INTRODUCTION TO
FLORICULTURE
Second Edition

CONTRIBUTORS

Allan M. Armitage

Douglas A. Bailey

William H. Carlson

Charles A. Conover

G. Douglas Crater

Dominic J. Durkin

William C. Fonteno

P. Allen Hammer

David E. Hartley

Royal D. Heins

August De Hertogh

Mark P. Kaczperski

Meriam G. Karlsson

R. Kent Kimmins

Anton M. Kofranek

Roy A. Larson

Alton J. Pertuit, Jr.

Marlin N. Rogers

Edward M. Rowley

Thomas J. Sheehan

Virginia R. Walter

Thomas C. Weiler

C. Anne Whealy

Richard E. Widmer

Gary J. Wilfret

INTRODUCTION TO
FLORICULTURE
Second Edition

Edited by
Roy A. Larson

Department of Horticultural Science
College of Agriculture and Life Sciences
North Carolina State University
Raleigh, North Carolina

ACADEMIC PRESS, INC.
Harcourt Brace Jovanovich, Publishers
San Diego   New York   Boston
London   Sydney   Tokyo   Toronto

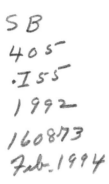

Academic Press, Inc.
San Diego, California 92101

United Kingdom Edition published by
Academic Press Limited
24–28 Oval Road, London NW1 7DX

Library of Congress Cataloging-in-Publication Data

Introduction to floriculture [edited by] Roy A. Larson. --2nd ed.
    p.   cm.
    Includes bibliographical references and index.
    ISBN 0-12-437651-7
    1. Floriculture.   I. Larson, Roy A.
SB405.I55   1992
635.9--dc20   91-24492
                                                                   CIP

PRINTED IN THE UNITED STATES OF AMERICA
92 93 94 95   9 8 7 6 5 4 3 2 1

Contents

2   Carnations

C. Anne Whealy

3   Roses

Dominic J. Durkin

4   Snapdragons

Marlin N. Rogers

II POTTED PLANTS

8 Bulbous and Tuberous Plants
August De Hertogh

9 Azaleas
Roy A. Larson

10 Potted Chrysanthemums
G. Douglas Crater

11 Gloxinias, African Violets, and Other Gesneriads

R. Kent Kimmins

12 Poinsettias

David E. Hartley

13 Easter Lilies

Thomas C. Weiler

14 Hydrangeas

Douglas A. Bailey

15 Cyclamen

Richard E. Widmer

16 Begonias

Meriam G. Karlsson and Royal D. Heins

17 Kalanchoe

Alton J. Pertuit, Jr.

18 Geraniums

William C. Fonteno

19 Other Flowering Pot Plants

P. Allen Hammer

20 Bedding Plants

William H. Carlson, Mark P. Kaczperski,
and Edward M. Rowley

List of Contributors

Numbers in parentheses indicate the pages on which the authors' contributions begin.

Allan M. Armitage (159), Horticulture Department, University of Georgia, Athens, Georgia 30602

Douglas A. Bailey (365), Department of Horticultural Science, North Carolina State University, Raleigh, North Carolina 27695

William H. Carlson (511), Department of Horticulture, Michigan State University, East Lansing, Michigan 48824

Charles A. Conover (569), Central Florida Research and Education Center, Apopka, Florida 32703

G. Douglas Crater (249), Department of Ornamental Horticulture, University of Tennessee, Knoxville, Tennessee 37901

Dominic J. Durkin (67), Department of Horticulture, Rutgers University, New Brunswick, New Jersey 08903

William C. Fonteno (451), Department of Horticultural Science, North Carolina State University, Raleigh, North Carolina 27695

P. Allen Hammer (477), Department of Horticulture, Purdue University, West Lafayette, Indiana 47907

David E. Hartley (305), Paul Ecke Poinsettias, Encinitas, California 92023

Royal D. Heins (409), Department of Horticulture, Michigan State University, East Lansing, Michigan 48824

August De Hertogh (195), Department of Horticultural Science, North Carolina State University, Raleigh, North Carolina 27695

Mark P. Kaczperski (511), Department of Horticulture, University of Georgia, Athens, Georgia 30602

Meriam G. Karlsson (409), School of Agriculture and Land Resources Management, University of Alaska Fairbanks, Fairbanks, Alaska 99775

R. Kent Kimmins (289), Department of Horticulture, Kansas State University, Manhattan, Kansas 66506

Anton M. Kofranek (3), Department of Environmental Horticulture, University of California, Davis, Davis, California 95616

Roy A. Larson (223), Department of Horticultural Science, College of Agriculture and Life Sciences, North Carolina State University, Raleigh, North Carolina 27695

Alton J. Pertuit, Jr. (429), Department of Horticulture, Clemson University, Clemson, South Carolina 29634

Marlin N. Rogers (93), Department of Horticulture, University of Missouri, Columbia, Missouri 65211

Edward M. Rowley (511), High Valley Nursery, Roosevelt, Utah 84066

Thomas J. Sheehan (113), Department of Environmental Horticulture, University of Florida, Gainesville, Florida 32611

Virginia R. Walter (551), Ornamental Horticulture Department, California Polytechnic State University, San Luis Obispo, California 93407

Thomas C. Weiler (333), Department of Floriculture and Ornamental Horticulture, Cornell University, Ithaca, New York 14853

C. Anne Whealy (43), Ball Seed Co., West Chicago, Illinois 60185

Richard E. Widmer (385), Department of Horticultural Science, University of Minnesota, St. Paul, Minnesota 55108

Gary J. Wilfret (143), Gulf Coast Research and Education Center, Institute of Food and Agricultural Sciences, University of Florida, Bradenton, Florida 34203

Preface

The first edition of *Introduction to Floriculture* was published more than ten years ago. Changes occur almost from month to month in this dynamic field, with all its innovations. Cultural techniques change, new cultivars and even new crops are introduced, new products and equipment become available, and consumer trends change. Geographical areas of production change as new sites become prominent, while long-established production centers lose their competitive edge. The elimination of physical and political barriers in Europe has already altered production and marketing situations on that continent. The initiation of new economic policies among selected European countries will result in further readjustments. Free trade agreements advocated by Mexican, United States, and Canadian government leaders should have pronounced consequences on commercial floriculture in all three countries. The importation of cut flowers from Central and South America into the United States can be expected to continually increase.

There also is a changing audience as growers with years of expertise retire and are replaced by a new generation of growers. The floriculture knowledge of the newcomers will not instantly equal that of their predecessors, but most will have a computer literacy that will surpass that of the previous generation. Constantly improving software programs will give them access to information and advice often gained previously by trial and error. Facsimile equipment has facilitated the transmission of information from teachers, researchers, and extension specialists to growers and vice versa.

Software programs, facsimile and telephone messages, and other avenues of communication can only be as good as fundamental sources of information enable them to be. It is a mission of the authors, editor, and publisher of the second edition of *Introduction to Floriculture* to produce such a fundamental, educational source. Many chapters in this edition are longer than those in the first edition as new information had to be added, but much information included in that edition was still too valid to delete. Geraniums have been discussed in a separate chapter, a promotion which the importance of geraniums in commercial floriculture surely justifies. Fourteen authors wrote chapters for both editions, but retirements primarily prompted the emergence of new authors. Crop expertise and the ability to communicate that knowledge were the necessary qualifications for authorship in this new edition just as they were for the first edition.

Authors were given the freedom to present their material using their own styles so the chapters are not all clones of one another. A general format was requested, and authors observed that framework. This approach should achieve a semblance of unity, while individuality also is respected.

It is a privilege and honor to edit a book written by a dedicated, intelligent corps of authors and to have the book published by a firm with the prestige and standards of Academic Press. Shirley Light of Academic Press has been extremely supportive and helpful, making the second edition a pleasant experience and not a chore.

I do wish to thank all the authors who were so willing to write chapters for a textbook. I hope the audience is as willing to read as the authors were to write. Writers and readers have a common bond called "knowledge." There is much to be told and learned about floriculture, and this book is written and published as that messenger of knowledge.

Roy A. Larson

I
CUT
FLOWERS

1
Cut Chrysanthemums

Anton M. Kofranek

Introduction to Floriculture, Second Edition
Copyright © 1992 by Academic Press, Inc.
All rights of reproduction in any form reserved.

I. HISTORY

The florists' chrysanthemum is a complex hybrid that, if grown from seed, segregates into many diverse flower forms. Most of the species in the lineage of present day cultivars are from China. These include *Chrysanthemum indicum* (a yellow single), *C. morifolium* (lilac and rose colors), and the Chusan daisy (species unknown); the latter was brought to England in 1843 by Robert Fortune and is thought to be one of the parents of spray or pompon chrysanthemums. Even before this date the British and Dutch were hybridizing chrysanthemums. In the United States, Elmer D. Smith began hybridizing for the florists' trade in 1889. He hybridized and named over 500 cultivars (Cathey, 1969; Langhans, 1964; Laurie *et al.*, 1979).

Commercial hybridization to improve cultivars continues today in America, Asia, and Europe. Selection is based not only on flower shape and color, but also on suitability of seedlings for year-round flowering programs for cool temperature tolerance and for postharvest qualities.

II. CLASSIFICATION

The genus *Chrysanthemum* has been reclassified to *Dendranthema*, and the florist species is *grandiflora* (Anderson, 1987), but literature cited in this chapter refer to it as *Chrysanthemum morifolium*. Readers should be aware of the change.

A. Inflorescence Forms

The florists' chrysanthemum is a composite inflorescence that has flowers borne on a receptacle or capitulum. The heads are borne on long peduncles in cymose clusters. The single inflorescences (daisy-like) have ray flowers (outside row), which are pistillate and disk flowers (the central ones or the "eye of the daisy"), which are bisexual and usually fertile. The receptacle is flat or convex and is surrounded by an involucre of bracts. Inflorescences are categorized on the basis of shape and form (Ackerson, 1957). Some of the most common inflorescence forms grown commercially and in the garden are illustrated in Fig. 1. They are described in the following section.

1. Singles—daisy-like—composed of one or two rows of outer pistillate flowers (ray) and flat bisexual flowers (disk) borne centrally (Fig. 1B).

2. Anemones—similar to the single form except the disk flowers are elongated and tubular—forming a cushion. Disk flowers may be the same color as or different color than the ray flowers (Fig. 1D).

3. Pompons—a globular head formed by short uniform ray flowers. The shape is considered formal; disk flowers are not apparent. The National Chrysanthemum Society (America) recognizes three distinct sizes: (a) small buttons, 1.5 inches

or less in diameter; (b) intermediate, 1.5 to 2 inches in diameter (Fig. 1C); and (c) large, 2 to 4 inches in diameter.

4. Decoratives—similar to pompons because they are composed mainly of ray flowers, but the outer rows are longer than the central flowers, giving the inflorescence a flat and informal shape. Sizes are mostly intermediate and large (Fig. 1E).

Fig. 1. Typical chrysanthemum inflorescence forms: (A) spider, (B) single, (C) intermediate pompon, (D) anemone, (E) decorative, (F) large-flowered incurve (standard chrysanthemum). (Courtesy of Yoder Brothers.)

5. Large flowered blooms—these are greater than 4 inches and are classified in many shapes. Disk flowers are not apparent in most of these forms: (a) Incurved double: globose and formal, with ray flowers similar in size to the disk flowers and that curve inward and toward the top (Fig. 1F). (b) Reflexed double: less formal and globose than the incurved double, with overlapping ray flowers curved downward, except for the ray flowers. (c) Tubular ray flowers: (i) spider-ray flowers, tubular and elongated in the outer rows but short in the center. The drooping outer row ray flowers are sometimes hooked on the ends (Fig. 1A). (ii) Fuji, similar to the spiderexcept the ray flowers may be shorter, droop less, and lack hooks on the ends. (iii) Quill, tubular ray flowers, long on the outside and short near the center, resembling feather quills. Ends of flowers are open and not flattened. (iv) Spoon, similar to the quill except the outer row flowers are open and are flattened, resemble a spoon. (d) Miscellaneous or novelty types, consisting of feathery, plume-like or "hairy" ray florets.

B. Commercial Use

Chrysanthemums are grown in two basic ways for cut flowers, depending on market demand.

1. Disbudded Inflorescences

All flower buds but the terminal one are removed to allow one inflorescence perstem to develop (Fig. 2). If the bloom is an incurved or reflexed form and is between 4 and 6 inches in diameter, it is usually referred to as a "standard." "Incurves" or"reflexes" smaller than 4 inches are known as "disbuds." Only cultivars that increase flower size markedly on disbudding are used. These may include fuji, spider, quill, and even certain single chrysanthemums, as well as the more usual incurved or reflexed double types.

2. Spray Inflorescences

The entire cyme is allowed to bloom but very often the central inflorescence (the oldest one) is removed about the time color begins to form in the ray flowers. Because this is the oldest bloom in the cyme (known as a spray), it will senesce before the lateral inflorescences if it is not removed. It also is larger than the surrounding blooms; therefore, it is removed so all remaining flowers will be more uniform in size. Spray chrysanthemums may have any of the inflorescence forms described above (i.e., singles, anemones, formals, decoratives, and the class of tubular ray flowers). Single forms perhaps are most popular at this time.

C. Photoperiod Response

Hybrid cultivars used for year-round flowering are short day (long-night) plants. Natural flowering occurs during several months in the fall. In the Northern Hemi-

Fig. 2. *Top*: The correct stage of development for lateral bud removal for disbudding a standard chrysanthemum (about the 29th short day). These laterals have peduncles that are long enough to "snap off" easily but have not become too long and woody at this stage. *Bottom*: The disbudding is accomplished (i.e., lateral buds and the peduncles have been completely removed) leaving only the terminal bud. Delaying this disbudding until lateral buds reach a morphological stage larger than that shown reduces the ultimate size of the terminal inflorescence at maturation.

sphere these cultivars have been classified as early (August to mid-October), midseason (mid-October to mid-November), and late (after mid-November) flowering.

They are now classified for natural or year-round flowering by response groups (i.e., 6- to 15-week cultivars). Cultivars in the 6-week response group require 6 weeks to reach harvest stage from the first inductive short day. Other response groups require 7 or more weeks, up to a maximum of 15 weeks to reach harvest stage from the first inductive short day. Most commercial cut chrysanthemums in the United States are in the 8- to 11-week categories. Plants in the longer response groups cost more to grow, but flower prices usually remain the same as the midseason type that may be grown in any season.

III. PROPAGATION

Plants are propagated by rooting terminal cuttings. These vegetative cuttings are removed from stock plants maintained under long-day conditions that inhibit flower bud formation (see Section VI). Terminal cuttings 3 to 4 inches long that are removed from stock plants can be placed directly into the rooting medium or may be stored at 35° to 38°F for several weeks in cartons lined with polyethylene to prevent desiccation. Some cultivars are not stored successfully in the winter because cuttings lack an adequate supply of carbohydrates. To enhance development of roots, basal ends of the cuttings are dipped in a talc containing 0.1 to 0.2% indolebutyric acid (IBA). The greenhouse temperature should be between 60° and 65°F and the rooting medium temperature between 65° and 70°F. About 500 and 600 cuttings per square yard are placed in the medium, depending on size of the lower leaf of the cultivar. Until rooting is accomplished a fine mist should be used intermittently during the daylight hours. Some propagators increase the mist frequency from 10 A.M. to 3 P.M. when light intensity and leaf temperatures are the greatest. The mist is usually turned off a day or two before cuttings are removed to "harden" them prior to shipping or planting. Cuttings are well rooted in 10 to 20 days, depending on the cultivar and time of year. Cuttings with roots slightly less than $\frac{3}{4}$ inch long are desirable; longer roots make planting difficult and more root damage can occur.

Almost any porous mixture that is not toxic can be used as a rooting medium. Perlite plus sphagnum peat moss is perhaps the most common medium because it is easily obtained, produces consistent results, and does not separate from the roots during shipment. Vermiculite, sand, fine coal cinders, scoria, pumice, and a sandy soil mixture have also been used as rooting media. When cuttings are shipped by common carrier, a light-weight medium is preferred to keep shipping costs as low as possible.

Total soluble salts below 15 milliequivalents per liter (mEq/liter) for the mist system does not affect rooting. However, magnesium should not exceed 70% (Paul, 1969). A high percentage of sodium (>67%) will cause "red root." Some calcium is necessary for good rooting. Applications of gypsum or ground limestone at rates of 5 to 7 pounds per 100 square feet can be broadcast over the surface of the rooting medium prior to inserting cuttings. This treatment will supply adequate calcium and will reduce the proportions (percentage) of sodium and magnesium if these cations are abundant in the irrigation water.

IV. SOIL PREPARATION

Chrysanthemums will grow in almost any soil if it is well managed. Plants are susceptible to several soil-borne pathogens (see Section XIII); therefore, those organisms must be controlled to ensure maximum growth. Assuming the rooted cuttings are pathogen-free, it may be feasible to grow the crop in a soil that has never been used for chrysanthemums previously and is hence probably free of the soil-borne organisms that affect the crop. The same soil can be used with caution for several successive crops until soil pathogens reach epidemic proportions. If and when this occurs steam pasteurization or chemical treatment is necessary to control soil-borne pathogens.

Chemical treatments are fumigants that control most soil pathogens or specific pathogens such as *Verticillium albo-atrum*. A combination of 2 parts chloropicrin to 1 part methyl bromide can be used for *Verticillium*. However, a general chemical soil treatment (e.g., 98% methyl bromide and 2% chloropicrin) is useful. Where steam heat is available, the soil can be treated as described by Mastalerz (1977).

Before the soil is treated with chemicals or steam, plant stubble either is removed or finely ground up and incorporated into the soil with a rototiller. Ingredients such as sphagnum peat moss, wood shavings, fir bark, or other locally available organic material plus gypsum, limestone, and superphosphate can be incorporated into the growing medium. Steaming clay soils can cause a temporary toxic condition to the roots because of an excessive buildup of ammonia or manganese. These excesses can be partially overcome by additions of gypsum or superphosphate after steaming or by allowing the soils to be fallow for several weeks until microorganisms convert ammonia to nitrate nitrogen (Mastalerz, 1977). Steaming sandy soils creates fewer problems. Soils that are chemically treated must be well aerated before cuttings are planted (a waiting period of 1 week).

Media in the bed is usually marked for planting by scratching the soil with a pronged tool at a preset spacing. Cuttings are planted at those marks or planted in the middle of squares with the support of wires or string. Often two cuttings are planted per square on the outer rows and only one per square in the interior of the bed.

V. MINERAL NUTRITION

Chrysanthemums have high requirements for both nitrogen and potassium. Maintenance of high levels of nitrogen during the first 7 weeks of growth is especially important. If moderate deficiencies develop during this early period, later nitrogen applications will not recapture the flower quality that will have been lost. The findings of Lunt and Kofranek (1958) showed that the quality of plants and flowers produced was ideal when the plants were fertilized early in their growth cycle. No further fertilization was necessary after the inflorescences reached a diameter of approximately $\frac{1}{2}$ inch. Later fertilization is wasteful and growers have to be concerned about groundwater quality. Too much nitrogen can also cause leaves to become brittle in some cultivars. There should be adequate nitrogen (4.5 to 6%) in the leaves for use by the flowers. During the first 80 days, chrysanthemum plants grow rapidly, and there is a large requirement for nitrogen (Lunt and Kofranek, 1958). During the last 20 days, only the inflorescence grows rapidly and mineral nutrients are translocated from the leaves. During the first few weeks the root systems of individual plants are not extensive throughout the soil, and there is low efficiency of nitrogen recovery. The efficiency increases, however, with time, and the greatest nitrogen requirement for all aboveground plant parts is between the 70th and 80th day. At maturity about 20 to 25% of the nitrogen of the aboveground plant portion is located in the inflorescence.

At one time, recommendations for applying fertilizers to chrysanthemums were to withhold fertilizers after planting rooted cuttings and until plants became established. These recommendations are no longer valid, as they were predicated on the assumption that the soil might have been high in salts from the previous crop. Soils with excessive salts should be leached before planting. Before the soil is treated for pathogens, certain fertilizers of low solubility should be incorporated into the soil. One possible recommendation is a moderate fertilizer application such as 1 pound of single superphosphate, 2 pounds of dolomitic limestone, and 2 pounds of urea–formaldehyde per 100 square feet. The urea–formaldehyde is a low-release nitrogen that at warm temperatures mineralizes 25 to 35% in 3 weeks, 35 to 50% in 6 weeks, and 60 to 75% before 6 months. Only 6 to 10% of the nitrogen is immediately available. A majority of the nitrogen becomes available to the crop when it is required by the crop. After cuttings are planted, they should be irrigated at once with a liquid fertilizer containing 200 ppm each of nitrogen and potassium. Liquid fertilizer should be applied at every irrigation. It is important, however, to analyze the soil at regular intervals for soluble salts and for a shift in pH. A pH between 5.5 and 6.5 is adequate and the (soluble salts) electrical conductivity of a saturated paste extract (ECe) should not exceed 2.5 millimhos per centimeter (mmho/cm). An ECe reading exceeding 2.5 mmho/cm indicates excess soluble salts, and the soil should be leached with plain irrigation water to reduce the excess salts. If the pH exceeds 6.5, an acid fertilizer (e.g., ammonium nitrate) will lower the pH; an alkaline fertilizer (calcium nitrate) will raise the pH. With this system of fertilization, there is no need for soil analyses for the individual cations and anions unless troubles are suspected. A soil analysis can be used for "trouble shooting" and is often used by some growers as a routine operation. The solubridge is an instrument used to measure the ECe of the soil, and a pH meter is used routinely to measure shifts in pH from the norm. These two measurements usually are adequate tests for a short-term crop such as chrysanthemums, but they should be used weekly.

Occasionally leaf disorders appear, indicating a problem in the soil. Leaf tissue analysis reflects the mineral status of the leaf more accurately than does a soil analysis. Tissue levels associated with the mineral status in the leaf appear in Table I. These levels may be useful as standards in solving mineral deficiency disorders.

Table I

Leaf Tissue Levels Associated with Adequate orDeficient Levels of Selected Essential Elements in *Chrysanthemum* × *morifolium* 'Good News'[a]

Element	Adequate range	Critical level	Level found in moderate or severe deficiency	Plant part effectively reflecting mineral status
Nitrogen (%)	4.5–6	4.0 upper limit 4.5 lower limit	1.5–3	Upper leaves
Phosphorus (%)	0.26(?)–5	0.26 upper limit(?)[b] 0.17 lower limit(?)[b]	0.10–0.21	Upper or lower leaves
Potassium (%)	3.5–10	2.75 upper limit 2.15 lower limit	0.2–2	Lower leaves
Calcium (%)	0.5–4.6	0.40 upper limit 0.46 lower limit	0.22–0.28	Upper leaves
Magnesium (%)	0.14–1.5	0.11(?)	0.34–0.064	Lower leaves
Sulfur (%)	0.3–0.75	0.25(?)	0.07–0.19	Upper leaves
Iron (ppm)		—[c]	35	Upper leaves
Manganese (ppm)	195–200	—[c]	3–4	Upper or lower leaves
Boron (ppm)	25–200 (240 is excessive)	20	18.1–19.5	Upper leaves
Copper (ppm)	10 ppm(?)	5	1.7–4.7	Middle leaves or leaves from lower axillary growth
Zinc (ppm)	7.26(?)	7	4.3–6.8	Lower leaves

[a] The adequate range for a given element extends from just above the critical level to the concentration at which toxic disorders develop. Except for boron, the upper limit of the adequate range is not well defined. Adapted from Lunt *et al.* (1964).

[b] Because phosphorus is redistributed within the plant at flowering, critical levels are estimates.

[c] Data are inadequate to estimate critical levels.

Many commercial grade fertilizers supply adequate microelements except for iron. Chelated iron can be supplied in liquid fertilizers at the rate of 3 to 5 ppm when fully diluted.

VI. VEGETATIVE GROWTH

A. Stock Plants

Two common spacings for stock plants are 4 × 5 inches and 5 × 5 inches. Long daylengths and liquid fertilizers are given from the day of planting to promote rapid vegetative growth. Plants are given a soft pinch (Fig. 3A) as soon as they are established to promote rapid development of shoots. It is not advisable to allow the new plant to grow to the size where the first pinch is large enough to be taken as the first cutting from that stock plant. This is tantamount to a hard pinch (Fig. 3B), which has the disadvantages of leaving too few nodes on the original plant and allowing the lower portion of the stem to become semiwoody before taking that cutting. Buds in the axils of leaves of semiwoody chrysanthemum stems do not grow as rapidly as do those of succulent stems. Of the three pinches shown in Figure 3, the soft pinch is best to promote rapid bud growth.

Cuttings should be taken as often as possible to keep the stock plant in a juvenile stage; premature flower buds are less apt to form on shoots that are actively growing. Usually stock plants produce flushes of growth (shoots) in the early stages because there is little light competition among shoots. Because of the

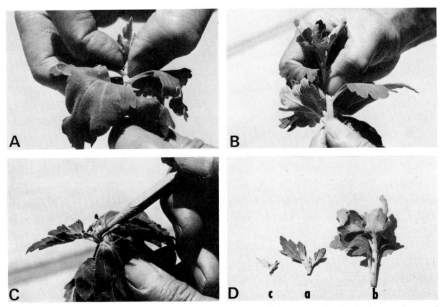

Fig. 3. (A) The "soft pinch" is most commonly used because buds that remain on the stem "break" (sprout) readily, because stem tissue is succulent and buds are reasonably large. As many as four to five breaks can be expected from a soft pinch during periods of high light intensity. (B) The "hard pinch" is made into semiwoody stem tissue. Note the bud just below the top thumbnail. These and lower buds are large but they do not "break" as rapidly as those buds that remain just below a "soft pinch" (A). A hard pinch is generally used to reduce plant size, especially in pots where some cuttings are taller than others. (C) The pencil point indicates where a "roll out" pinch is made. This pinch is only used on short plants or soon after cuttings are planted. The "breaking" (sprouting) of buds on the stem is less than satisfactory (i.e., not many shoots result from this type of pinch and bud growth is usually slower than from a soft pinch). (D) *Left to right*: Portions of stem tips removed from a "roll out pinch" (c); a "soft pinch" (a); and a "hard pinch" (b). Note the degree of leaf removal with each pinching method.

lower light flux during the winter months fewer axillary buds sprout after a pinch than in the summer. Later, in the 10th to 15th week after planting, plants become so dense that cuttings or shoots long enough (3 to 4 inches) for harvesting are available only on an irregular basis and only on the periphery of the stock plant. This limited availability can be attributed to the denseness of the stock plants and the limited light that exists in the center of the plants. When cuttings are taken, at least two leaves should remain on the stock plant just below the point where the cutting is removed. The two leaves are the photosynthetic surface and those buds in axils of the remaining leaves become part of the next flush of cuttings. If too many leaves remain at each harvest, the stock plant becomes too large and competition for light becomes a serious problem. Removal of shoots on a frequent basis, taken for cuttings, provides more light to the center and also removes shoot competition.

Stock plants are usually kept in beds for cutting production for 13 to 21 weeks. Leaving plants in beyond 13 weeks (about four to five flushes of cuttings) can result in premature budding of those cuttings removed for production (i.e., flower buds may form on those cuttings even under long-day conditions). Some cultivars form a flower bud more readily than others (Cockshull and Kofranek, 1985).

The longer a stock plant is grown, the more apt it is to produce premature flower buds. Some cultivars are prone to initiating premature flowers (e.g., 'Festival' and 'Mandalay'). Stock plants of these cultivars must be replanted more frequently to avoid this problem. A young plant (juvenile) is less apt to initiate flower buds than an older one. One way to keep the plant vegetative is to remove (take) cuttings frequently even if there is no demand for them. Long (older) shoots on older stock plants (beyond 13 weeks) are very likely to have premature flower buds (partially induced) even if supplementary lighting and temperature are adequate.

Supplementary lighting for the inhibition of flower initiation is more critical for stock plants than it is for production of flowering plants. Cuttings taken with premature buds are almost worthless because the resultant plants flower on short stems. Even if the rooted cuttings are destined to be pinched after planting and the flower bud is thus removed, the resultant vegetative shoots are not as vigorous as those that sprout from a completely vegetative cutting. The only reason for pinching a plant with a premature flower bud is to promote (stimulate) vegetative shoots; however, pinching may result in weak shoots or may even be unsuccessful because the lateral shoots may also have flower buds. Lighting to promote vegetative growth and to inhibit flowering on stock plants must, therefore, be complete. A minimum light intensity of 10 fc (0.11 klx) from incandescent lamps for 4 to 5 hours in the middle of the night during the winter and 2 hours during the summer is adequate for even the cultivars most insensitive to supplementary lighting. Cyclic lighting with incandescent lamps should be avoided on stock plants because cyclic lighting is marginal illumination for insensitive cultivars (see Section VII). As stock plants become older they are more likely to form flower buds if incandescent cyclic lighting is used. The use of fluorescent and low-pressure sodium vapor lamps is not well documented for stock plants.

B. Production Plants

Rooted cuttings in a vegetative state should be planted in a well-prepared moist soil, and then irrigated with a liquid fertilizer solution at once and illuminated at night to ensure a long-day effect from the first day. Spacing of cuttings in the bench orbed varies with season and cultivar and depends on whether plants are to be pinched or grown single-stem. Plants to be pinched are usually spaced 6 × 7 inches in summer and 7 × 8 inches or 7 × 9 inches in the winter. Interior-positioned plants are later pruned to two stems and outside plants are pruned to three stems per plant. Single-stem plants are usually planted 4 × 6 inches for summer and fall crops and 5 × 6 inches for winter crops. Some cultivars may require a wider spacing of 6 × 6 inches. These spacings are for beds that are approximately 42 inches wide.

More cuttings are required for single-stemmed plants than for pinched plants, but this additional cost can be overcome because the time to grow a single-stemmed crop is less than for a pinched crop. Pinching temporarily delays the growth of any plant.

It is necessary to maintain production plants in a vegetative state both for rapid growth and to attain desired stem length before flower induction is desired. Vegetative growth in hybrid chrysanthemums grown for year-round flowering is promoted by long-day conditions and the proper night temperature. A daylength greater than 14.5 hours for plants grown at 60°F is required to maintain a vegetative state. Plants are most effectively illuminated with incandescent lamps in the middle of the dark period, which breaks up the night into two short periods; this type of lighting is known as "night break lighting." Duration of night break lighting varies with season and latitudes because of the daylength. Recommendations given to growers for night break lighting is based solely on the daylength. Five hours may be recommended in northern latitudes (40° to 50°N) in the winter but only 4 hours are recommended between 25° and 40°N where the daylength is longer. Because daylengths in northern latitudes during the summer are so long, no night break lighting, or at the most 2 hours, is recommended to ensure a completely vegetative state. Within 15° of the equator, illumination is recommended all year round for at least 3 hours per night. The recommended light intensity for continuous lighting as a night break is between 7 and 10 fc [(0.077 to 0.11) klx, 0.30 to 0.43 Wm^{-2} PAR]. Experiments with 'Albatross,' however, a cultivar generally requiring higher light intensity or duration for the night break than most cultivars, showed that considerably less light intensity for the night break (3 to 5 fc, 0.13 to 0.22 Wm^{-2} PAR) was required during the winter than during the summer months (7 to 10 fc) (Sachs and Kofranek, 1979). The major difference between growing chrysanthemums in the greenhouse in the winter and summer is the daytime radiant energy (photosynthetically active radiation). Thus, the higher the daytime radiant energy, the greater the light energy required for an effective night break. This is the opposite of what has been generally understood or recommended. Moreover, these findings underline the importance of photo-synthetically active radiation (PAR), not merely daylength, for flower initiation and development (Cockshull, 1972). Cyclic lighting during the night break, proposed by Cathey and Borthwick (1961), conserves energy because plants are illuminated only 20% of the time. Their work recommended that incandescent lamps be operational for a 6 minute interval every 30 minutes during a 4 hour night break with a minimum of 5 fc (0.055 klx). An interval of 1 minute of light every 5 minutes, at 10 fc (0.11 klx) or more, was found to be needed for sensitive cultivars (Kofranek, 1963). A comprehensive review on cyclic lighting is offered by Mastalerz (1977). When long days are supplied by incandescent cyclic lighting employed over a long period (e.g., 7 weeks), flower inhibition may be marginal or incomplete. Plants grown under a cyclic lighting regime reached full bloom as much as 1 week earlier, after short days began, than those given a continuous 5 hour night break for the same period (Kofranek, 1963). This result clearly demonstrates that the floral inhibition was not complete in the latter stages of the long-day period because the cyclically illuminated plants were also shorter than those grown with continuous light and had fewer nodes to flower.

Low-pressure sodium lamps that have a narrow band of effective irradiation (589 nm) have been reported to be extremely efficient as a night break with four cultivars (Canham *et al.*, 1977).

When production plants are grown under long-day conditions, shoot apexes continue to produce leaves and nodes at the rate of two to four per week (plastochron). Lower leaves mature first, and internodes in the region 4 inches below the apex elongate rapidly. Depending on cultivar and season, a shoot will reach the appropriate length (about 14 to 20 inches) in a set number of days. More time is required in winter than in summer to produce the same number of leaves (nodes). Published chrysanthemum schedules list the time required from the pinch from planting time to the beginning of short days for floral induction. The need for the long days before the short-day cycles to obtain the final desired stem length is illustrated in Figure 4. If 36 inches of stem were required for the cultivar, the plant should be grown vegetatively for about 4 weeks in the winter to ensure that the minimum plant height is approximately 36 inches.

During the cold weather when greenhouses are tightly closed, the carbon dioxide content of the air can become depleted; chrysanthemums then respond favorably to carbon dioxide injection into the greenhouse atmosphere.

VII. FLOWERING OF PRODUCTION PLANTS

When plants have reached the desired stem length (about 14 to 20 inches), they are given a short-day treatment. Lights that provided long-days are turned off during the natural short-day period (winter), or the plants are covered with a blackout material during naturally occurring long days (summer). The blackout material can be either black sateen cloth with a minimum of 68 × 104 threads to

Fig. 4. Plants of 'Florida Marble', a 9-week cultivar, were given various numbers of long days (LD) before beginning short day cycles (SD): (A) No LD (SD on day of planting); (B) 1 week of LD, (C) 2 weeks of LD, (D) 3 weeks of LD, and (E) 4 weeks of LD. After the LD period they were given continuous SD to promote flowering. Rooted cuttings were planted on November 9 and the photograph was taken 77 days later. These plants are typical of winter-grown chrysanthemums in a region of low light (Davis, California): The node number above the soil surface to the oldest inflorescence (indicated by the arrow), and the height above the soil surface are as follows:

	A	B	C	D	E
Weeks of long days	0	1	2	3	4
Average number of nodes/stem before short day induction	21	24	26	30	34
Final plant height (inches)	21	26	32	38	43

the inch or black polyethylene. The blackout material is best applied for a minimum of 12 hours, which is possible with an automatic blackout system (such as 7 P.M. to 7 A.M.). Manually applied blackout usually begins at 4:30 P.M., to conform to the usual work hours, and is removed beginning at 8:00 A.M.. The sun is still high at 4:30 P.M. (especially with daylight savings time) and heat builds up under the black material. Excessive heat (above 85°F)) can cause a "heat delay" of flower initiation during the early inductive short days. (Whealy et al., 1987). It is preferable to wait until at least 5:30 to 6 P.M. in midsummer before covering plants to avoid heat problems under the black material and remove the material later in the morning. Cultivars do differ in their susceptibility to heat delay.

The blackout must be applied for at least 21 to 28 consecutive short days, if standard chrysanthemums are grown, and for a longer period (35 to 42 days) if spray-type chrysanthemums are grown (see Section IX). After 14 consecutive short days the capitulum of the inflorescence is completely formed, and the outer rows of florets are beginning to initiate. It is safe to skip the blackout 1 day a week after that stage of inflorescence development has been attained (i.e., after 14 to 19 consecutive short days for 'White Marble' or 'Polaris', respectively) (Kofranek, 1983). There may be 1 or 2 days' delay in flower maturity over those given the blackout every day, depending on temperature in the last days of development. High day and night temperatures that occur near maturity can hasten flowering up to 5 days but flower quality will diminish, compared to plants grown at optimum temperatures.

VIII. INFLORESCENCE INITIATION AND DEVELOPMENT

Hybrid chrysanthemums presently being grown for year-round flower production are short-day (long-night) plants when grown at minimum temperatures of 60°F. Post (1949) stated, based on work conducted at latitude 42°N, that a daylength of 14.5 hours (about August 15) was necessary for flower initiation, but that a shorter daylength of 13.5 hours (occurring about September 20) was required for flower bud development. Natural daylengths mentioned by Post include civil twilight (the light intensity when the sun is 6° below the horizon). He also noted that some cultivars initiated flower buds as late as the first week in September. Furuta (1954) found that late blooming cultivars required shorter daylengths to reach full bloom than did early flowering cultivars. Post and Kamemoto (1950) reported that an early cultivar, 'Gold Coast' (9 weeks), initiated flower buds in about 4 days on a short photoperiod and that a late cultivar, 'Vibrant' (14 weeks), initiated flower buds in 5 days. Doorenbos and Kofranek (1953) showed that after 24 continuous short days, floral apexes of 'Gold Coast' and 'Vibrant' were at the same morphological stage of inflorescence development. They concluded that early and late blooming cultivars initiated floral receptacles and flowers (inflorescence) at about the same rate, but that the difference in the final flowering time (9 versus 14 weeks) was attributed to the different rate of development of inflorescences after the 24th short day.

Various response groups (see Section II,C) have different critical (maximum) daylengths for inflorescence initiation and development (Table II). Cathey (1957) reported the following maximum daylength requirements for selected cultivars in several response groups at a minimum might temperature of 60°F.

The data in Table II show that a shorter photoperiod is required for inflorescence development than is necessary for floral initiation for cultivars of different flowering periods. In this same study, Cathey (1957) showed that the critical daylength for

Table II

Critical Daylength Requirements Based on Response Group

		Critical daylength requirements (hours/day)	
Cultivar	Response group (week)	Flower initiation	Flower development
'White Wonder'	6	16 hr	13 hr, 45 min
'Pristine'	8	15 hr, 15 min	13 hr
'Encore'	10	14 hr, 30 min	13 hr
'Fortune'	12	13 hr	12 hr
'Snow'	15	11 hr	10 hr

floral initiation and development of three cultivars was altered with a change in the minimum night temperature (Table III).

Extremely high night temperatures (an average of 85°F) will delay floral initiation (Furuta and Nelson, 1953; Whealy *et al.*, 1987). Low minimum temperatures (range: 55° to 35°F) at the beginning of the short days delay flower bud initiation from 1 to 49 days, depending on cultivar and duration (5 or 15 days) of low temperatures (Samman and Langhans, 1962). Lowering night temperatures to 50°F during the first 15 short days delayed flowering of the temperature-sensitive cultivar 'Lemon Spider' but had little effect between the 16th and 30th short day (Carow and Zimmer, 1977).

Many chrysanthemum cultivars were classified by Cathey (1954) into temperature categories of flowering response.

1. Thermozero cultivars—those that flower normally between 50° and 80°F. Flowering proceeds rapidly at 60°F. This category was suggested as best for year-round flowering.

2. Thermopositive cultivars—those that require night temperatures of at least 60°F. Flower buds may initiate but do not develop beyond the capitulum stage at lower temperatures. These cultivars may be grown for year-round flowering if the temperature is properly maintained.

3. Thermonegative cultivars—those that do not flower normally when night temperatures are over 60°F. Lower temperatures (50°F) may delay but do not inhibit initiation. These cultivars should only be grown when night temperatures can be controlled at 60°F or slightly lower. Summer culture should be avoided. This category includes the late season cultivars that are in the 13- to 15-week response groups ('Snow' in Table III, for example). The cost of growing chrysanthemums with such a long response period is prohibitive.

The light flux in the first 2 weeks of short days can alter initiation of the capitulum and flowers. Low light intensities during this period will delay initiation by several days (Cockshull, 1972). The optimum light intensity for growth and flowering should be above 3000 fc (32 klx or 132 Wm^{-2} PAR) for a large portion of the

Table III

Interrelationships of Temperature and Critical Photoperiod for Flower Initiation and Development of Three Chrysanthemum Cultivars

| Cultivar | Period to flower (weeks) | Temperature (°F) | Maximum light period required[a] | |
			Flower initiation	Flower development
'White Wonder'	6	50	13 hr, 45 min	13 hr, 45 min
		60	16 hr	13 hr, 45 min
		80	16 hr	12 hr
'Encore'	10	50	13 hr, 45 min	13 hr, 45 min
		50	14 hr, 30 min	13 hr
		80	15 hr, 15 min	12 hr
'Snow'	15	50	12 hr	12hr
		60	11 hr	10 hr
		80	10 hr	9 hr

[a] Experimental range of photoperiods from 9 to 16 hours. Adapted from Cathey (1957).

daylight hours. Additional information concerning the environmental control of chrysanthemum flowering can be found in several reviews (Cathey, 1969; Cockshull, 1972; Mastalerz, 1977; Machin and Scopes, 1978).

IX. YEAR-ROUND FLOWERING

Year-round flowering of chrysanthemums was first suggested by Post (1947), although experiments on early flowering were first conducted by Professor Laurie using blackcloth. Research to perfect flowering of a number of cultivars was conducted over several years by different workers and then later by Yoder Brothers of Barberton, Ohio. Flowering schedules are available from several commercial companies that provide rooted or unrooted cuttings on a year-round basis.

Growers scheduling year-round production or even a single crop must follow certain principles in order to grow high-quality flowers. Briefly, the essentials are the following.

1. Rooted vegetative cuttings are planted in beds having a well-prepared soil and are provided with mineral nutrients in a liquid fertilizer solution from the first day. During bright weather plants should be shaded to minimize wilting.

2. Rooted cuttings are properly spaced based on whether the plants will be pinched or grown single stem (see Section VI,B).

3. After plants are established, they can be pinched (Fig. 3) to induce branching for multiple-stemmed plants and later pruned to a desired number of stems per plant. Some plants are not pinched; these are known as single-stemmed plants.

4. When the single stems, or the lateral shoots of pinched plants, reach a given height, they are given short-days to induce flowering. Up to this point plants must be grown under long-day conditions to inhibit flower bud initiation.

5. Plants are then given short-day conditions (preferably a minimum of a 12-hour daylength) until induced flower buds develop to a stage where daylength does not affect flower development (see Section VII).

6. If the plant is grown for a single bloom, lateral flower buds should be removed beginning about the 28th short day or when it is possible to remove the buds readily (see Fig. 2). Generally, disbudding is a two-step operation, spaced about a week apart.

7. During periods of high light intensity the developing blooms that are beginning to show color should be shaded with cheesecloth or a light-shading material to prevent sun scald to developing blooms.

8. Flowers are harvested with the proper stem length and inflorescence development required by the wholesale market. Flower development within the bed is not always uniform, and a range of 5 to 10 days may be required for all flowers to reach the proper cutting stage. Less time is required in summer when warm temperatures hasten flower maturity. More time might be needed in winter when low daytime radiant energy delays flower development in some areas of the country.

9. After flowers have been harvested, plant stubble and roots are removed from the soil or can be rototilled into the soil in preparation for the next crop. Soil must be prepared by a certain date to keep the year-round schedule intact.

Certain fundamental principles regarding the cultivar being grown must be understood before one can develop a year-round schedule. Some of this information was discussed in the preceding section on flowering. The most important points are listed below and will assist in explaining the sample schedules in Figure 5.

1. Not all cultivars are suitable for year-round scheduling. Some may require night temperatures lower than those that can be attained during the summer months. These cultivars are typical of the 13- to 15-week response groups that require temperatures of 60°F or lower.

To develop quality flowers, these particular cultivars should only be used in greenhouses or outdoor growing locations where night temperatures will be cool during later stages of flower bud development. Cultivars should be chosen to produce the best quality on a year-round basis. Cultivars in the 9-, 10-, or 11-week response groups usually are the ones that produce marketable quality during warm summer weather but also during low light intensity periods of winter. Although the 6- to 8-week cultivars reach the harvest stage more rapidly than the 10-week cultivars, their overall quality is not considered commercially adequate. Assuming that the greenhouse night temperature will be maintained at 60°F, the 9-, 10-, or 11-week cultivars produce the best quality in the most reasonable time from the beginning of short days until full bloom. Ten-week cultivars were chosen for scheduling samples (Fig. 5) and for the following discussion.

2. Rooted cuttings may require more time to become established in the soil after planting in winter than in summer, assuming plants are not allowed to wilt excessively during periods of high light and heat in summer. Those plants destined for

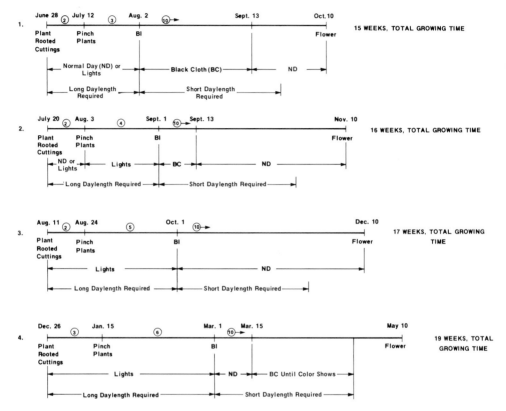

SAMPLE FLOWERING SCHEDULES FOR A 10-WEEK CULTIVAR

Fig. 5. Schedules for a 10-week cultivar during four periods having wide differences in light intensity. Definitions: Long daylength (LD) required from planting date to bud initiation date. Short daylength (SD) required from bud initiation date until the bud development is advanced enough to disregard the length of day. Lights: a night break with supplementary lighting to provide a long daylength to maintain a vegetative condition. Black cloth (BC): a blackout system to provide a short daylength to promote flowering. Natural daylength (ND): a daylength of adequate length at a given time of year to promote vegetative growth or flowering as required. Plant rooted cuttings: planting date for crops. Pinch plants: the date for removal of growing points to induce branching. Bud initiation date (BI): the day to change from a long daylength to a short daylength in order to promote flowering. Flower: a date to begin harvesting the blooms. Plants may bloom sooner if high temperatures prevail during short daylength cycle. The numerals 2, 3, 4, 5, 6, and 10 in circles refer to the number of weeks to complete a state of plant growth or development.

pinching will be ready for tip removal in as little as 10 days in the summer or as much as in 21 days in late fall or winter. Plants are provided with long days during this period by supplying supplementary lighting or with natural long days during the summer, depending on the latitude.

3. Pinched plants should be pruned as soon as possible to a desired number of shoots for the spacing provided. Early removal of excess shoots encourages rapid growth of the remaining shoots by reducing shoot competition.

4. Plants grown with single stems are planted closer together than pinched plants because of less shoot competition (see Section VI,B). The growth of single-stemmed plants is unchecked, and they reach the proper height for the cultivar earlier than pinched plants. A point worth repeating is that pinching always checks growth and that additional time is necessary for new shoots to develop after the removal of the shoot apex.

5. The time required from the pinching date to when shoots are long enough depends on available PAR. As many as 6 weeks will be required for plants pinched in October, November, and December at latitudes of 40° to 50°N to attain the proper shoot length, but only 4 weeks may be necessary at latitudes of 25° to 30°N during this same season. Daytime radiant energy at the northerly latitudes is more limiting for plant growth in winter than in the more southerly latitudes. During the late spring and early summer months of May, June, and July, as little as 3 weeks will be required to attain the proper shoot length at all the above-stated latitudes, because light intensity for vegetative growth is not limiting during this season. Some cultivars grow so rapidly during natural long days that fewer than 21 days are

Fig. 6. A shoot of 'Bright Golden Anne' with most of the large leaves removed to illustrate the size of a vegetative shoot above the most recently matured leaf. This shoot portion just above the leaf with the two white spots has 13 leaves, which range in size from one inch long to primordial leaves at the apex. If this cultivar is given a soft pinch (Fig. 3A), that portion above the marked leaf, about 13 leaves, is removed. 'Albatross,' however, has about 18 leaves in a shoot of similar size.

required to obtain proper shoot length. Chrysanthemum catalogs or experience will reveal these rapid growing cultivars.

6. Once proper shoot length is attained (usually 14 to 20 inches), the plant is given inductive short days for flower initiation and development. There are a given number of leaf initials in various stages of morphological development within the zone of expansion (subapical meristem and the region of elongation). The zone of expansion of certain cultivars has as many as 18 to 20 leaves between the most recently expanded leaf and the apical meristem ('Albatross'); other cultivars have as few as 12 to 14 leaves (Fig. 6) in various development stages ('Princess Anne'). About 4 to 5 inductive short days are required to change the vegetative apex into a terminal flower bud (Post and Kamemoto, 1950). In that period, leaf initials (perhaps only two to three) are still in the process of being initiated before the apex becomes reproductive and the inflorescence receptacle is finally initiated. Although a shoot may only be 14 inches long on the first inductive short day, it is reasonable to assume that at full bloom that shoot at maturity may be between 28 and 40 inches long. The internodes within the zone of expansion (Fig. 6), as well as those just below the most recently matured leaf expand, which results in a doubling of shoot length at full bloom. Some cultivars are more efficient in internode expansion than others and this difference would indicate that the former require fewer long days from the pinch to the start of short days. The actual number of nodes in the zone of expansion at the time of the first inductive short day may or may not be of significance in the shoot reaching a marketable length. Environmental conditions (light intensity and temperature) surrounding the plant and the genetic traits of the cultivar probably have more influence on ultimate shoot length by controlling the final internodal length than do the actual number of nodes within the zone of expansion when short days begin.

7. The minimum number of consecutive short days required to produce quality blooms depends on the cultivar and its culture. When a plant is grown for a single bloom as a standard, all buds except the terminal inflorescence are removed; this disbudding procedure (see Fig. 2), if conducted early (starting about the 28th short day) takes advantage of the "source-sink relationship" within the plant. The source of sugars in the leaves is directed to the sink (i.e., the single inflorescence at the terminus). Under such conditions, as few as 21 consecutive short days were required by 'Albatross' but 28 consecutive short days were required by 'Escapade' to develop a quality inflorescence (Kofranek and Halevy, 1974). When plants are grown as sprays (not disbudded) one should assume that more consecutive short days would be required to promote full bloom of a majority of the inflorescences because the sinks are in many locations on the flowering branches and individual inflorescences vary in their morphological development. (Note: On a cyme the most mature inflorescence is the uppermost one and the inflorescences are progressively less mature on the lower branches; inflorescences are formed basipetally.) Cultivars such as 'Marble' and 'Polaris' require 35 short days during the longest days to produce marketable sprays (Kofranek, 1983). Most chrysanthemum scheduling catalogs suggest continuing short days until inflorescences show color. This could be as late as the 49th short day for a 10-week pompon chrysanthemum.

8. Sample schedules of a 10-week cultivar grown as a pinched crop during four periods at latitude 42°N are given in Figure 5.

X. FLOWERING DURING THE NATURAL SEASON

With the advent of year-round flowering, fewer chrysanthemums are being grown during their natural flowering season. For the year-round grower natural flowering may not be convenient to fit into the year-round schedule. In the future, when fossil fuels are less available and more costly, year-round flowering may not be economically feasible. Then natural-season chrysanthemums may again be grown for approximately 3 months per year out-of-doors in mild climates or in greenhouses heated with as little fuel as possible.

Natural flowering is possible from early October until early January when the appropriate cultivars are chosen for flowering succession. Rooted cuttings are planted every week, starting in mid-June with the 7-week cultivars and ending in mid-August with the 15-week cultivars. As a general rule, rooted cuttings of each response group are pinched 2 weeks after planting (Table IV). Cuttings are easily established within 2 weeks of planting during these periods of high light intensity. Cultivars in the various response groups should be planted on the dates indicated in Table IV in regions of latitude 42°N and 1 week earlier in more northernly latitudes and 1 week later in the southern latitudes. A soft pinch to induce branching is given 2 weeks later for the flowering periods indicated; this is known as the "time pinch." The purpose of the time pinch is to improve the shape of the spray formation of pompon chrysanthemums (Fig. 7B). If the pinch is made too early, a premature crown bud (Fig. 7A) initiates and causes excessive peduncle elongation; if the pinch is made too late, a terminal bud initiates (Fig. 7C), surrounded by lateral buds on short peduncles.

Table IV

Natural Flowering Dates of Various Response Groups Grown in Greenhouses in the Northern United States and Southern Canada[a]

Response group (Week)	Approximate planting date	Approximate pinching date	Natural flowering date
7	June 25	July 9	Sept. 22–Oct. 9
8	July 2	July 16	Oct. 12–Oct. 21
9	July 9	July 23	Oct. 22–Nov. 1
10	July 16	July 30	Nov. 2–Nov. 11
11	July 23	Aug. 6	Nov. 12–Nov. 21
12	July 30	Aug. 13	Nov. 22–Dec. 4
13	Aug. 6	Aug. 20	Dec. 2–Dec. 11
14	Aug. 13	Aug. 27	Dec. 12–Dec. 26
15	Aug. 20	Sept. 3	Dec. 27–Jan. 5

[a] Plants are generally pinched 2 weeks after planting for the flowering dates shown. Adapted from Yoder Brothers (1968).

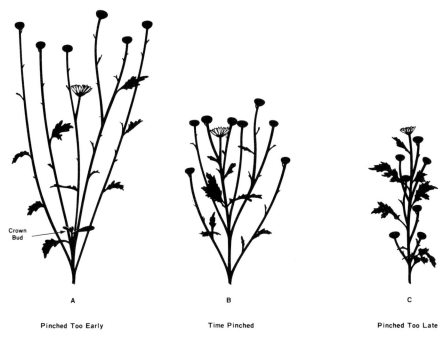

Crown
Bud

A   B   C

Pinched Too Early   Time Pinched   Pinched Too Late

Fig. 7. Natural flowering and spray formations of pompons as influenced by the date of pinching prior to the natural bud initiation date in the fall.

Although the rule is not rigid, the time pinch is made approximately 100 days before full bloom. The rationale is that the stem length from the resultant pinch attains the proper size for that cultivar, within a given response group, by the time it reaches its natural bud initiation date. The photoperiod (daylength) will then be short enough for floral buds to initiate on or about the date and will continue to develop properly and bloom within the dates indicated (Table IV). If planting and pinching are made earlier than indicated, such as 2 weeks earlier, the resultant shoots may initiate a crown bud (Fig. 8) before the natural bud initiation date. Because the daylength is not short enough to promote further flower bud development, the crown bud may not develop beyond the capitulum stage and will cause unwanted branching. A plant pinched too early (before the day indicated in Table IV) will usually develop a spray formation as shown in Fig. 7A with a crown bud low in the spray. Shoots below the crown bud elongate; later those shoots initiate a flower on or about the natural bud initiation date, thus forming an elongated spray. If plants are pinched at a later date than indicated, such as 2 weeks later, the resultant spray may resemble Fig. 7C. Those shoots reach maturity at a date later than the indicated natural bud initiation date; those buds, therefore, initiate and develop quickly to form a compressed spray. The peduncles below the inflorescence do not elongate adequately during the short days that follow the indicated bud initiation date, and the spray develops a compressed appearance. When the plant is pinched approximately 100 days before the flowering date, the shoots have ample time to reach both maturity and the proper length to produce a marketable stem. The flower bud then initiates on or about the natural bud initiation date and develops lateral buds with peduncles of a desirable length (Fig. 7B). The pinching dates may vary slightly among cultivars, even within similar response groups, because of genetic variability of peduncle lengths. For instance, 'Iceberg' has naturally long peduncles; therefore, pinching this cultivar late may produce an even better quality spray formation than if it were time pinched 100 days before bloom.

Planting, pinching, and early shoot development occur during periods of high radiant energy and high temperatures of summer. Growth is rapid and should not be checked by creating water and nutrient stresses. Bud initiation also occurs during periods of high light intensities and high temperatures, but subsequent bud development of midseason and late cultivars occurs during periods of low radiant energy and low night temperatures. During this latter period, blooming may be greatly

Fig. 8. A crown bud in the center of an inflorescence spray. The strap-shaped leaf growing below the bud and toward the right is typical of a crown bud. Buds such as these will never develop to anthesis because they lack true floral parts. Crown buds will cause branching when they form prematurely in shoots being grown for their vegetative state.

delayed by very low night temperatures (35° to 45°F). Cultivars of the late response groups (13 to 15 weeks), however, often require cool temperatures (48° to 55°F) for proper bud development. Earlier cultivars (9 to 11 weeks) require minimum temperatures of 53° to 54°F or greater for normal inflorescence development.

Standard chrysanthemums grown for the natural season should also be "time pinched," about 100 days prior to harvesting, to avoid the formation of premature crown buds that may not develop properly. A crown bud (Fig. 8), if selected (or "taken") as the one to become the marketable bloom, may be induced to develop by removing vegetative lateral buds below it. Crown buds, if selected or taken too early, are slightly larger, usually have paler colors, and develop on longer peduncles than those terminal buds selected at the proper time (i.e., after the natural bud initiation date). If buds are selected too early they may develop green centers because the daylength is marginal for development. Bud selection at a later date ensures that the days are amply short to promote proper bud development if the minimum temperatures are adequate. A more detailed explanation of flowering in the natural season is given in a paper by Kofranek (1983).

XI. IMPROVING INFLORESCENCE QUALITY WITH PHOTOPERIOD

Some spray chrysanthemums develop cymes or sprays resembling those shown in Figure 7C. The sprays can be made to open and be less compact as in Figure 7B by manipulation of the daylength (Post, 1950). When the plants have attained the proper height, they are given 12 short-day cycles to initiate the inflorescences on the entire spray. Then the plants are given an interruption of 10 consecutive long days (night break) and are later returned to short days to complete flower development. The long days elongate the peduncles below the inflorescences and improve the spray formation. Some cultivars do not require any manipulation because their peduncles are naturally long.

Flower diameter of standard chrysanthemums can be increased by subjecting disbudded chrysanthemums to long-day conditions after the 35th short day. This manipulation is called "after lighting" (Ben-Jaacov and Langhans, 1969). After lighting generally increases the length of ray flowers to increase the overall diameter of the inflorescence. Not all standards respond to this treatment, however.

XII. USE OF SELECTED GROWTH REGULATORS

A. Increasing Stem Length

Gibberellins can be used to increase the stem length of standard chrysanthemums. Concentrations as low as 1.5 to 6 ppm potassium gibberellate sprayed 1 to 3 days after planting and again 3 weeks later have resulted in significant increases in stem length without loss of stem quality under midwinter light conditions in California (Byrne and Pyeatt, 1976).

B. Reducing Peduncle Length of Standards

Daminozide (B-Nine) can be applied to standard chrysanthemums just after disbudding (see Fig. 2) to prevent excessive stem elongation from the uppermost leaf to the base of the flower ("neckiness"). One spray application (2500 ppm) to the point of runoff is usually adequate. Daminozide reduces cell division and elongation in the region just below the inflorescence, where elongation occurs rapidly at the time the florets are actively developing. Future use of daminozide is uncertain, however, because it is the same compound as Alar and the chemical has become controversial. Research efforts with new growth regulators, such as paclobutrazol (Bonzi) and uniconazole (Sumagic), have shown promise, but almost all of the experiments have been conducted on potted chrysanthemums rather than on cut crops.

C. Increasing Peduncle Length of Pompons

Some pompons develop short peduncles, and longer ones may be more desirable for floral arrangements. Peduncle elongation may be accomplished by spraying the tops of plants to runoff with 20 ppm gibberellic acid (GA) 4 weeks after the start of short days. This treatment is most effective during periods of high radiant energy, and it varies with cultivars. High concentrations or applications of GA later than the 4-week period may result in weakened inflorescences (Kofranek and Cockshull, 1985).

D. Root Initiation

The most common rooting hormone used for chrysanthemums is 0.1 to 0.2% IBA mixed with talc. Although it is possible to apply IBA in an aqueous solution, it is generally not recommended because of the possible spread of bacterial diseases in the solution.

E. Floral Inhibition

As little as 3 to 4 ppm ethylene has been reported to inhibit floral initiation during inductive short days (Tjia et al., 1969). Experiments by Cockshull and Horridge (1978) have been conducted using 2-chloroethylphosphonic acid, to determine whether or not ethylene can be substituted for long days in chrysanthemum floral inhibition.

F. Chemical Disbudding of Standard Mums

Many people have tried to find a way of eliminating or retarding growth of lateral flower buds chemically without injuring the terminal inflorescence. Moderate success has been achieved with materials such as HAN, naphthalenes, and 2,3-dihydro-5,6-diphenyl-1,4-oxanthin (UNI-P293) with selected cultivars (Zachariou-dakis and Larson, 1976). To date no chemical has been a satisfactory substitute for manual disbudding. Research results have been sporadic because of cultivar differences and environment. Lateral buds must be eliminated or retarded at a very early morphological stage (i.e., between the 14th and 20th short day). During this active growth period, the immature terminal inflorescence is extremely vulnerable to injury from chemicals. Any miscalculation of chemical application can destroy or seriously injure the terminal inflorescence. Leaves adjacent to those lateral buds that are retarded or eliminated are usually seriously distorted. If the terminal bud is not injured, there is a strong likelihood that the lateral buds will also be unaffected.

G. Chemical Pinching

The methyl esters of fatty acids (Emgard 2077 and Off-Shoot-O) were studied experimentally for chemically pinching chrysanthemums. Results have been variable and not as successful as with azaleas. These materials used at low rates (2%) can drip down the stem and girdle it. Because the hazards of using these chemical pinching agents are so great, they are not recommended for commercial use.

XIII. DISEASES

Chrysanthemums can be affected by fungi, bacteria, and viruses. Many of these diseases can be avoided by obtaining disease-free cuttings from propagators who culture-index their stock plants and cuttings and rigidly maintain "clean" production programs. Such disease-free plants are not resistant to most of the organisms, however, so growers must continue appropriate methods of control, such as the use of sterile media, sanitary greenhouse practices, and preventive spray programs. Major chrysanthemum diseases and their control will be discussed in this chapter but not as extensively as Horst and Nelson (1977), or McCain and Raabe (1990).

A. Fungal Diseases

1. Ascochyta Ray Blight

a. Pathogen: Ascochyta chrysanthemi, Mycosphaerella ligulicola. The conidia are spread by wind or splashing water from infected tissue. Disease severity is increased under humid conditions and temperatures exceeding 60°F. Flower buds may rot before opening; a distinctive feature is that the ray (or center) florets are affected first, becoming brown. Eventually the whole flower will be affected.

b. Control. Fungicides are available for its control but the first method of control should be an attempt to keep the flowers and foliage dry in the greenhouse and during the marketing phase. Disease symptoms sometimes are not apparent until flowers arrive at the market place.

2. Fusarium Wilt

a. Pathogen: Fusarium oxysporum. This disease has increased in importance. The fungus is soil-borne, and first infects the roots and then invades vascular tissue, causing severe wilting and eventual death of the plant. Warm temperatures and high relative humidity are conducive to disease development.

b. Control. The first line of defense against Fusarium wilt is the use of cultivars that are relatively resistant (Strider, 1985; Strider and Jones, 1985). Cuttings that have been culture-indexed to improve freedom from the organisms are also available. Fungicidal spray applications are ineffective as the disease is systemic but fungicidal drenches early in disease infection can be beneficial.

3. Gray Mold

a. Pathogen: Botrytis cinerea. Botrytis is favored by cool temperatures (50° to 60°F) and high relative humidity, and may sporulate on dead or dying plant tissue. It can also be severe at higher temperatures. Spots on petals appear water-soaked at first, and then are covered by countless mycelia and spores. Lower leaves can rot, and the organism may enter the stem and girdle it.

b. Control. The disease is particularly serious if old flowering plants are allowed to remain in the growing area. Plant debris should be removed. Sun scald or any other form of plant injury will also be an avenue of penetration by the fungus. Relative humidity should be kept as low as possible, and flowers and foliage should be dry before packing and shipping. Effective fungicides are also available for control.

4. Cottony Rot

a. Pathogen: Sclerotinia sclerotiorum. Sclerotia may germinate in soil, but airborne spores are readily spread from apothecia. Symptoms are similar to those caused by *Botrytis*. A cottony mass appears on affected tissue. Sclerotia (hard, black bodies) may form within the stem.

b. Control. Removal of plant residue or application of effective fungicides can reduce incidence of disease.

5. Powdery Mildew

a. Pathogen: Erysiphe cichoracearum. Deformed or puckered foliage is one of the first symptoms of infection by this organism. It appears first on the underside of the foliage, as white to gray powdery growth. The disease problem is most severe at high relative humidity.

b. Control. Heating and ventilating to reduce relative humidity helps reduce the severity of the problem. Adequate spacing of plants also is beneficial, as crowded plants will usually be exposed to higher humidities within the plant canopy. Crowding also makes fungicide spray applications more difficult.

6. Pythium Root Rot or Basal Stem Rot

a. Pathogen: Pythium species. *Pythium* is a soil-borne organism favored by excessive soil moisture. Spores are spread by contaminated soil, water, tools, and other implements. *Pythium* is a water mold, so it is particularly severe in poorly drained media. Sterilization of the medium will be beneficial in *Pythium* control, but only if serious efforts are made to maintain a sanitary growing area, with hose ends off the ground. Roots might be severely damaged or the stem near the soil line.

b. Control. Fungicides can be used to arrest or eliminate the pathogen. They are applied as drenches to the media, and repeat applications might be needed if the disease continues to spread. Pasteurization between crops might be required to reduce the risk of more disease problems with future crops.

7. Rhizoctonia Stem Rot

a. Pathogen: Rhizoctonia solani. This stem rot is caused by a soil-borne organism and is most destructive under moist, warm conditions. Infected plants wilt during the day, and eventually will not regain turgidity. The rot occurs on the stem just above the soil surface.

b. Control. Proper media preparation and sanitary greenhouse practices will help reduce the possibilities of infection. There also are effective fungicides to aid in control. Some of the most effective fungicides had long residual effects in the media, however, and were removed from the list of recommended pesticides.

8. Rust

a. Pathogen: Puccinia chrysanthemi. This disease occurs infrequently but can become serious when temperatures are 60° to 70°F, and free moisture is present on the foliage. Such conditions are ideal for urediospore germination. Small reddish-brown (rust) pustules appear as flecks on the underside of the leaves. The center of the pustule dies and becomes black. Airborne spores are produced.

b. Control. The disease will not be a problem if the relative humidity is kept low and the foliage is not allowed to remain wet for extended periods. Cultivars differ in susceptibility.

9. White Rust

a. Pathogen: Puccinia horiana. White rust occurs under the same conditions as described for common rust. Yellow spots first appear on the upper side of the leaf, and the center then turns brown. The lower leaf surface has small, waxy pustules that are buff to pink in color but later turn white.

b. Control. Control measures are similar to those described for common rust.

10. Septoria Leaf Spot

a. Pathogen: Septoria obesa and S. chrysanthemella. *Septoria obesa* is most common, but *S. chrysanthemella* has also been reported. This organism may remain in the plant debris for 2 years, so elimination of old plant parts is essential. The disease organism is spread by splashing water and is most troublesome when leaves remain moist for 12 or more hours. Leaf spots are circular to irregular in shape, usually develop from the base of the plant upward, and may coalesce. Spots are black, brown, or occasionally reddish.

b. Control. Low relative humidity and dry foliage will aid in control. Protection with a fungicide when symptoms first appear can also prevent the disease from escalating in severity.

11. Stemphylium and Alternaria Ray Blights

a. Pathogen: Stemphylium species and Alternaria species. Infections occur at temperatures ranging from 60° to 85°F but freestanding water on the petals is necessary for infection. Small, necrotic lesions form on ray petals. Lesions are reddish-brown on white florets, chocolate-brown on yellow florets, and light brown on pink florets.

b. Control. Freestanding water on the flowers should be avoided. Infected plants should be discarded. Adequate ventilation is required.

12. Verticillum Wilt

a. Pathogen: Verticillium dahliae or V. albo-atrum. These pathogens are soil-borne organisms that can remain in the growing medium for many years. Only one side of the plant might wilt, in contrast to overall wilting caused by Fusarium wilt and other vascular diseases. Leaves gradually become yellow and die, beginning at the base of the plant. Dried leaves remain on the stem. Infection usually begins in cool weather, and symptoms are obvious in warm weather.

b. Control. Media should be pasteurized at 140°F for 30 minutes with aerated steam or fumigated with an approved fumigant for 48 hours or more. Culture-indexed cuttings also are essential.

B. Bacterial Diseases

1. Bacterial Blight

a. Pathogen: Erwinia chrysanthemi. Bacterial blight is favored by high temperatures (80° to 90°F) and high humidity, and can be spread by infested hands, tools, equipment, or infected plant material. The first symptom is the gray-ish-colored foliage on some shoots. Wilting follows, particularly on bright days. The stem is easily crushed or may split, and the pith becomes jelly-like. Shoot tips turn black and exude droplets of liquids.

b. Control. Control can be achieved by starting with disease-free cuttings. Plants, when infected, should be destroyed immediately. There is no effective bactericide to recommend.

2. Bacterial Leaf Spot

a. Pathogen: Pseudomonas cichorii. This disease is most likely to occur during periods of rainy weather when the relative humidity remains high. Circular to elliptical spots appear on the foliage. These spots may increase in size or coalesce to form lesions on lower leaves. In severe cases, the bacteria enter leaf petioles and stems, and infected flower buds die prematurely.

b. Control. Cultivars do differ in degree of susceptibility. Efforts should be made to keep relative humidity as low as possible, but such a goal is not easy to achieve.

3. Crown Gall

a. Pathogen: Agrobacterium tumefaciens. Round growths or galls appear on the stem just below the soil surface, but sometimes are even found on leaves or stems. The disease, as with so many chrysanthemum diseases, is worse when moist conditions prevail.

b. Control. The disease can be spread from one crop rotation to another if galls or other infected plant parts remain in the growing medium. The medium should be steamed or fumigated between crops. This disease is now quite rare and is seldom encountered with current production practices. One suggested method of biological control is to dip cuttings or spray plants with *Agrobacterium radiobacter*.

C. Viral Diseases

1. Aspermy Flower Distortion

a. Pathogen: Chrysanthemum aspermy or tomato aspermy virus. This is a serious disease of chrysanthemums. It can be spread by aphids, tools, and handling. Flowers will be distorted and smaller than normal. Red, pink, and bronze petals can exhibit streaking or color break. Leaves usually are unaffected.

b. Control. The virus was particularly serious when chrysanthemums and tomatoes were grown in rotation in the greenhouse. Debris from either crop would affect the succeeding crop. Removal of infected plants, use of cultured cuttings, control of aphids, and proper greenhouse sanitary procedures decrease the chances for aspermy.

2. Chlorotic Mottle

a. Pathogen: Chrysanthemum chlorotic mottle virus. There are several forms of leaf mottling viruses such as vein mottle, dwarf mottle, necrotic mottle, and viruses B and Q (Gosling, 1984). The viruses can be spread through propagation of infected stock and during subsequent handling. Early symptoms of mottle can be confused with nutritional disorders.

b. Control. The major method of control is to obtain virus-indexed cuttings.

3. Chrysanthemum Mosaic

a. Pathogen. This virus is spread by aphids. Symptoms include mottled and malformed leaves, and will vary according to cultivar response. Mosaic alone might not be too serious, but its combination with other viruses can seriously affect plant growth and quality.

b. Control. Aphid control and virus-indexed plants are effective means of control.

4. Chrysanthemum Stunt Virus

a. Pathogen. This disease almost eliminated the chrysanthemum as a commercial crop in the 1940s. It only came under control when techniques were developed to index plants for the presence or absence of the virus. Infected plants are unduly small (stunted) and might reach anthesis about a week earlier than scheduled.

b. Control. Virus-free stock plants and cultured cuttings are essential, and any suspicious looking plants should be discarded immediately. The virus can be spread by propagation knives, other tools, or hands.

5. Tomato Spotted Wilt Virus

a. Pathogen. This serious virus is spread by western flower thrips. Many plant species are susceptible, and it is probably the most devastating virus to affect floriculture crops since chrysanthemum stunt almost destroyed the chrysanthemum industry in the 1940s. Ring patterns appear on the foliage of some cultivars. Leaf necrosis and distortion are noticeable and may be on only one side of the plant. Necrotic streaks might also appear on the stems.

b. Control. Virus-free plant material should be purchased, and then precautions should be taken to avoid later infection. Weeds in and around the greenhouse should be eradicated, as they can harbor the vector. Sticky traps should also be installed in the greenhouse to detect the presence of thrips. Because other flower crops are subject to tomato spotted wilt virus, they should also be purchased from reputable sources and monitored regularly. Screening of ventilators might be needed to exclude thrips from the greenhouse. Effective insecticides might also be part of the pest control program, particularly in field production.

XIV. PESTS

Included in this section are true insects, spider mites, slugs, and nematodes [disease-causing organisms are also pests, and their chemical control is regulated by the Environmental Protective Agency (EPA)]. No specific chemical control materials are recommended here because such recommendations can vary from one state to another, and pesticides are frequently introduced to or removed from marketing channels. One should make certain that any pesticides used on chrysanthemums have label clearance.

A. Insects

1. Sucking Insects

These include numerous species of aphids, leaf hoppers, mealy bugs, tarnished plant bugs, thrips, and whiteflies. They suck plant juices and may distort leaves and flowers.

2. Chewing Insects

Beet army worms, cabbage loopers, chrysanthemum gall midge, corn-ear worms, cutworms, garden millipedes, serpentine leaf miner, and spotted cucumber beetles are among the chewing insects that can be troublesome in chrysanthemum production.

B. Spider Mites

These mites suck plant juices, reducing leaf color and plant vigor. Mites were about the first pests to exhibit resistance to pesticides, and no one group of miticides can be relied on to give adequate control.

C. Slugs and Snails

There are various species that chew flowers and leaves. These pests are nocturnal so either the damage they cause or the slime trails they leave behind them will indicate their presence during the day.

D. Nematodes

1. Leaf Nematodes

These are microscopic, nonsegmented round worms. They are spread through stomates by splashing water. They cause dark green to brownish angular lesions in leaves. The progression of the injury on the plant is upward.

2. Root Nematodes

These pests suck juices from roots and cause root galls, thereby weakening plants. They are soil-borne or can be transmitted to uninfested soil on contaminated plants or soil.

XV. PHYSIOLOGICAL DISORDERS

Chrysanthemum crops can be affected by disorders that cannot be attributed to disease organisms, insects, or any of the other pests mentioned earlier. These disorders are abiotic or physiological in nature. Symptoms of several disorders and possible causes are listed in Table V; however, one must evaluate the cultural practices employed to determine the most likely cause of the problem. Written records of temperatures, application dates, and rates of fertilizer and pesticide applications, as well as soil and foliar analyses are very, very helpful in making these evaluations.

XVI. POSTHARVEST HANDLING

Flowers should be cut about 4 inches above the soil line to avoid cutting into woody tissue that is less likely to absorb water. Foliage from the lower third of the stem is stripped at harvest, and stems are immediately placed in acidified water containing a surfactant. The concentration of the surfactant (wetting agent) should be about 0.01% on a volume basis. Effective acidifiers can be either citric acid (300 ppm) or hydroquinoline citrate (60 to 200 ppm). The latter material has some biocidal features (Durkin, 1981).

To maximize mature flower and leaf longevity, the procedure to follow is to hydrate the cut stems quickly with the surfactant in acidified water and then to transfer those stems to a solution with a biocide and possibly some sugar. An effective biocide is a quaternary ammonium compound such as Physan 20 (200 ppm) or bleach water (sodium hypochlorite). The latter can be as little as 4 ppm, but it soon volatilizes as chlorine gas. Sugar should not exceed 1.5%, because leaves of some cultivars may be damaged. However, the benefits to flower longevity from added sugars is insignificant for mature flowers.

After cut stems are hydrated, they are graded into units appropriate for the market. Spray mums are graded into 8 to 12 ounce bunches usually containing a minimum number of stems. Floral sprays are wrapped in a conelike plastic sleeve to prevent damage and to facilitate packing. Standards or disbuds are graded in groups of 10 or 12, depending on the market. Tissue paper is usually placed between flowers, especially for spider or Fuji mums, to prevent intertwining of ray flowers.

Mature flowers can be wrapped in plastic and stored dry for 6 to 8 weeks at 31°F. After storage, stems are cut and placed in tepid water (96°F) in a cool location (40° to 45°F) to rehydrate the cut stems (Post and Fisher, 1952).

Standard chrysanthemums can be harvested in an unopened stage, when the inflorescences are only 2 to 4 inches in diameter with only a few outer ray florets unfurled. This practice enables growers to remove crops from the greenhouse earlier than previously practiced, and increases the number of crops produced annually in the same growing area. It also enables outdoor producers to harvest flowers prior to a threatened early frost. More flowers also can be packed in the

Table V

Visual Symptoms and Possible Causes of Cut Chrysanthemum Problems

Symptoms	Possible causes
Stunted growth with small leaves	Nutrient deficiency, especially nitrogen Excess soluble salts in media Overwatering Underwatering (water stress) Low temperature during vegetative period; plant develops slowly Low light intensity during vegetative period Viruses (see Section XIII on diseases) Root nematodes
Undesirable branching (i.e., interruption of vegetative growth) including initiation of premature flower buds	Shoot apexes of some cultivars might be damaged by emulsifiable concentrates used in some pesticides Crown bud initiates but does not develop Calcium or boron deficiency Tarnished plant bug injury to shoot apex
Occasional leaf wilting at any stage of growth	Plants not irrigated Sunny days following an extended period of cloudy weather Particularly noticeable on plants infected with *Verticillium* disease Cold soil temperature, combined with warm air and leaf temperatures Reduced water absorption, increased transpiration
Chronic wilting at a later stage of development	*Verticillium* in specific stems Recent injured root system from an excessive fertilizer application Recent injured root system from an excessive irrigation (soil saturated for long durations)
Interveinal chlorosis of leaves	Media pH too high (alkaline) Iron or manganese deficiency Poor root system (many possible causes) Heavy infestation by spider mites Soil nematodes
Light green leaves (general chlorosis)	Nitrogen deficiency Sulfur deficiency (rare) Viruses (see Section XIII) High light intensity coupled with high temperatures (usually summer) Low temperatures, 'Marble'
Marginal leaf burn (necrosis)	Potassium deficiency Excessive soluble salts in media Boron in irrigation water; leaves accumulate boron Poor root system causing water stress (many possible causes)

Table V *(continued)*

Symptoms	Possible causes
Marginal leaf burn (necrosis) *(cont.)*	Verticillium fungus causing water stress Insecticide damage
Bronze-colored foliage	Phosphorus deficiency Low temperature as plants mature
Bronze coloration on lower leaf surface	Potassium deficiency Smog damage
Brittle leaf	Nitrogen level excessive as plants become mature
Leaf spots (other than those caused by pathogens)	Excessive daytime radiant energy Insect or insecticide damage
Death of lower foliage	Shading caused by dense spacing Pathogens, such as Botrytis or Verticillium wilt Foliar nematodes
Bleached petals (especially bronze and pink cultivars)	Temperature too high during floral development Stunt virus
Injured dark-colored petals	Sun scald, excessive flower temperatures
Intense colors of pink and bronze petals	Low night temperature during flower development
Pink colorations in white petals	Low night temperature during flower development Overmature flowers
Quilling of florets	Lower than recommended night temperatures Low light intensity during flower development Low storage temperatures, less than 38°F, for immature flowers
Green centers in standard flowers	Early selection of buds (terminal inflorescences during normal season production) Long photoperiod interrupting early stages of inflorescence initiation
Distorted inflorescence	Crown bud initiates but does not develop normally because of marginal short days, improper temperatures, or both Aster yellows Heat delay Tarnished plant bug injury after the 14th short day
Delayed or no floral induction	Photoperiod too long, which can be caused by light pollution at night, holes in blackout material Improper night temperatures during floral initiation Heat delay in summer Radiant energy too low during the day during flower initiation Ethylene pollution at time of initiation Tarnished plant bug injury

Fig. 9. *Left to right*: Inflorescences matured in the greenhouse on an intact plant (GH), and detached buds opened with a 3% sucrose solution in a lighted room; buds were cut in stage 2 at a 2 inch diameter or were cut in stage 3 at 3 to 4 inch diameter. Six days were required for complete opening of the cut buds of 'Albatross' in a sugar solution. This experiment was conducted in the winter when the light intensity in the greenhouse was poor. Note the flatness of the inflorescence allowed to develop in the greenhouse (left); the central flowers (florets) developed poorly under low light conditions. Those cut buds opened in the 3% sugar solution had well-developed central florets that resulted in a globular inflorescence; the sugar solution was the carbohydrate substrate necessary for growth of the immature florets. (From Kofranet *et al.*, 1972. Photograph courtesy of Florists' Publ. Co.)

shipping containers. Someone in the marketing channel must have proper facilities and expertise to open the flowers, however, so quality is not impaired. Stems are placed in a solution of sugar and biocide. Several biocides suggested (Marousky, 1971; Kofranek and Halevy, 1972; Sacalis, 1989) are effective, but their use cannot be recommended in the United States because they lack EPA approval. The flower opening room should be 70°F, with constant fluorescent lighting at an intensity of 1.1 klx, 100 foot candles, or 3.2 Wm^{-2} PAR. Quality of flowers opened in this manner often exceeds those that are allowed to mature in the greenhouse or field and then shipped (Kofranek *et al.*, 1972) (Fig. 9).

REFERENCES

Accati-Giribaldi, E., Kofranek, A. M., and Sachs, R. M. (1977). "Relative efficiency of fluorescent and incandescent lamps with inhibiting flower induction in Chrysanthemum morifolium 'Albatross.'" *Acta Hortic.* **68**: 51–58.

Ackerson, C. (1957). "The Complete Book of Chrysanthemums." *Amer. Garden Guild.* Garden City, N.Y.: Doubleday.

Anderson, N. O. (1987). "Reclassification of the Genus *Chrysanthemum* L." *HortScience.* **22**(2): 313.

Ben-Jaacov, J., and Langhans, R. W. (1969). "After-lighting of chrysanthemums." *N.Y. State Flower Grow. Bull.* **285**: 1–3.

Byrne, T. G., and Pyeatt, L. E. (1976). "Gibberellin sprays to increase stem length of 'May Shoesmith' chrysanthemums." *Florists' Rev.* **157**(4077): 31–32, 75.

Canham, H. E., Cockshull, K. E., and Hand, D. W. (1977). "Night-break lighting with low pressure sodium lamps." *Acta Hortic.* **68**: 63–67.

Carow, B., and Zimmer, K. (1977). "Effects of change in temperature during long nights on flowers in chrysanthemums." *Gartenbauwissenschaft* **42**(2): 53–55.

Cathey, H. M. (1954). "Chrysanthemum temperature study. B. Thermal modifications of photoperiods previous to and after flower bud initiation." *Proc. Am. Soc. Hortic. Sci.* **64**: 492–498.

Cathey, H. M. (1955). "Temperature guide to chrysanthemum varieties." (*N.Y. State Flower Grow. Bull.* **119**: 1–4.

Cathey, H. M. (1957). "Chrysanthemum temperature study. F. The effect of temperature upon the critical photoperiod necessary for the initiation and development of flowers of *Chrysanthemum morifolium*." *Proc. Am. Soc. Hortic. Sci.* **69**: 485–491.

Cathey, H. M. (1969). Induction of Flowering—Some Case Histories, edited by L. T. Evans, pp. 268–290. Australia: MacMillan.

Cathey, H. M., and Borthwick, H. M. (1961). "Cyclic lighting for controlling flowering of chrysanthemums." *Proc. Am. Soc. Hortic. Sci.* **78**: 545–552.

Cockshull, K. E. (1972). "Photoperiodic control of flowering in the chrysanthemum." In *Crop Processes in Controlling Environments*, edited by A. R. Rees, K. E. Cockshull, D. W. Hand, and G. Hurd, pp. 235–250. New York: Academic Press.

Cockshull, K. E., and Horridge, J. S. (1978). "2-Chloroethylphosphonic acid and flower initiation by *Chrysanthemum morifolium* Ramat in short and long days." *J. Hortic. Sci.* **53**: 85–90.

Cockshull, K. E., and Kofranek, A. M. (1985). "Long-day initiation by chrysanthemum." *HortScience.* **20**(2): 296–298.

Doorenbos, J., and Kofranek, A. M. (1953). "Inflorescence initiation and development in an early and late chrysanthemum variety." *Proc. Am. Soc. Hortic. Sci.* **65**: 555–558.

Durkin, D. J. (1981). "Factors affecting hydration of flowers." *Acta Hortic.* **113**: 109–117.

Farnham, D. S., Thompson, J. F., Hasek, R. F., and Kofranek, A. M. (1977). "Forced air-cooling for California flower crops." *Florists Rev.* **161**(4162): 36–38.

Furuta, T. (1954). "Photoperiod and flowering of *Chrysanthemum morifolium*." *Proc. Am. Soc. Hortic. Sci.* **63**: 457–461.

Furuta, T., and Nelson, K. S. (1953). "The effect of night temperature on development of chrysanthemum flower buds." *Proc. Am. Soc. Hortic. Sci.* **61**: 548–550.

Gosling, S. G., ed. (1984). *The Chrysanthemum Manual of the National Chrysanthemum Society.* London: National Chrysanthemum Soc.

Halevy, A. H., Byrne, T. G., Kofranek, A. M., Farnham, D. S., Thompson, J. F., and Hardenburg, R. E. (1978). "Evaluation of postharvest handling methods of transcontinental truck shipments of cut carnations, chrysanthemums, and roses." *J. Amer. Soci. Hortic. Sci.* **103**(2): 151–155.

Horst, R. K., and Nelson, P. E. (1977). "Diseases of chrysanthemums." *Cornell Univ. Info. Bull.* **85**.

Kofranek, A. M. (1963). "Experiments continue on cyclic lighting for greenhouse mums." *Florists' Rev.* **133**(3434): 23–24, 55.

Kofranek, A. M. (1983). "The minimum number of short day exposures for producing high quality spray chrysanthemums." *Acta Hortic.* **147**: 199–210.

Kofranek, A. M., and Cockshull, K. E. (1985). "Improving spray formation of pompon cultivars with gibberellic acid and intercalated long days." *Acta Hortic.* **167**: 117–124.

Kofranek, A. M., and Halevy, A. H. (1972). "Conditions for opening cut chrysanthemum flower buds." *J. Am. Soc. Hortic. Sci.* **97**(5): 578–584.

Kofranek, A. M., and Halevy, A. H. (1974). "Minimum number of short days for production of high quality standard chrysanthemums." *HortScience.* **9**: 543–544.

Kofranek, A. M., Shepard, P., and Kubota, J. (1972). "Seasonal bud opening of 'Albatross' mums." *Florists' Rev.* **151**(3902): 22–23, 58–61.

Langhans, R. W., ed. (1964). *Chrysanthemums, A Manual of Culture, Insects, and Economics of Chrysanthemums.* Ithaca, New York: N.Y. College of Agriculture

Laurie, A., Kiplinger, D. C., and Nelson, K. A. (1979). *Commercial Flower Forcing.* 8th ed. New York: McGraw-Hill.

Lunt, O. R., and Kofranek, A. M. (1958). "Nitrogen and potassium nutrition of chrysanthemum." *Proc. Am. Sci. Hortic. Sci.* **72**: 487–497.

Lunt, O. R., Kofranek, A. M., and Oertli, J. J. (1964). "Some critical levels in *Chrysanthemum morifolium* cv. 'Good News.'" In *Plant Analysis and Fertilizer Problems.* Edited by C. Bould, P. Prevot, J. R. Magness, vol. IV, pp. 398–413. Geneva, New York: W. F. Humphrey.

McCain, A. H., and Raabe, R. D. (1990). *California Greenhouse Disease Management.* Oakland, CA: Univ. of Calif.

Machin, B., and Scopes, N. (1978). *Chrysanthemums, Year-Round Growing.* Poole, Dorset U.K.: Blandford Press Ltd.

Marousky, F. J. (1971). *Handling and Opening Bud-cut Chrysanthemum Flowers with 8-Hydroxyquinoline Citrate and Sucrose.* U.S. Dept. Agric. Mark. Res. Rep. No. 905.

Mastalerz, J. W. (1977). *The Greenhouse Environment.* New York: Wiley.

Paul, J. L. (1969). "Water quality and mist propagation." *Int. Plant Prop. Soc.* **18**: 183–186.

Post, K. (1947). "Year round chrysanthemum production." *Proc. Am. Soc. Hortic. Sci.* **49**: 417–419.

Post, K. (1949). *Florist Crop Production and Marketing.* New York: Orange-Judd.

Post, K. (1950). "Controlled photoperiod and spray formation of chrysanthemums." *Proc. Am. Soc. Hortic. Sci.* **55**: 467–472.

Post, K., and Fischer, C. W., Jr. (1952). "Commercial storage of cut flowers." *N.Y. State Ext. Bull.* **853**: 1–14.

Post, K., and Kamemoto, H. (1950). "A study on the number of short photoperiods required for flower initiation and the effect of interrupted treatment on flower spray formation in two commercial varieties of chrysanthemum." *Proc. Am. Soc. Hortic. Sci.* **55**: 477–482.

Rij, E., Thompson, J. F., and Farnham, D. S. (1979). *Handling, Precooling and Temperature Management of Cut Flower Crops for Truck Transportation.* U.S. Dep. Agric. ATT-W-5.

Sacalis, J. N. (1989). *Fresh (Cut) Flowers for Design. Postproduction Guide I.* Columbus, OH: Ohio State Univ.

Sachs, R. M., and Kofranek, A. M. (1979). "Radiant energy required for the night break inhibition of floral initiation as a function of daytime light input in *Chrysanthemum × morifolium* Ramat. *Hort-Science.* **25**(5): 609–610.

Samman, Y., and Langhans, R. W. (1962). "Interaction of temperature and photoperiodism in *Chrysanthemum morifolium.*" *Proc. 15th Int. Hortic. Cong., Nice, France, 1958.* **2**: 400–411.

Strider, D. L. (1985). "Resistance of chrysanthemum cultivars to Fusarium wilt: A compilation." *N.C. Flower Growers' Bull.* **29**(4): 1–5.

Strider, D. L., and Jones, R. K. (1985). "Fusarium-free chrysanthemum cuttings are available." *N.C. Flower Growers' Bull.* **29**(4): 5–7.

Strider, D. L., and Jones, R. K. (1986). "Susceptibility of Chrysanthemum to bacterial leaf spot and bud blight caused by *Pseudomonas cichoril.*" *N.C. Flower Growers' Bull.* **30**(2): 22–24.

Tjia, B. O. S., Rogers, M. N., and Hartley, D. E. (1969). "Effects of ethylene on morphology and flowering of *Chrysanthemum morifolium* Ramat." *J. Am. Soc. Hortic. Sci.* **94**: 35–39.

Whealy, C. A., Nell, T. A., Barrett, J. E., and Larson, R. A. (1987). "High temperature effects on growth and floral development of chrysanthemum." *J. Am. Soc. Hortic. Sci.* **112**(3): 464–468.

Wienke, J. (1968). "Time to avoid crown buds." *Yoder's Grower Circle News.* **69**.

Zacharioudakis, J. N., and Larson, R. A. (1976). "Chemical removal of buds of *Chrysanthemum morifolium* Ramat." *HortScience.* **11**: 36–37.

2

Carnations

C. Anne Whealy

Introduction to Floriculture, Second Edition
Copyright © 1992 by Academic Press, Inc.
All rights of reproduction in any form reserved.

I. INTRODUCTION

*D*ianthus caryophyllus L., family Caryophyllaceae, is native to the Mediterranean area. The genus name comes from the writings of Theophrastus about Dios Anthos, the flower of the gods. Linnaeus chose the species name, caryophyllus, after the genus of clove, as the fragrance from carnations is reminiscent of clove. The name carnation probably comes from its use by the ancient Greeks as a coronation flower.

Today's commercial carnations are the products of more than 200 years of breeding. Carnations are flowered year-round and have a wider color range, larger flowers, and stronger stems than their wild ancestors. Most commercial cultivars are diploids as tetraploids are less productive although tetraploid flowers are larger.

Extensive breeding efforts continue worldwide to provide carnations of higher quality and productivity (Sparnaaij and Demmink, 1983). Present-day breeding objectives include: increasing flower color range; offering new flower forms; extending low light and wider temperature tolerance; eliminating disbudding; improving branching on sprays; increasing yield distribution, postharvest longevity, stem strength, compactness, and resistance to insects and diseases; reducing calyx splitting; and enhancing fragrance.

Carnations have cymose inflorescences, hence they can be cultivated as either standard carnations or sprays. Standard carnations are produced by removing all the lateral flower buds and leaving the terminal flower. Sprays are formed by removing the terminal flower bud, which results in lateral flower bud development.

II. WORLD PRODUCTION CENTERS

A. United States

Sales of cut carnations in the United States continue to increase annually. In 1988, 935 million stems of standard carnations and 24 million bunches of spray carnations were sold in the United States. It is estimated that by 1990, one billion stems of standard carnations, worth $150 million at wholesale, will be purchased. The majority of the carnations sold in the United States are imported.

Originally, the major domestic carnation production area was in New England (Kingman, 1983). With the advent of air cargo transportation, production moved to Colorado and California where higher light intensity was available during the winter months. Today, Colorado and California producers face the same competitive challenges from Colombia that they posed to the New England producers 40 years ago. In 1975, domestic production peaked at 600 million stems annually and has continued to decline since then. This decline is due in part to the increase in energy costs and the increase of highly competitive Colombian imports. Imported carnations in the United States totaled 162 million flowers from Colombia in 1975, less than 20% of the consumption. In 1988, 95% of the standard carnations and 75%

of the sprays sold in the United States were imported from Colombia (Anon., 1989b).

B. Latin America

The export carnation industry has prospered in Latin America because of ideal climatic conditions, affordable labor, and the ability to ship the product to global customers. Argentina, Chile, Colombia, Costa Rica, the Dominican Republic, Ecuador, Mexico, and Peru export cut carnations to North America and Europe (Daum, 1983). Ideal climatic factors contributing to the success of the exported carnation industry include high light intensity, a mild climate requiring only a minimal amount of protection, and uniform temperatures and daylength. Labor costs in Latin America average only $10 per day including benefits—insignificant compared to North American or European labor costs. Limitations to Latin American production do exist, however, such as accessibility, availability of water and energy, and political and economic concerns.

Colombia is the largest producer of carnations. In 1987, Colombian producers exported 1.27 billion stems worth $71 million to world flower markets. The primary market for Colombian carnations is the United States, which imported 883 million stems worth $49 million in 1987. The European community imported 254 million stems, worth $15 million, of Colombian carnations in 1987.

Floricultural products rank third in Colombia's agricultural exports. The carnation is the most important ornamental crop in Colombia, accounting for 41% of the total Colombian cut flower exports in 1987. In 1987, there were 2400 acres in standard carnations and 250 acres in sprays. The Colombian industry consists of over 100 producers employing more than 30,000 people. Obviously, the carnation industry in Colombia is an important source of foreign currency and employment.

The ideal conditions for standard carnation production exist on the high Andean plateau of Sabanade Bogota, 8500 to 12,000 feet above sea level. Night temperatures range between 40° and 45°F, and day temperatures range from 58° to 68°F, year-round. This area has extremely high light intensity, a constant photoperiod of 12 hours, and high organic matter content in soils having a pH of about 5.5 to 6.0.

Carnations produced on the plateau are grown under open plastic structures without winter heating. Production of carnations is in ground beds, which increases the problems with soil-borne diseases. Indeed, *Fusarium* has proven to be a problem in Colombian carnation soils. This has led the Colombian industry to invest in costly pathogen screening and selection programs. Crop diversification is the most economical technique for control, which may indirectly lead to other Latin American countries becoming more competitive carnation producers.

The rapid growth of the Colombian industry has resulted in the United States cut flower producers' success in the imposition of import duties on Colombian flowers; this, too, will indirectly favor other countries' industries. Colombian producers face increasing competition in the future. Spray carnation production is growing in Costa Rica, Peru, and Mexico. Venezuela is encouraging flower production in response to developments in the oil industry. Exacerbating the situation, Colombian producers have experienced a 30 to 35% increase in production costs per year since 1980, concomitant with a 12% devaluation in the Colombian peso. The key to the Colombian producers' continued success will be their ability to become more efficient in order to maintain a competitive edge over other Latin American producers.

C. Europe

The carnation is third in popularity behind the rose and the chrysanthemum in Europe. Germany is the world's largest importer of standard carnations. In 1987, Germany imported 1 billion carnations worth $272 million, mostly from Holland. The largest European producers are Italy, Spain, France, and Holland (Sparnaaij, 1983). Most of the Spanish, French, and Italian product is grown for high-demand local markets.

During the summer and fall, the demand in Western Europe is met by local production; however, in the winter and spring, standard carnations are imported from Colombia, France, Israel, Italy, Kenya, and Spain. In the overlapping months of April/May and November/December there is considerable competition among carnation producers. Colombia supplies the majority of Europe's imported standard carnations, although Spain has made impressive increases in production in the last 10 years.

Similar to the United States, a trend is evident in Europe where the growers in the northern European countries are increasingly competing with imports. Since 1970, European carnation production has declined with the exception of Holland and Spain. This decline can be attributed to an increasing variety of cut flowers in the marketplace, problems with maintaining and stimulating demand, increasing labor and energy costs, urbanization of carnation production areas, and aggressive competition.

Standard carnation production in Holland has remained static from 1960 to 1980, while the total number of carnations produced has increased 3.5 times because of the increase in the popularity of spray carnations. Although the amount of carnation production area in Holland has declined, the Dutch have increased production by building energy-efficient structures and by establishing central auctions (Hoogendoorn and Sparnaaij, 1987), so that the producers can concentrate solely on production because they are assured of a realistic price for their product. In 1987, Holland exported 656 million carnations at a value of $248 million, most going to German markets.

D. Israel

In 1988, there were 570 acres in spray carnation production and 75 acres in standard carnation production in Israel. In general, standard carnations are grown for the local market. Israeli producers supply 80% of the spray carnations for the

European winter market, the majority going to Holland (Mor and Halevy, 1983). In 1987, Israel exported 305 million stems worth $28 million.

Most Israeli carnation production is in unheated polyethylene-covered structures. Rapid expansion of the industry occurred in 1975 with the area doubling annually until 1979, when the industry reached a plateau. This stabilization was the result of increasing production costs and the devaluation of European currencies against the United States dollar. Decreasing profitability has led to a decrease in the number of producers and an increase in productivity.

E. Kenya

The carnation is the primary cut flower crop in Kenya. In 1987, Kenyan producers exported 156 million stems worth $13.4 million. Most of this product was exported to Germany.

Spray carnations are grown outside at an altitude of 6000 feet and standard carnations are grown at 9000 feet under polyethylene (Cox, 1987). Night temperatures range from 40° to 57°F and day temperatures from 77° to 95°F. Total production in Kenya is 400 acres with sprays representing 350 acres. The majority of this production is concentrated at one operation.

III. VEGETATIVE GROWTH

A. Propagation

Cuttings can be purchased from specialty propagators who produce culture- and virus-indexed cuttings. Some growers purchase clean cuttings and produce their own stock in a mother block program. The mother block is kept isolated from production and is used only for cutting production. However, for most producers it is not economically advantageous to maintain stock plants.

Typically, cuttings are harvested with four to five nodes, 4 to 6 inches in length, with a mass of 0.35 ounces. Cuttings are broken, not cut, to prevent pathogen transfer. Cuttings are stored in polyethylene-lined boxes, but the polyethylene is not sealed. Cuttings are stored dry as cuttings stored wet are more susceptible to pathogens.

Cuttings can be stored for several weeks at 32°F. Increasing the storage temperature reduces the time the cuttings can be stored; however, storing cuttings at temperatures above 32°F improves root initiation (van de Pol and Vogelzang, 1983).

Media and benches should be steam-pasteurized so that they are free of pathogens. Attention to sanitation is essential to successful propagation of carnation cuttings. Rooting of carnations occurs within 2 to 4 weeks depending on the cultivar and medium temperature. At a 60°F medium temperature, cuttings root in about 3 weeks. Increasing the medium temperature to 70°F decreases rooting time to 2 weeks. Rooting hormones, such as napthaleneacetic acid (NAA), enhance rooting. It is advisable to use powder, rather than liquid rooting hormones, to reduce the chances for the spread of pathogens. Full light and mist are recommended for optimal and rapid rooting.

B. Planting and Media

Planting dates depend on market demand. Plantings are typically made between April and September in the Northern hemisphere and from September to April in the Southern hemisphere. It is common for standard carnations to be grown for 2 years, with half of the production area planted each year. Spray carnations are usually replanted each year with plantings staggered every 3 to 4 months to even out production. With ground beds or in field culture, a 1-year cropping schedule is recommended to reduce pathogen potential during the second year. Raised beds are preferred over ground beds for improved pathogen and insect control, labor usage, aeration, and nutritional concerns.

Excellent drainage and aeration are prerequisites for producing high quality carnations. Benches or beds should be well drained to prevent the buildup of excess water and soluble salts. Sphagnum peat can be added to improve the porosity of the medium. Over time, however, peat decomposes and can actually reduce porosity. For maximum porosity, coarse inert media such as calcareous gravel, coarse sand, perlite, fired clay aggregates, or volcanic scoria can be used successfully, but irrigation frequency is higher and uniformity of watering must be maintained. Ideally, the pH of the medium should be maintained between 6.0 and 6.8.

Fumigation with methyl bromide/chloropicrin or steam pasteurization is necessary prior to planting to eliminate pathogens, nematodes, and weed seeds. However, because carnations are susceptible to bromide toxicity, methyl bromide should only be used on sandy soils with adequate aeration and leaching prior to planting. Aerated steam is preferred over steaming at higher temperatures as energy costs are reduced and more beneficial organisms survive this treatment.

Cuttings should be planted at the same level as they were in propagation. Cuttings that are planted shallow establish faster due to increased aeration and are less susceptible to stem rot caused by *Rhizoctonia*. If 25% of the cuttings fall over after the initial watering, they have been planted at the proper depth. A fungicidal drench should be made at planting time to reduce fungal infection. Cuttings should be misted or syringed for the first few days following planting to maintain turgor.

C. Spacing and Support

Spacing depends on the cultivar, light level, timing, and pinching programs. High-density spacing is used for 1-year crops, and more generous spacings are used for 2-year crops to maintain high yields. For spray or 1-year standard carnations, cuttings should be planted at a spacing of 4 × 6 inches or a plant density of six plants per square foot. Two-year crops are planted on 6 × 6 inch or 6 × 8 inch spacings or four and three plants per square foot, respectively.

Support wires reduce the incidence of weak and crooked stems, which decrease quality and productivity. Wires also reduce wind damage in field production. Layers of box wire are stretched over the bed prior to planting and serve as marking guides for planting. As the crop grows, layers are raised. Plastic mesh, string, and bamboo canes can be used for upper layers, but metal wire is required for lower layers. On standards, four to six levels of wire may be required for 1-year crops and up to eight on 2-year crops. The lower layer should be 6 inches above the soil line with successive layers 8 inches apart. Labor is required to continuously maintain the plants within the wire support.

D. Pinching and Pruning

Pinching is done by hand to reduce the possibility of transmitting pathogens by using knives or clippers. Most spray-type cultivars break freely without pinching, but some require a soft pinch. Most standard cultivars are not freely branching and are pinched by using one of four pinching systems that affect time of flowering, yield, and quality.

The "single pinch" method results in the most rapid flowering, but all stems flower about the same time. After the plants have become established, about 3 to 4 weeks after planting, the shoot apex is pinched, leaving four or five lateral shoots to develop. Plants are pruned to four breaks on outside rows and to three on interior rows. The amount of pruning influences both yield and grade. Permitting the maximum number of shoots to develop increases yield, but mean flower grade suffers, whereas pruning decreases total yield, but more flowers of higher grades are produced.

Carnations can also be pinched using a method called a "pinch and a half." After the initial pinch, one-half of the breaks from the first pinch are pinched a second time after they have attained size, about 5 to 7 weeks after the first pinch. Although this method results in lower yields on the first harvest compared to the single pinch method, it permits more evenness of production over time.

In the "double-pinch" method, breaks from the first pinch are pinched when they are 4 to 5 inches in length, approximately 5 weeks after the first pinch. This method is not commonly used because production is extremely high at one time. Weak stems after harvest are also a problem.

A fourth method, "single pinch plus pull pinches," is similar to double-pinching except the second pinch consists of only removing the growing tip rather than a harder pinch. The larger shoots are pull-pinched for 2 months. This method evens out production for an extensive cropping.

If optimal production temperatures cannot be maintained during the summer, the plants are topped to 12 to 16 inches or gradually pruned back. Severely pruning the plants eliminates unwanted summer production and renews plants for production in the second year. Pruning debris should be removed to prevent problems caused by pathogens.

E. Irrigation

Although overhead irrigation can be used until flower buds are visible, surface irrigation systems are commonly used throughout carnation production. The types of irrigation systems used include the Gates sprinkler system, plastic soaker hoses, drip irrigation, and furrow and flood irrigation. Drip lines covered with white plastic mulch are quite popular in Holland (Van den Heuvel, 1987). The white plastic increases the utilization of available light and provides the benefits of decreased evaporation and greenhouse humidity, increased irrigation efficiency, and enhanced upper root growth.

For carnations the optimal moisture for the medium should be between 300 to 500 cm tension. Under low light levels, tensions less than 300 cm produce soft elongated growth and low flower quality. Toning the plants by withholding water until they are almost wilted increases stem strength on plants grown under low light. Water tensions greater than 500 cm under high light intensities result in poor flower quality, smaller flowers, and hard growth. Irrigations should be frequent once flowers begin to develop to enhance flower color, size, and longevity.

F. Nutrition

The optimal nitrate level in the medium is between 25 and 40 ppm. During low light conditions, nitrate, rather than ammoniacal nitrogen, is applied for toning and for stronger stems. However, excess nitrogen causes weak stems in carnations. Low nitrogen levels inhibit flower bud initiation. Steam sterilizing at 212°F may cause ammonium toxicity. Nitrate fertilizers should be used until nitrifying bacteria have been reestablished.

The optimal potassium level in the medium for carnations is between 25 and 40 ppm. Low potassium levels reduce yield, grade, stem strength, and longevity. Calyx tip dieback is associated with potassium deficiency.

Phosphorus deficiency reduces growth and causes stunted plants. The optimal phosphorus level in the medium is 5 to 10 ppm and is attained by adding superphosphate or triple superphosphate. This fertilizer is added and mixed into the soil before adding limestone to optimize phosphorus availability.

Optimal medium calcium levels are 150 to 200 ppm. Calcium deficiency appears as crescent-shaped necrotic lesions 1 to 2 inches from the leaf tip and as calyx scorch. Depending on medium pH, dolomitic limestone or gypsum is added to attain the desired calcium level.

Boron deficiency increases calyx splitting (Adams et al., 1979) and induces bud abortion. Less than optimal foliar boron levels (less than 20 to 25 ppm) causes shortened internodes, clubbiness, distorted flower buds, and "witches-broom" symptoms. Boron deficiency is more evident with carbon dioxide injection and varying light levels. Boron toxicity appears as leaf tip necrosis on young leaves.

The duration and temperature of the steam pasteurization, low pH (below 5.5), or poor drainage may lead to manganese toxicity on carnations. Manganese toxicity reduces carnation yield and quality and is diagnosed by stunted and hard growth, leaf distortion, and a distinctive purplish leaf tip burn on the lower leaves. Reducing the length of steaming, using aerated steam (140°F), increasing the soil pH, and adding chelated iron will reduce the incidence of manganese toxicity. The recommended foliar manganese level for carnations is about 35 ppm.

Soluble salt levels should be kept low while maintaining optimum fertility. High salinity levels will decrease yield, fresh flower weight, and mean flower grade. A specific conductivity of 2.5 millimho/cm, using the saturated paste extract, is optimum.

G. Temperature

Temperatures must be adjusted to light intensity to enhance grade and growth. Temperatures should be reduced with decreasing solar radiation and short photoperiods during the winter. Production temperatures should increase in the spring with increasing duration and amount of ambient light.

Warmer temperatures are required for producing high quality cuttings than for producing high quality flowers. Moreover, because low temperatures enhance reproductive development on cuttings, stock plants are grown at temperatures higher than for flowering plants. The optimum day temperature for young plants is 5° to 9°F higher than for flowering plants. Spray carnations will tolerate greater temperature extremes than standard carnations.

H. Carbon Dioxide

Carbon dioxide levels affect both growth and quality. Low levels of carbon dioxide, 100 to 150 ppm, in closed greenhouses during the day inhibit growth as the rate of photosynthesis becomes equal to the respiration rate. Thus, ventilating or adding supplemental carbon dioxide is necessary to maintain adequate growth and enhance quality.

Carbon dioxide injection is done during the day, and the amount injected varies with light level. The greenhouse carbon dioxide level should be maintained at 300 to 500 ppm on cloudy days and at 750 to 1500 ppm on sunny days. Carbon dioxide injection (1000 ppm) can increase yields 30 to 35%, reduce the time between harvesting, and increase percent dry weight. Increased yield and reduced time between harvests adds up to increased and more uniform production. An increase in percent dry weight is manifested in enhanced stem strength and flower longevity.

With supplemental carbon dioxide, irrigation and fertility levels must be increased to ensure that optimal growth is maintained. Light level and temperature must also be optimal with the addition of carbon dioxide into the greenhouse environment. Slightly higher temperatures are advantageous with carbon dioxide injection as benefits are not realized until the day temperature reaches 68° to 72°F.

IV. FLOWERING

The strategy in carnation production is to maximize productivity during periods of high demand. Timing of flowering can be predicted as long as the environmental conditions are monitored and adjusted accordingly. Time to flower varies with photoperiod, temperature, light intensity, and cultivar. Hence, flowering is programmed by planting date, temperature, pinching method, light level, photoperiod, and cultivar response.

A. Disbudding

Disbudding of standard carnations is done as soon as lateral buds can be easily removed and should be done on a continuous basis. For this reason, it is the most labor-intensive aspect of standard carnation production but is essential to maintain flower size and quality. Most standard carnation growers remove lateral buds 12 to 16 inches down the stem, or to six nodes below the terminal flower bud.

In spray carnations, the terminal flower bud is removed to give a more uniform and open spray. This can also be done with standard carnations. Removing the terminal bud on standards will allow the laterals to flower at about the same time and increases yield, but reduces flower grade.

B. Light and Temperature

Dianthus caryophyllus is a long-day plant that naturally flowers in the summer. Breeding has resulted in carnations being perpetually flowering, but photoperiod, light intensity, and temperature can modify the flowering response.

Carnations have been classified as quantitative long-day plants (Blake, 1957) as long days promote and short days delay flower initiation. Photoperiod does not affect subsequent flower development. Long-day treated plants flower sooner and more uniformly, but have fewer nodes and longer internodes than plants produced under short days. More rapid bud initiation translates to reduced harvest period and total crop time. Lighting, however, does not enhance quality or increase flower size. Most cultivars respond to 2 to 3 weeks of lighting. The effectiveness of the long-day treatment depends on the cultivar, the shoot position, and the stage and rate of development (Sparnaaij *et al.*, 1987).

Although there are cultivar differences in response to photoperiod, the critical photoperiod for most spray and standard carnations is about 13 hours. Long days can be given as continuous, as an extension, or as a night-break treatment. Cyclic lighting is as effective as continuous lighting on standards and sprays provided that it is given throughout the dark period and the total amount of light per dark period is similar (Van de Hoeven, 1987).

Long-day treatments inhibit lateral branching, so stock plants are produced under short days to increase lateral shoot production and cutting quality, and to ensure the production of vegetative cuttings. Carnation cuttings from short-day treated stock plants root better than plants from long-day stock plants, however, long days during propagation promote rooting (Pokorny and Kamp, 1960). After pinching, short days are given to promote branching and are continued until 70% of the shoots have 10 visible nodes (Healy and Wilkins, 1983). Long days are only effective after a certain critical leaf number has been attained. Long days are provided for one to several weeks to promote flower initiation and to increase lateral stem length. After long days, plants flower in about 12 to 16 weeks depending on temperature, cultivar, and light intensity. Short days are then given as extended lighting will delay or even completely inhibit the formation of new shoots.

The rate of flower initiation in carnation is also affected by irradiance (Bunt et al., 1981). The amount of light has had a tremendous influence on determining the world's carnation production centers. A minimum amount of light to sustain carnation growth is 21.5 klx.

There is an interaction between amount of ambient light and the number of long days necessary for flower initiation (Bunt and Powell, 1983). Twenty to thirty long days are required for low light levels and 7 to 14 long days under high light conditions. Similar to the effect of photoperiod, light level had little effect on the rate of flower development. This response is related to the effect of light intensity on photosynthetic assimilate supply. High light increases flower size, number of petals, and stem diameter. High plant densities or low light intensity reduce the number of flowering shoots, flower quality, and dry and fresh weight (Mastalerz, 1983). Supplemental lighting during low light periods can increase growth and development and can enhance flowering.

Temperature affects the rate of growth and flower development, productivity, quality, and longevity. Temperatures less than optimal will improve flower quality, but yields will be lower, development will be slower, and the incidence of malformed flowers and splitting is increased. Temperatures higher than optimal enhance flower development and productivity but cause weak stems, small flowers, and poor flower color.

Flower bud initiation is most rapid and most uniform at temperatures less than 60°F and is delayed by temperatures above 60°F. Subsequent flower development is promoted by higher temperatures, but the amount of light determines the optimal temperature for flower development. Supraoptimal temperatures, above 90°F, delay development. Low finishing temperatures increase harvesting time.

C. Calyx Splitting

Calyx splitting is a major problem in standard carnations as flowers are asymmetrical and value is reduced (Fig. 1). Carnation flowers with split calyxes are worth only 20 to 25% as much as flowers with intact calyxes. Cultivars vary in susceptibility to splitting. Those with short and broad calyxes are less likely to split than those with long and narrow calyxes. Splitting is attributable to conditions that favor the proliferation of petals and accessory flowers or conditions that prevent normal calyx expansion.

Low temperatures cause an increase in the production of extra growth centers inside the calyx, but the calyx is not able to contain these extra petals or petaloids and splits. Rapid temperature drops at night enhance growth center proliferation. A gradual reduction in night temperature (over several hours) will not cause splitting. Carnation buds are most sensitive to this rapid temperature drop approximately 3 to 5 weeks before harvest or when buds are beginning to open. One night at 40°F will greatly increase the incidence of splitting. Warm night temperatures (55° to 60°F) are not conducive to splitting. Moreover, wide variations in day versus night temperatures also favor splitting. Day and night temperatures should not vary more than 18°F to reduce the incidence of calyx splitting.

Both the boron and nitrogen contents in carnations are important factors in calyx splitting (Adams *et al.*, 1979). Low nitrogen, high ammoniacal nitrogen, or low boron levels enhance splitting by attributing to weak calyxes. Higher nitrate to ammoniacal nitrogen ratios during low light periods are recommended to reduce splitting.

Fig. 1. A major problem in standard carnation production, flowers with split calyxes are asymmetrical and value of the stem is reduced 75 to 80%.

Small rubber bands and clear tape have been used to maintain calyx integrity. Calyxes can be banded when the bud shows a small opening. Banding is often done when carnations are grown in nonheated outdoor or open-house production where temperature control is not possible.

Precisely controlled temperatures and optimal fertility levels can virtually eliminate calyx splitting. Additionally, choosing cultivars that are less prone to splitting will reduce this problem.

V. PESTS

A. Pathogens

Bacterial wilt in carnations is caused by *Pseudomonas caryophylli* in a synergistic relationship with *Corynebacterium*. Bacterial wilt is most common on older plants and is most severe at high temperatures (75° to 95°F). The earliest symptom of bacterial wilt is wilting; the wilt affects one or more branches or the entire plant. Leaves are dull and grayish, with subsequent chlorosis and death. Vascular tissue degraded by bacterial wilt appears frayed and yellowish. The diseased tissue is sticky and this stickiness distinguishes bacterial wilt from other pathogens.

Pseudomonas is a soil-borne bacterium that enters plant roots and stem bases through wounds or openings. Thus, bacterial wilt can be controlled by steam-pasteurizing the medium and using clean cuttings from culture-indexed stock plants. Tools and hands should be cleaned and disinfected prior to handling cuttings or established plants, and all diseased plants should be destroyed.

Symptoms of Fusarium wilt caused by *Fusarium oxysporum f. dianthi* are similar to those of bacterial wilt. *Fusarium oxysporum* is regarded as the most devastating fungal pathogen affecting carnations. Warm climates and poorly drained soils are conducive to *Fusarium* infection.

The Sim cultivars are very susceptible to *Fusarium*, however, resistant spray and standard carnation cultivars are commercially available. Newly planted cuttings are more susceptible to wilt caused by *Fusarium* than are established plantings.

Fusarium enters the plant through the roots and develops in the vascular tissue. Plants appear distorted due to the wilt. The distortion and the absence of the stickiness of the infected tissue differentiate wilt caused by *Fusarium* from bacterial wilt. Leaves of plants infected with *Fusarium oxysporum* turn gray and then straw yellow. The vascular tissue becomes brown.

Pseudomonas carophylli and *Fusarium oxysporum f. dianthi* can survive in the soil for 5 to 10 years. Like *Pseudomonas*, the development of *Fusarium* is promoted at soil temperatures of 75° to 95°F. Poor aeration and drainage and overwatering are also conducive to *Fusarium* infection.

Plant calcium levels have been shown to be related to *Fusarium* infection, because calcium-deficient carnations are more sensitive to *Fusarium* and the severity of infection is higher with reduced calcium content (Blanc *et al.*, 1983). As a binding component of cell walls, calcium decreases the susceptibility of the cell walls to enzymatic pathogenic degradation. Carnation cultivars that are more resistant to *Fusarium* tend to accumulate more calcium than more susceptible cultivars.

It is extremely difficult to eliminate this pathogen from ground beds as it can survive in the subsoil where steam or fumigants do not penetrate. Commonly, carnations are grown in raised benches or in artificial media to eliminate this source of contamination. However, spores are transmitted by wind and can infect raised benches or artificial media.

The incidence of *Fusarium oxysporum* can be reduced by integrating steam pasteurization, fumigating fields, growing in raised beds or artificial media, increasing the pH and calcium levels of the medium, reducing nitrogen levels, lowering temperatures, maintaining good sanitation practices, establishing fungicidal drenches, and planting clean cuttings of resistant cultivars.

Disease suppression by antagonistic bacteria in certain soils shows promise for *Fusarium* control in carnations (Garibaldi and Gullino, 1987). Soils with naturally occurring suppressive bacteria are added as amendments to sterilized carnation beds to reduce the incidence of wilt due to *Fusarium*. Dipping the roots of carnations into a soil suspension with suppressive bacteria prior to planting has also been shown to be effective in reducing *Fusarium* infection (Yuen *et al.*, 1983).

Fusarium roseum f. cerealis causes Fusarium stem rot or stub dieback on carnations and is a concern during carnation propagation and harvesting. *Fusarium roseum* survives as a soil saprophyte and can be transmitted by air currents. This fungus enters the stubs left after flower harvest and causes dieback and subsequent girdling of main branches. On cuttings, Fusarium stem rot causes basal stem rot, and reddish lesions are evident at the cutting base.

Planting cuttings too deeply will cause lower leaf abscission, which provides an entrance for the pathogens. High soil moisture and soil temperatures over 75°F contribute to Fusarium stem rot susceptibility. Cuttings from stock plants maintained with high nitrogen levels are also more susceptible.

Prior to cutting harvest, stock plants should receive low levels of nitrogen and high levels of phosphorus and calcium to reduce susceptibility of cuttings. Steam pasteurization, clean tools, removal of plant debris, good drainage, low medium temperatures, drenching of newly planted carnations, and regular fungicide sprays on stock plants will reduce the incidence of Fusarium stem rot.

Fusarium bud rot, causal agent *Fusarium tricinctum*, is transmitted by mites. The fungus causes rot of carnation buds, and mites can be found in decayed tissue. Miticides are effective in reducing the spread of this pathogen and infected buds should be removed.

Rhizoctonia stem rot, caused by *Rhizoctonia solani*, is common in newly planted carnations. Plants infected with *Rhizoctonia* will appear pale green and wilted. Brown stem lesions are evident at the soil level, and basal stems are water-soaked and may be girdled.

Plant debris, unclean tools and hands, and nonpasteurized or nonfumigated soils will increase the probability of *Rhizoctonia* infection. Planting cuttings too deeply and compacting the medium also contribute to *Rhizoctonia* infection. Warm, moist conditions favor disease development.

Sclerotia of *Rhizoctonia solani* can survive for long periods of time in the soil and in plant debris. Fungicide drenches are effective, and, as with other pathogens, sanitation and removal of diseased plants are recommended.

Alternaria blight, leaf spot, or branch rot is caused by *Alternaria dianthi*. Early symptoms are small purple spots on the lower or upper surface of lower foliage. A yellow-green halo then appears around the spots. Eventually, the spots expand and coalesce, and the interiors of the lesion appear sunken and brownish-gray. Lesions may also appear on the stem, eventually girdling it. Clusters of black spores are evident on infected leaves and stems. Like *Fusarium oxysporum*, *Alternaria dianthi* can penetrate plant tissue directly.

Alternaria dianthi thrives in dead plant material and can be spread by overhead or splashing water. Temperatures in the range of 75° to 95°F promote *Alternaria* growth.

Overhead watering should be avoided and temperatures maintained for optimal carnation, and not *Alternaria*, development. Fungicides will control *Alternaria*, but benomyl increases the susceptibility of carnations to *Alternaria* attack.

Carnation rust, caused by *Uromyces dianthi* or *Uromyces caryophyllinus*, is quite common on carnations. Long, narrow, reddish-brown lesions appear on both sides of the leaves, on stems, and on flower buds. Leaf curl is evident and plants are stunted.

Like *Alternaria*, rust requires freestanding water for development. Keeping the foliage dry, venting, and maintaining optimum temperatures reduces rust infection.

Botrytis cinererea causes Botrytis flower blight of carnation. The blight is common during the storage of cut carnations and affects the petals. High humidity causes moisture to collect in the opening flower and provides a favorable environment for the development of the pathogen. If high humidity is maintained, gray spore masses will be evident. Petal edges appear water-soaked and eventually entire petals and flowers are affected.

Reducing humidity by venting and heating when carnations are beginning to show color will inhibit Botrytis flower blight development. Watering should be done in the morning to prevent high moisture conditions. Fungicides can be sprayed before harvest to reduce infection during storage and transport of cut flowers.

Phialophora wilt, caused by *Phialophora cinerescens*, is evident as a pale green discoloration of the leaf margins that may subsequently turn red. Wilting and chlorosis are later symptoms. Stem cracking may also occur. Cool soil temperatures favor *Phialophora* development and spread. Steam pasteurization is effective in reducing *Phialophora*.

Heterosporium echinulatum causes carnation ring spot or fairy ring. The symptoms are raised circular lesions with central depressions surrounded by a yellowish ring. Keeping the foliage dry, avoiding overhead water, and applying fungicide will inhibit carnation ring spot problems.

Zygophiala (greasy blotch) dissolves the leaf cuticle and causes small oily, radiating patterns to develop on leaves and stems of carnations. High humidity favors greasy blotch. Ventilation and fungicides are effective control measures.

Although carnations do not typically die from viral infection, reduced flower quality and production can severely affect marketability and profitability. The four most common carnation viruses are carnation streak virus (CSV), carnation mosaic virus (CMV), carnation mottle virus (CMoV), and carnation ringspot virus (CRSV). It is not unusual for carnations to be infected simultaneously with more than one virus (Lommel *et al.*, 1983).

Carnation streak virus (CSV) symptoms are yellow or reddish spots paralleling the leaf veins. Spots become more prevalent and lower leaves may become chlorotic and die. Carnation mosaic virus (CMV) is characterized by leaf mottle, irregular light green blotches on the leaves, and vein paralleling in the flowers. Carnation mottle virus (CMoV) is the most common carnation virus and is recognized by a faint leaf mottle and streaking on the flowers. Carnation ringspot virus (CRSV) shows as irregular yellow or gray spots. Leaf streaking is often observed, and leaf margins may be wavy or irregular in shape.

Sanitation can help to control these viruses. Vectors, such as the green peach aphid (*Myzus persicae*) that transmits CRSV and CMV, should be controlled. Viruses in carnations can also be transmitted through vegetative propagation and contaminated harvesting tools. Disinfection of hands and sterilization of tools reduce virus transfer.

Viruses can be kept in check by using clean cuttings from certified virus-free stock plants. Elaborate virus indexing programs are employed by specialist propagators to ensure production of virus-free carnation cuttings. Indexing involves subjecting carnations to heat therapy (100°F dry heat for 2 months) to inhibit the movement of viruses to the meristem. After the treatment, meristems are excised, cultured, and then tested for the presence of virus with indicator plants or serological tests such as ELISA. Shoot tips or meristems are excised from certified clean plants and propagated for the nucleus stock block. The specialty propagator continuously renews the certified nucleus block. The plants are continuously checked for trueness to type, productivity, pathogens, and viruses. Plants are then selected to provide cuttings for increase stock blocks.

B. Insects and Mites

A known vector of CRSV and CMV, the green peach aphid, *Myzus persicae*, can be the most serious insect pest on carnations. Aphids also reduce plant vigor and secrete honeydew, which reduces quality. A number of species of thrips are common on carnations. Thrips nymphs and adults suck plant sap and cause distortion, streaking, and spotting of flowers, foliage, and stems. Leaf miner maggots (*Liriomyza*) tunnel inside the leaves, leaving serpentine mines.

Two-spotted spider mites, *Tetranychus urticae*, cause speckling of leaf undersides as they remove plant sap. Severely infested carnations can succumb to spider mites. *Aceria paradianthi*, the carnation bud or shoot mite, causes distorted and stunted new growth. These Eriophyidae mites are extremely small and live at the bases of leaves and stems and under the calyxes.

In field production or ground beds, nematodes, especially *Meloidogyne* species, can cause root injury and enhance the transmission of wilt pathogens.

Integrating chemicals with cultural and biological controls is an economical and environmentally sound pest control strategy. Reduced temperatures inhibit pest population increases. Destroying all weeds and plant debris in and around the production area, which can harbor mites, aphids, and thrips, will also decrease pest populations.

VI. POSTHARVEST HANDLING

Carnations are extremely responsive to postharvest treatments. Longevity can be increased two to three times with various postharvest treatments. However, postharvest procedures can account for 30% of the total cost of production.

Longevity is also contingent on production conditions. Because longevity is dependent on the carbohydrate status of the stems at harvest, temperature, light, and carbon dioxide levels influence longevity. Thus, bolstering the percent dry matter will increase longevity.

A. Harvesting

Flowers are harvested with sharp knives or pruning shears. On standard carnations two to three nodes and on spray carnations three to four nodes are left on the shoots for the next flowering. Flowers should be cut in the early morning when plants are turgid. Standard carnations are harvested as open flowers or in the bud stage. Spray carnations are harvested with two flowers open and the rest showing color. Flowers are handled carefully to avoid breakage and bruising. It is important to expose flowers to a 40° to 48°F environment as soon as possible to reduce plant temperature. Precooling the flowers maintains quality and increases longevity.

B. Storage

Carnations can be stored longer than almost any other cut flower crop without a decrease in quality or longevity. Peak demands can be met by storing carnations over periods of time. Long-term cold, dry storage is commonplace in carnation production, however, low pressure or hypobaric storage can also be used for carnation flower storage.

Cold storage and shipping environments should be maintained at 33° to 34°F and provide good air circulation and high relative humidity (90 to 95%). At these low temperatures the production of ethylene is inhibited, thus the rate of senescence is reduced.

Cardboard shipping and storage cartons are lined with polyethylene and newspaper. Newspaper is placed on top of the polyethylene, in between layers of flowers, and on top of the flowers to absorb condensation and excess moisture and to prevent bruising. Flowers should not be in contact with the polyethylene, and it should not be sealed. Cross cleats are used in the boxes to prevent the bunches from shifting during transport. The cartons are precooled to 32°F prior to closing the cartons.

An air space should be maintained around the cartons in storage and during shipping to maximize air circulation. Moreover, shipping cartons should have holes to ensure that the cold air can flow freely through the carton to avoid ethylene buildup in the carton. The air is forced through the vent holes in the cartons at a flow rate of 600 to 900 feet per minute, and the polyethylene and newspaper should not restrict the flow.

The water balance in the cut carnation should be at an optimal level for maximum longevity. Moisture stress and low humidity can lead to reduced longevity and petal burn. At the other extreme, excessively high humidity or freestanding water on the flowers and leaves can promote *Botrytis* and other petal blight diseases. Preventing condensation is the best method to avoid petal blight problems. Petal blights not only reduce quality but also increase ethylene evolution from infected flowers.

C. Ethylene

Carnations are sensitive to endogenous and exogenous ethylene. Therefore, carnations should not be stored with fruits, vegetables, cuttings, or cut flowers that release ethylene gas. Exposure of carnation flowers to ethylene accelerates senescence and "sleepiness" and reduces water uptake (Halevy *et al.*, 1983). Senescence is characterized by a decrease in fresh and dry weight, a climacteric rise in respiration and ethylene production, and an increase in membrane permeability. Sleepiness refers to the upward curling, cupping, or bending of the petal margins to the extent that the flower is partially closed. Besides exposure to ethylene gas, sleepiness can also be caused by calcium deficiency, lack of water uptake or dehydration, or high temperature during storage.

The stage of flower development influences the response of the carnation to ethylene (Camprubi and Nichols, 1978). With increasing stage of flower development, carnations become increasingly sensitive to ethylene. Buds or immature flowers are relatively tolerant of ethylene and are increasingly so at low temperatures.

D. Stage of Development

Carnations harvested in tight bud or partially open stages store for longer periods of time and have greater longevity than flowers harvested when fully open (Reid *et al.*, 1983). Tight buds produce less ethylene, have lower respiration rates, and are more resistant to fungal pathogens. Open flowers can be stored at low temperatures for a maximum of 2 to 3 weeks. Buds can be stored for longer periods of time, such as 6 to 10 weeks. However, for maximum longevity, buds are not stored for longer than 4 to 5 weeks. It has been demonstrated that carnations harvested in the tight bud stage and preconditioned in silver thiosulfate and sucrose can be successfully stored dry for 20 to 24 weeks at 32° to 34°F.

There are additional benefits to harvesting buds rather than open flowers. Cutting at the bud stage helps to meet peak demands, increases yield, salvages flowers that would be discarded before replanting, and decreases the time between croppings (Mynett et al., 1983). Moreover, buds are easier to handle and can be packed tighter than open flowers, hence more buds can be stored per unit area, and transportation costs are lower.

E. Postharvest Treatments

Silver is a powerful and specific inhibitor of ethylene action, hence silver delays flower senescence. Although silver nitrate can be used to enhance carnation longevity, silver thiosulfate (STS) is preferred due to the lower phytotoxicity potential. Carnation buds are conditioned with STS and sucrose before storing to improve longevity. A 20-hour STS pulse treatment can increase flower longevity three to four times compared to nontreated carnation flowers, and longevity is similar to nonstored flowers. Pulsing temperatures of 68° to 80°F maximize uptake of STS. Because of environmental concerns with silver, however, other inhibitors of ethylene action or synthesis, such as cytokinins, are being tested and have shown promise as carnation pretreatments.

Adding sucrose (10%) to the STS pulse or conditioning solution further increases carnation quality and longevity. Sucrose serves as an energy source and helps maintain membrane structure, function, and favorable osmotic potential.

Fungicides and bactericides are also used in the pulsing and conditioning solutions to reduce fungal and bacterial infections (Farnham et al., 1978). Bacterial or fungal contamination of the solution reduces water uptake by plugging the vascular system. Distilled or deionized water is recommended because of the problems with water quality and bacterial content.

Flowers cut in the bud stage are subsequently opened in bud opening solutions at 76° to 80°F. Development is accelerated by this treatment compared to opening in the greenhouse, and quality is typically higher than flowers cut at an open stage. Bud opening solutions contain high sucrose levels (10 to 12%) that promote opening and enhance longevity. Generally, STS and a biocide are added to the opening solution. Nonionic surfactants or detergents are sometimes added to reduce surface tension. Successful opening of buds requires constant light, at an intensity at least 10.8 klx, and 70°F. Tight buds require 4 to 6 days to open, while more mature buds harvested with the petals straight up require only 2 to 3 days to open.

F. Grading

After low temperature storage, long-term transport, or both open flowers are graded, bunched, recut, and rehydrated in a warm preservative (Fig. 2). Bunches of standard carnations are sold 25 stems to the bunch, and spray carnation bunches have 35 opened or partially opened flowers on seven to ten stems.

Carnations are graded on measurable characteristics such as stem length, stem strength, and flower diameter. Other characteristics such as the presence of pathogens; insects; damage caused by either pathogens, insects, or both; foliar or flower blemishes; sleepiness; malformed flowers; crooked stems; split calyxes; and faded or off-color flowers are determined judgmentally.

VII. SUMMARY

In the last 20 years, the carnation industry has taken on a more global appearance. Market share continues to decline in developed countries and to shift to undeveloped countries. Third World countries in Latin America, Asia, and Africa are emerging as carnation exporting giants. Increasingly, political and economic

Fig. 2. After opening, standard carnations are graded, bunched, and stems recut prior to sale.

concerns will become important in these countries. As the industry moves into the 21st century, there are increasing demands for horticultural technology to be adaptable to the world's new carnation producers. Technology will be tailored to their particular environments, soils, pathogens, and insect pests; to increased productivity and predictability of production; and to continuity of selection and breeding programs for these production centers.

Moreover, the future of the carnation industry depends on increasing market demand and the development of new markets. As more new products come onto the market, the position of the carnation may decline, but if new types of *Dianthus* are offered, the demand for carnations may increase.

REFERENCES

Adams, P., Hart, B. M. A., and Winsor, G. W. (1979). "Some effects of boron, nitrogen and liming on the bloom production and quality of glasshouse carnations." *J. of Hort. Sci.* **54**: 149–154.

Anonymous. (1965). *Carnation Manual*, 10th ed. New York: F. C. Gloeckner and Co., Inc.

Anonymous. (1968). "Yoder carnations." *Grower Circle News* **67**: 1–15.

Anonymous. (1969). "Yoder carnations." *Grower Circle News* **79**: 1–15.

Anonymous. (1970). "Yoder carnations." *Grower Circle News* **90**: 1–15.

Anonymous. (1988). "Statistical information on the international production and trade of fresh cut flowers." Liaison Committee for Information on Cut Flowers, Bezuidenhoutseweg, The Netherlands.

Anonymous. (1989a). *Floriculture Crops, 1988 Summary.* Agricultural Statistics Board, National Agricultural Statistics Service, United States Department of Agriculture, Washington, D.C.

Anonymous. (1989b). *Ornamental Crops National Market Trends.* Agricultural Marketing Service, United States Department of Agriculture, Washington, D.C.

Besemer, S. T. (1980). "Carnations." In *Introduction to Floriculture*, edited by R. A. Larson. New York: Academic Press.

Blake, J. (1957). "Photoperiodism in the perpetual flowering carnation." In *Report of the 14th International Horticulture Congress of the Netherlands.* Wageningen, The Netherlands: H. Veenman and Zonen.

Blanc, D., Tramier, R., and Pallot, C. (1983). "Calcium nutrition and its effect of the receptivity of carnation to *Fusarium oxysporum* f. sp. dianthi." *Acta Hort.* **141**: 115–123.

Bunt, A. C., and Cockshull, K. E. (1985). "*Dianthus caryophyllus.*" In *Handbook of Flowering* edited by A. H. Halevy, vol. 2, pp. 433–440. Boca Raton, FL: CRC Press.

Bunt, A. C., and Powell, M. C. (1983). "Quantitative response of the carnation to solar radiation, long days and supplementary light source." *Acta Hort.* **141**: 139–150.

Bunt, A. C., Powell, M. C., and Chanter, D. O. (1981). "Effects of shoot size, number of continuous-light cycles and solar radiation on flower initiation in the carnation." *Sci. Hort.* **15**: 267–276.

Camprubi, P., and Nichols, R. (1978). "Effects of ethylene on carnation flowers (*Dianthus caryophyllus*) cut at different stages of development." *J. of Hort. Sci.* **53**: 17–22.

Cox, R. J. (1987). "Carnation production in Kenya." *Acta Hort.* **216**: 43.

Daum, P. L. (1983). "Carnation trends in Latin America." *Acta Hort.* **141**: 245–248.

Farnham, D. S., Udea, T., Kofranek, A. M., Halevy, A. H., McCain, A. H., and Kubota, J. (1978). "Physan-20, an effective biocide for conditioning and bud opening of carnations." *Florists' Rev.* **162**(4190): 24–60.

Garibaldi, A., and Gullino, M. L. (1987). "Fusarium wilt of carnation: Present situation, problems and perspectives." *Acta Hort.* **216**: 45–54.

Halevy, A. H. (1987). "Recent advances in postharvest physiology of carnations." *Acta Hort.* **216**: 243–254.

Halevy, A. H., Borochov, A., Faragher, J. D., Harel, R., and Mayak, S. (1983). "Physiological changes in carnation petals as affected by storage and transport." *Acta Hort.* **141**: 213–220.

Healy, W., and Wilkins, H. (1983). "Photoperiod control of lateral branching and flower production in carnations." *Acta Hort.* **141**: 151–156.

Holley, W. D., and Baker, R. (1963). *Carnation Production.* Dubuque, Iowa: Wm. C. Brown.

Hoogendoorn, C., and Sparnaaij, L. D. (1987). "International developments in production and consumption of carnations." *Acta Hort.* **216**: 159–164.

Kingman, R. (1983). "The carnation industry in the United States." *Acta Hort.* **141**: 249–252.

Lommel, S. A., Stenger, D. C., and Morris, T. J. (1983). "Evaluation of virus diseases of commercial carnations in California." *Acta Hort.* **141**: 79–88.

Mastalerz, J. (1983). "Supplementary irradiation or dusk to dawn lighting for cropping carnations at several population densities." *Acta Hort.* **141**: 157–163.

Mor, Y., and Halevy, A. H. (1983). "Carnation trends in Israel." *Acta Hort.* **141**: 253–259.

Mynett, K., Nowak, J., Rudnicki, R. M., and Goszczynska, D. (1983). "The yield of carnation flowers cut in the bud stage." *Acta Hort.* **141**: 197–202.

Pokorny, F. A., and Kamp, J. R. (1960). "Photoperiodic control of growth and flowering of carnations." *Ill. State Florists Assn. Bull.* **202**: 6–8.

Reid, M. S., Kofranek, A. M., and Besemer, S. T. (1983). "Postharvest handling of carnations." *Acta Hort.* **141**: 235–238.

Sparnaaij, L. D. (1983). "Carnations in Europe." *Acta Hort.* **141**: 261–266.

Sparnaaij, L. D., and Demmink, J. F. (1983). "Carnations of the future." *Acta Hort.* **141**: 17–23.

Sparnaaij, L. D., Demmink, J. F., and Koeshorst-van Putten, H. J. J. (1987). "Factors influencing the reponse to photoperiod (LD treatment) during the winter in an autumn-planted carnation crop." *Acta Hort.* **216**: 303–311.

van den Heuvel, J. (1987). "Carnation growing techniques in The Netherlands in the period 1980–1987." *Acta Hort.* **216**: 339–342.

van de Hoeven, A. P. (1987). "The influence of daylength on flowering of carnation." *Acta Hort.* **216**: 315–319.

van de Pol, P. A., and Vogelezang, J. V. M. (1983). "Accelerated rooting of carnation 'Red Baron' by temperature pretreatment." *Acta Hort.* **141**: 181–188.

Yuen, G. Y., McCain, A. H., and Schroth, M. N. (1983). "The relation of soil type to suppression of Fusarium wilt of carnation." *Acta Hort.* **141**: 95–102.

3

Roses

Dominic J. Durkin

I. INTRODUCTION

Rose flower production for the United States market was once limited to the Northeastern United States but improvements in environmental control and transportation have allowed the industry to expand across the country, and into South and Central America. In seeking the best combination of price and quality, rose flower production can be found at both the most primitive and the most advanced levels, depending on environmental conditions, production costs, and distance to markets. The cut rose is easy to transport, and its high value is used increasingly by developing countries to provide hard currencies.

There are plantings in South America under plastic with no supplemental heat where crop time may be more than 70 days at any time of the year while crop time may be as short as 35 to 40 days in the United States, Canada, and Holland at warmer temperature under supplemental lighting and added carbon dioxide. Other technologies such as rockwool culture can be found in areas with poor soils or, more commonly, in areas of long-standing production where soil-borne diseases and salinity make soil an unprofitable medium. The main competition occurs between areas having good rose climate, low wages, and a large labor supply and areas of high environmental and labor costs, but with proximity to markets that makes for easier flower quality maintenance. This competition will continue for the foreseeable future, permitting a more economical product to be marketed but requiring continued focus on improved production efficiency and more effective quality maintenance.

II. HISTORY

Cut rose production was developed to bring the garden to the residents of large urban centers. Production sites were restricted primarily by the transportation system available; in the United States, they were found along the Atlantic Coast, along the railroad lines serving the country, and in the Great Lakes area. After World War II and the growth of the airline industry, cut rose production spread rapidly to areas of the country with superior climates, featuring bright winter sun and night temperatures of about 60°F.

The rose is the strongest component of domestic cut flower production, increasing from $54 million in 1970 to $181 million in 1988. During the same period, the number of producers declined in nearly all states but California, and the average production value per grower increased about five times from $157,000 to $700,000, based mainly on increased return per flower. Imported roses, as a proportion of total sales, increased from 0.2% in 1970 to 42% in 1988. Plantings in Kenya, Mexico, Morocco, Spain, and Zambia suggest that this trend will continue. The red rose remains the top color choice in the United States and Japan, but in Europe red roses constitute less than 40% of total sales. There is increased interest in lavender and bicolored flowers, and serious efforts are being made to increase rose flower fragrance.

III. MORPHOLOGY

Flowers are evident in the tip of rose shoots approximately 2 weeks after axillary bud growth begins, when the shoot is less than 2 inches long. Not all flower buds develop to maturity, for there is an intense competition among the growing organs

Fig. 1 Blind shoots. (A) Entire blind shoot. (B) Close up of blind shoot. (C) Point of shoot tip abortion.

for products of photosynthesis and the root system, such as water, nutrients, and hormones. As a result, blind shoot production is common, particularly under reduced light (Fig. 1). When two buds develop after a cut, a blind shoot from the second bud below the cut is more often the rule than the exception. One of the major benefits of supplemental lighting is the increase of marketable flowers from the second buds.

Flowers might be solitary, as in the hybrid tea, or they might be multiple, as in the floribunda, but, depending on season and plant vigor, it is not unusual for the hybrid tea rose stem to have several short axillary flowering shoots at maturity.

The morphology of the rose shoot is interesting; there is a clear variation in leaf shape, in the shape of the axillary buds, and in the rate at which the axillary bud grows (Fig. 2). The rose shoot at maturity has from 12 to 20 nodes, including those with scales and leaves. Following from the point of origin on the parent shoot, there are one to four or five strap-shaped scales, part of the original axillary bud. The next one to two nodes may have three leaflet compound leaves, the next four to ten leaves will have five leaflets and occasionally seven leaflets on vigorous shoots; there will be one or two 3-leaflet leaves, a scale-like leaf or two on some varieties, and finally the flower bud. From one to six of the buds immediately below the flower will produce side shoots, short flowering shoots of limited value that must be removed in the greenhouse or in the grading area (Fig. 3). Axillary buds under this zone will be pointed while buds close to the stem base will be more rounded. Buds near the top of the five-leaflet leaf range are quicker to develop, there is greater likelihood that buds will develop at the second node below the cut, and it is more likely that the second bud will produce a flower than when a cut is made closer to the stem base. However, flowers will generally be cut to a lower and thus slower bud to obtain the long stem length that brings a premium price. When roses are pinched to increase shoot numbers or to time the crop for a specific market, a pinch to the fourth or fifth 5-leaflet leaf from the stem base will be used to produce a faster crop with more double breaks (Fig. 4).

IV. CULTIVAR DEVELOPMENT

There are many rose breeding programs around the world. Objectives include resistance to diseases such as powdery mildew and Botrytis flower mold, plant productivity, ease of timing for market planning, increased flower size and stem length, flower aroma, and longer vase life. Crosses are made in the spring and rose "hips" are harvested when the color changes to red or orange. The hip contains an average of six achenes, each comprised of a very hard pericarp and the true seed. After harvest, the achene is scarified to enhance water entrance and is usually stratified in flats containing moist sphagnum peat moss for 2 weeks at 68°F before transfer to 39°F for at least 3 to 4 weeks. When some seeds germinate, the flat is transferred to a temperature of 60° to 70°F for germination of the remaining seeds.

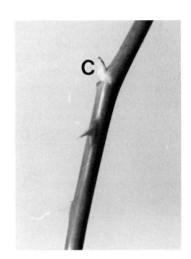

Fig. 2.   Rose shoot morphology. (A) Shoot with flower. W, three leaflet leaf; X, five and seven leaflet leaves. (B) Axillary bud shape, midshoot. (C) Axillary bud shape, upper shoot.

Fig. 3. Side shoot growth. A, terminal flower bud; S, side shoots. Note that lower side shoots are longer than upper shoots. Terminal flower bud will be larger when side shoots are removed early in development.

Seedlings are transplanted to a well-aerated mix for the first flowering. Selected seedlings are increased by budding for tests in many production areas. It is the rare seedling that progresses from this stage to naming and release in less than 6 years; from a seedling population of 150,000, three or four might be named and patented. The patent protects the breeder rights to a cultivar under an international convention in effect since 1961. Usually, the grower pays a royalty on each patented cultivar for the life of the plant. In effect, the producer rents the plants for the duration of productive life and in this way, the rights of the breeder carry over to any sport produced.

V. VEGETATIVE PROPAGATION

To preserve the specific characteristics of a rose cultivar, it is necessary to increase numbers by vegetative means, usually by budding. Grafting is still done, and for rockwool culture, rooted cuttings are often as good as grafted or budded plants.

Fig. 4. Hard and soft pinches. (A) Hard pinch made after stem matured. Note branch angle and double breaks. (B) Soft pinch was made while the stem was immature. This pinch generally produces one break but the diameter of the stem is usually larger than the parent stem. Note leaf form. Leaves developed from bud scales.

A. Cuttings

Cuttings are made at any time of the year; shoots are allowed to mature for 7 to 10 days after the usual harvest stage to increase food storage and rooting response. The cutting can be one, two, or three nodes but only the top leaf is retained to delay growth of the upper axillary bud until rooting is well advanced and to provide food for the developing shoot. The base of the cutting is lightly dusted with a rooting compound and cuttings can be stuck at a spacing of $1\frac{1}{2}$ inches in the row and 3 inches between rows. The medium should have good water retention and excellent aeration; a mix comprised of an equal volume of peat moss and horticultural grade vermiculite exhibits these characteristics. Rooting in 3-, 4-, or 5-inch pots facilitates handling and planting. Rooting occurs quickly at a media temperature of 70° to 75°F, and misting should be used to reduce water loss. The mix should contain fertilizer because some growth will occur before planting; this could include limestone and epsom salts, a minor element mix, and low levels of 5–10–5 fertilizer and potassium nitrate. A high nitrate soluble fertilizer at the rate of 2 pounds per 100 gallons can be used after rooting. Decrease in misting frequency should begin when roots are observed, in 11 to 14 days, to harden the plants for planting. The entire process should be complete in 4 to 5 weeks at the above medium temperature.

B. Budded Plants

Budded plants are still the most popular. They are larger and can produce quality flowers in a relatively short time. There is variation in the rootstock, with *Rosa manetti* being most important in the United States, while *R. canina* and *R. indica* (called *R. odorata* in the United States) are more important in Europe and Israel. Budding in the United States is done in California, Arizona, and Texas, while much budding is also done in Spain and Israel. It is important to find an area of low rainfall during the harvest season to harden the plants for harvest and storage. After the field is leveled and furrowed for irrigation to a depth of 18 to 27 inches, it is fumigated 3 to 5 weeks later with methyl bromide and chloropicrin for control of Verticillium wilt, soil pests, and weeds. By October 1, the soil is furrowed for irrigation. The field is treated with herbicide by the end of October and if needed, a preplant 10–34–0 fertilizer is placed to the side. Rootstock from virus-free blocks are cut back to 30 inches, bunches of 50 canes are collected and placed in 0.5% sodium hypochlorite solution for 15 minutes for crown gall control. Canes are sawed into 9-inch lengths and debudded of all but the top two buds; the stem bases are dipped in indolebutyric acid (IBA) to promote rooting, and cuttings are planted from late October to mid-December. Cuttings are watered after planting, all are rooted by early February, and budding takes place about 8 weeks later.

Shoots of the desired variety are allowed to mature on the plant after flowering, leaves are removed but petioles may be left for ease of handling the buds. The "budsticks" are soaked in a fungicide preparation, wrapped, and stored at 36°F. The bud is cut from a budstick, inserted into a T cut made in the bark below the growing rootstock shoot, and the union is wrapped with a rubber band. In 3 weeks, the rootstock shoot is cut through one-third of the way, about 1 inch above the bud. Six weeks after that, when the grafted bud has produced a flower, the rootstock shoot is removed completely.

Plant digging begins in mid-December after the tops are pruned to 9 inches. Plants are cleaned of soft growth and graded by the size of the understock, canes, and extent of the root system. Once cleaned and graded, they are packed in plastic-lined corrugated cardboard boxes for storage at 31° to 32°F.

C. Grafted Plants

Plant reproduction by grafting is a minor part of propagation today. Costs for greenhouse space are high, plant care is expensive, labor is not available for plant maintenance, and grafted plants require a long growing period before quality flowers are produced. Grafting still is a significant propagation method in Germany and Holland where winter light conditions are unfavorable for flower production. Known as winter grafting, the technique consists of grafting a softwood scion with one leaf on a bare-root dormant seedling understock, normally *R. canina inermis*. More than 1 million plants were produced in Holland in 1988 by a technique called "stenting." This consists of a graft of a softwood scion with a single leaf on a softwood understock cutting. Placed under mist, the graft union and rooting occur concurrently, in approximately 3 weeks producing plants with potential for earlier and larger basal shoots, and greater flower production. Temperatures of 75°F hasten the grafting process, but misting must be used to prevent desiccation. When the graft union is complete, temperature and humidity are reduced over a period of 7 to 10 days to acclimatize the plants to greenhouse conditions.

D. *In Vitro* Plants

Plant production from *in vitro* culture has some potential for improving the rose business. Thus far using this technique has been expensive, the plants have a high proportion of mutations, and though vigorous, they produce main branches of a smaller diameter and shorter stems than those plants produced by budding or by "stenting."

VI. PLANT CULTURE

The greenhouse rose is a perennial crop with a potential productive life of 8 to 10 years. Decisions made before planting will have consequences for many years. Cultivar selection and soil preparation are fundamental decisions because they are not correctable during crop life. The aim is to start with a good, productive soil, and through positive cultural practices, delay the inevitable loss in soil quality brought on by compaction, subsidence, salinity, nematodes, and soil-borne diseases.

A. Soil

In the traditional rose greenhouse, plants were grown in a bench containing 5 to 6 inches of soil. Shallow soils tend to retain more water than deeper soils of the same structure. Roots have a high need for oxygen, and because oxygen and water compete for the same pore space, retention of more water results in a lower soil oxygen level. Plants in such soils will be less vigorous, more prone to chlorosis, and slower to rebound from heavy pruning or cropping. In short, they will need careful attention to cultural practices such as mulching, irrigation frequency, salinity control, and the control of soil pests and diseases.

Soil preparation for production in benches begins weeks before replanting to ensure that salinity levels are low, that nutrients are at the proper levels, and that the soil pH is between 5.5 and 6.3. In addition to problems of chlorosis, poor root growth and short stems indicate problems in soil compaction and limited aeration. Plans should be made to add materials to form channels of large pores for quick restoration of soil air after irrigation. Amendments used for this purpose include coarse sand, sphagnum peat moss, haydite, scoria, and a variety of organic residues, such as various seed hulls, woodchips, sawdust, manure, and straw. In most cases, amendments equal to 40 to 50% of soil volume should be used, particularly in heavy clay soil, but in practice, materials are more commonly added in smaller amounts after several plantings, accepting the modest improvement in plant growth for the lesser cost and effort. For these soils, use of a mulch provides a gradual supply of organic materials that, if not contributing to improved soil structure, at least minimizes the loss in structure over time. Steam pasteurization of shallow soils is highly recommended to control soil insects, nematodes, diseases, and weeds.

Soil preparation in ground beds of good agricultural soil is less demanding. After restoring nutrient levels and correcting salinity and pH, organic levels may be increased by addition of some of the materials mentioned above. The soil may be pasteurized with steam or treated with methyl bromide and chloropicrin, or it may be replanted after tilling.

B. Planting

Planting can be done at any time of year, but plant availability and the flower market are usually decisive. A common scheme in the United States is to plant after one of the holiday periods, Christmas, Valentine's Day, Easter, or Mother's Day. This provides planting times of about January 1, February 20, April 10, or May 15. The early planting will allow flowering for Easter, and the next for Mother's Day.

Spacing is approximately one plant per square foot with three rows in a bed. Particularly for early planting, plants should be placed in the greenhouse for 2 to 3 days to initiate root growth in a process called "sweating out." Boxes are opened and plants are periodically misted to maintain a warm, humid atmosphere. Planting is done when root growth is observed. Special care is taken to see that roots are spread and that the plants are no deeper than they were in the field. Many cultivars produce roots from the scion when planted with the bud union under the soil. This allows "own root" development within a year, with the loss of rootstock benefits. For plants in heavy soil, deep planting also reduces air supply to the roots, which reduces plant take-off and vigor.

It is common for newly planted roses to receive two or three irrigations, 1 day apart, before the normal irrigation schedule is started to ensure that the soil is wet and that salts and chemical residues are leached from the soil.

Stored rose plants may have some developed shoots at planting. As this occurs more often after lengthy storage, such as late in the spring when stress conditions are intense, it may be wiser to remove shoots at planting to preserve plant moisture. Shoot growth will be slower as it depends on development of less mature buds, but the risk is greater of excess water loss before roots are established. After planting at any season, high humidity should be maintained around the plants. This can be done by hand misting, but it is expensive and difficult to perform on schedule. High pressure mist and fogs are better and because humidity is so important in the production of quality roses, some automatic humidification should be included in the production system. Irrigation schedules should be based on plant growth status and solar radiation. Rose stem length and flower size are very sensitive to soil moisture; with all things considered, it is better to irrigate when the soil still retains moderate amounts of water as judged by feel or by tensiometer readings of 100 to 300 cm tension.

The rose shoot must be held in an upright position during growth, otherwise the stem will bend and the long stem rose will be cut to a lower grade. Support has traditionally been by string to a central metal stake but this labor intensive operation has given way to the use of plastic or metal mesh, 3 to 7 inches on a side. The first network is installed 16 to 20 inches above the soil and additional layers will be 12 to 14 inches apart. Shoots are periodically guided to an upright position.

C. Nutrition

Most roses are grown using a liquid fertilization schedule based on soil tests, especially of pH and salinity, and an occasional foliar analysis. A pH range of 5.5 to 6.3 allows moderate solubility of the micronutrients. Salinity or soluble salts should be low at planting to allow for the increases following soil pasteurization, but once plants are established, aggressive nutrition practices should keep salts at a level about intermediate between normal and excessive. Within the flowering cycle, lower salinity at cropping will reduce root loss normal in this period. Following bud break, salinity levels can be increased. In some programs, it is common to make a dry fertilizer application at this time, often with calcium or potassium nitrate at 1.75 pounds per 100 square feet.

Soil pH is a result of many interacting influences but for the rose producer the most important are the alkalinity of the water, the nature of the mineral soil, and the composition of the fertilizer solution. Water sources with high levels of calcium and magnesium (called "hard water") also contain high levels of the bicarbonate ion. Consequently, they have a double alkalizing effect. Such waters react with the acid in the soil solution until much of the bicarbonate is converted to carbonic acid and then to carbon dioxide and water. This process drives soil pH upward, and the accumulation of calcium and magnesium extends this process further. Use of acid residue fertilizers can retard this process but more growers now use acids, most commonly phosphoric or nitric acid, injected directly into the water at a rate calculated to offset the alkalizing effect of the bicarbonate ion.

Rose fertilizer solutions contain nitrogen, potassium, magnesium, and iron as regular constituents, and periodically, a micronutrient supplement of boron, copper, manganese, molybdenum, and zinc will be added. Base levels of nitrogen are in the range of 160 to 200 ppm (11 to 14 milliequivalents per liter) with a milliequivalent ratio of nitrate to ammonium nitrogen of 5–1 to 10–1. Base levels of potassium are about 150 ppm (4 milliequivalents per liter). There is evidence that production under supplemental lighting requires increased nitrogen to approximately 300 ppm. Potassium nitrate is a common source of both nutrients while other nitrogen sources are ammonium nitrate, calcium nitrate, and ammonium sulfate. Liquid systems using any sulfate or phosphorus source with calcium will need a double-headed injector as the combinations produce calcium sulfate or calcium phosphate, both nearly insoluble salts. Regular use of ammonium fertilizers will decrease soil pH, while regular use of nitrate fertilizers will have the opposite effect. Long-term pH control is also managed with periodic dry applications of limestone or dolomitic limestone where supplements of magnesium are needed. Besides potassium nitrate, potassium may be added as the chloride or the sulfate salt. Potassium sulfate is preferred as the potential for salinity and toxicity are much less when compared to potassium chloride. Salinity control is also managed by careful selection of fertilizer sources, the use of appropriate fertilizer concentrations, and periodic applications of water alone.

D. Temperature

Growth rate of the rose is very responsive to temperature; as temperature is increased, growth rate increases. Flower quality is affected in the opposite direction; as temperature is increased, petal number and flower bud size decreases. Generally, roses are grown at 62°F night temperature with a day temperature of 70°F under low light conditions and 80°F under high light. Flowers will take about 40 to 48 days to develop at these temperatures. Selection of day and night temperatures represents a compromise between numbers of flowers harvested and flower quality. Where the growing environment is better, higher temperatures may produce more flowers of a better quality, whereas in poorer conditions, higher temperatures will produce some increase in flowers with poorer quality and increased blindness. Night temperatures for roses are considerably lower in many parts of the world, frequently near 44°F. Roses produced at these temperatures will take approximately 75 days to flower from a previous harvest if day temperatures are also low, but with day temperatures of 81° to 86°F, flowering may take only 55 to 60 days. Flower size will be large and stem length can be very long under cool night conditions.

Where roses are cropped so that all flowers are harvested in a short period, bud break can be accelerated by high temperatures, up to 81°F for the first 3 weeks as long as humidity is maintained at a level of 60% or above. After the 3-week period, temperatures should be reduced to normal to provide for flower quality. There has been much interest in manipulating night temperatures to allow fuel saving without loss of growth rate but no recipe has consistently produced the energy savings without a delay in flowering.

E. Ventilation

Greenhouses must have air replacement for a number of reasons. First, in most environments, sunlight produces a temperature rise in the greenhouse that can exceed the desired day temperature. Excess heat is dissipated with the exit of hot air from the top of the greenhouse. Even where heat accumulation is not a serious problem, carbon dioxide levels are reduced by photosynthesis and must be replenished by ventilation. Carbon dioxide can also be added to the greenhouse air, either by burning propane or by use of liquid carbon dioxide. Optimum carbon dioxide levels in the 900 to 1000 ppm range increase growth rates and flower quality.

Air replacement is also used to manage greenhouse humidity as a general approach to control powdery and downy mildews and *Botrytis*. Relative humidity of the air is an important factor in the development of fungal spores. If the greenhouse is closed to conserve heat, the relative humidity of the air increases as the temperature falls. In most environments, the air becomes saturated, water will condense on the leaves and flowers providing exceptional conditions for rose diseases. In most seasons, this can be avoided by providing adequate ventilation as the temperature falls to allow exchange of the humid greenhouse air for the drier outside air. During seasons when the outside temperature is near the night temperature, it may be necessary to combine periodic heating with ventilation to decrease the likelihood of saturated air. There is some evidence that air movement of 30 feet per minute markedly reduces problems of *Botrytis*. When this is provided by the horizontal air flow system, it also provides uniform greenhouse temperature and ease of distributing greenhouse fumigants.

F. Light

The light requirement of the rose is complicated by the effect of light on transpiration and the effect of transpiration on plant growth and quality. Generally, rose growth and quality increase with increasing light to about 1200 μmol sec^{-1} m^{-2} photosynthetically active radiation (PAR). This is about 6000 foot-candles (fc) and is close to the maximum winter light in most of the United States. It must be remembered that increased PAR will be accompanied by an increase in all solar radiation to produce higher greenhouse temperature and decreased relative humidity. High pressure mist and fogging systems are an essential part of humidity management to allow use of the available light in spring, summer, and fall. Daytime humidity of 60 to 80% is optimum for rose production. Greenhouse shade is required in the summer and for periods in spring and fall, depending on location and light levels.

In the more northerly parts of the world, where winter light is of low intensity and short duration, there is widespread use of lighting for the rose crop. High-pressure sodium (HPS) lights are used to produce 50 to 100 μmol sec^{-1} m^{-2} PAR (250 to 500 fc). With night temperatures up to 68°F, growth is more rapid, the number of flowers is increased, and plant vigor is excellent. As mentioned earlier, nutritional and irrigation requirements are increased, but carbon dioxide addition is not needed at this light intensity.

G. Plant Growth Management

Regardless of the plant type, it is necessary to delay flowering to increase the amounts of foliage and roots so that high quality flowers can be produced. There are two methods, depending mainly on the season and timing requirements for the next good market. In the first case, after the new plants are trimmed of damaged wood and weak shoots, shoots are allowed to grow until the flower bud is apparent in the shoot tip. Shoots are pinched to the uppermost five-leaflet leaf. Not all shoots will resume growth, as there is competition for water and nutrients, but perhaps half the shoots on a vigorous plant will grow; the other shoots will be shaded out by the developing canopy and die within the year. High quality flowers from the second growth flush will be marketed, and usually flowers will be cut below the point of origin on the parent shoot to provide longer stems. The second method, used when the market is poor, is to allow the shoots to grow until pea-size flowers are produced. Stems are pinched to the upper five-leaflet leaf and axillary shoots are removed in a practice called "deshooting." When done regularly on a 2- to 3-day schedule, the root system becomes very extensive, the food from photosynthesis accumulates, and the number of renewal canes (bottom breaks) increases substantially. Shoot vigor, particularly at the base of the plant, is important because stem length and growth rate are associated with thick stems. Because the producing plant for the next year will originate from these vigorous shoots, there is some assurance of high quality flower production.

H. Harvesting

Roses are cut two times per day during much of the year and three times in the summer when development rates and markets warrant. The stage at cutting varies with cultivar, season, and distance to market, but generally the rose is cut at the tightest stage at which the flower will open in plain water (Fig. 5). Tight roses will often open in preservative solution but because preservative use is still not wide-spread, it is better to make the test in water with roses comparable to those being shipped. For the nearby market, the more open rose will develop better and make a better presentation. There is the belief among retail florists that more open roses do not last, but the opposite is more often true.

The position of the cut on the shoot will vary with market needs for stem length and on the height of the plants, the season, and the leafiness of the plants. As a general rule, the cut will be made to the first 5-leaflet leaf above the point of origin of the flowering stem, but it will be higher at the time of a holiday pinch or if the plants have suffered a heavy leaf drop. In deep winter conditions without supple-mental lighting, the cut will be made at the second 5-leaflet leaf. Following a pinch to the third to fifth 5-leaflet leaf, the cut will be made under the pinch to the next leaf in the event of a single break and in the event of a double break, the cut will be made to just below the first flower and when the second flower develops (a few days to a week later), it will be undercut to the next five-leaflet leaf. During the spring season when plants are vigorous, flowering shoots may be cut to the "knuckle," just above the node of origin, to gain additional stem length. Usually, two or three shoots will develop from this cut, originating from buds in the axils of the original bud scales (Fig. 6).

Fig. 5. Flower stages. *Left to right*: a, too tight; b, for distant market; c, for local market; d, better for local market; e, open too far.

Fig. 6. Knuckle cut on a flower from a soft pinch. A, node at which pinch was made; B, point of cut; C, shoots originating from axillary buds of the original bud scales.

The rose is cut with a sharp knife or with shears. Some shears are constructed to hold the rose while it is being moved to the arms or to a basket. It was common for roses to be cut and held in the arm until a circuit of a bed was made, but for roses grown in long beds, cropping systems where an entire bed may be cut in 7 days, and long stem flowers, damage to flowers can be serious. Now most cut roses are placed in a basket for protection. Once a bundle is cut, it is wrapped in plastic sheeting or in plastic netting to protect the flowers during handling. The bundle may be placed in a hydrating solution in the greenhouse to offset the loss of water from the foliage, or it may be moved directly to the storage area. There is an advantage to put the rose into solution as soon after harvest as possible but in most greenhouses, bucket sanitation is primitive and the gain in some small amount of water uptake is more than offset by the inoculation of the clean roses with bacteria that will reduce vase life (see Section IX).

I. Crop Planning

An efficient business must not only produce excellent flowers at a saleable price, but it must also manage the production to satisfy the market throughout the year.

This takes thorough planning, a knowledge of the cultivar characteristics, anticipation of seasonal effects, and some good fortune. The planning will involve production timing for key holidays, seasons, and special sales. It also involves an awareness of growth rates by cultivar and season to allow proper timing and an awareness of how many pinches are needed to meet a production goal. Several questions arise in scheduling rose production. How long does it take for the cultivar to flower again after a previous crop? If one made 100 pinches on October 27, how many shoots would produce marketable flowers for Christmas? What would be the range in flowering time? What proportion of the Christmas crop might return to flower for Valentine's Day? How much variation can one find from year to year? It should be apparent that a keen mind, good records, and an awareness of weather patterns are important for this planning effort.

J. Pruning

If flower harvest is always made above the origin of the flowering stem, the plants will grow too tall. There are two ways to reduce plant height. First, as a flower is cut, rather than cutting it at the first 5-leaflet leaf, the stem on which the flower is growing is cut to 24 inches above the soil after the first year of growth and to 6 inches higher after each succeeding year. The advantages of this system are there is a continuous supply of flowers and these plants will produce two to three more flowers per plant and the plant root system will remain in better condition with the supporting foliage. The disadvantages are that the harvest is slower, the greenhouse is left in disarray because the lower stem is pruned into the walkway, and the decisions of where to make the cut require considerable experience.

In the second pruning system, the plants are pruned with hedge cutters at the 24 inch level. Because a major loss of roots can be expected following such a foliage loss, salinity should be reduced to the low to medium levels before pruning. Additionally, to limit xylem sap flow from the cut canes and prevent stem dieback, it is useful to prune near the end of the watering cycle. The soil can be further dried to prevent sap flow but this hardens the plant and delays bud break. After pruning, all dead and weak wood is removed, and it is a good time to spray for spider mites. Bud break can be accelerated by warm temperature and high humidity.

Where summer markets are limited, the flush of growth is sheared periodically to increase numbers of growing shoots and to stimulate development of basal shoots for plant renewal.

VII. INSECTS, RELATED PESTS, AND DISEASES

There are a host of insects and related pests attacking greenhouse roses. The rose grower has to be acquainted not only with the pests and their control, but with the host of regulations that have developed to make pesticides safer for the operator and others. A good control program starts with the employees, because they have so many opportunities to inspect the plants each day. Early detection not only lessens the chance of the pest spreading throughout the greenhouse, but at low populations the pests are easier to control and are less likely to develop resistance to pesticides.

The second goal of a good pest control program is to know the life cycle of the pest, the growth stages when control by spraying is not possible, the array of available control measures, and the chemical families so that rotation of materials can be practiced to minimize development of resistance.

The third part of the pest control program is to acquaint employees with the health and safety problems of the pesticides. They should understand that effective and careful application may be the best approach to pesticide health and safety by reducing application frequency. Pesticide recommendations can be obtained through the Cooperative Extension Service and through professional rose organizations.

A. Insects and Related Pests

1. Spider Mites

The two-spotted spider mite (*Tetranychus urticae*) remains the most important pest on greenhouse roses. Its life cycle consists of eight stages from egg to adult, including three quiescent stages of insensitivity to miticide. At day temperatures of 75° to 80°F and night temperature of 65°F, it may pass through all stages in less than 13 days. Feeding principally on the underside of the leaf, mites leave pinpoint chlorotic spots that turn the leaf bronze when the population is high. Mites are most easily detected along the south side of the greenhouse and at the end of beds where temperatures are high. They can easily be distributed throughout a rose planting during one flower harvest. Control is more difficult than for other pests because of the large populations involved and also because of their ability to develop resistance to pesticides. The pest feeds mostly on the underside of the leaf, and the eggs are laid there, so it is crucial that miticide coverage be adequate there. Special spray nozzles have been designed for mite control. Even moderate mite infestations reduce foliage size, cause leaf drop, and restrict stem elongation.

2. Aphids

There are at least three species of light to dark green aphids that feed on stems, leaves, and flowers. Aphids are easier to detect from the cast skins and the black sooty fungus that grows on the "honeydew" than by direct observation. In the greenhouse, all aphids are females. They give birth to live young, which can then begin reproducing in 7 to 8 days.

3. Thrips

Two species of thrips are found in greenhouses, the Eastern thrips (*Frankliniella tritici*) and the Western thrips (*Frankliniella occidentalis*). In the general life cycle, eggs are laid in the flower bud and go through four nymph stages to maturity. During development, nymphs drop off the plant onto the soil two times before maturing. The entire life cycle takes about 20 days. Adults are winged and very mobile. Control is difficult because of the protection afforded by the flower and the fact that the nymphs spend two periods in the soil.

4. Whiteflies

The greenhouse whitefly (*Trialeurodes vaporariorum*) is becoming a serious pest of the greenhouse rose. Eggs laid on the underside of the leaf hatch in 10 days, and remain on the leaf for the next 10 to 15 days. Control is difficult because the pest is naturally resistant to pesticides at many of the immature stages. Weed cleanup in and around the greenhouse will help in control, as will the exclusion of tomato, pepper, and other annual plants from the vicinity of the greenhouse. There is considerable interest in use of screens to exclude whiteflies and thrips from the greenhouse.

B. Diseases

1. Powdery Mildew (Sphaerotheca pannosa var. rosae)

The most common disease of roses in greenhouses, powdery mildew, attacks leaves, stems, and flowers. Spores are wind-borne from fruiting bodies to new leaves and can germinate in 3 hours. Following penetration of the tissue, the fungus develops along the outside of the tissue. Ultimately, fruiting bodies develop to complete the cycle.

Like many fungi, the development of mildew is related to optimal temperature and humidity. Nighttime conditions of low temperature (60°F) and high humidity (90 to 99%) favor spore germination, infection, and spore production. High temperatures (80°F) and low humidity (35 to 70%) favor spore maturation and release. To some extent then, control can be managed with careful control of heating and ventilation at the end of the day, when humidity increases as temperatures drop, and in the morning when a slow temperature rise will delay spore release. Concurrent use of heat and ventilation can allow replacement of humid inside air with drier outside air. There is also some evidence that in most greenhouses, night temperature fluctuations can result in cyclic leaf wetting thereby providing conditions for spore germination. Devices to measure condensation events provide guidance for when greenhouse heating is required. Additional control is provided by sublimated sulfur and by spraying with appropriate fungicides.

2. Downy Mildew (Peronospora sparsa)

A disease enhanced by cool temperature and high relative humidity, downy mildew is found in rainy periods when relative humidity exceeds 85%. After spore germination, the fungus grows within the leaf (in contrast to the surface growth of powdery mildew) and, depending on the humidity, spores may be seen on the underside of the leaf. In usual conditions, spores are not seen, but the leaves on the current shoot may show dark irregularly shaped spots; leaf drop may be heavy and may include leaves from the current growth. Heat and ventilation will aid in control but in seasons when the outside and inside temperatures are similar, spraying will also be required.

3. Botrytis Blight (Botrytis cinerea)

Botrytis or gray mold is a common disease of stored plants and flowers and, under cool and humid conditions, is a serious problem of roses in transit. Spore germination requires up to 12 hours of continuous freestanding water on plant surfaces. Proper heat and ventilation are important control measures in the greenhouse and research shows that spore germination is inhibited by air movement of 30 feet per minute. Conditions such as the frequent movement of flowers from cold storage to a warm grading area or from cold storage to poorly refrigerated trucks are the most common ways to obtain the long-term wet surface needed for infection.

VIII. PHYSIOLOGICAL DISORDERS

A. Bullheads

Bullheads are distorted flowers characterized by shorter petals that give the rose a flattened appearance. Common in some cultivars grown at cool temperatures, it is also seen in flowers distorted by thrips, by excess stem vigor as in the case of renewal shoots, and in some cultivars that have a tendency to produce bullheads.

B. Blind Shoots

Rose shoots initiate flowers at an early stage of development, but blind shoots abort most flower buds soon thereafter. Blind shoots increase in periods of low light, but the absolute number of blinds may be highest after pruning. More blinds are produced from the lower buds on a stem, and it is common to see a high proportion of blinds near the end of a crop. Increased carbon dioxide and light levels help to minimize numbers of blinds during the winter period.

C. Leaf Drop

Leaf drop is not uncommon in the rose. Some common causes are shading from foliage above, inconsistent irrigation and nutritional practices, plant age, and, occasionally, a reaction to a pesticide application. Additionally, heavy infestation of mites and powdery mildew will accelerate leaf aging and cause leaf drop.

D. Leaf Distortion

Rose leaves grow to maturity in 10 to 12 days from the embryonic size in the bud. Spray applications during this period can damage the leaf and prevent normal growth, producing distortions. Water stress, usually caused by high light intensity and low humidity, causes a tissue burn. In some of the new, very tight greenhouse structures during the winter, greenhouse temperatures do not require the use of venting. It is not unusual to see, on a single stem, a succession of normal leaves followed by one to four small distorted leaves, and then a return to normal size leaves. This problem is thought to result from temporary limitation in the uptake of micronutrients during periods of low transpiration.

E. Volatiles

The rose is sensitive to a large number of chemicals, from mercury metal to mercuric compounds in paints and fungicides, to ethylene gas, ammonia, Treflan herbicide, sulfur dioxide, and many of the phenoxy-type herbicides. As a general rule, if a chemical has not been shown to be safe on roses, assume that it is not!

F. Sugar Damage

Roses produced under supplemental high density discharge (HID) lighting and carbon dioxide additions have been shown to transpire excessively, a fact that might help in some problems (see Section VIII,D) but that causes problems in the handling of the cut flower. Although it is recognized that roses require preservatives if the potential vase life is to be attained, use of sugar-containing preservatives has caused considerable sugar burn on these roses. It seems that the rate of water entry into the leaf cells is fast compared to the rate of sugar entry. With rapid water loss from the leaves, the sugar concentrates outside the cell until cell death by dehydration occurs. Similar problems can be observed in rose handling for roses produced during periods of vigorous growth as in the spring season.

IX. POSTHARVEST HANDLING

Flower life varies with the cultivar and the season; life of the flower on the plant is perhaps the best index of potential vase life. There are two principal differences between the cut rose and the rose on the plant; the rose on the plant has a secured water source limited only by negligence in soil irrigation. The rose developing on the plant also has the benefit of a continual food supply, either directly from its own leaves or from storage areas in the roots or stems. Death of the rose flower on the plant is almost always by petal drop.

Problems for the cut rose begin at harvest; if cut too early, the rose may contain little stored food. The correct harvest stages are difficult to define as cultivars vary greatly, but in general, the rose should be cut at the stage at which the flowers will continue to develop for the final consumer.

There are two short-term problems of the cut rose; water loss increases at harvest, and as water moves into the leaves from the stem, air enters the stem base. The first problem causes wilting of the flower while the second limits the rate of uptake when flowers are placed in water. Hydration of the cut flowers in deep, acidified water usually solves the harvest water problems. Two common acidifiers are citric acid at 350 ppm and aluminum sulfate at 200 to 600 ppm. The former is a faster hydrating agent, but sanitation is easier to maintain with aluminum sulfate solutions.

Once hydrated, the long-term performance of the cut rose depends on microbe control. The most common problem involves bacteria in the water, but there are problems with yeast and fungi as well. Control starts with the use of clean containers and either use of long-term sanitation measures like slow release chlorine compounds or with frequent replacement of the water. Use of sugar containing preservatives is not recommended at the production level; storage temperatures are too low for significant uptake, and sugar can make the microbe control problem more difficult. In contrast, preservatives are recommended whenever storage temperatures exceed 38° to 40°F, usually at the retail level and always at the consumer level. As water quality is variable, tests should be run to determine which of the available preservatives is most effective.

Roses are graded by length, depending on market preferences starting at about 9 inches and running to 30 inches. Grading is frequently done by machine after initial screening for malformed or crooked flowers (Fig. 7). Roses are packed in bunches of 25 in the United States and 20 in Europe. Care must be taken to avoid tight wraps as the flowers will swell with continued hydration, and they will bruise. All stems should be cut after wrapping so that at later marketing stages, approximately 1 to 2 inches of stem can be removed from each flower without unwrapping the bunch. Flowers that receive little or no recutting will be slow to hydrate, and their performance in the vase is uncertain. For long-stem roses, it is common to wrap the stem bases with string or rubber bands. This wrap should be placed at least 4 inches from the stem base to allow for easy recutting.

The flower box must be large enough to accommodate 400 to 600 flowers, rigid enough to protect the flowers, and insulated well enough to keep out the subfreezing temperatures of winter and to limit heat gain during the rest of the year. Cardboard boxes, often waxed to limit water loss and occasionally sprayed with polyurethane to limit heat gain, are used. Other insulators in use are newspaper, fiberglass liners, or Styrofoam sheets. Flower bunches must be immobilized, sometimes with wooden slats cleated to the outside of the box or with rubber bands that are held by plastic pieces that fit through the sides of the boxes. Bunches are commonly iced for additional heat control but where Botrytis flower

Fig. 7. Peduncle form. A, normal, straight peduncle; B, short, crooked peduncle often resulting from excessive plant vigor.

mold is a problem, some shippers use sealed ice or one of the many thermal gels. Temperature control is easier if the flowers are completely cold at packing but when that is not feasible, forced air cooling systems are available to cool packed flowers before shipping.

In the usual case, roses are shipped to market as soon as the grading process is completed but sometimes supply gets ahead of demand. If the need is for a 1 to 2 day storage period, flowers may be held in packed boxes by the grower or by the wholesaler. For longer holding, flowers may be placed into dry storage immediately after harvest. Because these flowers will be dry at the start of storage, conservation of tissue water is paramount. The flowers are placed flat down in a plastic-lined box, the box is sealed, and stored at 31° to 32°F. It may take several hours for the flowers to cool to 31°F, and it is not uncommon for the heads of some cultivars to turn upward, producing crooked roses of lesser value. In an alternative method, roses can be stored in plastic-lined drums with stems in an upright position. In either case, the results are more satisfactory if the cooling is accomplished quickly. There is less water loss and fewer hydration problems later, less time for reaction to the inevitable ethylene accumulation from closely packed flowers, and greater conservation of stored food. Though there is no apparent change in the stored flowers, flower performance in the vase is significantly reduced by dry storage of more than 5 days. Moreover, even when flower life is not diminished by dry storage, flower form may be seriously affected. Flowers of some cultivars may exhibit full opening in as little as 4 hours after 7 days of dry storage. Conversely, flowers of other cultivars do not open. Though the variety of problems is frustrating, they all seem related to ethylene sensitivity.

Dry stored roses will be more difficult to hydrate. Stems should be cut, preferably under water with removal of 1 to 2 inches of the old stem base. If stems are cut in air, they should be placed in 10 to 12 inches of warm (105°F) water containing citric acid. Leaves and flowers should be crisp in 30 to 60 minutes and the roses can be graded and packed.

REFERENCES

Augsburger, N. D., and Powell, C. C. (1986). "Greenhouse ventilation: Some basics on controlling humidity." *Roses Inc. Bull.* Sept., 20–23.

Brown, W. (1989). "Water, pH and nutrient availability." *Roses Inc. Bull.* Dec., 28–35.

Brugger, M. F., Short, T. H., and Bauerle, W. L. (1987). "An evaluation of horizontal air flow in six commercial greenhouses." *Roses Inc. Bull.* Oct., 32–41.

Bryne, T. G., and Doss, R. P. (1981). "Development time of 'Cara Mia' rose shoots as influenced by pruning position and parent shoot diameter." *J. Am. Soc. Hort. Sci.* **106**: 98–100.

Bryne, T. G., Doss, R. P., and Tse, A. T. Y. (1978). "Flower and shoot development in the greenhouse roses, 'Cara Mia' and 'Town Crier' under several temperature-photoperiodic regimes." *J. Am. Soc. Hort. Sci.* **103**: 500–502.

Darlington, A. (1990). "Relative humidity control and plant growth." *Roses Inc. Bull.* Jan., 24–27.

Dixon, M. A., Butt, J. A., Murr, D. P., and Tsugita, M. J. (1988). "Water relations of the cut greenhouse rose: The relationship between stem water potential, hydraulic conductance, and cavitation." *Scientia Hort.* **36**: 109–118.

Durkin, D. (1988). "Studies on the handling of cut rose flowers." *Roses Inc. Bull.* May, 53–61.

Halevy, A. H., and Mayak, S. (1979). "Senescence and post harvest physiology of cut flowers, Part 1." *Hortic. Rev.* **1**: 204–236.

Halevy, A. H., and Mayak, S. (1981). "Senescence and post harvest physiology of cut flowers, Part 2." *Hortic. Rev.* **3**: 60–143.

Hammer, P. E., and Marois, J. J. (1989). "Nonchemical methods for postharvest control of *Botrytis cinerea* on cut roses." *J. Am. Soc. Hort. Sci.* **114**: 100–106.

Hanan, J. J. (1986). "Yield and quality of 'Samantha' roses in three inert media." *Roses Inc. Bull.* Oct., 31–33.

Hanan, J. J., Holley, W. D., and Goldsberry, K. L. (1978). *Greenhouse Management.* New York: Springer-Verlag.

Hasegawa, P. M. (1980). "Factors affecting shoot and root initiation from cultured rose shoot tips." *J. Am. Soc. Hort. Sci.* **105**: 216–220.

Horst, R. (1983). *Compendium of Rose Diseases.* St. Paul, Minnesota: Published by The American Phytopathological Society.

Hughes, H. E., and Hanan, J. J. (1983a). "Effects of salinity in water supplies on rose production: Experiment 1." *Roses Inc. Bull.* March, 16–19.

Hughes, H. E., and Hanan, J. J. (1983b). "Effects of salinity in water supplies on rose production: Experiment 2." *Roses Inc. Bull.* March, 20–22.

Langhans, R., ed. (1987). *Roses: A Manual of Greenhouse Rose Production.* Haslett, Michigan: Roses Inc.

Laurie, A., Kiplinger, D. C., and Nelson, K. S. (1979). *Commercial Flower Forcing.* New York: McGraw Hill.

Lawson, R. H., and Horst, R. K. (1987). "Avoiding cane injuries may prevent rose canker." *Roses Inc. Bull.* April, 40–44.

Marois, J. J., Macdonald, J., and Redmond, J. (1987). "Biological control of botrytis on rose." *Roses Inc. Bull.* March, 47–60.

Marois, J. J., MacDonald, J., and Redmond, J. (1988). "Quantification of the impact of environment on the susceptibility of *Rosa hybrids* to *Botrytis cinerea*." *J. Am. Soc. Hort. Sci.* **113**: 842–845.

Marois, J. J., Macdonald, J., Tanner, L., Wagner, S., and English, J. T. (1989). "Control of *Botrytis cinerea* on rose: Microclimate effects on disease development." *Roses Inc. Bull.* Sept., 42–48.

Mastalerz, J. W. (1977). *The Greenhouse Environment.* New York: Wiley.

Mor, Y., and Zieslin, N. (1987). "Plant growth regulators in rose plants." *Hort. Rev.* **9**: 53–66.

Mor, Y., Johnson, F., and Faragher, J. D. (1989). "Preserving the quality of cold-stored rose flowers with ethylene antagonists." *Hortscience.* **24**: 640–641.

Nelson, P. V. (1981). Greenhouse Operation and Management. Reston, VA: Reston.

Pasian, C. C., and Lieth, J. H. (1989). "Analysis of the response of net photosynthesis of rose leaves of varying ages to photosynthetically active radiation and temperature." *J. Am. Soc. Hort. Sci.* **114**: 581–586.

Post, K. (1959). *Florist Crop Production and Marketing.* New York: Orange-Judd.

Powell, C. C., and De Long, R. E. (1989). "Studies on the chemical and environmental control of powdery mildew on greenhouse roses." *Roses Inc. Bull.* Nov., 47–55.

Reid, M. S., Evans, R. Y., Dodge, L., and Mor, Y. (1989). "Ethylene and silver thiosulfate influence opening of cut rose flowers." *J. Am. Soc. Hort. Sci.* **114**: 436–440.

Rupp, L. A., and Bugbee, B. (1990). "High intensity rose production using rooted cuttings." *Roses Inc. Bull.* Feb., 41–51.

Skirvin, R. M., Chu, C., and Young, H. J. (1986). "The tissue culture of rose," In *Handbook of Plant Cell Culture*, vol. 5, edited by Evans, D. E., Sharp, W. R., Ammirato, P. V., Yamada, Y., and Bajaj, Y. P. S. New York: Macmillan.

Smitley, D., and D'Angelo, T. (1988). "Biological and integrated control of spider mites in rose production." *Roses Inc. Bull.* March, 57–64.

van de Pol, P. A., and Breukelaar, A. (1982). "Stenting of roses: A method for quick propagation by simultaneously cutting and grafting." *Scientia Hort.* **17**: 187–196.

4

Snapdragons

Marlin N. Rogers

Introduction to Floriculture, Second Edition

I. INTRODUCTION

Ancestors of the greenhouse snapdragon grew originally in the Mediterranean region, where they were tender, summer-flowering perennials. Cultivated varieties were first grown in the United States as open pollinated, inbred lines for flowering during long-day periods in the field or greenhouse, but about 1926, Mr. Frank Volz, Chevoit, OH introduced a cultivar 'Chevoit Maid' that flowered during winter (Lindstrom, 1966).

The first F_1 hybrid greenhouse-forcing snap, 'Christmas Cheer,' was introduced by Fred and Helen Windmiller in Columbus, OH in 1938 (Ball, 1952). This was quickly followed by many other F_1 types such as 'Maryland Pink' (Fred Winkler), and 'Mary Ellen' and 'Dorcas Jane' (J. S. Yoder). These hybrid types combined the early flowering characteristics of the winter greenhouse forcing types with the strong vegetative growth characteristics of F_1 hybrid plants. George J. Ball, Inc. and Yoder Brothers, Inc. have been the principal introducers of new F_1 hybrid cultivars for greenhouse forcing since about 1950.

A. Classification of Greenhouse-Forcing Cultivars

Greenhouse-forcing types are classified into four response groups based on their growth and flowering responses in relation to temperature and daylength. Group I cultivars (winter and early spring group) are highly reproductive cultivars and flower quickly at 50°F night temperatures during the short, dark days of midwinter in northern growing areas. Group II (late winter and spring group) cultivars have resulted from crosses of more vegetative inbreds, and flower with good quality at 50°F night temperatures, but they require more crop time than Group I cultivars. Group III (late spring and fall group) cultivars are extremely slow to flower at 50°F, but perform much better at 60°F night temperatures under longer days and higher light intensities. Cultivars in Group IV (summer group) are reproductive only at high night temperatures (60°F or higher) and at 50°F are blind (nonflowering). They are used for midsummer flowering (Duffett, 1960). A summary of recommended flowering periods for cultivars from each of the response groups in both northern and southern growing areas is shown in Table I. The dividing line between northern and southern areas would be about 38°N latitude—roughly a line from Washington, D.C. through Cincinnati, OH and Kansas City, MO.

B. Timing in Different Areas

As can be seen from the summary timing information given in Table I, there is considerable variation in the time required to produce a crop at different seasons and in different growing areas. Detailed studies have been completed on the timing required in both southern (e.g., Alabama) (Sanderson and Martin, 1984) and northern (e.g., Pennsylvania) (White, 1961) producing areas. Timing representative of growing conditions in Missouri is shown in Figure 1.

Table I

Recommended Times for Flowering and Approximate Time to Flower of Different Snapdragon Response Groups for Northern and Southern Growing Areas in the United States

Response group	Northern United States		Southern United States	
	Best flowering period	Seed to flower (days)	Best flowering period	Seed to flower (days)
I	December 1–February 15	130	Not recommended for southern United States	
II	February 15–May 1	155	December 15–April 1	130
	November 1–December 1	100	—	—
III	May 1–July 1	110	April 1–June 1	95
	September 15–November 1	85 ,	November 1–December 15	85
IV	July 1–September 15	85	June 1–November 1	85

The point of practical importance is that beginning about August 1 in mid-America, plantings made 1 week apart will later flower at intervals of 2 to 3 weeks apart. If one is attempting to time plantings so that one can cut a crop every other week, the sowings will have to be made every 4 to 5 days for crops seeded during August and early September, following which light levels are decreasing consistently. During the spring, as light levels are increasing, plantings made about 3 weeks apart will mature about 2 weeks apart.

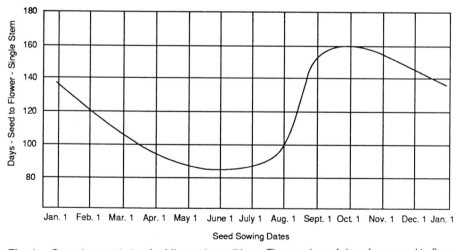

Fig. 1. Snapdragon timing for Missouri conditions. The number of days from seed to flower (salable maturity) for single-stem snapdragons planted at different times during the year.

C. Economic Importance

Snapdragons were included as an individual flower crop for the first time in the 1959 Agricultural Census of Horticultural Specialty Crops, and at that time they were ranked as the seventh most valuable cut flower crop produced—being eclipsed by roses, carnations, pompon chrysanthemums, standard chrysanthemums, gladiolus, and *Cattleya* orchids. At that time snapdragons made up 3.2% of the total wholesale value of the flowers produced in the United States—about $4.5 million out of $142.5 million total. The most important states involved in their production were Pennsylvania, New York, Ohio, Massachussets, Michigan, New Jersey, Indiana, Illinois, Minnesota, and Maryland.

Snapdragons experienced declines in production and value, bottoming out in the late 1970s. Since that time, they have staged a strong comeback, until in 1984 they had doubled in total wholesale value compared to 1979, and showed a 50% increase in total number of stems sold, and an increase in price per stem of over 30%. The leading states in their production are now (in order): California, Florida, Pennsylvania, New York, Ohio, Massachussets, New Jersey, Michigan, North Carolina, and Maryland. Half the total production now comes from the three top states, indicating a southerly movement of the prime production area.

Snapdragons are popular with retail florists, and they could be much more important commercially than these figures would suggest. They are fragile cut flowers that do not ship well for long distances, but if more high quality stems were grown locally, many more flowers could be sold. They are ideal for local market production. They fit in beautifully with the current market desires for greater variety in the kinds of cut flowers available for commercial use.

II. CROP PRODUCTION

A. Propagation

Before the early 1900s, snapdragons were vegetatively propagated almost exclusively, but suffered severe losses from snapdragon rust, transmitted from one crop to the next on infected cuttings. To combat this problem, growers began growing seed propagated, inbred lines and then later moved to F_1 hybrid types. Seedlings are easily grown or may be secured from specialist propagators. A number of factors are important in securing high germination percentages.

Seeds can be infected internally with several fungi (e.g., *Alternaria alternata, Phyllosticta antirrhini*, and *Stemphyllium botryosum*) all of which are pathogenic to snapdragons and can result in poor germination percentages. Infection is more common when seed is produced under high humidity conditions outdoors, than when it is grown in a well-ventilated greenhouse. Treatment with seed fungicides such as thiram is not effective in overcoming the problem and appears even to be detrimental to germination of healthy seed (Harman *et al.*, 1973). Most greenhouse-forcing snapdragon seed is greenhouse-grown today, and the above disease problems are only rarely encountered.

Snapdragon seedlings are highly susceptible to *Pythium* and *Rhizoctonia*, so the seed should be sown in pathogen-free media. It is worthwhile to prepare the germinating medium in the flat, and then steam flat and medium simultaneously. Seed can be sown as soon as the medium is cool. If a fungicidal drench is applied, use Banrot (6 ounces per 100 gallons) rather than captan or ferbam, because these have been shown to reduce snapdragon seed germination (Strider and Jones, 1984). A loose, open, well-aerated medium is necessary for good root development and ease of transplanting.

Seeds germinate well under a mist propagation system during the summer when medium temperatures can be maintained at 68° to 70°F. At other seasons, when the mist might cool the medium to suboptimum temperatures, better germination will be obtained by covering the newly seeded flats with a polyethylene sleeve to retain moisture. Optimum germination of snapdragon seeds occurs at temperatures of 65° to 70°F and is also enhanced by light (Atwater, 1980).

Seedlings should not be too crowded in the seed flat. A spacing of about 300 seedlings per square foot is recommended. Growing them as individual seedlings in plug trays is even better (Larson *et al.*, 1985). After germination is complete, cooler, drier conditions result in stocky growth. High light intensities are desirable. Supplemental lighting hastens growth of the seedlings to transplanting size (Petersen, 1955) (see Section II,F).

Snapdragon seedlings are one of the easiest kinds of plants to store, should this become necessary. They can be held for up to 6 weeks at 33° to 40°F if provided with fluorescent light at 2.7 klx for 14 hours daily. The flat should be enclosed in a polyethylene sleeve to prevent excessive drying of the medium during the storage period. (Kumpf *et al.*, 1966).

B. Plant Culture

"Unchecked growth" results if flat-grown seedlings are transplanted when the first set of true leaves has developed and the small plants are about 1 inch tall. At this stage in their development, transpirational water losses are small, and the roots readily absorb the moisture required by the plants. Plug-grown seedlings are even better and should provide 100% stands with no misses. Adding $2\frac{1}{2}$ to 5 pounds of microsized Osmocote (16–9–12) per cubic yard of medium in the plug trays produced high-quality seedlings that grew into early-flowering, top-quality flower spikes at maturity (Larson *et al.*, 1985).

A bench spacing of 15 to 20 square inches per stem is recommended for winter crops (3 × 5 inches or 4 × 5 inches), and 12 to 15 square inches (3 × 4 inches or 3 × 5 inches) for summer crops. Should a pinched crop be planned, individual plants would be planted farther apart, and the number of branches to be retained would be calculated to give about the same amount of space per flowering stem.

If only a single crop of flowers is desired from a given planting, two tiers of support that can be moved upward as the crop grows will be adequate. If successive crops of flowers are to be harvested from the same planting, it is necessary to have four to six tiers of support 6 to 8 inches apart, to enable the new stems to remain erect. Because snapdragons exhibit strong negative geotropism, maintaining flowering stems in a vertical plane is necessary to produce straight-stemmed cut flowers. Welded wire mesh or nylon mesh is used almost exclusively today. Support fabrics with mesh sizes 4 × 5 inches, 6 × 6 inches, or 6 × 8 inches are all suitable for this purpose.

Snapdragons grow best in open, porous, well-drained, and well-aerated growing media. Because snapdragons need good soil aeration, proper preparation of the growing medium prior to planting is important (Miller, 1957). Unamended field soil does not result in as good plant growth as soil to which leaf mold, peat moss, or other sources of organic matter have been added. Severe soil compaction also causes stunted plant growth, weak root systems, and delayed flowering.

Excellent quality snaps have been grown in media ranging from a completely soilless medium of a 1:1 mixture of peat:perlite, to a 1:1:1 mixture of peat:soil:perlite or sand, to a 3:1 soil:sand mix (Hanan and Langhans, 1962), or in inert, loose rockwool (Lee *et al.*, 1986), provided soil watering, aeration, and nutritional control were properly maintained.

C. Nutrition

Snapdragons have lower quantitative nutritional requirements than crops such as chrysanthemums, roses, or poinsettias. No statistically significant differences in plant growth resulted between plants grown in soil-based media at Spurway soil test values over the range of 1 to 10 ppm phosphorus, or 5 to 30 ppm potassium; there were differences, however, between 2 to 5 ppm and 25 to 50 ppm nitrate levels (Rogers, 1951), with stronger growth occurring at the higher nitrogen levels.

If nitrogen levels in the medium are low at the time of benching seedlings, fertilization with soluble nitrogen should begin promptly to insure the development of thick, heavy stems (Boodley, 1962). Overfertilization with nitrogen, however, can lead to excessive grassiness and should be avoided. Sanderson (1975) recommended using about one-half the normally recommended 200 ppm nitrogen and potassium levels if snapdragons are to be given constant liquid fertilization. Young plants respond better to nitrate than to ammonium forms of nitrogen, so the actual nitrogen source should be selected with this response in mind (Haney, 1961).

Detailed studies of nutritional deficiency symptoms have been published (Oertli, 1970); the observations are summarized in the following paragraphs.

Nitrogen deficiency causes stunted plant growth and an overall yellow-green color of the foliage. Few side branches are produced and stems tend to be thin, hard, and wiry.

Phosphorus deficiency also causes stunted plant growth, and the young leaves become very dark green in color. In some cases, purpling, especially of the undersides of older leaves, is noted. Young leaves may have the tips recurved downward and inward toward the central stem.

Potassium deficiency causes symptoms in snapdragons somewhat different from those in most plants. The initial response is an interveinal chlorosis of the young leaves similar to the classic iron deficiency symptom. This is followed later by the development of the symptoms normally associated with potassium deficiency (i.e., necrosis of tips and margins of the older leaves on the plant).

Calcium deficiency in snapdragon first affects the developing root system, because calcium is an element essential for the formation of primary cell walls in meristematic areas. Seedlings placed in calcium-deficient solutions died very quickly and had thin, poorly branched, and poorly developed root systems. Older, more developed plants placed under calcium-deficient conditions began to wilt after 1 or 2 weeks and died shortly thereafter. Symptoms similar to the iron chlorosis noted previously for potassium deficiency were also seen by Oertli (1970).

The symptoms of magnesium deficiency in snapdragons are more typical of the classic pattern. Interveinal chlorosis and necrosis appear first on the older leaves of the plant while the young, newly developing leaves remain relatively healthy in appearance. Tips of the older leaves have been reported to curl downward and tips of younger leaves to curl upward. High levels of ammonium nitrogen antagonize magnesium uptake (Haney, 1961), so growers should consider this as one possible cause of a magnesium deficiency problem. Because snapdragons appear to have a higher than normal magnesium requirement (Dunham et al., 1956), and because its uptake is also antagonized by high levels of calcium and potassium, low to medium levels of these latter cations in the growing medium should also be maintained for balanced nutrition of the plant.

Sulfur deficiency causes pale, yellow-green upper leaves in which the main veins are lighter than the rest of the leaf blade. As the deficiency continues, the symptom pattern progresses downward. A fine, interveinal chlorosis may be seen on the older leaves, while the young leaves are more uniformly yellow-green.

Iron chlorosis of snapdragon appears first as interveinal yellowing of the youngest, most recently developed leaves and progresses to nearly complete loss of all green color from these leaves. Small axillary shoots that develop on the main stem may be almost completely white. Such symptoms may appear even in the presence of adequate supplies of iron in the growing medium if any interference with root growth or function occurs, such as might be caused by excessive levels of soluble salts, improper pH levels, calcium deficiency, root rot problems, overwatering, or inadequate soil aeration. Iron in most plants cannot be accumulated for future use, but must be taken up daily to meet the daily needs of the plant. Any interruption in the ability of the plant to do this can result in the appearance of iron deficiency symptoms in the young leaves.

Boron deficiency has also been reported as a problem in snapdragon culture (Furuta, 1960). Lack of boron inhibits meristematic activity, and a common symptom will be death or abortion of the terminal growing point (in snapdragons, the terminal flower spike) followed by the growth and development of bypassing shoots from axillary buds further down the stem (Fig. 2). Boron deficiency symptoms can be precipitated by overapplication of calcium-containing fertilizers or lime, due to an antagonistic interaction between the two ions in the soil (Carmichael, 1968). Because boron is a trace element, only small quantities are needed. Normal recommendations for application to snapdragons would be a maximum of 1 ounce of household-grade Borax per 100 square feet of bench area per year, which may be broken up into two or three applications at lower dosage rates.

Other nutrient deficiency symptoms have occurred so rarely in normal snapdragon production that they can usually be disregarded.

Fig. 2. Boron deficiency symptoms (*right*) are made more severe by the antagonistic effects of calcium in the nutrient solution. (From Carmichael, 1968.)

D. Moisture and Aeration

Because the pore spaces in the growing medium are filled with either air or water, these two environmental factors need to be considered together. Many growers have noted that if snapdragons are grown at relatively high moisture levels, they often succumb to a root rot and wilt problem as the plants near maturity. Because of this, it has often been recommended that the crop should be grown on the "dry side" to prevent such problems.

Short-term waterlogging of snapdragon plants in the seedling stage can cause damage that will show up as decreased growth and quality of flower spikes produced at maturity (Miller, 1957). When potted seedlings were completely submerged in water for a period as short as 3 days, significant reductions in the grade of the flowering stems occurred. This was probably caused by the damage to the existing root system of the plant, and by the necessity of generating a new functional root system after the period of flooding.

A major complicating factor in attempting to study the effects of relative air–moisture balances in the growing medium in which snapdragons are produced is the almost universal presence in most growing situations of *Pythium*, pathogens which cause root rot. Although severity of root rot problems due to *Pythium* may be reduced in some other kinds of plants by reducing soil moisture levels, the degree of control of the disease achieved in snapdragons from less frequent watering was not sufficient to overcome the simultaneous decrease in total plant growth resulting from moisture stress (Hanan and Langhans, 1962, 1963).

Further exhaustive studies in this area (Hanan *et al.*, 1962, 1963) showed that if *Pythium* is rigorously excluded from the growing medium, snapdragons can be grown to perfection in a wide variety of media over a range of soil moisture contents (24 to 34% by volume, or from 1 to 10 inches of mercury tension on a tensiometer) and when air fills between 45 and 55% of the total pore space in the root zone. Moisture contents in excess of 40% by volume reduced final product quality. When moisture levels became that high in some growing media, oxygen diffusion rates began to approach levels that had been shown previously (Stolzy *et al.*, 1961) to limit initiation of new roots in snapdragons. These oxygen diffusion rates were somewhat lower than the critical rate for tomatoes, which suggests that snapdragons are somewhat more tolerant of low soil aeration than some other crops.

Although growing snapdragons at high moisture levels and with minimum moisture stress maximizes total growth and flower grade, it does cause a reduction in the percentage dry weight and a probable reduction in the keeping quality of the product. At the same time, the efficiency of water usage is reduced (Hanan, 1965). Conversely, with coarse textured soils, lateral movement of water by capillarity is slow, which can result in high water stress in the plants grown in them unless they are watered frequently. This means that the commercial grower has to maintain fairly critical balances in this area for optimum results. If *Pythium* is eliminated, and if growing media are deep enough (6 inches or greater) and porous enough to offset the deleterious effects of the perched water table at the bottom of the bench, the plants can be watered daily during periods of high solar radiation when moisture stress is maximum, for optimum growth and quality. Today's commercial snapdragon producer should probably adopt this system to ensure consistent success with the crop.

E. Temperature

Seeley (1965) indicated only small differences in the rate of growth and development of snapdragons between growing medium temperatures of 52° to 58°F and 68° to 73°F. A later study, however, was carried out under more rigidly controlled environmental conditions in a growth chamber rather than in a greenhouse (Rutland and Pallas, 1972). This showed that soil or root temperatures of 50°F, when compared to those of 77°F, greatly reduced the ability of the plant to maintain high levels of transpiration and resulted in increased plant moisture stress, stomatal closure, and probably reduced photosynthetic rates. Marked resistance to passage of water through the roots is known to occur at temperatures lower than 70°F. This is undoubtedly a major reason for midmorning wilting of snapdragon foliage on healthy, well-watered plants on clear winter mornings when light levels are high enough to encourage high transpiration rates before soil temperatures warm up in a 50°F night temperature greenhouse. Cultivars that maintained higher leaf hydration values because of earlier stomatal closure under moisture stress produced higher quality cut flowers (McDaniel and Miller, 1976) under southern summer greenhouse conditions than cultivars that had less control of moisture loss. Cut flower quality would probably be improved by use of soil warming treatments if soil temperatures were low enough to cause any appreciable plant wilting.

Although most plants probably have different optimum temperatures at different stages in their overall development, this relationship has been studied more intensively for snapdragons than for most other plants (Miller, 1962; Tayama and Miller, 1965). They have shown that optimum growing temperatures for snapdragons, based on dry weight accumulation of the top of the plant, decreased from a high of about 81°F for young seedlings held at constant day and night temperatures to a low of about 59°F for plants approaching flowering. A series of plants grown at a constant day temperature of 59°F but at various night temperatures, showed optimum night temperatures varying from 68°F for young plants to night temperatures of 55°F for plants near flowering.

The causes for these differences were later shown to be changes in the ratio of leaf dry weight to total plant dry weight as plants aged. Young plants had high leaf dry weight (tissue capable of carrying on photosynthesis) in relation to total dry weight (tissue capable of carrying on respiration). In young plants photosynthesis is the dominant process; in older plants respiration dominates. High temperatures, then, would increase the rate of photosynthesis in young plants to a greater extent than respiration, and this would result in increased rates of dry matter accumulation. In older plants with a lower ratio of leaf area to total bulk, lower temperatures would tend to retard respiration (the dominant process in such plants) more than they would photosynthesis and, thus, would result in maximum rates of dry matter accumulation (Tayama and Miller, 1965).

Some authors have suggested that reducing night temperatures after dark cloudy days and increasing them after bright, sunny days might increase the growth rates and the quality of some flower crops. Miller (1960) attempted this procedure with snapdragons. He found that reducing night temperatures after cloudy days had no measurable effect on size or quality of flower spikes but that increasing night temperatures from the "normal" 50° to 60°F after bright days resulted in slightly earlier flowering but slightly smaller flower spikes. Differences were probably not great enough to be commercially important.

Snapdragons have long been grown as a cool greenhouse crop with a night temperature of 50° to 52°F being considered optimum. Group I or group II cultivars should be selected for growth under such conditions in the northern greenhouse during midwinter months. Another alternative has also been suggested (Duffett, 1961); to grow group III cultivars at 60°F night temperatures in the same greenhouse with year-round chrysanthemums. Quality was good and timing of the crop was similar to that required for a single-stem chrysanthemum crop grown during the same season. Many group I or group II cultivars grown at these warm night temperatures in Missouri, however, were found to have soft, short stems and poor overall quality, and were considered unsalable (Rogers, 1961).

F. Light

Snapdragons were long-day, summer-flowering plants until 1926 when 'Chevoit Maid,' the first winter-flowering cultivar, was introduced (Lindstrom, 1966). Since that time many cultivars have been introduced, some that flower better in winter and others that are more satisfactory for summer flowering.

Early research indicated that snapdragons responded to daylength, flowering earlier under long days. Haney (1953) suggested giving young snapdragon plants short days by pulling black cloth during September and October to prevent premature flowering of late fall and early winter crops caused by unusually bright, warm fall weather. Increased stem length also resulted from this practice. These findings were confirmed by Rogers (1958), who used other cultivars developed at a later date.

The response of snapdragons to daylength differs from the daylength response of chrysanthemums or poinsettias (Maginnes and Langhans, 1967a). Instead of inhibiting or promoting flowering as was true for the latter two crops, daylength treatments exerted quantitative effects on snapdragons, and plants have now been classified as quantitative long-day plants in their response to daylength (Hedley, 1974). Long days hasten flowering in most cultivars; short days retard flowering but do not completely prevent it in most cases, with the possible exception of some of the group IV cultivars.

It also became evident that snapdragons reacted most markedly to daylength treatments when plants were between 2 and 8 inches high (Haney, 1953), when they had 10 to 12 leaves (Maginnes and Langhans, 1967a), or were 5 to 7 weeks after seeding (Rogers, 1958; Maginnes and Langhans, 1967a). Long-day treatments applied prior to the beginning of and during this light-sensitive stage reduced the number of leaves produced by the plant, hastened flower bud initiation, shortened stems, and also hastened flower development to salable maturity. Short-day treatments applied at the same stage of development caused opposite effects. Although temperature modified responses of some cultivars, it was judged to have a rather minor influence in the plant's response, except as it affected overall growth rates (Maginnes and Langhans, 1967a).

Four-hour exposures from 10 P.M. to 2 A.M. to flashing light from incandescent lamps at 0.1 to 0.27 klx intensities were shown to be as effective in providing long-day treatment for snapdragons as a continuous 4-hour light break during the same period, if the light was on at least 10 or more seconds each minute. At 5 seconds per minute, plant response was not quite as marked as at longer exposures (Maginnes and Langhans, 1967b).

In addition to responding to length of day as a facultative long-day plant, snapdragons have also been shown to respond photosynthetically to supplementary lighting at intensities and durations higher than needed to elicit photoperiodic responses. Petersen (1955) showed marked increases in dry weight of young snapdragon plants following all-night supplementary lighting with fluorescent lamps at about 10 lamp watts per square foot with the tubes 8 to 10 inches above the plants.

Snapdragons have also been shown to benefit from higher midwinter light levels achieved by reflecting additional natural sunlight onto the greenhouse benches using bright aluminum foil reflectors placed on the north side of benches oriented east–west (Carpenter, 1964).

Because high-intensity supplemental lighting is relatively expensive to apply, it is recommended that such treatments be carried out only during the small plant stage, before plants have been transplanted at final spacing to the cut flower benches (Flint, 1958). However, if young seedlings are kept in small pots before benching to make it possible to apply economical supplementary lighting treatments, the plants soon benefit more from transplanting to the final bench than from continued lighting (Rogers, 1960). This pivotal point appears to be when the plant height reaches approximately 4 inches (Rogers, 1958).

In still another study, Rogers (1961) found that providing snapdragons with 1.1 to 3.3 klx of supplementary lighting, using mercury vapor lamps for 3 weeks after benching in a 50°F night temperature greenhouse, gave final plant quality and timing results comparable to growing the plants at approximately 60°F night temperature without supplementary lighting. Plant growth was hastened about equally by either warmer growing temperatures or added light.

G. Carbon Dioxide Enrichment

Although roses and chrysanthemums have been shown to respond dramatically to carbon dioxide enrichment of the greenhouse atmosphere, especially at slightly elevated day temperatures (Shaw and Rogers, 1964), snapdragons were found to be much less responsive in midwinter in Missouri. Little difference in stem length, stem weight, or flower spike length resulted from carbon dioxide enrichment (1200 to 1500 ppm) at either 50° to 60°F night temperatures and higher day temperatures. Flowering was accelerated about 10 days, however, compared to control plants that did not receive additional carbon dioxide. Koths (1964) also reported only slight benefits from carbon dioxide enrichment at 400 to 500 ppm when both treated and untreated plants were grown in Connecticut at 50°F night temperature.

Lindstrom (1966) grew cultivars representing groups I, II, III, and IV at approximately 60°F night temperatures and at carbon dioxide levels of 1500 and 4000 ppm in Michigan. Under such growing conditions, cultivars from groups I or II were not considered salable, either with or without additional carbon dioxide. Groups III or IV cultivars, however, flowered sooner and were of higher grade when grown under conditions of carbon dioxide enrichment, and were considered to be of excellent quality. There were not enough additional benefits, however, from 4000 ppm CO_2 over 1500 ppm to justify the additional cost of adding it.

Duffett (1968) recommended the use of 60°F night temperatures and carbon dioxide enrichment to 800 ppm for routine winter culture of this crop under the low light conditions of northern Ohio. Because the cost of carbon dioxide enrichment came to only $0.04 to 0.05 per square foot per year, the 2 to 6 weeks saving in bench time for each crop made its use a highly profitable innovation. Carbon dioxide levels higher than 800 ppm were conducive to excessively heavy stems and grassy growth in that area.

Under the milder winter conditions of North Carolina, where much more ventilation is required for daytime temperature maintenance than in the North, carbon dioxide enrichment at either 750 or 1500 ppm was found to have only slightly beneficial effects for snapdragon production (Nelson and Larson, 1969). It would seem then that carbon dioxide enrichment to about 800 ppm with night temperatures higher than 52°F are commercially desirable practices only when the later flowering response groups are used, and predominantly in the low-light, colder growing areas of the northern United States.

III. PLANT PROTECTION AND PEST CONTROL

A. Air Pollution Problems

Of the various air pollutant gases that can injure plants, ethylene (Fischer, 1950) probably is the most damaging to snapdragons. This gas causes shattering or premature dropping of florets. Haney (1952) conclusively demonstrated that this problem is controlled by a single gene, dominant for shattering. In the development of new hybrid cultivars, one test given routinely by some breeders is an ethylene test to separate shattering from nonshattering cultivars.

Ethylene gas can be produced as a result of incomplete combustion of fossil fuels in gas- or oil-fired unit heaters or carbon dioxide generators in tightly sealed plastic greenhouses that do not have properly sized air intakes. To prevent such problems, growers using such plastic growing structures should provide at least 1 square inch of cross-sectional area of fresh air intake ducting for each 2000 BTU heat output from the fuel burning units (Rogers et al., 1969).

B. Diseases

Several reviews, some well illustrated, on the subject of diseases have been published (Williamson, 1962; Nelson, 1962; Dimock, 1958; Forsberg, 1958). The most common and serious diseases of greenhouse snapdragons vary from one growing area to another. Botrytis blight, for example, is a very serious disease in the cool, moist, low winter light area of northeastern United States, but is rarely seen in the midWest. Excellent greenhouse sanitation in which all dead and fallen leaves are quickly removed from the growing area is one key to prevention of this disease, because the causal fungus grows and sporulates freely on such dead and decaying organic matter.

Powdery mildew, downy mildew, and rust are likewise much more troublesome in moist, humid growing areas. Benomyl, applied at recommended rates, provides effective control of powdery mildew. The spores of the fungi causing Botrytis blight, downy mildew, and rust require the presence of liquid moisture films on the leaves for germination and infection to occur. Growers who are able to control relative humidity levels below the saturation point in their greenhouses will essentially prevent the occurrence of all three diseases. Horizontal air flow, internal air circulation systems, and the use of heating and ventilation controls that bring in cool, dry outside air when internal relative humidity levels become too high, will prevent moisture condensation on foliage and the start of such high-moisture mediated diseases.

Several diseases, previously unreported as occurring on snapdragons, have been noted. *Alternaria* species and *Helminthosporium* species have been found infecting snapdragon flowers in Florida, producing symptoms very similar to *Botrytis* flower infection. Instead of producing masses of gray mold mycelium and spores on the surface of rotted tissue typical of *Botrytis*, however, *Alternaria* produced abundant masses of black spores on tissue, and *Helminthosporium* sporulated on the infected tissue but did not develop dense spore masses characteristic of the other two fungi (Engelhard, 1971).

Snapdragon leafspot, caused by *Cercospora antirrhina*, has been a serious problem for many North Carolina growers for over 20 years (Porter and Aycock, 1967). All greenhouse cultivars tested were about equally susceptible. The fungus may survive in dried infected leaves for over 14 months and in leaves in or on the soil for over 3 months. For these reasons, careful sanitation to remove potential inoculum sources is recommended as a primary control measure. Use of fungicidal sprays as protectants prior to inoculation would also seem prudent for growers who have experienced this problem in the greenhouse.

C. Insects

In general, the snapdragon is relatively free of insect pests, but several species may invade plantings. Aphids, spider mites, and looper larvae are the most common. Because it is illegal according to the Federal Environmental Pesticide Control Act of 1972 as amended, to use any pesticide in any way inconsistent with the product label, only materials specifically labeled for the crop and for the pest concerned may be used. Consult the regularly updated chemical control recommendations provided by several of the leading state florists associations (e.g., Ohio, Florida, Michigan, California) for information about chemicals effective and legal for use on specific snapdragon pests.

With increasingly stringent state and federal regulation of chemical pesticides and with the rapid development of insect and mite resistance to more and more chemicals, florists are going to be forced to implement other kinds of pest control procedures (Parella, 1989). Integrated pest management procedures are being used more and more widely by commercial florists. This means that many more nonchemical pest control procedures will be used, such as greenhouse sanitation (e.g., removal of weeds in walks, under greenhouse benches, and in the outdoor areas immediately surrounding the greenhouse), insect exclusion from the greenhouse (e.g., prevention of entry from outside by installation of insect-proof screening of all air intakes and entries), and the integration of biological pest control methods (e.g., use of insect pathogens such as *Bacillus thuringiensis*, insect growth regulators such as kinoprene or Enstar, or the programmed introduction of insect parasites, predators, or both—e.g., *Chrysopha carnea* or lacewing or *Phytosieulius persimilis*, which is a predacious mite) into greenhouse crop cultural procedures. Chemicals will still be used where it is possible to integrate them into the picture without destroying these other primary pest control procedures, but we will cease to depend on them as the primary or only method for control of insects and mites, as we have tended to do for the past 30 to 40 years.

The use of the biological control procedures mentioned is new to growers in the United States and will require the completion of much research to develop the biological information needed about both pests and beneficial insects before we can recommend effective commercial pest control programs. Such programs are far-advanced in Europe, and many commercial greenhouse flower and vegetable growers are already producing high quality crops with a minimum use of chemical pesticides.

IV. CARE AND HANDLING OF THE FINISHED PRODUCT

A. Harvesting and Grading

Traditionally, snapdragons have been harvested when the florets on the lower third of the spike are open. Because the cut flowers are subject to tip curvature if stems are placed other than vertically (negative geotropism), they must always be held in an upright position. Attempts have been made in the past to overcome this problem by chemical (growth regulator) treatments, but none have been successful enough to have been adopted commercially. Cut flowers have been shipped in upright containers such as gladiolus hampers to prevent crooked tips.

The Society of American Florists (SAF) has established standard grades for snapdragons based on stem length, stem weight, and number of open florets per stem (Table II). Material to be marketed under these grade standards must have reasonably straight stems, and clean, uniform foliage free of insect or disease injury. Foliage on the lower third of the stem should be removed.

B. Long-Term Storage

The earliest work on long-term holding of snapdragons was done by Mastalerz (1953). This involved the use of 31°F temperature and dry-pack storage. Under these conditions some snapdragon cultivars could be held for up to 3 weeks without appreciable loss of quality or vase life (compared to that of freshly harvested flowers). It was necessary to cut spikes before the oldest floret was more than 1 week old, because florets that opened following storage were often faded in color.

Snapdragons that are cut when only one or two florets are open can now be opened successfully in floral preservative solutions containing 8-hydroxyquinoline citrate (8-HQC) and sucrose (Raulston and Marousky, 1971). Chrysanthemums, gladiolus, *Strelitzia*, and carnation buds react similarly, and these can open successfully even after prolonged holding of cut buds in low-temperature dry-pack storage conditions (Kofranek, 1976).

Another development that may have future potential in this area is the use of hypobaric storage. Burg (1973) found that snapdragons could be held at <2 in mercury pressure for up to 6 weeks in good condition, compared to the 3-week maximum for the best methods used previously. The equipment necessary for this

Table II

Minimum Specifications for SAF Standard Grades of Snapdragons[a]

Grade name	Label color	Weight per spike (gm)		Minimum open flowers per stem (number)	Minimum stem length (cm)	Stems per bunch (number)
		Minimum	Maximum			
Special	Blue	71	113	15	91	12
Fancy	Red	43	70	12	76	12
Extra	Green	29	42	9	61	12
First	Yellow	14	28	6	46	12

[a] The term *utility* is not a grade within the meaning of these standards. Material that does not grade because of crooked stems, foliage injury, or abnormal growth may be marketed under the designation *utility* (white label). The term *unclassified* is not a grade within the meaning of these standards but is provided as a designation to show that no definite grade has been applied to the lot.

technology, however, has been expensive to acquire and difficult to operate, and hypobaric storage has not as yet been brought into any widespread commercial use in the floral industry.

C. Postharvest Physiology

Few cut flower crops are more responsive to good postharvest treatment than snapdragons. Freshly cut flower spikes of most cultivars will have a vase life of about 1 week in tap water or distilled water. When the best combination of flower preservatives is used, vase life can be increased two or three times.

Larsen and Scholes (1966) and Raulston and Marousky (1971) found that longest vase life, greatest number of florets opening, and greatest increase in spike length after cutting occurred when flowers were held in a solution containing 300 ppm of 8-HQC + 1.5% sucrose. The former researchers also found the addition of 25 ppm Alar (n-dimethylaminosuccinamic acid) to be beneficial. Johnson (1972) got best results from a solution of 300 ppm 8-HQC + 0.5% sucrose. Both light and floral preservatives are crucial for proper development of floret color in florets that open after harvest (Marousky and Raulston, 1970). Regardless of the solution used, spikes held in darkness produced little anthocyanin and were poorly colored. In the light, those spikes held in 8-HQC + sucrose produced much more intensely colored florets than those held in tap water. Light (2.15 klx) incident on the developing floret at the time of opening was critical for anthocyanin production.

Self-generated ethylene gas can be a prime cause of early senescence in cut snapdragons. One of the reasons for excellent results with hypobaric storage is the constant removal of trace quantities of ethylene from the storage atmosphere. Pretreatment of cut snapdragon stems for 20 hours immediately after harvest in a solution containing silver thiosulfate (STS) and sucrose inhibits ethylene action and added about 6 days vase life compared to distilled water controls (Nowak, 1981). The highly toxic silver ion, however, has not yet been cleared for commercial use in postharvest treatment of cut flowers in the United States.

Another approach to control of ethylene problems has involved use of chemicals to suppress ethylene formation (Wang et al., 1977). In this study, two analogs of rhizobitoxine and sodium benzoate were tested to determine the relationships between their effects on ethylene production by flowers and keeping quality. Both ethoxy and methoxy analogs of rhizobitoxine significantly reduced ethylene production and increased vase life. Like hypobaric storage, however, this treatment has also not yet gained commercial acceptance.

REFERENCES

Atwater, B. R. (1980). "Germination, dormancy and morphology of the seeds of herbaceous ornamental plants." *Seed Sci. Tech.* **8**(4): 523–573.
Ball, V., ed. (1952). *The Ball Red Book, 8th ed.*, West Chicago, Illinois: Geo. J. Ball, Inc.

Boodley, J. W. (1962). "Fertilization." In *Snapdragons*, edited by R. W. Langhans, pp. 28–34. Ithaca, New York: N.Y. State Flower Grow. Assoc., Inc.

Burg, S. P. (1973). "Hypobaric storage of cut flowers." *HortScience.* **8**: 202–205.

Carmichael, O. E. (1968). "Boron Toxicity of Flowering Plants." Masters thesis, University of Missouri, Columbia, Missouri.

Carpenter, W. J. (1964). "Response of snapdragons and chrysanthemums to supplemental reflective sunlight." *Proc. Am. Soc. Hortic. Sci.* **84**: 624–629.

Dimock, A. W. (1958). "Snapdragon diseases common in New York." *N.Y. Flower Grow. Bull.* **145**: 2–3.

Duffett, W. E. (1960). "Response groups and varieties for year-round snapdragons." *Ohio Florists' Assoc. Bull.* **371**: 4–5.

Duffett, W. E. (1961). "Grow these snaps in 60° greenhouses along with mums." *Yoder Grower Circle News* **7**: 3.

Duffett, W. E. (1968). "Culture of greenhouse snapdragons." *Ohio Florists' Assoc. Bull.* **466**: 5–7.

Dunham, C. W., Hamner, C. L., and Asen, S. (1956). "Cation exchange properties of the roots of some ornamental plant species." *Proc. Am. Soc. Hortic. Sci.* **68**: 556–563.

Engelhard, A. W. (1971). "Botrytis-like diseases of rose, chrysanthemum, carnation, snapdragon, and King aster caused by *Alternaria* and *Helminthosporium*." *Proc. Fla. State Hortic. Soc.* **83**: 455–457.

Fischer, C. W., Jr. (1950). "Production of a toxic volatile by flowering stems of common snapdragon and calceolaria." *Proc. Am. Soc. Hortic. Sci.* **55**: 447–454.

Flint, H. L. (1958). "Snapdragon lighting." *N.Y. State Flower Grow. Bull.* **145**(1): 3–5.

Forsberg, J. L. (1958). "Snapdragon diseases." *Ill. State Flor. Assoc. Bull.* **186**: 5–8.

Furuta, T. (1960). "Test boron deficiency in snapdragons at Auburn." *Florists' Rev.* **126**(3244): 25.

Hanan, J. J. (1965). "Efficiency and effect of irrigation regimes on growth and flowering of snapdragons." *Proc. Am. Soc. Hortic. Sci.* **86**: 681–692.

Hanan, J. J., and Langhans, R. W. (1962). "Soil aeration—Progress report." *N.Y. State Flower Grow. Bull.* **198**: 1–2, 6.

Hanan, J. J., and Langhans, R. W. (1963). "Soil aeration and moisture controls snapdragon quality." *N.Y. State Flower Grow. Bull.* **210**: 3–6.

Hanan, J. J., Langhans, R. W., and Dimock, A. W. (1962). "Soil aeration and the *Pythium* root rot disease of snapdragon." *N.Y. State Flower Grow. Bull.* **195**: 1–6.

Hanan, J. J., Langhans, R. W., and Dimock, A. W. (1963). "*Pythium* and soil aeration." *Proc. Am. Soc. Hort. Sci.* **82**: 574–582.

Haney, W. J. (1952). "Snapdragon shattering." *Mich. Florist.* **258**: 24.

Haney, W. J. (1953). "Daylength manipulation to time snapdragons." *Natl. Snapdragon Soc. Bull.* **2**: 1–3, 12.

Haney, W. J. (1961). "Snapdragon culture." *Mich. Florist.* **366**: 25–26, 29.

Harman, G. E., Heit, C. E., Pfleger, F. L., and Braverman, S. W. (1973). "Snapdragon blight—A serious problem caused by seedborne fungi." *Plant Dis. Rep.* **57**: 592–595.

Hedley, C. L. (1974). "Response to light intensity and day-length of two contrasting flower varieties of *Antirrhinum majus* L." *J. Hortic. Soc.* **49**: 105–112.

Johnson, C. R. (1972). "Effectiveness of floral preservatives on increasing the vase-life of snapdragons." *Florists' Rev.* **149**(3868): 47, 95–97.

Kofranek, A. M. (1976). "Opening flower buds after storage." *Acta Hortic.* **64**: 231–237.

Koths, J. S. (1964). The effect of CO_2 enriched greenhouse atmosphere on growth of snapdragons. *Mich. Flor.* **399:** 15.

Kumpf, J., Horton, F., and Langhans, R. W. (1966). "Seedling storage." *N.Y. State Flower Grow. Bull.* **244**: 1–3.

Larsen, F. E., and Scholes, J. F. (1966). "Effects of 8-hydroxyquinoline citrate, N-dimethyl amino succinamic acid, and sucrose on vase-life and spike characteristics of cut snapdragons." *Proc. Am. Soc. Hortic. Sci.* **89**: 694–701.

Larson, R. A., Thorne, C. B., and Milks, R. R. (1985). "Growth of geranium and snapdragon 'plugs' fertilized with controlled release micro-fertilizer and exposed to root-zone heating." *N.C. Flower Growers Bull.* **29**(3): 12–16.

Lee, C. W., Goldsberry, K. L., and Hanan, J. J. (1986). "Production of cut snapdragons in rockwool." *CO. Greenhouse Grow. Assoc. Bull.* **438**: 1–2.

Lindstrom, R. S. (1966). "Snapdragons—60°F and CO_2." *Florists' Rev.* ·**139**(3591): 18–19, 51–54.

Maginnes, E. A., and Langhans, R. W. (1967a). Photoperiod and flowering of snapdragon. *N.Y. State Flower Grow. Bull.* **260**: 1–3.

Maginnes, E. A., and Langhans, R. W. (1967b). "Flashing light affects the flowering of snapdragons." *N.Y. State Flower Grow. Bull.* **261**: 1–3.

Marousky, F. J., and Raulston, J. C. (1970). "Enhancement of snapdragon floret color with light and floral preservatives." *HortScience* (Abstract). **5**: 355.

Mastalerz, J. W. (1953). "Low temperature conditioning of snaps." *Natl. Snapdragon Soc. Bull.* **1**: 3.

McDaniel, G. L., and Miller, M. G. (1976). "Transpiration of snapdragon under southern summer greenhouse conditions." *HortScience.* **11**: 366–368.

Miller, R. O. (1957). "Snaps need good drainage." *N.Y. State Flower Grow. Bull.* **140**: 1–3.

Miller, R. O. (1960). "Growth and flowering of snapdragons as affected by night temperatures adjusted in relation to light intensity." *Proc. Am. Soc. Hortic. Sci.* **75**: 761–768.

Miller, R. O. (1962). "Variations in optimum temperatures of snapdragons depending on plant size." *Proc. Am. Soc. Hortic. Sci.* **81**: 535–543.

Nelson, P. (1962). "Diseases," In *Snapdragons*, edited by R. W. Langhans, pp. 70–80. Ithaca, New York: N.Y. State Flower Grow. Assoc.

Nelson, P. V., and Larson, R. A. (1969). "The effects of increased CO_2 concentrations on chrysanthemum (*C. morifolium*) and snapdragon (*Antirrhinum majus*)." *N.C. Agric. Exp. Stn. Techn. Bull.* **194**: 1–15.

Nowak, J. (1981). "Chemical pre-treatment of snapdragon spikes to increase cut-flower longevity." *Scientia Hort.* **15**(3): 255–262.

Oertli, J. J. (1970). "Nutrient disorders in snapdragons." *Florists' Rev.* **146**(3773–3780), 20–21, 28–29, 29, 51, 23, 28, 65, 24.

Parella, M. P. (1989). "Where are we in the development of integrated pest management." *Am. Floral Endowment Res. Rep.* **1**(1): 1–8.

Petersen, H. (1955). "Artificial light for seedlings and cuttings." *N.Y. State Flower Grower. Bull.* **122**: 2–3.

Porter, D. M., and Aycock, R. (1967). "Snapdragon leafspot caused by *Cercospora antirrhina*." *N.C. Agric. Exp. Stn. Techn. Bull.* **179**: 1–31.

Raulston, J. C., and Marousky, F. J. (1971). "Effects of 8–10 day 5°C. storage and floral preservatives on snapdragon cut flowers." *Fla. Flower Grow.* **8**(2): 4–10.

Rogers, M. N. (1951). "Greenhouse Soil Fertility Analysis and Interpretation." Masters thesis, University of Missouri, Columbia, Missouri.

Rogers, M. N. (1958). "Year around snapdragon culture. 1. Effects of lighting and shading snaps seeded during the summer months." *MO State Florists News* **18**(3): 3–7.

Rogers, M. N. (1960). "Direct benching vs. potting snapdragon seedlings." *PA Flower Grow. Bull.* **118**: 3–5.

Rogers, M. N. (1961). "The reactions of varieties of different response groups grown during the winter months at night temperatures of 60°F, 50°F, and 50°F, with supplementary lighting." *Natl. Snapdragon Soc. Bull.* **13**: 1–6.

Rogers, M. N., Hartley, D. E., and Tjia, B. O. S. (1969). "Ethylene prevents normal photoperiodic response in *Chrysanthemum morifolium*." *Florists' Review.* **143**(3716): 19–21, 87–90.

Rutland, R. B., and Pallas, J. E., Jr. (1972). "Transpiration of *Antirrhinum majus* L. in relation to radiant energy in the greenhouse." *J. Am. Soc. Hortic. Sci.* **97**: 34–37.

Sanderson, K. C. (1975). "Fertilization, watering, temperature, light and photoperiod." *Florists' Rev.* **156**(4038): 17, 59–61.

Sanderson, K. C., and Martin, W. C. (1984). "Evaluation and scheduling of snapdragon cultivars." *Ala. (Auburn) Agr. Exp. Sta. Bul.* **468**(revised): 1–27.

Seeley, J. G. (1965). "Soil temperature and the growth of greenhouse snapdragons." *Proc. Am. Soc. Hortic. Sci.* **86**: 693–694.

Shaw, R. J., and Rogers, M. N. (1964). "Interactions between elevated carbon dioxide levels and greenhouse temperatures on the growth of roses, chrysanthemums, carnations, geraniums, snapdragons, and African Violets. Various flowers." *Florists' Rev.* **135**(3941): 19, 37–39.

Stolzy, L. H., Letey, J., Szuszkiewicz, T. E., and Lunt, O. R. (1961). "Root growth and diffusion rates as functions of oxygen concentration." *Soil Sci. Soc. Am. Proc.* **25**: 463–467.

Strider, D. L., and Jones, R. K. (1984). "Captan and ferbam drenches to control damping off may cause injury to some seedlings." *N.C. Flower Grow. Bull.* **28**(1): 12–14.

Tayama, H. K., and Miller, R. O. (1965). "Relationship of plant age and net assimilation rate to optimum growing temperature of the snapdragon." *Proc. Am. Soc. Hortic. Sci.* **86**: 672–680.

Wang, C. Y., Baker, J. E., Hardenburg, R. E., and Lieberman, M. (1977). "Effects of two analogs of rhizobitoxine and sodium benzoate on senescence of snapdragons." *J. Am. Soc. Hortic. Sci.* **102**: 517–520.

White, J. W. (1961). "Timing snapdragons." *Penn. Flower Grow. Bull.* **125**: 5–7.

Williamson, C. E. (1962). "Root diseases and soil sterilization," In *Snapdragons*, edited by R. W. Langhans, pp. 62–69. Ithaca, New York: N.Y. State Flower Grow. Assoc.

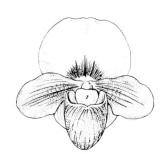

5

Orchids

Thomas J. Sheehan

Introduction to Floriculture, Second Edition
Copyright © 1992 by Academic Press, Inc.
All rights of reproduction in any form reserved.

113

I. INTRODUCTION

The cultivation of orchids is not new. Confucius (551–479 B.C.) mentioned orchids in his writings. He speaks of the fragrance of lan (orchids) in the home, indicating that the Chinese were using orchid flowers to decorate their homes (Withner, 1959). However, the evolution of orchid culture from the hobbyist to commercial production was very slow. The early Greeks and Romans looked to the orchids more for their medicinal than for their aesthetic qualities. It was not until the 1700s that interest in orchids really began to develop. During the early 1700s, sea captains, missionaries, and botanists began introducing orchids into Great Britain from all parts of the world. Plants were often brought back as gifts for their sponsors or benefactors. As these exotic plants bloomed, they helped stimulate additional interest in dispatching collectors to distant corners of the globe. Yet it was not until 1821, when Conrad Loddiges and Sons started growing orchid plants commercially at their nursery in Hackney, near London, that the orchid industry was born. Conrad Loddiges started producing flowering orchid plants for sale to the landed gentry, who could afford the glasshouses needed to grow them.

Almost another century passed before commercial production of orchids for cut flower sales came into vogue. In 1913, the Sun Kee Nursery opened in Singapore to produce spray-type orchids for cut flower sales. This nursery is still in production and has 30 acres devoted to *Arachnis, Aranda,* and *Aranthera* orchids. Most of the flowers are exported to Europe. Some of the earlier growers in the United States were Pitcher and Manda, in South Orange, New Jersey; Lager and Harrell, in Summit, New Jersey; Baldwin, in Mamaroneck, New York; and Linden in New York City.

Cut flowers are still widely grown in many areas of the world. *Cymbidium* flowers, for example, are produced mainly in California, New York, and Australia. Individual *Cymbidium* ranges in California may cover over 20 acres, with most plants flowering during late winter and early spring. However, flowers are available on a year-round basis, with production from the Southern Hemisphere supplementing California and New York production.

Dendrobium hybrids are being grown mainly in Hawaii, Thailand, and Singapore. Thailand presently is the largest exporter of *Dendrobium* sprays, with over $10 million in sales annually.

Singapore, Malaysia, and Thailand also export large quantities of other cut orchid sprays, with most of the production being exported to Europe, especially West Germany.

Although there are many major cut flower production areas around the world, in the United States growers are again producing plants. The demand for plants by the hobbyist has made plant sales more profitable than growing orchids for cut flowers.

The latest census figures indicate that the value of all orchid sales in the United States was $14,041,000 (1979), with Hawaii now producing $4,168,000 worth of orchids annually (1988).

II. BOTANICAL INFORMATION

Taxonomically, orchids are a unique group of plants. They are vastly different vegetatively, yet all species can be tied together by their floral characteristics as members of this huge family. The Orchidaceae contain over 800 genera and over 25,000 known species of monocotyledonous herbaceous perennial plants. Plants may be erect growers (monopodial) or prostrate (sympodial)(Fig. 1), with a few tree climbers (*Vanilla*). Although the majority of orchids are so-called green plants, there are a few saprophytes and leafless plants in the family. The stems may have one or more swollen internodes (pseudobulb) and have one to many leaves.

The parallel-veined leaves, either thick and leathery or thin, soft, and often pleated, come in a variety of shapes from linear to oval to orbicular. They are arranged alternatively along the stem.

The flowers are very distinctive and range in size from 1/8 to 18 inches in diameter. They come in every color and include many bicolor and tricolor flowers (*Cattleya bicolor* and *Vanda tricolor*). Some orchids have no odor, whereas others may have a wide variety of fragrances. Two highly fragrant orchids are *Maxillaria tennufolia* and *Aerides odorata*.

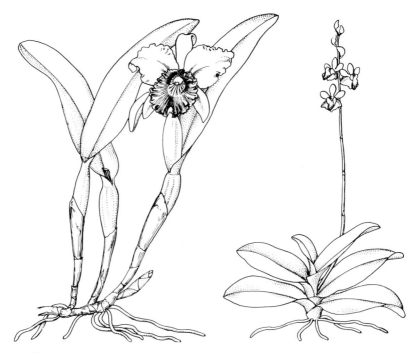

Fig. 1. The two major growth habits are sympodial (*Cattleya* on *left*) and monopodial (*Doritis* on *right*).

There are five distinguishing characteristics that separate orchid flowers from all others in the plant kingdom. These floral characteristics are:

1. *Zygomorphic flowers*—The orchid flower is a special type of irregular flower in that it has bilateral symmetry (*zygomorphic*). It can be cut in one plane and will divide into equal halves. Cutting the flower in any other plane would result in two unequal pieces (Fig. 2).

2. *Pollen*—The pollen of orchid flowers is agglutinated into small packets called *pollinia* (Fig. 3) that are removed by insects in the act of pollination. The number of pollinia per flower varies in different genera from two to eight with one species, *Brassavola cucullata,* having twelve. The number of pollinia, and their arrangement within the flower, can often be used in identification of the flower (e.g., *Cattleya* has four, whereas *Laelia* has eight).

3. *Column*—The reproductive structures of the orchid flower (anther and pistil) have been fused together in a waxy unit called the column (gynandrium). Within the column, there is a canal leading from the stigmatic surface to the ovary. The column may bear one fertile stamen, represented by an anther terminal on the column (*Cattleya*) or by two lateral anthers situated midway along the sides of the column (*Paphiopedilum*). The stigmatic surface is on the under side of the column (Fig. 4).

4. *Rostellum*—On the underside of the column between the anther cap and the stigmatic surface lies the rostellum. The *rostellum* is a gland formed at the apex of the style and often appears as a beaklike projection between the stigma and another cap (Fig. 4).

Fig. 2. A typical zygomorphic flower is exemplified by the *Cattleya*. It can be cut in only one plane to produce two equal halves.

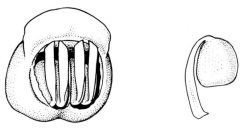

Fig. 3. Pollen of orchids is agglutinated into packets called pollinia. The figure on the *left* shows four pollinia with the anther cap. A single pollinium is on the *right.*

The rostellum performs two unique functions. First, it acts like a dam separating male and female portions of the flower, thus preventing self-pollination. The second function is that of a gland, wherein it disperses a viscid substance on the back of any insect that comes in contact with it. As an insect forces its way toward the base of a flower to partake of the nectar, its back rubs the rostellum, and a bead of gluey material is applied. As the insect backs out, the glue comes in contact with the stipe (caudicle) of the pollinia, and the insect takes the pollinia to the next flower it visits, thus ensuring cross-pollination.

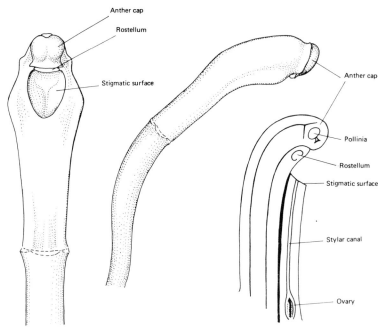

Fig. 4. The reproductive structures of the orchid flower are fused to form the column. *Left,* ventral view of column showing anther cap at the apex and stigmatic surface. *Center,* column side view. *Right,* a diagrammatic sketch of the column.

5. *Seeds*—Orchid flowers produce copious amounts of seed. A single orchid seed pod may contain between 500,000 and one million minute seeds (Fig. 5). These seeds, unlike corn or peas, contain no endosperm and are often called "naked seed." Because the seeds contain no endosperm, they cannot germinate in the wild without the aid of a fungus, whereas under laboratory conditions, they have to be germinated aseptically with all the necessary chemicals supplied via the germination medium.

Any plant that has four or more of the aforementioned characteristics in the flower would belong to the orchid family. Once one becomes familiar with the basic characteristics of orchid flowers, it becomes easy to identify unknown flowers as orchids (Sheehan and Sheehan, 1985).

III. PROMINENT GENERA GROWN AS CUT FLOWERS

A wide variety of orchid genera are grown as cut flowers (Fig. 6). The number will vary from country to country and, in some cases, area to area within a country, depending on climatic conditions. The following are some of the better known genera, and they are shown in Figure 6.

A. *Cattleya*

A genus native to Central and South American tropics, *Cattleya* has over 50 species and thousands of hybrids. Plants, species, and hybrids can be selected to provide a grower with flowers every month of the year. Some species and their hybrids are photoperiodic responders and can be flowered twice in 1 year (Hager, 1957). The colors range from white through various shades of lavender, yellow, and red. Bicolors, white with purple lips and yellow with purple lips, are available. Flower size ranges from 2 to 6 inches.

B. *Cymbidium*

Cymbidium is native to Asia and the Philippines. The species and hybrids grown are the cool types requiring 50°F night temperatures for flowering. Flowers (3 to 5

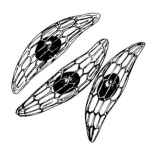

Fig. 5.   Typical orchid seeds.

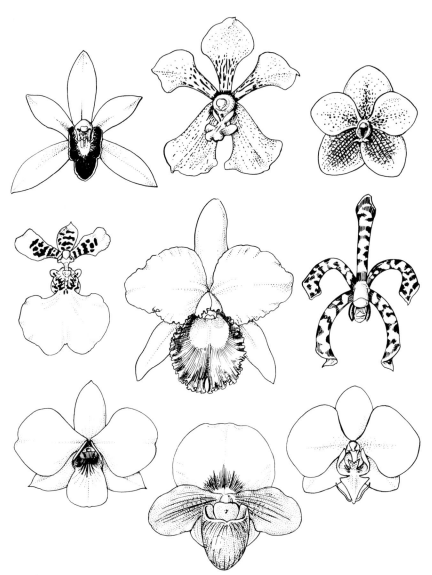

Fig. 6. Orchid genera grown as cut flowers. From left to right, *top row: Cymbidium, Vanda, Ascocenda. Middle row: Oncidium, Cattleya, Arachnis. Lower row: Dendrobium, Paphiopedilum, Phalaenopsis.*

inches) are grown basically for spring trade (Easter and Mother's Day), when their pastel shades are most fitting. However, flowers are available year-round because growers in Australia supply the northern markets during the Australian spring when plants in the United States are vegetative.

C. *Phalaenopsis*

These orchids, native to Asia, the Philippines, and Indonesia, are very popular for use in wedding bouquets. White flowers, hybrids of *Phalaenopsis amabilis,* are available year-round, as these plants can be maintained in flower continually. Pink and other colors are available in fall and spring.

D. *Dendrobium*

These natives of the Western Pacific basin are widely grown for their long-lasting sprays of cut flowers. Thailand, Singapore, and Hawaii are the largest producers of *Dendrobium* sprays. A typical Thailand spray is cut with seven flowers and seven buds. Most of the production in Thailand and Singapore is shipped to West Germany. This large genus, with a wide variety of flower colors, sizes, and shapes, has great untapped potential for cut flower production, and new hybrids developed at the University of Hawaii are very promising for year-round flowering.

E. *Vanda*

These natives of southeast Asia have long been popular plants. Probably the best known is *Vanda* 'Miss Joaquim,' which has been used in Hawaiian leis for many years. It also has been widely used as a promotional flower. *Vanda* flower sprays are now being grown in Singapore, Thailand, and Hawaii. The latter are available throughout the year in a variety of sizes and colors.

F. *Ascocenda*

These hybrids of *Vanda* and *Ascocetrum* resemble miniature vandas and have an excellent shelf life. Presently, most of the cut flower production is in Thailand, and most flowers are shipped to West Germany. Yields as high as 15 spikes per square foot per year have been reported.

G. *Arachnis* and Its Hybrids [*Aranthera (Arachnis* × *Renanthera*) and *Aranda (Arachnis* × *Vanda*)]

These southeast Asia natives are widely grown as cut flowers in Singapore and Malaysia. They are grown in open fields with minimal care and produce up to 12 spikes per plant per year. One acre is capable of producing between 300,000 and

360,000 spikes per year, depending on the number of plants per acre. A variety of colors is available.

H. *Oncidium* 'Golden Showers'

A hybrid of Central and South American plants, this *Oncidium* is a very popular cut flower used in West Germany. Most of the production is in Singapore where the plants flower all year. The sprays of dainty, bright yellow flowers are excellent for use in flower arrangements.

I. *Paphiopedilum*

The lady slipper orchids, natives of southeast Asia, have long been popular cut flowers in Europe and the northern parts of the United States. Most of those cultivars grown are hybrids of species, such as *Paphiopedilum insigne,* and require cool nights (50°F) for best flowering, so production is limited to the more temperate climates. However, interest in the warm types, such as *P. nivium* and *P. callosum,* should lead to the introduction of more warm-growing types, which could be produced commerically as far south as Florida.

Other genera are cut and sold on a limited basis in tropical and subtropical areas. These may be native plants such as the *Eulophia* sold in the markets in Nairobi, Kenya. *Phaius* spikes have also been sold as cut flowers by nurseries growing these plants for landscape use in central and southern Florida. Undoubtedly, many other genera, especially native plants, are cut and sold elsewhere in the world. Most of these, however, are not grown commercially for the flowers.

Today, these and many more genera, especially small-flowered species of little importance as cut flowers, are grown as pot plants for sale to hobbyists. Many orchid nurseries are predominantly in business to supply plants, from seedings in flasks all the way up to large, flowering-sized plants in pots or baskets.

IV. PROPAGATION

Orchids, like most floricultural crops, may be propagated either sexually or asexually. Because most orchids do not come true from seed, once a hybrid or a clonal selection has been made, then all further propagation must be by asexual means to be sure offspring will be true to type. Orchids are asexually propagated by several means not commonly used to increase other floricultural crops. Some unique characteristics of the orchids themselves make this possible.

A. Seed Germination Technique

Orchid seeds are very small. If put end to end, it would take 50 seeds to make a line 1 inch long. The seeds are not only small but also lack endosperm and hence are difficult to germinate. In their native habitats, germination takes place only when certain fungi are present, which supply sugars to a germinating seed until such time as the seedling has sufficient chlorophyll to produce its own sugars and sustain itself.

As early as 1903, scientists, such as Bernard (1903) at the Paris Botanic Garden, were trying to germinate orchid seeds. Both he and Burgeff (1909) concluded that the fungus was necessary to germinate orchid seeds in the laboratory. In the early 1920s, Knudson showed that the fungus was converting starch in the medium to sugar and that the seeds used the sugar for germination. Knudson substituted sugar for starch and got excellent growth without the fungus (1922). Just to prove his point, Dr. Knudson grew an orchid plant from seed to flower in a 3 gallon flask without any fungus. This research was the basis for the development of the industry as we know it today. Orchid seeds now are as easy to germinate and grow as any other floricultural crop, but they do require some special handling.

Knudson's work led to the development of Knudson's "C" Solution for orchid seed germination (1923). This formula has been the basis for most of the solutions used to germinate orchid seeds today (Table I).

A grower may either make up the medium or buy a prepared medium. The latter is simply mixed with water and autoclaved. It is ready to use as soon as it cools.

A variety of containers may be used. Old square milk bottles or new orange juice bottles are very popular, but any size of glass container can be used. Erlenmeyer flasks are also very popular.

When the medium is prepared, each chemical is weighed out and dissolved separately in a liter of water. Sugar is dissolved, and agar is added at the end. The solution should be heated until the agar is dissolved. The pH is adjusted to 5.0 to 5.2, and the medium is poured into containers and autoclaved at 15 to 20 pounds for 15 minutes. Sufficient medium should be poured into the container to provide a layer of medium approximately 0.5 inch deep.

After autoclaving and cooling, containers are ready for seed sowing. Sowing should be done in a clean area. All surfaces should be cleaned and the room free

Table I

Knudson's "C" Solution

Compound	Formula	Amount
Calcium nitrate	$Ca(NO_3) \cdot 4H_2O$	1.00 gram
Monobasic potassium phosphate	KH_2PO_4	0.25 gram
Magnesium sulfate	$MgSO_4 \cdot 7H_2O$	0.25 gram
Ferrous sulfate	$FeSO_4 \cdot 7H_2O$	0.025 gram
Ammonium sulfate	$(NH_4)_2SO_4$	0.50 gram
Manganous sulfate	$MnSO_4 \cdot 4H_2O$	0.0075 gram
Distilled water	H_2O	1000 cm^3
Agar	—	18 grams
Sucrose	$C_{12}H_{22}O_{11}$	20 grams

of dust. Before sowing, the seed must be cleaned. Seeds should be placed in a vial and then covered with a 10% solution of calcium hypochlorite (10% Clorox). The container is closed and shaken vigorously. The seed should be treated for 5 to 10 minutes to be sure it is clean. A good rule of thumb is to sow the seeds when they turn yellow. If one waits too long, seeds will turn white and will have lost most of their viability.

Before the seed is sown in a container, gather all materials and light a Bunsen burner or alcohol lamp. Open the container, flame the neck, pick up the wire loop, heat it, and then pick up a loop full of seed and transfer the seed to the flask; flame the neck and replace the stopper. Gently shake the flask to distribute the seeds evenly over the surface of the medium.

Depending on the genus sown, the seeds will turn green after only a few days or after a few months. Flasks should be checked for contamination as it is most apt to occur within 3 to 5 days after sowing. Little contamination should occur after the first week. Seedling cultures should be maintained at a minimum of 70° to 72°F (Post, 1949) with a maximum light intensity of 1.6 klx in a greenhouse. In the laboratory, 16 hours at 1.0 klx of light at 70°F will be sufficient for good growth.

Seedlings will remain in the flask for possibly a year before they are transplanted to community pots. Some growers, depending on the genus being grown, may reflask the seedlings after several months and then at the second transplanting use community flats.

B. Vegetative Propagation

1. Cuttings

Most monopodial orchids (*Vanda, Arachnis*) can be propagated by tip cuttings. Usually, orchid cuttings are much larger than the 3 to 4 inch cuttings used for many floricultural crops. *Vanda* tip cuttings (Fig. 7A) are usually 12 to 15 inches tall and bear up to 12 leaves and usually a few aerial roots. Cuttings can be potted and will grow without being put in a propagation bed. *Arachnis* cuttings are usually 18 to 24 inches in length (Fig. 7B). They, too, will have aerial roots and can be lined out directly in the field or potted up.

Some monopodial and sympodial orchids produce offsets. Those genera such as *Dendrobium* and *Epidendrum* produce offsets in an axil of a leaf. The offsets root while still attached to the plant (Fig. 7C). Once four or more roots have formed, the offset can be snapped off, potted up, and grown as any transplant with little retardation of growth. Many offsets on reed-type *Epidendrum* plants are often in flower when snapped off, and if properly handled, will not lose any flowers after they are potted up.

Two genera, *Phalaenopsis* and *Phaius,* can be increased by using flower stalk cuttings. Although both are increased by this method, the techniques used are entirely different. A typical *Phaius* flower stalk will have seven or more nodes between the lowest flower and the base of the stem (Watkins, 1956). Each of these

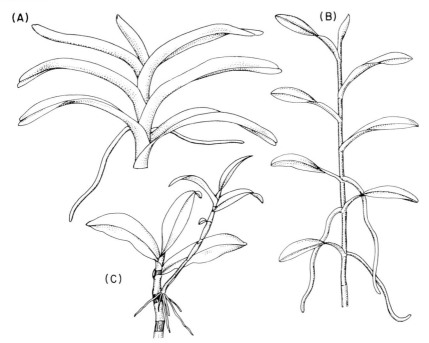

Fig. 7. Vegetative cuttings. (A) *Vanda* tip cutting. (B) *Arachnis* tip cutting. (C) *Dendrobium* offset.

nodes is covered with a leafy bract that protects a small bud. After the last flower is gone, the spike is ready to harvest. The spike is removed as close to the pseudobulb as possible. The top portion is then cut off just below where the first flower was, leaving a cane 15 to 18 inches long. The canes are laid on their sides in a flat of moist sphagnum moss, with the ends of the canes covered with the moss to prevent drying out. After 2 to 3 months, a small plant will arise at each node. Once the plant has three or four roots, it can be snapped off and potted up, and in 2 to 3 years will be a flowering-sized plant.

Phalaenopsis flower spikes must be propagated aseptically (Rotor, 1949), as are seeds. The spike is ready to propagate as soon as the last flower is gone. Like *Phaius,* it will have about seven nodes, each containing a bud. In most cases, however, the top node, just below where the first flower opened, will contain a flower bud and the bottom node frequently will not have a viable bud. Thus, only the center nodes should be used. The spike is cut 1 inch above and below the node. The bract is carefully dissected to remove all rough edges that could harbor fungus spores. The trimmed stem segments are placed in 10% Clorox for 10 to 12 minutes. The segments are removed and the ends are recut with a sterile knife. Each segment is placed in a test tube containing Knudson's "C" solution. After 3 months, a plant will start forming at the node. When it has formed two to three roots, the plant can be removed from the agar and potted up.

2. Division

Cattleya and the other sympodial orchids are propagated by division of the parent clump. This is usually accomplished on plants that have six or more pseudobulbs. The rhizome is cut between the third and fourth pseudobulbs and both sections are potted up as individual plants. Because most *Cattleya* plants produce only one new leaf per year, most plants are divided every 3 years. Genera such as *Paphiopedilum* and *Cymbidium* can be divided more frequently, as a division containing only one fan of leaves or one pseudobulb is all that is necessary to increase these plants.

3. Mericloning

Mericloning is a relatively new technique whereby, under aseptic conditions, one can have within 1 year 1 million plants identical to the parent plant. Like seedlings, they will take several years to flower.

This technique, first discovered by Morel in 1960 and later modified and refined by others, was first used to free plants of viruses. Now it can be used, however, to increase plants rapidly (Arditti, 1977).

The technique is similar for most orchids and should be done in an aseptic room or area in a laboratory. A rapidly growing shoot is selected. In the case of *Cattleya* or *Cymbidium,* the shoot is usually 1 to 2 inches long. The shoot is severed from the plant with a sharp knife. The shoot is cut as close to the base as possible. The severed shoot is placed into 10% Clorox for 10 to 15 minutes to sterilize it. Three or four of the outer leaves are removed after sterilizing. The shoots are sterilized in 5% Clorox for 8 to 10 minutes. The remaining two to three leaves are then removed to expose the growing point and leaf primordia. The shoots are again sterilized in 3% Clorox for 3 to 5 minutes. The leaf primordia is removed, and a small cube of tissue is cut out and dipped in 1% Clorox and plated out on a multiplication medium such as that of Murashige–Skoog. The tissue can then be divided, flasked, and reflasked until the desired number of explants is available. As previously mentioned, a single meristem can be increased to 1 million in less than 1 year. Under commercial laboratory conditions, they seldom increase an explant to more than 20,000 individuals, before going back to the parent plant for a new meristem. The exact number will be governed by the stability of the clone.

Once the required number of plants has been obtained, the explants can be treated as normal orchid seedlings and carried through to flowering as with any seedling population. The mericloned plants will all be identical to the parent plant, whereas the seedlings differ at time of flowering.

This brief description just outlines the process. A much more comprehensive review has been prepared by Arditti (1977), and the reader should refer to that article for additional details.

V. PLANT CULTURE

There is no difference in the culture of orchid plants, whether they are grown for pot plant sales or for cut flowers. In general, orchids being grown for cut flower sales will be repotted every 2 to 3 years. At this time, they may be divided or tip cuttings taken. The divisions, if in excess of the needs of the cut flower grower, are then sold as flowering plants.

A. Greenhouse Production

1. Media

The type of medium used for growing orchids will vary, depending on whether the orchid is an epiphyte or a terrestrial orchid. Epiphytic orchids (e.g., *Cattleya*) are found growing on trees in their native habitats. Hence, the medium should be similar to that on which they grow in their native habitat. Most epiphytic orchids (*Cattleya, Phalaenopsis, Dendrobium, Vanda*) can be grown in *Osmunda* fiber, tree fern fiber, tree fern-redwood chips, fir bark, sphagnum moss, and even cinder-like materials such as Solite and Holite.

a. Osmunda Fiber. This fiber is the root of the *Osmunda* ferns native to eastern United States. There are two types, the soft, light brown fiber and the dark brown, wiry form. *Osmunda* decomposes slowly and will last 2 to 4 years before decay sets in. It contains 2 to 3% nitrogen, which is released slowly as the fiber decomposes. Plants growing in *Osmunda* should be fertilized with a 1:1:1 ratio fertilizer.

b. Tree Fern (Tree Fern-Redwood Chips). This fiber comes from the stems of tree ferns. They are shredded or sawed into plaques. The stiff, wiry fibers are very long lasting; some last 5 to 7 years before decay occurs. A 1:1:1 ratio fertilizer should be used for plants growing in tree fern.

c. Barks (Mainly Fir Barks). These barks are by-products of the lumber industry. The chips are graded as large, medium, or fine. Large chips are used in 8 inch or larger pots. Medium is the most widely used grade for 1.5 to 7 inch pots, and the fine chips are used for thumb and community pots. Fir barks contain little or no nitrogen, and although slow to decay (3 to 4 years), they are broken down by a myriad of microorganisms. These microflora are capable of obtaining nitrogen at the expense of the plant, and a grower must compensate for this loss. Plants grown in bark are fertilized with a 3:1:1 ratio fertilizer.

d. Aggregated Materials (e.g., Solite, Holite). Materials that are inert and capable of holding moisture can also be used for epiphytic orchids. Most aggregates are about the same size as medium grade bark and are ideal for 3 to 6 inch pots. Because the aggregate does not decompose, it will last for years. It can be used over and over again, but should be cleaned and sterilized before being reused. This procedure prevents the transfer of disease organisms from one plant to another. A 1:1:1 fertilizer should be used on plants grown in aggregates. Actually, these materials are very similar to volcanic rock.

e. Sphagnum Moss. Interest in sphagnum moss as a medium has increased, mainly due to the introduction of New Zealand sphagnum moss. This sterilized medium is becoming more widely used.

Terrestrial orchids (*Cymbidium, Phaius*) are found growing in rich, organic soils in their native habitats. Any well-drained potting mixture containing 50% or more organic matter can be used. Peat, shavings, and sand (1:1:1, v/v) and even 100% peat moss have been recommended.

Studies by Poole and Sheehan (1977) have indicated that both terrestrial and epiphytic orchids can be grown in a mixture of peat and perlite (1:1, v/v) with excellent results. The use of this medium for growing orchids in the future looks very promising. Growers who have tried this medium say adjustment of watering to prevent overwatering is the most critical phase at this time.

A number of synthetic media are on the market today. None are widely used at this time but may play a role in the future.

B. Environmental Controls

1. Watering

It has often been said that more orchids are killed by improper watering than by any other factor. Watering and the quality of water used on orchids are the most important environment factors involved in orchid culture.

Like any floricultural crop, orchids should be watered thoroughly and then not watered again until the surface of the potting mixture begins to become dry. The number of days it takes for the surface of the potting mixture to become dry will be governed by climatic conditions, types of container and medium, age of the medium, and size of the plant in the pot. The best practice is to group orchids by pot size, type, and medium, thus ensuring that all plants are properly watered.

Quality of water is as important as the quantity of water applied. Fortunately, *Cattleya* orchids can be watered with water having a pH range of 4 to 9 (Northern, 1970), and hard or soft water appears to have little effect on the growth of orchids. Hard water, however, should not be used in an overhead sprinkler system because the orchid leaves will soon be covered with a thin film of calcium crystals, resembling white salt, on the leaves.

The most important factor to consider is the level of soluble salts in the water. Water having soluble salts of less than 125 ppm is excellent to use, 125 to 500 ppm is good, 500 to 800 ppm should be used with caution, and water with salts above

800 ppm should not be used. Most city water would fall within the good to excellent range and can be used. Well water should be analyzed to be sure it is safe to use on orchids.

2. Temperature

The temperature regimes used will be governed by the genera grown. *Cymbidium* orchids require 50°F night temperatures to produce flowers. Thus, 50°F night temperature and a range of 70° to 75°F day temperature would be ideal. During the summer, plants can survive 90°F temperatures but should not be exposed to them for long periods of time. *Cattleya* species and hybrids thrive best at night temperatures between 60° and 65°F. *Phalaenopsis,* especially the white-flowered cultivars, grow best with 65°F night temperatures and up to 80°F during the day. Pink *Phalaenopsis* cultivars flower better when the night temperatures are 55° to 60°F. When several orchid species are grown in one house, a compromise of temperature has to be reached and that usually means 60°F night temperature and up to 80°F during the day. If *Cymbidium* is included in a mixed group, then the night temperature will have to be dropped to 50°F to ensure flowering on *Cymbidium;* however, it will delay flowering of *Cattleya* and *Phalaenopsis.*

3. Fertilization

Orchids should be fertilized every 2 weeks for maximum growth. This, of course, assumes they are being provided the proper light, temperature, and water. The ratio of fertilizer used will vary with the medium. *Osmunda,* tree-fern, terrestrial mixtures, Holite, and Solite should be fertilized with a 1:1:1 ratio fertilizer such as 10:10:10 at the rates of 1lb/100 gallons of water applied to 400 square feet of growing area. Fir barks or other barks should be fertilized with a 3:1:1 fertilizer, 30:10:10 at the rate of 1lb/100 gallons of water. Bark media require additional nitrogen to offset that required by the myriad of microorganisms breaking down the bark in the containers. However, because bark decomposes very slowly, there is no danger of a rapid release of nitrogen tied up by microorganisms during the time the medium is in the pot.

Slow-release fertilizers have been successfully used on orchids. If they are used, the same fertilizer ratios will apply, and the same amounts should be used. Some growers prefer a combination of slow-release and liquid fertilizer, which provides a more uniform supply of nutrients over a longer period of time.

Slow-release fertilizers are safe to use and are less apt to be damaging if an accidental overdose is applied. However, this does not mean that care should not be taken to be sure the proper amounts of any given fertilizer are applied. The old adage, "If a pound is good, two pounds is better," does not hold true.

4. Light

Light, like many other cultural factors, will vary depending on the orchid genus under cultivation. *Phalaenopsis* plants thrive at 1.6 to 1.9 klx and *Cattleya* at 2.6 to

3.9 klx; *Cymbidium* will grow under full sun. Therefore, it becomes necessary to shade some orchid greenhouses to ensure that the proper amount of light will be available for good orchid growth. Any greenhouse shading compound can be used. The compound should be applied in late spring as the days are becoming brighter. In the fall in northern climates, the shading compound should be allowed to wear off, as the low light intensities encountered in the winter months will reduce plant growth. In the South or in Florida, California, and Hawaii, however, shade will be required all year, as winter radiation in these areas will often be four times that of the northern United States.

Some cultivars or orchids, notably *Cattleya trianaei* and *Cattleya labiata* and their hybrids, can be manipulated photoperiodically (Rotor, 1952), and the same plant can be flowered twice in 1 year. As an example of how it is done, the following schedule was developed by Hagar (1957).

Cattleya labiata
> Light: June 5 to October 12; temperature, 65°F night temperature (NT) (standard chrysanthemum lighting)
> Normal days: October 12 to December 15
> Harvest flowers: December 15 to 20
> Light: December 15 to April 1
> Shade: Use standard chrysanthemum blackcloth April 1 to June 5; temperature, 65°F NT
> Harvest flowers: June 5 to 10

Many *Cattleya labiata* and *C. trianaei* hybrids can be manipulated in the same manner, doubling flower production without increasing the number of plants.

The time to flowering of *Cattleya* can also be shortened by temperature manipulation in case the crop is running late. If bud size is known and the numbers of days left are counted, then the proper night temperature can be selected (Table II) (Hager, 1957).

5. Flowering

There are so many species, cultivars, and hybrids of orchids available that it is possible to have flowers the year round just by the selection of cultivars. *Cattleya* and *Vanda* flowers are relatively easy to obtain year-round, and a grower needs to select only a few species and their hybrids to have flowers all year (Table III).

Table II

Number of Days to Flower

	Night temperature (°F)		
Bud size (inches)	55	60	70
1	67	50	36
1–2	53	38	28
2–3	42	24	18
3–14	30	20	12

Table III

Selection of Some *Cattleya* and *Vanda* Species to Provide Year-Round Flower Production

Genus and species or hybrid	Months in flower											
	Jan.	Feb.	Mar.	Apr.	May	June	July	Aug.	Sept.	Oct.	Nov.	Dec.
Cattleya												
gigas					X	X	X	X				
labiata									X	X		
mossiae				X	X							
trianei	X	X	X									
luddenmanniana			X	X	X	X	X	X				
Vanda												
coerulea	X	X									X	X
dearei						X	X					
sanderiana								X	X	X	X	
merrillii							X	X				
Miss Joaquim	X	X	X	X	X	X	X	X	X	X	X	X

Some orchid plants, such as *Phalaenopsis amabilis* and its white hybrids, can be kept in flower all year. It is possible to harvest three or more crops of flowers from an individual plant. The primary flower spike is produced in the fall and will flower for a long period of time. Some spikes have been known to produce flowers for over 2 months. Once the last flower has been harvested, the top of the spike is removed just below the point of attachment of the first flower bud. A secondary flower spike will be produced in approximately 8 weeks, and when it has finished, it too can be cut off, just below the first flower, and a tertiary spike will form. Frequently, the tertiary spike will still be in flower when the next primary spike is being produced in the fall.

Standard *Cymbidium* orchids require cool nights for flower production. These plants then primarily flower in the spring. Selection of hybrids is very important, as the range of flowering can be spread from December to May, or even longer. Most growers prefer midspring and late spring flowering cultivars because they are very popular corsage flowers for Easter and Mother's Day.

Dendrobium, especially *Dendrobium phalaenopsis* and its hybrids, are excellent plants to produce sprays of flowers for late fall, winter, and early spring. The flowers are very suitable for corsages, and two or three flowers make an outstanding corsage. These plants will flower all year in the tropics.

A grower, by carefully selecting orchids, can have a wide variety of sizes, shapes, and colors, not to mention some very fragrant orchid flowers available year-round. Orchid flowers come in many colors of the rainbow, and many have two or even three colors on an individual flower.

VI. FIELD PRODUCTION IN THE TROPICS

Today, in tropical countries, there is a rapidly increasing interest in growing flowers, and many of these countries are growing spray-type orchids for export. At present, the major production area is in southeast Asia and is centered around Thailand and Singapore. Hawaii and Sri Lanka are also developing production areas.

The majority of spray orchids are field grown with little or no protection and with minimal maintenance. In Singapore, field-grown orchids are called "ground or-chids" although they are epiphytes. The base of the stem is in the ground when they are set out in the field.

The size of the farms varies from small units of 2 to 3 acres to over 100 acres. During a visit to Malaysia in 1971, the author saw one field of *Aranthera* 'James Storie' that covered almost 60 acres on the side of a hill. As an illustration of the size of this industry, Thailand exports over $10 million (United States) of cut orchids annually, and the majority are *Dendrobium* 'Mme. Pompadour.'

Actually, there are two distinctly different methods used for orchid production in the tropics. Most *Arachnis, Aranthera, Aranda, Vanda* and *Renanthera* species and hybrids are grown in full sun in the field on trellises. *Dendrobium,* some *Arandas*

('Wendy Scott,' Christine #1'), *Ascocenda,* and *Oncidium* cultivars are grown in pots, often under 40 to 50% lath shade.

The culture of *Arachnis* 'Maggie Oei' can serve as a good example of the typical "ground orchid" grown in the tropics. Native woods are used to build trellises in the fields prior to planting. Uprights (about 2 inches × 2 inches) are spaced in rows, approximately 8 feet apart with rows usually 3 feet apart. Cross bars are spaced about 2 feet apart (Fig. 8). Spacing the trellises in this manner will allow the grower to plant between 25,000 and 30,000 cuttings per acre.

After trellises are in place, tip cuttings 24 to 30 inches long are planted. Cuttings will have aerial roots and may even have flower spikes developing. A trench, usually 4 to 6 inches deep, is dug along the row. Individual cuttings are spaced 6 to 8 inches apart along the row. The base of the cutting is placed in the trench, and the top is tied to the first cross arm. After the row of cuttings is tied in place, the trench is filled with native soil, covering the lower 3 to 4 inches. In 4 to 6 months, the cuttings are considered flowering-size plants (although some flowers may have been cut all during this growth period), and spikes will be harvested from these plants for 18 months. At that time (2 years from cuttings), plants are 8 feet tall, and sprays are difficult to harvest. The grower will then take tip cuttings from these plants and replant the field or expand his production. If a grower is interested in expanding production, he can leave the old plants in the field after taking cuttings, and in 6 months he will be able to remove another set of cuttings. This time, the

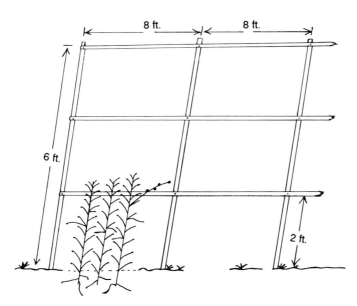

Fig. 8.   Trellis used for growing orchids in the field. (Not drawn to scale.)

harvest may be 2 to 2.5 times the original harvest, as each of the plants cut the first time will produce two to three secondary growths.

Arachnis plants and their hybrids are fast growing, producing a new leaf every 2 weeks and up to 12 flower spikes per stem per year in tropical areas. Hence, an acre of *Arachnis* 'Maggie Oei' is capable of producing 300,000 to 360,000 spikes per year. This, of course, is ideal production, but the plants have to be replanted every 2 years, which will reduce production, because they produce only 18 months out of every 2 years. Thus, actual production per year is reduced to 225,000 to 275,000 spikes per acre per year. Even if only half of these spikes were salable (112,000 to 137,000) and sold for 10 cents per spray, the return would still be $11,200 to 13,700 per acre per year.

Aranda hybrids, such as 'Wendy Scott' and 'Christine', *Ascocenda* hybrids, *Oncidium* 'Golden Showers' and *Dendrobium* hybrids, are all grown in pots and are generally grown under approximately 50% lath shade in the tropics.

Although these orchids are all grown in pots, culture varies from genus to genus. *Aranda* 'Wendy Scott' or 'Christine' are handled very similarly to *Arachnis* 'Maggie Oei'. Tip cuttings 15 to 18 inches tall, with aerial roots, are taken. Two stakes are then placed upright into a 6 to 8 inch clay pot. A few rocks are placed in the bottom of the pot to hold plants in place. As soon as the roots adhere to the sides of the pot, plants will support themselves. Growers are also using various media, such as coconut husks, and are adding chemical fertilizers, such as 10–10–10, at the rate of 1 pound per 100 gallons of water and applying it as a normal watering every 2 weeks.

Like *Arachnis,* these cuttings are considered flowering-size plants after 4 to 6 months, and flowers are harvested for 18 months. At this time, *Aranda* plants are becoming too tall for the pots; tip cuttings are taken and the cycle is started all over again.

Dendrobium plants in southeast Asia are often grown in cut-up coconut husks. Husks are fitted into the pot like a jigsaw puzzle and hold the rhizome in place. Plants can remain in the pot for 3 or 4 years before they need to be divided and repotted.

Potted dendrobiums are usually grown on raised benches. Camp and Philipp (1976) suggest using benches 4 × 90 feet long. A 200 × 210 foot shade house would have 58 benches and could house 21,244 or more plants. The two most widely grown *Dendrobium* cultivars in Hawaii are 'Louis Bleriot', which produces 5 to 15 spikes per year depending on the age of the plant, and 'UH 44', with 6 to 24 spikes. The lower figure is for the first year, and the last figure for the fourth year. If the 200 × 210 foot house were half in 'Louis Bleriot' and half 'Jacqueline Thomas' ('UH-44'), a yield of 18,553 dozen could be expected; if they sold for $2.53 per dozen, the gross sales would be $46,939.09, leaving a net return of almost $10,000.

Singapore has always been the leader in cut spray-type orchids, but the industry is constantly changing. The increasing demands for lands for urban use has led to a decline in both the number of farms as well as the size of the farms. There were approximately 800 acres producing 25,000,000 spikes in 1981 when the industry reached its peak. Since then there has been a decline, dropping to 400 acres in 1985.

Most farms in Singapore are small. In 1985, there were 289 farms approximately 0.66 acre in size, 48 at approximately 0.88 acre, 38 at 1.32 acres, 37 at 1.75, and 7 at 3.5 acres or larger. A continual decline in size and number can be expected in the future. There has also been a similar decline in price structure. When *Aranda* 'Christine' was introduced in 1972, the financial return per spray was twice what it was in 1986. A similar trend is also evident in *Aranda* 'Noorah Alsagoff'.

The four main exports in 1960s and 1970s were *Arachnis* 'Maggie Oei', *Aranda* 'Wendy Scott,' *Aranthera* 'James Storie,' and *Oncidium* 'Golden Showers'. In 1986, prices were reduced by approximately 50%. This steady decline has made it uneconomical for most growers to grow 'Maggie Oei' or 'Wendy Scott'. However, one item that helps keep consumer interest high is the constant introduction of new hybrids by Singapore growers. The leading varieties are *Aranda* 'Christine'; *A.* 'Noorah Alsagoff'; *A.* *'Tay Swee Keng'; A.* 'Kooi Cheow'; *Mokara* 'Chark Kuan'; *Dendrobium* 'Tay Swee Keng'; *D.* 'Mary Trowse'; and *Oncidium* 'Golden Showers.' Arandas are the leading export flower with *Oncidium* 'Golden Showers' a close second. 'Golden Showers' has been popular for over 2 decades, and this can be attributed to its yellow color. Because this is presently the best spray of yellow flowers and yellow is important in Japan, it should play an important role for some time to come (Goh and Karaljion, 1989). Thailand produced more *Dendrobium* sprays than any other country.

Because the aforementioned plants flower all year in tropical parts of the world, flowering is not a problem, as there is always a constant supply with some genera (*Dendrobium*) having a peak of flowering in August.

Although the majority of the spray-type orchids produced today are sold in Europe, they have great potential for sale in other parts of the world, because there are few spike flowers readily available except gladiolus and snapdragons.

VII. INSECTS, DISEASES, AND PHYSIOLOGICAL DISORDERS

The numbers of pests that attack orchids in cultivation are relatively few compared to those that attack orchids in the wild. In general, with proper care and sanitation there is little potential for pests to build up. However, environmental conditions that are conducive to good orchid growth are also favorable for development of orchid pests. The grower should constantly monitor the plants for any signs of disease and insect problems.

A grower should be especially watchful when new plants are introduced into the range. All new plants should be isolated for at least 1 month. Careful checking and application of preventative sprays will provide the protection needed to prevent introduction of unwanted organisms. The following brief descriptions cover the major pests encountered. A more detailed and illustrated description can be found in the AOS handbook on "Orchid Pests and Diseases" (Pridgeon, 1990).

A. Insects and Related Pests

1. Armored Scales

Boisduval scale (*Diaspis boisduvalii*) is the most common armored scale found on orchids. The first indication of a Boisduval scale attack is often a yellow spot on the upper surface of the leaf. The small circular scales are frequently attached to the underside of leaves, in tight cracks, or under pseudobulb sheaths, and often go undetected until large colonies have built up. Males gather in great numbers and look like cotton on the leaf surface.

2. Soft Scales

Brown soft scales (*Coccus hesperidum*) are small oval scales, usually greenish-brown in color. They usually attack leaves, but in severe infestations may be found on stems. There have been reports of ants introducing brown soft scale into greenhouses, where they will pasture them on plants to obtain honeydew.

3. Mealybugs (Pseudococcus *and* Ferrisia)

Mealybugs move about more than scales, and their white appearance is caused by a secretion that covers their bodies. Underneath the covering, they may look pink or yellow. These, too, secret honeydew and attract ants. When there are large infestations, sooty mold may develop on the leaves.

4. Mites

The two most common mites on orchids are the *Phalaenopsis* mite (*Teniupalpus pacificus*) and the two-spotted mite (*Tetranychus urticae*). These minute, sucking insects are hard to detect, but the telltale signs of their feeding give them away. As they suck the juices from the cells, empty cells reflect light, and soon little silver dots show up all over the leaves. In heavy infestations, webbing will appear on the plants. These are very prolific pests, and in dry, hot weather will build up enormous populations in a very short time.

5. Cockroaches (Periplanta)

These pests can be very harmful to orchids. They will eat the tips off young roots and also buds and newly opened flowers. Because they are active only at night, they often go undetected. Although the damage they cause is very similar to that of snails and slugs, they should be immediately suspected when no slime trails are evident.

6. Snails and Slugs (Slytommatophora)

Snails and slugs are also nocturnal feeders, often hiding in slits in the bottom of orchid pots during the day. They will eat root tips, young leaves, flower buds, and even open flowers. If slime trails are detected in the early morning hours, it is evident that snails and slugs are present and need to be brought under control.

B. Disease Pests

Diseases, like insects, can be a serious problem with orchids, sometimes just marking the foliage and making the plants unsalable, or sometimes destroying plants completely. Probably the most serious at present are the viruses that attack orchids, because they are often hard to detect and no reliable control measures are available.

Orchid diseases have been thoroughly discussed by Burnett (1985) and only some of the major disease problems will be mentioned here.

There are a number of cultural practices that can be used to help prevent disease organisms from building up. Water should be kept off plants as much as possible. Plants should be watered in the morning. All tools used in cutting orchid flowers and in propagating should be disinfested. Insect infestation should be avoided or controlled as much as possible, and all new plants should be isolated for up to 3 months, just in case they harbor a disease organism. All of these practices will help reduce the incidence of disease in any greenhouse range.

The most commonly encountered diseases are petal-blight, black rots, bacterial rots, viruses, and a wide variety of minor leaf spots. The latter, although disfiguring, do not usually destroy plants.

1. Petal-Blight (Botrytis cinerea)

As the name denotes, petal-blight attacks flowers of a variety of orchids (*Cattleya, Phalaenopsis, Dendrobium, Oncidium, Vanda*). This disease is most prevalent during cool, moist weather. It is typified by small, brown, circular spots on the petals and sepals of flowers. In advanced stages, the entire flower may be covered with spore masses. All infected flowers should be cut off and destroyed. Keeping water off flowers and having good air movement help control this disease.

2. Black Rots (Pythium ultimum *and* Phytophthora cactorum)

Black rots are widespread throughout the world, and because one fungus is prevalent in cool weather (*Phytophthora*) and the other in warm, humid weather (*Pythium*), they can occur at any time during the year. *Pythium* is the most dangerous, as it spreads throughout the plant at a very rapid rate. The symptoms of both fungi are similar, with infected parts turning black, sometimes with a yellow margin. These fungi usually attack leaves, but may attack pseudobulbs and stems, as well. They attack a wide variety of orchids. Care should be taken to keep foliage dry when these fungi are present.

3. Viruses (Cymbidium *Mosaic Virus and* Odontoglossum *Ringspot Virus*)

Viruses are potentially the worst enemies of orchids. First, there are no known control measures and, second, they are easily transmitted on cutting tools during propagating and cutting of the flowers. Whether dividing plants or cutting flowers, one should sterilize all tools between use on each plant to prevent spread of viruses.

Some viruses (*Cymbidium* mosaic virus, CMV) will develop chlorotic streaks in leaves, which eventually become pitted dark areas. These are easy to detect. However, color-break virus is visible only when plants are in bloom and color break can be detected in the flowers. The rest of the time, the plant looks healthy. This virus may be spread during propagation because virus symptoms are not evident. Because the plant appears clean, tools may not be sterilized and the virus is, thus, spread. Again, it is important that all tools be sterilized between uses on plants. A saturated solution of trisodium phosphate or lime water with a pH of 12 or higher should be used. Dipping tools in either of these solutions will help to keep viruses from spreading.

When a plant is suspected of having a virus, it should be isolated and checked by the local agricultural experiment station. If the results indicate presence of a virus, the plant should be isolated or destroyed (Wisler, 1989).

4. *Leaf Spots* (Colletotrichum, Cercospora, Gloeosporium, Phyllostictina)

Leaf spots are often found on orchids. Some cause tip dieback on leaves, whereas others cause a variety of spots, yellow to black sunken areas, on leaves. Most leaf spot organisms will not kill the leaves, but once the spots form, they will be present as long as the leaves remain on the plants, making the leaves unsightly.

C. Control of Insect Pests and Diseases

The rapidly changing picture in the pest control field makes it very difficult to give recommendations for control of most pests. Regulations vary from country to country and even between states within a country. Therefore, it is best to contact the local agricultural authorities to obtain the best control method available in an area, because some pesticides are not cleared for orchids.

D. Physiological Disorders

There are two physiological disorders occuring in orchids that are often confused with diseases. These disorders are difficult to identify and may require expert assistance to detect.

Calcium deficiency (Poole and Sheehan, 1970) occurs in a variety of orchids, especially during warm weather when plants are actively growing. Young leaves or new leads of *Cattleya* will turn black, usually commencing near the tip and progressing toward the stem. Immature leaves may turn completely black and drop off, or in severe cases, the young pseudobulb will turn black. As the black area moves down the leaf, it is preceded by a thin yellow halo. Applications of calcium to the medium will prevent this.

Mesophyll cell collapse occurs most frequently in *Phalaenopsis* plants in winter months (Sheehan and McConnell, 1978). The surface of one or more leaves on the plant will become pitted, and the pitted areas turn yellow to tan to black, finally looking like a virus infection. The next leaf, however, will develop normally, indicating it is not a virus. This disorder occurs on cold nights (temperatures below 45°F) or when cold water is applied to leaves that are near maturity. The malady can be prevented by using water that is at the same temperature as the greenhouse atmosphere or by keeping the greenhouse above 60°F at night.

VIII. HARVESTING AND HANDLING CUT FLOWERS

A. Grading

As with many other floricultural crops, there are not standardized grades for orchid flowers. Prices will often be governed by size. For example, a box of *Cymbidium* orchid flowers may contain 6, 8, or 12 flowers and is often sold at a flat price per box. Assuming $12 per box, the cost of the individual flower will range from $1 to $2 between the smaller and larger flowers. The same flower may be even cheaper if it is sold by the spray and may only be 50 to 75 cents per flower.

In the case of *Cattleya* flowers, both size and color are considered in pricing. A white flower will cost more than a purple flower or a white flower with a purple lip, when flowers are of equal size.

Attempts have been made to standardize some flowers. In Thailand, a No. 1 *Dendrobium* 'Mme. Pompadour' flower spike has seven open flowers and seven buds. *Aranthera* 'James Storie' is graded as a single spray or a branched spray, but no limits are put on size of the spray. *Arachnis* 'Maggie Oei' sprays usually have one to three flowers open and four or more buds.

Grading is left primarily to the grower and will often vary from country to country. In general, grading is done mainly on length of the flower spike, flower number, and size and arrangement of flowers on the spike. In some cases, the number of lateral branches on the inflorescence is also taken into consideration.

B. Cutting Flowers

Unlike many of our floricultural crops, orchid plants are kept for many years, so harvesting flowers is one activity in cultivation that is extremely important to growers. Actually, it is much more important than most growers will admit. Perhaps more viruses are spread when orchid flowers are cut than in any other way. Consequently, it is essential that all orchid flowers be harvested to avoid the spread of viruses. Harvesting can be done in a number of ways. One recommendation is for the grower to use disposable surgical blades. A person harvesting can carry a pocketful of blades, and, after harvesting the flowers from an individual plant, change knife blades and go on the next plant. These blades can be sterilized and reused.

Cutting tools (e.g., knives) should be dipped between the cutting of each plant.

A solution of trisodium phosphate or a saturated lime solution (pH 12) is good for disinfecting knives.

In general, orchid flowers do not mature until 3 to 4 days after they open, so it is important to know how old blooms are before harvesting. Flowers cut before they mature will not hold up and may wilt before they reach the wholesaler.

Spray-type orchids present no problem. Each flower opens 1½ to 2 days apart. If three or more flowers are open on the spike, the lower flower is mature and can be harvested. Some growers allow the entire spike to open before they harvest, to ensure that the flowers are mature.

Cattleya flowers are a little more difficult to handle as it is very difficult, when there are large numbers of plants in flower, to determine when any one flower opened. Frequently, more than one flower will open on the same spike in one day. One of the easiest ways to keep track is to have the grower go through the house each morning and insert a colored golf tee, colored plastic label, or other such label in each pot where there is a flower opening. By using a different color each day, the grower knows exactly when each flower opened. This system makes it easy to send any person into the house to cut flowers, for example, only from the pots with red markers.

When individual *Cattleya* and *Cymbidium* flowers for market are cut, the peduncle should be immediately inserted into a tube of water (orchid tubes with rubber or plastic covers). The tubes are generally filled from a pan of water that has been standing in the greenhouse overnight and has reach ambient temperature.

Spray-type orchids are cut and often shipped dry to the market. In Hawaii and Singapore, some *Dendrobium* and *Aranda* growers immerse the entire spray of flowers in water for 15 minutes before packing and shipping. Still other growers will wrap a little moist cotton around the base of stems before shipping and place 12 stems in a clear plastic bag.

C. Packaging

There are as many ways to package orchid flowers as there are cultivars grown. As previously mentioned, *Cymbidium* flowers in small tubes are packed 6, 8, or 12 in glassine-fronted boxes, ready to be made into corsages. Spikes of *Cymbidium* flowers are often packaged 100 flowers to a box. *Cattleya* flowers are packaged in standard florist boxes. Tubes are taped to the bottom of the box and shredded wax paper is placed around flowers to protect sepals and petals in transit.

Hawaiian *Dendrobium* is packaged 4 dozen sprays per box. The standard box is 30 × 10 × 7 inches. In Singapore, growers will pack up to 12 dozen stems of *Arachnis* 'Maggie Oei' in almost the same size box. The sprays actually have to be compressed in order to put the lid on the box. Nevertheless, when the box is opened in the market, the sprays spring back to their uncompressed state.

D. Storage

Orchids, unlike many cut flowers, do not store for any length of time at 31°F. Flowers start turning brown in 3 days at this temperature and lose their salability very rapidly.

Because most orchid flowers are long-lived on the plants, up to 3 or 4 weeks, growers will often leave them on the plants until they are needed. If they must be cut and stored, they should be stored at 42° to 45°F. At this temperature, most orchids can be safely stored for a 10- to 14-day period. If orchids are not at their peak, then storage time will be less.

E. Shipping

Because orchids are very durable flowers, they can be safely shipped long distances. It is not uncommon to ship orchid flowers halfway around the world. Orchids are shipped almost daily from Singapore and Bangkok to many cities in western Europe and arrive in excellent condition. If they are properly handled, as mentioned under harvesting and packaging, they should be very salable on arrival.

There are reports of some shipments of orchids arriving with flowers in a very bleached condition. Flowers such as lavender *Vanda* 'Miss Joaquim' appear to be a dirty white. This can occur when the pollinia have accidently been removed from the flower and flowers have been shipped in a sealed plastic bag. If one pollinium has been removed, that flower begins to produce ethylene, which soon builds up in the container and bleaches the flowers. Punching a few holes in the bag, as for apple bags, will eliminate most of this problem.

F. Consumer Care

Orchid flowers, although long-lived, should be handled as other cut flowers to ensure maximum shelf life. Immediately on arrival, individual flowers are removed from the tubes. Tubes may even be empty when the flowers arrive, depending on the length of time they have been in transit. The lower 0.5 inch of the peduncle is cut off, and the flower is inserted into a fresh tube of water with preservative added. If the flower is made up in a corsage, the corsage is placed in the refrigerator at night and will last for many days, as it can be worn evening after evening.

Spray-type orchids should be handled in the same manner as gladiolus or chrysanthemums. The basal 1.0 inch of the stem is cut on arrival, placed in warm water at 100°F with a floral preservative, and hardened off at 42°F. When used in an arrangement, a preservative is placed in the water to prolong shelf life.

REFERENCES

Arditti, J. (1977). *Orchid Biology.* Ithaca, New York: Cornell Univ. Press, pp. 203–293.

Bernard, N. (1903). "La germination des orchidees."*G. R. Acad. Sci. Paris,* **137:** 483–485.

Burgeff, H. (1909). *Die Wurzelpilae der Orchideen, ihre Kulture und ihr Leben in der Planze.* Jena: Fischer.

Burnett, H. (1985). "Orchids," In *Diseases of Floral Crops,* vol. 2, pp. 283–350, edited by D. L. Strider. New York: Praeger Pub.

Camp, S. G., and Philipp, P. F. (1976). "The economics of growing Dendrobium on Oahu for mainland export." *Hawaii Agric. Exp. Stn. Dept.* paper 37, p. 1–12.

Davidson, O. W. (1960). *Proceedings of the World Orchid Conference, 3rd,* pp. 224–233.

Dekle, G. W., and Kuitert, L. C. (1968). "Orchid insects, related pests and control." *Fla. Dept. of Agric. Bull.* **8:** 1–28.

Goh, C. J., and Karalgian, L. G. (1989). "Orchid industry in Singapore." *Econ. Bot.* **43**(2): 24l–254.

Hager, H. (1957). *Proceedings of the World Orchid Conference, 2nd,* pp. 130–132.

Knudson, L. (1922). "Non-symbiotic germination of orchid seeds." *Bot. Gaz.* (Chicago) **73:** 1–25.

Northern, R. T. (1970). *Home Orchid Growing.* New York: Litton.

Poole, H. A., and Sheehan, T. J. (1970). "Effects of levels of phosphorus and potassium on growth, composition and incidence of leaf-tip die-back in *Cattleya* orchids." *Proc. Fla. State Hortic. Soc.* **82:** 465–469.

Poole, H. A., and Sheehan, T. J. (1973). "Leaf-tip die-back of cattleyas—what's the real cause?" *Am. Orchid Soc. Bull.* **42**(3): 227–230.

Poole, H. A., and Sheehan, T. J. (1977). "Effects of media and supplementary micro element fertilization on growth and chemical composition of *Cattleya.*" *Am. Orchid Soc. Bull.* **46**(2): 155–160.

Post, K. (1949). *Florist Crop Production and Marketing.* New York: Orange-Judd, pp. 663–717.

Pridgeon, A. M. (1990). *Handbook on Orchid Pests and Diseases.* West Palm Beach: American Orchid Society, pp. 1–108.

Rotor, G. B. (1949). "A method of vegetative propagation of Phalaenopsis species and hybrids." *Am. Orchid. Soc. Bull.* **18**(8): 738–739.

Rotor, G. B. (1952). "Daylength and temperature in relation to growth and flowering of orchids." *Cornell Exp. Stn. Bull.* **885.**

Sheehan, T. J. (1966). *Proceedings of the World Orchid Conference, 5th,* pp. 95–97.

Sheehan, T. J. (1975). "Floricultural potential in Kenya, Part I & II." *FAO KEN/528,* pp. 1–47 and 1–39.

Sheehan, T. J. (1985). "Recent advances in botany, propagation, and physiology of orchids," In *Hort Reviews,* vol. 5, edited by J. Janick. Westport, CT: AVI Publishing Co.

Sheehan, T. J., and Sheehan, M. R. (1985). *Orchid Genera Illustrated.* Ithaca, NY: Cornell Univ. Press, pp. 1–207.

Watkins, J. V. (1956). *ABC of Orchid Growing,* 3rd ed. Englewood Cliffs, NJ: Prentice-Hall, pp. 25–26.

Wisler, G. C. (1989). *How to Control Orchid Viruses.* Gainesville, FL: Maupin House, pp. 1–119.

Withner, C. L. (1959). *The Orchids,* New York: Ronald Press, pp. 1–648.

6

Gladiolus

Gary J. Wilfret

Introduction to Floriculture, Second Edition
Copyright © 1992 by Academic Press, Inc.
All rights of reproduction in any form reserved.

I. INTRODUCTION: HISTORICAL BACKGROUND

The modern gladiolus (family Iridaceae; genus *Gladiolus*) cultivars (Fig. 1) offer a diversity of colors, shapes, and sizes available in few other flowering plants. They are used as landscape plants in the home garden, as specimens for exhibtion, and as cut flowers. Gladiolus flowers can be all colors except true blue, although some of the violet shades appear to be very near to blue in subdued light. Floret shapes can be round, triangular, flat, hooded, or orchid-like, and petals can be plain, ruffled, laciniated, recurved, needle-pointed, or deeply crinkled. Florets range from miniatures about 1 inch across and widely spaced on thin, single, or multibranched stems to the towering 6 foot giants with florets 7 or more inches in diameter in a two-rowed formal display. A grower can choose any combination of these characteristics and can be almost sure of finding a matching cultivar.

Very few flowers match the complex ancestry of *Gladiolus* and a revision of the South African species (Lewis *et al.,* 1972) has further complicated the understanding of its development. *Gladiolus* species were recognized over 2000 years ago growing in the fields of Asia Minor and were called "corn lilies." The European species were cultivated at least 500 years ago. Prior to 1730 the major garden species in England were *Gladiolus communis, G. segetum,* and *G. byzantius,* the latter being introduced in 1629 from Constantinople. With the establishment of trade routes from England to India via the Cape of Good Hope, several South African species were sent to England, starting in 1737. The species *communis, carneus (blandus),* and *cardinalis* were the prominent types grown prior to 1880 and, because they are sexually compatible, many natural hybrids were cultivated (Buch, 1972). The first *Gladiolus* was hybridized by W. Herbert, Dean of Manchester, who crossed *G. cardinalis* and *G. carneus (blandus),* whose progeny were fertile, and *G. tristis* and *G. recurvus,* whose offspring were sterile. The first important gladiolus hybrid was made in 1823 at Colville's Nursery in Chelsea, England, where *G. tristis* var. *concolor* was pollinated by *G. cardinalis* to produce the *colvillei* hybrids, which soon became the most important types for growing under glass for spring flowering. The cross that led to the development of present *Gladiolus* was made in 1837 by H. Bedinghaus of Belgium, who pollinated the parrot gladiolus *G. natalensis (psittacinus)* with *G. oppositiflorus* to give the summer flowering *Gandavenesis* hybrids.

The use of gladiolus as cut flowers in North America developed from the European hybrids and in 1870 up to 10,000 spikes per day were shipped to New York from local fields. When Dr. E. F. Palmer of Canada introduced 'Picardy' in 1932, the gladiolus cut-flower industry of the South and West was born. This cultivar was a blend of the *gandavenesis* and *primulinus* types and was the first that could be cut in the bud stage and shipped to distant markets where the florets would open. Field production of 'Picardy' and its progeny soon eliminated greenhouse production of gladiolus. Soon after World War II L. Butt of Canada introduced 'Rufmins,' which were small-flowered nonprim types, the most famous of which was 'Crinklette.' These are now commercially known as 'Pixiola' types (Wilfret, 1979). A few of the

Fig. 1. A modern exhibition and cut flower gladiolus cultivar, 'T-210.' Note symmetry, floret arrangement, and placement.

modern hybridizers of gladiolus cut flowers in the United States are C. Fischer of Minnesota, H. Turk of Oregon, E. Frazee of California, and G. Wilfret of Florida.

The gladiolus cut-flower industry in the United States peaked in the 1950s and demand and production steadily declined through the mid-1980s. Although the number of spikes sold has been on the increase through 1988, the quantity of spikes sold has not attained the 1950s level. Several reasons have been given for this decline, such as the flower's association with funerals, the limited use of large arrangements, or the emergence of other favored flowers, such as snapdragons. For whatever reason, the industry in 1988 was concentrated in seven states (United States Department of Agriculture, 1989), where over 199 million spikes, grown on 6200 acres, were sold. The wholesale value of these spikes was $33.9 million. Year-round production continued in California on more than 1750 acres that produced over 27 million spikes. Most flower production was in Florida, where 3200 acres yielded 110.8 million spikes, mostly from October through May. Limited production from North Florida yields spikes during May–June and September–

October. Summer production was primarily in Illinois, Michigan, New Jersey, and New York. Hawaii has started growing gladiolus year-round, but production is low (400,000 spikes). Another separate industry is corm production and sales, where at least 40 million corms are sold each year for cut-flower and landscape use, with a wholesale value close to $10 million.

II. TAXONOMY

The genus *Gladiolus* of the Iridaceae is represented by 180 species (Lewis *et al.,* 1972). It is found throughout Africa and the Mediterranean area, with the greatest concentration in southern Africa. Two species are endemic to Madagascar and 15 are found in countries bordering the Mediterranean. The Cape species are primarily diploids ($2n=30$) whereas the European species are polyploids ($2n=60$–130), indicating a southern origin of the genus. Modern hybrids, designated as *G. grandiflorus,* are a complex of at least 11 species, several of which are represented by different color forms or botanical varieties. *Gladiolus* is a herbaceous plant that develops from axillary buds on a corm. The leaves overlap at the base and may number from 1 to 12. The inflorescence is a spike and originates as a terminal axis. Florets number up to 30 or more and are tubular with flower parts in threes. Individual florets are enclosed in two green spathe valves. The pistil consists of a three-lobed stigma, a simple unbranched style, and an inferior ovary. The capsule contains between 50 and 100 ovules that mature within 30 days after fertilization. Florets are either bilateral or radially symmetric.

Gladiolus cultivars are classified by use of a three-digit number, (adopted by The North American Gladiolus Council). The first digit indicates floret size, the second digit basic color, and the third digit depth of color. The second digits from 0 to 8 correspond to a progression from yellow-green through orange, red, purple, and violet. White is given a value of 0. In addition, many gladiolus cultivars have a smoky hue and have been assigned a number of 9, depending on the base color. An odd third digit indicates the presence of a conspicuous mark (CM) in the lip or throat. The diameter of the lowest floret, without spreading or flattening of the petals, determines the size.

III. PROMINENT CULTIVARS

With the ease of hybridization of *Gladiolus* (Wilfret, 1974), over 10,000 cultivars have been recorded. Any extensive listing of cultivars would soon be obsolete but several have persisted over the years as cut flowers and need to be mentioned (Jenkins, 1963b; Wilfret, 1970). In the winter-growing areas of Florida, major cultivars grown are 'White Friendship,' 'T-210' (white); 'Friendship,' 'Daydream' (pink); 'Summer Rose,' 'Wine & Roses' (rose); 'Red Majesty,' 'Intrepid,' 'T-126' (red); 'Jester,' 'Jester Gold,' 'Goldfield' (yellow); 'T-704,' 'Beverly Ann' (lavender); 'Jenny

Lee,' 'Dr. Magie' (salmon); and 'Peter Pears' (orange). These cultivars are grown because they can be harvested in the tight bud stage and will open at the retail store. They are also less daylength sensitive and less temperature responsive than most cultivars and produce long spikes with at least 16 florets regardless of the season. In the northern growing areas, some of the cultivars used are: 'Trader-horn,' 'Florida Flame,' 'Wigs Sensation' (red); 'Vega,' 'White Prosperity,' 'Morning Bride' (white); 'True Love,' 'Rose Supreme' (rose); 'Saxony,' 'Tiger Flame' (orange); 'Golden Sceptre,' 'Golden Harvest,' 'Ice Gold,' 'Nova Lux' (yellow); and 'Greenbay' (green). In addition, several growers have invested in their own breeding programs and have developed cultivars for their exclusive use.

IV. PROPAGATION

A. Cormel Production

Gladiolus corms (Table I) are propagated from cormels that grow in clusters on outgrowths (stolons) between mother and daughter corms. Cormels are usually graded into three sizes: large, ≥0.4 inch diameter; medium ≥0.2 and <0.4 inch; and small <0.2 inch. Number of cormels per gallon ranges from 2700 for the large to 16,000 for the small. One gallon of cormels weighs approximately 4.4 pounds. Most commercial growers use only large cormels for planting stock production. Cormel stocks should be chosen carefully to prevent the spread of disease into developing corms and preferably should only be saved from "mother" blocks of corms that have been grown in fumigated soil and carefully rogued. In addition, cormels should be treated in a hot water solution to eradicate latent fungi, insects, and nematodes. Within the last few years, use of tissue culture has

Table I

Grades of Gladiolus Corms Developed by the North American Gladiolus Council

Description		Size (diameter, inch)
Large		
Jumbo		>2.0
No. 1		>1.5 to ≤2.0
Medium	Flowering Stock	
No. 2		>1.25 to ≤1.5
No. 3		>1.0 to ≤1.25
Small		
No. 4		>0.75 to ≤1.0
No. 5	Planting Stock	>0.5 to ≤0.75
No. 6		>0.4 to ≤0.5

been demonstrated to be feasible with *Gladiolus* to provide a source of clean stock for future propagation (Simonson and Hildebrandt, 1971; Wilfret, 1971; Ziv *et al.,* 1970).

The so-called hot-water treatment of cormels (Forsberg, 1961; Milholland and Aycock, 1965) has been modified to include the addition of fungicides in the solutions to complement the action of the hot water (Magie, 1971, 1975). Basically, this treatment consists of a 30-minute immersion of pretreated cormels in a suspension of benomyl (0.75 pound per 100 gallons of water) plus an additional fungicide, held at 127° to 131°F. Cormels used should have been dug during the warm months and stored at 75° to 90°F for 8 weeks prior to treatment. Two days prior to treatment, cormels should be covered with warm water (90°F) to soften the husks. Any cormels that float should be discarded. Cormels enclosed in mesh sacks should be submerged in the hot suspension for 30 minutes and then plunged for 10 minutes in cool running water. Treated cormels should be air-dried in thin layers in sterilized trays and then placed in cold storage (35° to 39°F) until planted. Dormancy of larger cormels is usually broken within 4 months of treatment. Root bud swellings indicate that cormels are ready to be planted. It is a good practice to soak cormels in water for 2 days just prior to planting to ensure uniform sprouting.

Land preparation consists of soil fumigation, maintenance of a soil pH of 5.8 to 6.5, and incorporation of approximately 250 pounds per acre of a 10–4.4–8.3 (nitrogen, phoshorus, potassium) or equivalent dry fertilizer. Gladioli require full sun and a well-drained soil for optimum growth. Moist cormels are planted in single rows in 4 to 4.5 inches wide furrows spaced 24 to 30 inches apart. Cormels should be covered with about 3 inches of soil, which is then leveled and compressed. A suitable herbicide is applied, based on soil type and cultural conditions (Gilreath, 1985, 1986a,b). About 40 large cormels are planted per foot of row (Magie *et al.,* 1966). The soil must be kept moist initially to obtain good germination, and then soil moisture should be reduced gradually to obtain optimum growth. Monthly applications of a dry or liquid fertilizer are necessary to obtain maximum yield within 5 to 6 months after planting (Waters, 1965). Small corms are dug with a modified potato digger. Yields of about 30 corms larger than 0.5 inch diameter can be dug per foot of row in a good operation when large cormels are used. Corms from 0.5 to 1.0 inch diameter are called "planting stock" and are used for the production of flowering-size corms.

B. Planting Stock Production

Planting stock treatment is similar to that of cormels except the temperature of the fungicide suspension is decreased to about 115°F, and the time the corms are submerged is limited to 15 minutes (Magie, 1975). Small corms <1.0 inch in diameter are planted in one or two rows per bed at a depth of 2.5 to 3 inches. Number of corms planted per foot of row ranges from 15 to 25, depending on corm size. The soil should provide adequate moisture and nutrition for good growth but should not be allowed to be saturated with water. Irrigation should cease 2 to 3 weeks prior to harvest to prevent rotting of corms in the field and to facilitate cleaning of the new corms. Removal of flower spikes improves corm size, but many growers allow the first floret to open to observe purity of the stock and allow roguing of odd plants. Yields of 480,000 flowering size corms (≥1.0 inch in diameter) or more can be expected per acre (Magie *et al.,* 1966). Short-day conditions at and after flowering are favorable for cormel formation, and research has shown that foliar application of gibberellin just after the emergence of spikes from the leaves promotes cormel production (Imanishi *et al.,* 1970). Corms should be cleaned and dipped in a fungicide solution within 2 days of digging to obtain maximum effect of the fungicide. Corms dug in the warm part of the year are dormant and require 3 to 4 months of cold storage (35° to 39°F) to break this dormancy. An alternative method to break dormancy is to expose the corms to ethylene chlorohydrin (Denny, 1938; Jenkins *et al.,* 1970). Corms and cormels, following at least 1 week of cold storage, are sealed in 1 gallon containers that contain 0.14 ounce of 40% ethylene chlorohydrin solution and are then held for 3 to 4 days at room temperature (73°F). Another method is to soak cormels in a 3% solution of ethylene chlorohydrin for 3 to 4 minutes and then to seal them in a tight glass container for 24 hours at 73°F. Corms and cormels planted immediately after treatment will sprout within 3 to 4 weeks. The major endogenous growth inhibitor controlling cormel germination has been identified as abscisic acid and can be regulated by cold storage plus 6-benzyladenine or by prolonged cold storage alone (Ginzburg, 1973).

V. FLOWERING STOCK CULTURE

A. Field Production

Gladiolus produce the best flower spikes when grown in deep, well-drained sandy loam soils, but they can be grown in sandy soils having less than 1% organic matter if proper cultural conditions are practiced. A heavy clay soil with poor drainage should be avoided whenever possible, as the gladiolus root system is easily damaged by excessive soil moisture. Soil fumigation with Vorlex (35 gallons per acre) or methyl bromide–chloropicrin (350 pounds per acre) is used for maximum flower production, but many growers rely on a 3 to 4 year crop rotation system to reduce buildup of soil-borne diseases. Size of corms used for flower production is dependent on the planting season. Many northern growers are able to produce marketable spikes in the summer from No. 2 and No. 3 corms. While under short-day, cool-night winter conditions, southern growers plant No. 1 or Jumbo corms. Corms are spaced in furrows at 4 to 6 per foot of row, with row spacing dependent on soil type and irrigation method. Single rows spaced 2.5 feet apart are adequate in heavier soils whereas sandy soils require wider spacing in order to maintain the integrity of the beds. In Florida, growers use either single rows on 4.5 foot centers or two rows per bed spaced 1.3 feet apart with beds on 6 foot centers. Corm density for single rows on 2.5 and 4.5 centers is 70,000 and 38,000 per acre, respectively. Double rows in beds on 6 foot centers can hold approximately 55,000 corms per acre. Large corms are planted 6 to 9 inches deep while medium corms are covered with only 5 to 6 inches of soil. Irrigation methods used are primarily open ditches or subtiles in Florida and overhead in other areas of the country. The latter method is used for three purposes: (1) to provide adequate moisture, (2) to prevent freezing of very early and very late crops, and (3) to reduce sunburning of florets. Floret count per spike can be greatly reduced when plants are grown excessively dry, especially from time of emergence to the second-leaf stage (Halevy, 1965). High-temperature and high-light intensities are well tolerated by gladiolus (Beijer, 1962), but blind plants or reduced floral bud counts occur during short-days, low-light intensity, and cool night temperatures (Shillo and Halevy, 1966). Plants are especially sensitive at the two-leaf stage, which is approximately the time of floral bud initiation. Various chemicals have been applied to stimulate rooting and flower formation (Halevy and Shilo, 1970; Zimmerman, 1938), but none have been shown to be advantageous.

B. Nutritional Requirements

Fertilizer requirements for rapidly growing plants vary with climatic conditions, irrigation method, and soil type. In sandy soils, it is necessary to provide fertilizer frequently, especially during the rainy season. In some heavier loam soils, little or no fertilizer is required for flower production (Stuart and McClellan, 1951; Van Diest and Flannery, 1963; Woltz, 1955a), as the large supply of inorganic and organic nutrients present in large corms is sufficient. In fact, the benefit of fertilizer is often seen only during the second season.

Nitrogen deficiency can be manifested as a reduction in the number of spikes and the number of florets per spike as well as the typical pale green foliage. Symptoms of phosphorus deficiency are dark green upper leaves and a purple coloration in the lower leaves. A lack of potassium causes a reduced floral bud count, shortening of the flower stem, delay in flowering, general yellowing of older leaves, and interveinal yellowing of younger leaves. Several symptoms of minor element deficiency have been reported for gladiolus (Woltz, 1957, 1965, 1976). Calcium deficiency can cause cracking of the spike generally below the second or third floret,and more severe cases are evident as bud blasting or bud rot. Magnesium deficiency causes interveinal chlorosis of older leaves whereas iron deficiency is manifested by interveinal chlorosis of new leaves. Deficiencies of boron cause cracking of leaf margins, deformed leaves, and stunted inflorescenses. Brown tips of leaves and spathes have been associated with fluoride toxicity (Brewer et al., 1966;Jenkins, 1963a; Woltz, 1957), but similar symptoms may result from anything that injures the root system, such as close cultivation, disease, nematodes, and waterlogging of soil.

The nutritional requirements of gladiolus vary depending on prior fertilization of the mother corm, but in general a gladiolus crop grown on sandy soils should have

100 to 150 pounds nitrogen (supplied partly as nitrate and partly as ammonium), 100 to 200 pounds phosphorus pentoxide (P_2O_5)(from superphosphate), and 125 to 200 pounds potassium oxide (K_2O)(from potash) per 2.5 acres (Woltz, 1955b, 1965, 1976). Secondary nutrients, such as calcium, magnesium, iron, and boron, may be applied as fritted traced elements during land preparation. At least four applications of fertilizers are advisable: (1) preplant incorporated, (2) side-dressed at the two- to three-leaf stage, (3) side-dressed at the "slipping" stage when the inflorescence emerges from leaves, and (4) side-dressed about 2 weeks after flowering to develop the new corm and cormels (Wilfret, 1970). Corms are dug 6 to 8 weeks after flowering and are given postharvest handling similar to planting stock.

VI. PEST CONTROL

A. Insects, Mites, and Nematodes

Gladiolus are excellent host plants for many insect pests (Kelsheimer, 1956; Magie and Cowperthwaite, 1954; Magie and Poe, 1972; Short, 1976). Several species of aphids attack gladiolus, including green peach aphid (*Myzus persicae*), potato aphid (*Macrosiphum solanifolii*), and melon aphid (*Macrosiphum gossypii*). These sucking insects damage developing foliage and flowers and transmit many virus pathogens. Aphids are controlled effectively with organic pesticides such as dimethoate, malathion, or endosulfan. Unsightly scars on florets are often caused by both the gladiolus thrips (*Taeniothrips simplex*) and common flower thrips (*Frankliniella* species). Use of diazinon, monocrotophos, or acephate in conjunction with good weed management will keep these under control. Primary sources of thrip infestations for new plantings of gladiolus originate from infested corms or volunteer plants (Schuster *et al.*, 1984). Water at 115°F was as effective in reducing the number of thrips on corms as insecticide preparations when applied as corm dips but could reduce the number of spikes produced. Either oxamyl or acephate could be used as corm dips at 75°F to reduce thrips. Loopers (*Trichoplusia ni* and *Pseudoplusia includens*), armyworms (*Spodoptera frugiperda, S. eridania,* and *S. exigua*), cutworms (*Feltia subterranea* and *Prodenia dolichos*), and corn earworms (*Heliothis zea*) feed on gladiolus foliage and flowers. There are three stages in the crop cycle when these larvae are more damaging: (1) at plant emergence to the two-leaf stage, (2) at the slipping stage, and (3) just prior to the opening of the lowest floret. Regular spray programs of *Bacillus thuringiensis* (a bacterial pathogen of lepidoptera larvae), monocrotophos, and trichlorfon will control these organisms. Two species of mites (*Tetranychus urticae* and *T. bimaculatus*) have been identified in gladiolus but generally do not create a major problem in commercial plantings (Engelhard, 1969). Nematodes, particularly those causing root-knot (*Meloidogyne* species), are controlled by hot-water treatment of corms and cormels and by soil fumigation (Overman, 1962, 1969).

B. Diseases

Diseases of gladiolus can be divided into those of the neck, leaf, and flower, and those of the corm and roots (Jenkins *et al.,* 1970; Magie and Cowperthwaite, 1954; Magie, 1957, 1967; Magie and Poe, 1972). Botrytis blight (*Botrytis gladiolorum*) can damage both leaves and flowers. It develops primarily in cool, wet weather and is evident as small brown or gray spots on one side of the leaf but may progress to both sides when advanced. Flower symptoms are small to large water-soaked spots on petals that may develop into a gray mold. Control is by sprays of maneb (with zinc) and benomyl. Curvularia blight (*Curvularia trifolii* f. species *gladioli*) attacks young leaves during warm, humid weather and may develop in flowers. It is particularly destructive on young cormels, where it destroys the plant at the soil line. Sprays with maneb and chlorothalonil are effective in its control. Bacterial leaf and neck rots (*Pseudomonas marginata* and *Xanthamonas gummisudans*) are especially destructive in warm, rainy seasons. No effective bacteriacides have been found to control these rots, but sanitation is helpful in reducing their spread. Stromatinia dry rot (*Stromatinia gladioli*) occurs in cool, moist weather and is evident as yellow-brown tissue about the corm with a sharp, moldy odor. Small black sclerotia are usually visible between the leaf bases. Hot-water treatment of corms and cormels, soil fumigation, and incorporation of dichloran in the planting bed help to keep this disease under control. Infested soils should be avoided during cool seasons.

Fusarium corm rot (*Fusarium oxysporum* f. species *gladioli*) is the most destructive disease of gladiolus (Forsberg, 1955; Magie, 1971, 1985). The fungus can exist as latent infections in the corm and cause storage rot, deformed and blind plants, and floret disfigurement. This fungus is a worldwide problem, and no host resistance has been found although tolerant lines are available (Wilfret and Magie, 1979; Wilfret, 1981, 1983). Control measures include hot water treatment of cormels, fungicide dips of corms, and soil fumigation. Other corm rots can be caused by *Curvularia, Stromatinia,* and *Botrytis* (soft rot). Postharvest fungicide dips of corms will generally keep these under control. Other fungi and bacteria have been reported on gladiolus, but they normally are not a serious problem in commercial operations.

In addition to the fungal and bacterial diseases, gladiolus can be infested by a number of viruses. Cucumber mosaic, tomato ringspot, tobacco ringspot, and bean yellow mosaic viruses have been reported on gladiolus (Bing, 1972; Beute *et al.,* 1970). Symptoms of these include leaf and spathe chlorosis, flower mottling, spike distortion, and plant stunting. Control is only by propagation of disease-free stock, roguing of infested plants, and proper insect control.

C. Weeds

Chemical weed control is essential for commercial operations, and herbicides are applied both pre- and postemergence of the crop. Because most herbicides are specific for soil types and prevalent weed populations, no one chemical can be used universally. Several herbicides have been reported as safe on gladiolus corms, such as Alachlor, Alachlor+CIPC, Trifluralin, 2,3,5,6-Tetrachloro-1,4-Benzenedicarboxylic Acid, Dimethyl Ester (DCPA), Oryzalin, Pronamide, and Diuron (Bing, 1977, 1978; Jenkins *et al.*, 1968; Raulston and Waters, 1971; Waters, 1967; Waters and Raulston, 1972; Gilreath, 1984). Small corms and cormels are more sensitive to these chemicals and proper care should be exercised in the use of any herbicide (Gilreath, 1986b).

VII. HARVESTING AND HANDLING FLOWERS

A. Harvest

Gladiolus spikes can be harvested from 60 to 100 days after planting, depending on the cultivar and time of year (Jenkins, 1963; Jenkins *et al.*, 1970; Wilfret, 1970). Spikes are cut in the tight-bud stage with two to three leaves remaining on the stem and from one to five floral buds showing color. Care is taken to not damage leaves remaining on the plant as these are needed for development of the new corm and cormels (Compton, 1960; Hussein *et al.*, 1962; Wilfret and Raulston, 1974). Spikes are bundled in groups of 100 and sent to the packing house for grading. They are transported in an upright position to prevent curving or crooking of stems. Some growers stand the cut spikes in water or a floral preservative and transport them in refrigerated trucks from the field to the packing house.

B. Grading

Spikes are sorted into five grades based on overall quality, spike length, and floret number (Table II). When graded, they are bunched in units of 10 and held together by rubber bands. They are then held upright in cold storage (39° to 42°F) until packed. Many growers stand the spikes in a floral preservative to prevent desiccation prior to and after grading. Studies have shown that sucrose pulsing of stems prior to storage of 7 or 10 days resulted in greater floret opening (Bravdo *et*

Table II

Grades of Cut Flowers Used in Florida by Commercial Gladiolus Growers

Grade	Spike length (inch)	Number florets (minimum)
Fancy	>42	16
Special	>38 to ≤42	14
Standard	>32 to ≤38	12
Utility	≤32	10

al., 1974; Fahoomand et al., 1980; Kofranek and Halevy, 1976; Mayak et al., 1973; Halevy and Mayak, 1974; Mor et al., 1981).

C. Packing, Sorting, and Shipping

Graded spikes are usually stored less than 24 hours before they are packed and shipped to the markets. They are held at a minimum temperature of 39°F, as many cultivars will not open well when stored at a lower temperature. The 10-spike bunches are packed without water in hampers made of fiberboard or wood which measure 13 inches wide and deep and from 42 to 51 inches tall. Fifteen to 24 bunches per hamper are wrapped in Kraft paper or polyethylene for protection from sudden temperature fluctuation, bruising, and moisture loss. The hampers are then stored at 39°F until shipped. Spikes may remain in these hampers from 3 to 7 days prior to use by the retailer, with quality decreasing with increased number of days of storage. Most gladiolus flowers are shipped in refrigerated trucks or by air to wholesalers or directly to retailers. They may be in transit from 1 to 3 days, depending on the distance to market. Spikes must be held in an upright position to prevent distortion of the stems because of the expression of negative geotropism.

D. Consumer Care of Flowers

Spikes should be removed immediately from the hamper on arrival. The basal 1 inch of each stem should be cut off, and stems placed in a floral preservative that contains at least a carbohydrate source and a bacteriacide. Flower life can be extended by 3 to 5 days by using a floral preservative instead of water (Kofranek and Paul, 1974; Marousky, 1968a,b, 1969, 1971). Water quality is important also and should be low in soluble salts (Waters, 1966, 1968) and free of dissolved fluorides (Marousky and Woltz, 1971; Spierings, 1970; Waters, 1968). As little as 0.25 ppm fluoride has been shown to damage some gladiolus cultivars (Marousky and Woltz, 1971). Flowers should be opened at a moderate temperature (70° to 73°F) with some light but not in direct sunlight. Once flowers have opened or are in the desired arrangement, they should be stored at 39° to 43°F until displayed. Vase life of these flowers varies from 5 to 10 days, depending on the cultivar and room temperture.

REFERENCES

Beijer, J. J. (1962). "The forcing of Gladioli in the hothouse." Ann. British. Gladiolus Soc. **1962:** 22–25.
Beute, M. K., Milholland, R. D., and Gooding, G. V. (1970). "A survey of viruses in field-grown gladiolus in North Carolina." Plant Dis. Rep. **54:** 125–127.
Bing, A. (1972). "Virus." In The World of the Gladiolus edited by N. Koenig and W. Crowley. Maryland: Edgewood Press, pp. 182–191.
Bing, A. (1977). "Preemergence weed control in gladiolus, cormels, 1976." North Amer. Gladiolus Coun. Bull. **130:** 62–65.

Bing, A. (1978). "Preemergence weed control in gladiolus cormels, 1977." *North Amer. Gladiolus Council. Bull.* **133:** 55–56.

Bravdo, B., Mayak, S., and Gravieli, Y. (1974). "Sucrose and water uptake from concentrated sucrose solutions by gladiolus shoots and the effect of these treatments on floret life." *Can. J. Bot.* **52:** 1271–1281.

Brewer, R. F., Guillemet, F. B., and Sutherland, F. H. (1966). "Effects of atmospheric fluorides on gladiolus growth, flowering, and corm production." *Proc. Amer. Hort. Sci.* **88:** 631–634.

Buch, P. O. (1972). "The Species." In *The World of the Gladiolus,* edited by N. Koenig and W. Crowley. Maryland: Edgewood Press, pp. 2–7.

Compton, O. C. (1960). "Effects of leaf clipping upon the size of gladiolus corms." *Proc. Amer. Soc. Hort. Sci.* **75:** 688–692.

Denny, F. E. (1938). "Prolonging, then breaking, the rest period of Gladiolus corms." *Contrib. Boyce Thompson Inst.* **9:** 403–408.

Engelhard, A. W. (1969). "Bulb mites associated with diseases of gladioli and other crops in Florida." *Phytopathology* (Abstract) **52:** 1025.

Fahoomand, M. B., Kofranek, A. M., Mor, Y., Reid, M. S., and Awak, A. R. E. (1980). "Pulsing *Gladiolus hybrida* 'Captain Busch' with silver or quaternary ammonium compounds before low temperature storage." *Acta Horta.* **109:** 253–258.

Forsberg, J. L. (1955). "Fusarium disease of gladiolus: Its causal agent." *Ill. Nat. Hist. Surv. Bull.* **16:** 447–503.

Forsberg, J. L. (1961). "Hot water and chemical treatment of Illinois-grown gladiolus cormels." *Ill. Nat. Hist. Surv. Biol. Notes* **43:** 1–12.

Gilreath, J. P. (1984). "Chemical weed control in flowering gladiolus." *Proc. Fla. State Hort. Soc.* **97:** 297–299.

Gilreath, J. P. (1985). "Cypressvine morning-glory control in gladiolus." *HortScience.* **20**(4): 701–703.

Gilreath, J. P. (1986a). "Response of gladiolus, statice and gypsophila to residues of preemergence herbicides." *Proc. Fla. State Hort. Soc.* **99:** 5–278.

Gilreath, J. P. (1986b). "Preemergence weed control in gladiolus cormels." *Weed Science* **34:** 957–960.

Ginzburg, C. (1973). "Hormonal regulation of cormel dormancy in *Gladiolus grandiflorus.*" *J. Exp. Bot.* **24:** 558–566.

Halevy, A. (1965). "Irrigation experiments on Gladiolus." In *The Gladiolus* edited by P.C. Vasaturo, New Hampshire: Maxfield Press, pp. 129–136.

Halevy, A. H., and Mayak, S. (1974). "Improvement of cut flower quality, opening and longevity by preshipment treatments." *Acta Hort.* **43:** 335–347.

Halevy, A. H., and Shillo, R. (1970). "Promotion of growth and flowering and increase in content of endogenous gibberellins in *Gladiolus* plants treated with the growth retardant CCC." *Physiol. Plant.* **23:** 820–828.

Hussein, M. F., El-Gamassy, A. M., and Serry, G. A. (1962). "Effects of number of leaves at flower cutting on the yield of 'Snow Princess' and 'Bloemfontein' gladiolus corms and cormels." *Agr. Res. Rev. Cairo.* **40:** 1–9.

Imanishi, H., Sasaki, K., and Oe, M. (1970). "Further studies on the cormel formation in gladiolus." *Bull. Univ. of Osaka, Series B.* **22:** 1–17.

Jenkins, J. M. (1963a). "Influence of different plant nutrients upon brown tip of gladiolus." *Plant Dis. Rep.* **47:** 976–977.

Jenkins, J. M. (1963b). "Some characteristics of commercial gladiolus varieties." *North Amer. Gladiolus Council. Bull.* **75:** 37–38.

Jenkins, J. M., Chambers, E.E., and McGee, F.G. (1968). "Chemical weed control in Gladiolus." *Weed Sci.* **16:** 86–88.

Jenkins, J. M., Milholland, R.D., Lilly, J.P., and Beute, M.K. (1970). "Commercial gladiolus production in North Carolina." *N.C. Agr. Ext. Circ.* **44B:** 1–34.

Kelsheimer, E. G. (1956). "Insects and other pests of gladiolus and their control." *Fla. Agr. Exp. Stn. Circ. S-91.*

Kofranek, A. M., and Halevy, A. H. (1976). "Sucrose pulsing of gladiolus stems before storage to increase spike quality." *HortScience.* **11:** 572–573.

Kofranek, A. M., and Paul, J. L. (1974). "The value of impregnating cut stems with high concentrations of silver nitrate." *Acta Hort.* **41:** 199–206.

Lewis, G. J., Obermeyer, A. A., and Barnard, T. T. (1972). "Gladiolus—A revision of the South African species." *J. S. Aft. Bot.* **10** (Suppl.).

Magie, R. O. (1957). "Soil fumigation in controlling gladiolus Stromatinia disease." *Proc. Fla. State Hort. Soc.* **70:** 373–379.

Magie, R. O. (1967). "Bacterial neck rot of gladiolus in Florida." *North Am. Gladiolus Council Bull.* **89:** 99–100.

Magie, R. O. (1971). "Effectiveness of treatments with hot water plus benzemidazoles and ethephon in controlling fusarium disease of gladiolus." *Plant Dis. Rep.* **55:** 82–85.

Magie, R. O. (1975). "The hot water treatment for gladiolus propagation." *GladioGrams.* **17:** 4–6.

Magie, R. O. (1985). "Gladiolus." In *Diseases of Floral Crops,* edited by David L. Strider, Vol. 2., pp. 189–226. New York: Praeger Pub.

Magie, R. O., and Cowperthwaite, W. G. (1954). "Commercial gladiolus production in Florida." *Fla. Agr. Exp. Stn. Bull.* **535.**

Magie, R. O., and Poe, S. L. (1972). "Disease and pest associates of bulb and plant." In *The World of the Gladiolus* edited by N. Koenig and W. Crowley, Maryland: Edgewood Press, pp. 155–181.

Magie, R. O., Overman, A. J., and Waters, W. E. (1966). "Gladiolus corm production in Florida." *Fla. Agr. Exp. Stn. Bull.* **664A.**

Marousky, F. J. (1968a). "Influence of 8-hydroxyquinoline citrate and sucrose on vase-life and quality of cut gladiolus." *Proc. Fla. State Hort. Soc.* **81:** 415–419.

Marousky, F. J. (1968b). "Physiological role of 8-hydroxy-quinoline citrate and sucrose in extending vase-life and improving quality of cut gladiolus." *Proc. Fla. State Hort. Socl.* **81:** 409–414.

Marousky, F. J. (1969). "Conditioning gladiolus spikes to maintenance of fresh weight with pretreatments of 8-hydroxyquinoline citrate plus sucrose." *Proc. Fla. State Hort. Soc.* **82:** 411–414.

Marousky, F. J. (1971). "Effects of temperature, container venting, and spike wrap during simulated shipping and use of floral preservative on subsequent floret opening and quality of gladiolus." *Proc. Trop. Reg. Am. Soc. Hort. Sci.* **15:** 216–222.

Marousky, F. J., and Woltz, S. S. (1971). "Effect of fluoride and a floral preservative on quality of cut gladiolus." *Proc. Fla. State Hort. Soc.* **84:** 375–380.

Mayak, S., Bravdo, B., Guilli, A., and Halevy, A. H. (1973). "Improvement of opening of cut gladioli flowers by pretreatment with high sugar concentrations." *Scientia Hort.* **1:** 357–365.

Milholland, R. D., and Aycock, R. (1965). "Propagation of disease-free gladiolus from hot-water treated cormels in southeastern North Carolina." *N.C. Agr. Exp. Stn. Tech. Bull.* **168.**

Mor, Y., Hardenburg, R. E., Kofranek, A. M., and Reid, M. S. (1981). "Effect of silver-thiosulfate pretreatment on vase-life of cut standard carnations, spray carnations, and gladiolus after a transcontinental truck shipment." *HortScience.* **16**(6): 766–768.

Overman, A. J. (1962). "Effective use of soil nematicides for gladiolus." *Proc. Fla. State Hort. Soc.* **74:** 382–385.

Overman, A. J. (1969). "Gladiolus corm dips for root-knot nematode control." *Proc. Fla. State Hort. Soc.* **82:** 362–366.

Raulston, J. C., and Waters, W. E. (1971). "Use of herbicides in ornamental flower production under sub-tropical conditions." *Proc. Trop. Reg. Amer. Soc. Hort. Sci.* **15:** 229–238.

Schuster, D. J., Price, J. F., and Magie, R. O. (1984). "Gladiolus thrips: Reduction on gladiolus corms following insecticidal and hot water dips." *J. Agr. Entomol.* **1**(2): 147–154.

Shillo, R., and Halevy, A. H. (1966). "The effect of low temperature on the flowering of Gladioli." In *The Gladiolus,* edited by P.C. Vasaturo, New Hampshire: Maxfield Press, pp. 239–245.

Short, D.E. (1976). "Pest control guide for commercial flower crops in Florida." *Univ. of Fla. Ext. Entomol. Rep.* **50.**

Simonson, J., and Hildebrandt, A. C. (1971). "*In vitro* growth and differentiation of *Gladiolus* plants from callus cultures." *Can. J. Bot.* **49:** 1817–1819.

Spierings, F. (1970). "Injury to gladiolus by fluoridated water." *Fluoride Quant. Rep.* **3:** 66–71.

Stuart, N. W., and McClellan, W. D. (1951). "Effect of nutrient supply and fertilizer practices on Gladiolus growth in the greenhouse and field." *Gladiolus Mag.* **15:** 2.

United States Department of Agriculture. (1989). *Floriculture crops—1988 Summary.* U. S. D. A. Crop Reporting Board SpCr **6**-1 (89).

van Diest, A., and Flannery, R. L. (1963). "The nutritive requirements of gladiolus in New Jersey soils." *Proc. Am. Soc. Hort. Sci.* **82:** 495–503.

Waters, W. E. (1965). "Nutrient requirements of gladiolus cormels on sandy soils of Florida." *Proc. Soil and Crop Sci. Soc. of Fla.* **25:** 59–63.

Waters, W. E. (1966). "The influence of post-harvest handling techniques on vase-life of gladiolus flowers." *Proc. Fla. State Hort. Soc.* **79:** 452–456.

Waters, W. E. (1967). "Influence of herbicides on gladiolus flower and corm production in Florida." *Proc. South Weed Conf.* **20:** 171–178.

Waters, W. E. (1968). "Relationship of water salinity and fluoride to keeping quality of chrysanthemums and gladiolus cut flowers." *Proc. Am. Soc. Hort. Sci.* **92:** 633–640.

Waters, W. E., and Raulston, J. C. (1972). "Weed control." In *The World of the Gladiolus,* edited by N. Koenig, and W. Crowley, Maryland: Edgewood Press, pp. 150–154.

Wilfret, G. J. (1970). "A critical evaluation of the commercial gladiolus cultivars grown in Florida." *Proc. Fla. State Hort. Soc.* **83:** 423–427.

Wilfret, G. J. (1971). "Shoot-tip culture of gladiolus: An evaluation of nutrient media for callus tissue development." *Proc. Fla. State Hort. Soc.* **84:** 389–393.

Wilfret, G. J. (1974). "Gladiolus breeding." In *Breeding Plants for Home and Garden—A Handbook,* edited by F. McGourty, Jr. *Brooklyn Bot. Gard. Rec.* **30:** 35–38.

Wilfret, G. J. (1979). "Pixiola: A new cut flower for Florida." *Proc. Fla. State Hort. Soc.* **92:** 313–316.

Wilfret, G. J. (1981). " 'Florida Flame'—A red gladiolus for a cut flower." *Fla. Agr. Exp. Stat. Circ.* **S-274:** 8.

Wilfret, G. J., (1983). " 'Dr. Magie'—A salmon gladiolus for a cut flower." *Fla. Agr. Exp. Stat. Circ.* **S-298: 11.**

Wilfret, G. J., and Magie, R. O. (1979). " 'Jessie M. Conner'—A landscape gladiolus for Florida." *Fla. Agr. Exp. Stat. Circ.* **S-265:** 7.

Wilfret, G. J., and Raulston, J. C. (1974). "Influence of shearing height at flowering on Gladiolus corm and cormel production." *J. Amer. Soc. Hort. Sci.* **99:** 38–40.

Woltz, S. S. (1955a). "Effect of differential supplies of nitrogen, potassium, and calcium on quality and yield of gladiolus flowers and corms." *Proc. Amer. Soc. Hort. Sci.* **65:** 427–435.

Woltz, S. S. (1955b). "Studies on nutritional requirements of gladiolus." *Proc. Fla. State Hort. Soc.* **67:** 330–334.

Woltz, S. S. (1957). "Nutritional disorder symptoms of gladiolus." *Florists Exch.* **129,** 17–20.

Woltz, S. S. (1965). "Fertilizing gladiolus." *Fla. Flower Grower.* **2:** 1–5.

Woltz, S. S. (1976). "Fertilization of gladiolus." *GladioGrams.* **21:** 1–5.

Zimmerman, P. W. (1938). "Adventitious roots with P-indolebutyric acid." *Contrib. Boyce Thompson Inst.* **10:** 5–14.

Ziv, M., Halevy, A., and Shilo, R. (1970). "Organs and plantlets regeneration of Gladiolus through tissue culture." *Ann. Bot.* **34:** 671–676.

7

Specialty Cut Flowers

Allan M. Armitage

Introduction to Floriculture, Second Edition
Copyright © 1992 by Academic Press, Inc.
All rights of reproduction in any form reserved.

159

I. INTRODUCTION

Specialty cut flowers may be defined as any cut flower other than roses, carnations, and chrysanthemums. Warren-Aumen (1980) described 55 minor cut crops and at least that many more are grown as cut flowers or foliage somewhere in the world. Although usually thought of as "minor" cut flowers, *Gypsophila paniculata* (baby's breath), *Limonium* (statice), and *Gladiolus* are of major importance in the cut flower industry. Actual dollar value of specialty cut flowers is unknown but their importance has risen dramatically in the last 5 years. While the wholesale volume of carnations, chrysanthemums, and roses remained level or declined between 1987 and 1990, the value of "other cut flowers" rose 56% during that period (USDA, 1989, 1991). A conservative wholesale value of specialty cut flowers is approximately $173 million, greater than that of standard and miniature carnations, standard and pompon chrysanthemums, and sweetheart roses put together ($123 million) (USDA, 1991). Although the major cut crops are seldom produced outdoors, the resurgence of interest in specialty cut flowers has taken place in the field. For this reason, they are also known as "field flowers" or "summer flowers," although numerous specialty cut flowers (e.g., *Gypsophila, Limonium, Lilium,* and *Salvia*) are also forced in greenhouses during peak demand periods.

Part of the reason that specialty flowers have increased in popularity is because of their introduction to the United States by overseas growers and distributors, particularly by the Dutch and South Americans. American growers found their product in more demand. This increase in demand resulted in the emergence of better quality stems, greater production volume, and many more growers. California, Hawaii, and Florida are the leading producers of specialty cut flowers, accounting for the majority of acreage devoted to specialty cut flowers, but many states have increasing numbers of producers. Cut greens are a major product in Florida, accounting for over 91% of the United States production (USDA, 1989).

All growers of cut flowers are concerned about imports. While import tariffs are not fair and wages earned by many overseas workers are impossible to compete against, the American growers must be realistic. Imports will continue, but if we grow the highest quality, provide the best service, and find our own market niches, we can remain profitable. There is no doubt that demand for specialty flowers will continue; imports will increase, but so will domestic production.

The quality of the American product is steadily improving, postharvest storage and handling have been constantly updated, transportation is available, and florists and mass market outlets are demanding more specialty flowers. Growth of the specialty cut flower industry is inevitable. The most limiting factor, however, to the growth of the American industry is the lack of promotion and marketing skills necessary to compete with overseas producers. Growers must be "market oriented" rather than "production oriented." That is, the demand should dictate the mix and volume of the species produced. A mixed market for the sale of specialty cut flowers exists and Seals (1990) discusses three niche markets that can be utilized. First, the grower may sell directly to consumers through farmers' markets, pick your own, catering services, hotels, and restaurants. Although markup is high, volume is low.

A more lucrative market includes retail outlets, such as traditional florist shops, mail order catalog companies, craft or other specialty shops, and supermarkets (mass market). In this market, the assortment of items need not be as large, however, the profit per item decreases. The most rapidly growing segment of cut flower buyers is supermarket floral buyers. According to one survey, 65% of chain supermarkets regularly carry cut flowers.

The wholesale market (auctions, brokers, commissioned wholesalers, growers' cooperatives) is the third outlet and offers the greatest volume of available markets but the smallest profit per time. Although some wholesalers are progressive, this is often the most difficult market for the small grower and specialty grower to penetrate.

Numerous markets are available to growers, and one or all may be at their disposal. Organizations, such as Roses, Inc., perform an important function to rose growers, and recently American growers of specialty cut flowers formed the National Association of Specialty Cut Flower Growers (ASCFG). The objective is to provide production and marketing information for American grown flowers. As of 1991, there are approximately 500 members throughout the country.

There is no lack of potential crops to be grown for the specialty cut flower market. Species of annuals, perennials, bulbs, grasses, and woody stems may all be economically produced. The grower must decide what can be produced profitably depending on the local environment and the type of markets—consumer, retail, or wholesale available. Because most crops are grown outdoors or with minimal protection, growers cannot all produce the same flowers, and more importantly, flowers mature at different times throughout the country. A crop in Florida may peak in February and March, while the same crop may be marketed in April and May in Michigan, creating gaps in the market for local growers.

Proper postharvest treatment of cut stems is arguably the most important aspect of cut flower acceptance in the market today. If the consumer spends money on cut flowers instead of food, chocolates, or wine, the stems must not disappoint the consumer by senescing rapidly. Confidence in the product is the most important sales tool; if lost, the market will stagnate. The American grower has a significant advantage over foreign growers in that their product should reach the wholesale and retail outlets more quickly. If handled properly between the field and the marketplace, the potential vase life of stems can be realized. Use of high quality water and preservatives in the field and packing area; clean buckets, where appropriate; providing silver thiosulfate (STS), sugars, and acidification treatments; adequate cold storage facilities; and packing to ensure quality, not volume, in the shipping box are but a few of the ways harvested stems can fulfill the confidence of the consumer. For without consumer confidence, there is no market.

Excellent reviews (Halevy and Mayak, 1981; Evans and Reid, 1990), manuals (Sacalis, 1989), and books (Nowak and Rudnicki, 1990; Vaughn, 1988) are available on postproduction treatments of cut flowers.

II. PRODUCTION OF CUT FLOWERS IN THE FIELD

A. Site Selection

Site selection for specialty flowers is extremely important (Fig. 1). Some of the considerations are given in the following sections.

1. Land with Arable Soils

Without fertile, well-drained soils, the addition of large amounts of compost and organic matter or raised beds must be continuously practiced.

2. Availability of Good Quality Water

All plants are susceptible to drought and saline water. The availability of good quality water is probably the most important consideration in the production of specialty cut flowers.

3. Land Adaptable to Equipment

Relatively flat land makes bed development, seeding, fertilizing, and harvesting easier and the subsequent implementation of equipment more feasible.

4. Land with Some Shaded Areas

Some crops benefit from shade, particularly during the afternoon, and an area of natural shade is most useful. Shade not only reduces light intensity, often

Fig. 1.   Field of specialty cut flowers.

resulting in longer flower stems, but also reduces temperature. Some specific benefits of shading include greener foliage and taller stems, particularly in bulbous crops.

5. Well-Drained Land

The majority of plants used for cut flowers require good drainage. This is especially true in areas of mild winters where poor winter drainage results in severe losses of perennial crops.

B. Influence of Shade

A number of species benefit from shade. Shade is useful in at least two ways. (1) It reduces soil and air temperature significantly. Temperate zone species that do not tolerate high summer temperatures benefit most from the application of shade. This is particularly important in the South (zone 6b and south) (Fig. 2). Examples of crops that benefit from application of shade in this manner are *Centaurea americana*, *Centaurea moschata*, and *Echinops ritro* (Table I). Too much shade, however, may significantly reduce yield quality for many crops (Table I). (2) Shade results in the elongation of flower stems, particularly in bulbous crops. For example, the addition of 55% shade cloth significantly increased scape length of *Anemone coronaria* and *Zantedeschia* hybrids under field conditions in Georgia (Table I). Although stem length was increased with the addition of shade in all genera, yield of *Eryngium* and *Zantedeschia* significantly declined when too much shade (>55%) was applied.

1. Spacing

Species that are transplanted are generally spaced 9 × 9 inches, 9 × 12 inches, 12 × 12 inches, or 12 × 15 inches. Some plants may be spaced as close as 6 × 6 inches to increase yield per square foot but stems are weaker because of a lack of air movement and light. Large plants (e.g., *Salvia leucantha*) should be grown at least 18 inches and up to 2 feet apart. Direct-sown plants can be thinned on germination to appropriate spacing but often survival of the fittest ensures "natural spacing." In general, the closer the spacing, the greater the yield per square foot, but less yield per plant occurs. Conversely, poor yield per square foot and high yield per plant occur at wider spacing. Table II shows how spacing affects the yield of *Salvia leucantha* (an annual), *Achillea* × 'Coronation Gold' (a perennial), and *Liatris pycnostachya* (a cormous species).

2. Single Planting versus Multiple Planting

In annual crops, one time planting produces flowers continuously, but often the stems of the third or fourth harvest are shorter and thinner than those harvested in the first or second flush. Multiple plantings, however, result in a constant supply of first and second flush flowers, and after the first or second harvest, plants may be discarded. The overall yield per plant is reduced with multiple plantings, but high

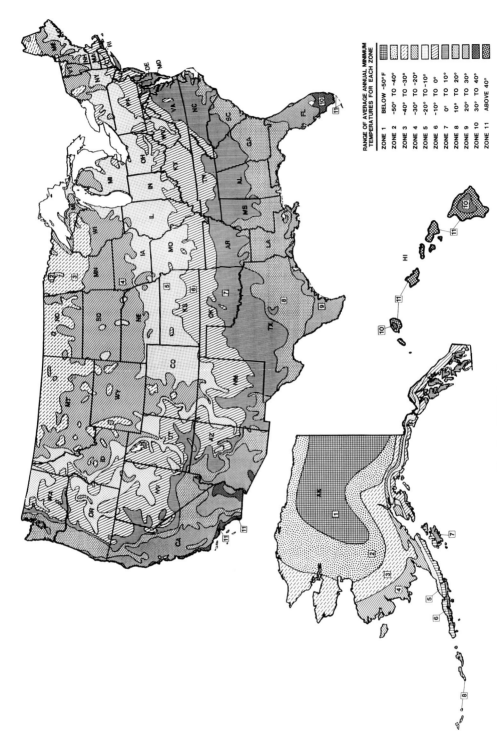

Fig. 2. USDA hardiness zone map.

Table I

Effect of Shade on Yield and Stem Length of Field-Grown Crops[a]

Crop	Shade (%)	Yield (stems/plant)	Scape (flower stem length) (inches)
Echinops ritro 'Taplono Blue'	0	4.7	33.0
(3rd year)	55	6.8	43.0
	67	5.8	36.0
Eryngium planum	0	8.6	35.6
(2nd year)	55	4.2	34.4
	67	0.4	43.6
Zantedeschia 'Majestic Red'	0	4	14.2
(2nd year)	55	4	23.9
	67	1	25.0

[a]Adapted from Armitage (1989e).

quality stems are maintained throughout the season. Crops such as celosia and sunflower may be planted at biweekly or monthly intervals to ensure long, straight stems. Some Zinnia cultivars, however, should only be planted once for best yield (Healy and Aker, 1988).

Another reason for multiple plantings is to ensure that high quality stems will be available when the market is strongest. If a strong market occurs, but only secondary or tertiary stems are being harvested, the potential price to the growers will not be realized.

Table II

Influence of Spacing on Yield of Field-Grown Cut Flower Species[a]

Crop	Distance between plants (feet)	Stems/plant	Stems/square foot
Salvia leucantha	2	124	31.6
	3	165	17.7
	4	195	10.2
Achillea × 'Coronation Gold'	1	47	46.5
	2	74	18.6
	3	119	12.1
	4	158	10.2
Liatris pycnostachya	1	7	6.5
	2	8	1.9
	3	9	0.9
	4	10	0.7

[a]Adapted from Armitage (1987).

C. Crop Selection

Crops should be selected based on market demand and the environment available to the grower. The market constantly changes, therefore, the crop mix should be studied annually and changed appropriately. The environment, however, is relatively constant. Growers will only be frustrated trying to grow species not adapted to their environment. Growers in Georgia will not only have a different crop mix than growers in Wisconsin, but planting and harvesting dates of similar crops will also differ. For example, *Centaurea macrocephala*, Armenian basket flower, is an excellent cut flower for northern growers but is poorly adapted for growers in the South. On the other hand, *Achillea filipendulina*, fern leaf yarrow, is well adapted to the South. Poor crop selection results in poor quality product and loss in market confidence.

Annuals, perennials, and bulbs are used for cut flowers, but categories of plant taxa may change based on environment. Delphiniums and baby's breath are annuals in Florida and southern Georgia but are treated as perennials in most other parts of the country.

III. SPECIES USEFUL FOR SPECIALTY CUT FLOWER PRODUCTION

A. Annual Species

Many species are available as cut flowers, and the majority are propagated from seed. Articles by Healy and Aker (1988) and Laushman (1989) on cut flower production of annuals should be consulted. A number of advantages and disadvantages to growing annuals are listed in Table III. Some examples of field production are shown for statice (Fig. 3) and stocks (Fig. 4). Table IV lists annual species that are useful for cut flowers. Table IV is adapted from research at the University of Georgia and of Seals (1989) and Kieft (1989). Many additional cultivars exist that are not mentioned. Consult seed producers or plant brokers for availability of cultivars.

Table III

Some Advantages and Disadvantages of Growing Annuals for Cut Flowers

---

Advantages
    Flowers are usually "cut and come again," resulting in high yield
    Plants may be transplanted to the field at different times to extend season
    Seeds, in general, are not difficult to germinate
    Some species may be directly sown in the field, eliminating transplanting costs

Disadvantages
    Harvesting is very labor intensive
    Market price may be low on common annual species
    Competition could be severe

Fig. 3. A field of statice *(Limonium sinuata)*. Brothers Brothers, Watsonville, CA.

Fig. 4. Stock *(Matthiola incana)*. Produced by Park Ave., Ventura, CA.

Table IV

Annual Species Used for Cut Flowers

Genus and species (common name) and common cultivar	Flower color	Height (feet)	Planting method	Germination Temperature (°F)	Time (days)	Uses
Ageratum houstonianum (flossflower)						
'Blue Horizon'	Blue	2–3	Transplant	70–73	10–20	Fresh, dry
Other cultivars: 'Cut Wonder' has fuzzy blue flowers in dense clusters						
Agrostemma githago (corn cockle)						
'Purple Queen'	Purple	3–4	Direct sow	70–73	10–20	Fresh
Comments: Better for northern climates, does not flower well in the South after June 30. Can be used as a winter crop in Florida						
Other cultivars: 'Milas' bears lilac-pink flowers with white centers						
Amaranthus caudatus (love-lies-bleeding)						
	Red	3–4	Direct sow	70–73	7–10	Fresh, dried
Other cultivars: var. *viridis* produces green-yellow tassels						
Other species: *A. cruentus* (prince's feather) has red spires on 2 feet tall stems. 'Red Cathedral' bears blood-red spires on 3–4 feet tall stems and dark red foliage. 'Green Thumb' is a cultivar of *A. hybrida* with green tassels.						
Ammi majus (false Queen Anne's lace, white dill)						
'Snowflake'	White	3–3.5	Direct sow	68–72	14–21	Fresh, dried
Comments: Workers should use gloves when handling as plants can cause dermatitis in sensitive people. The flower color is a "cleaner" white than Queen Anne's lace. Has been extremely successful in southern states. Tolerates cold to approximately 25°F						
Ammobium alatum (winged everlasting)						
'Grandiflorum'	White	2–2.5	Direct sow	68–72	7–10	Fresh, dried
Comments: Excellent for drying, becoming more popular every year						
Antirrhinum majus (snapdragon)						
Rocket series	Various	2–3	Transplant	75–72	10–14	Fresh
Other cultivars: 'Madame Butterfly', double flowered; Supreme series, double flowered, Liberty series, strong compact stems in many colors, many others available. Best planted in the fall in the south for spring harvesting.						
Calendula officinalis (calendula)						
'Pacific Beauty'	Yellow/Orange	2–2.5	Transplant	60–65	10–14	Fresh
Comments: Grow as winter/early spring crop in South and Midwest						
Other cultivars: 'Kabloyna' series, many others available.						
Callistephus chinensis (China aster)						
Perfection series	Various	2–5	Transplant	68–75	10–14	Fresh
Comments : Significant advances have taken place in the breeding of China asters. Yield and disease resistance still need improvement. See Healy and Aker (1988)						
Other cultivars: Many others available, including 'Compliment,' 'Emperor,' 'Matador,' and 'Matsumoto' series with 3–5 inch wide flowers and 2.5–3 feet tall stems in mixed colors						
Carthamus tinctorius (golden safflower)						
'Goldtuft'	Gold orange	3–5	Transplant	64–72	10–14	Fresh, dried
Caryopteris incana (bluebeard). Also known as *C.* × *bungei*						
	Lavender blue	2–4	Transplant	70–78	14–21	Fresh

Table IV (continued)

Genus and species (common name) and common cultivar	Flower color	Height (feet)	Planting method	Germination Temperature (°F)	Time (days)	Uses

Comments: Plants are perennial south of zone 7; woody at the base. See Armitage (1989a). C. × clandonensis is cold hardy to zone 4 (see Fig. 2) and is grown as a perennial. Although flowers are not as ornamental as C. incana, significant potential also exists with that hybrid

Catananche caerulea (cupid's dart)

	Blue	2–2.5	Transplant	66–73	8–14	Fresh, dried

Comments: Plants are short-lived perennials but best treated as annuals.

Celosia cristata (cockscomb)

Chief series	Various	1–2.5	Transplant	70–73	7–14	Fresh, dried

> *Other cultivars*: 'Pampas Plume' mix is a feathered form useful for cut flowers. Transplant from last frost every 2 weeks for best quality stems. 'Rocket' and 'Sparkler' series bear plume-type flowers. Staggered planting dates are better than a single plant date.

> *Other species*: *Wheat celosia* has tremendous potential. It grows 4 feet tall, is floriferous in bronze wheat shades, and stems are terrific for dried or fresh uses.

Centaurea americana (American basket flower)

'Jolly Joker'	Lilac-rose	2–4	Direct sow	64–72	10–14	Fresh

Comments: The flowers are spider-like. Plants perform best under cool nights and low humidity

Centaurea cyanus (bachelor's buttons)

'Boy' series	Various	2–3	Direct sow	64–72	10–14	Fresh

Comments: Plants will reseed the following year in most areas

> *Other cultivars*: 'Frosty' stands about 2 feet tall and bears mix-colored double flowers tinged with white

Centaurea moschata (sweet sultan)

'Imperialis'	Various	2–2.5	Direct sow	60–70	10–14	Fresh

Comments: Does not produce high quality blooms south of zone 7b east of the Rocky Mountains (see Fig. 2). Can be produced in winter months as a cool season crop (Oct–March) in Zones 8–11

> *Other cultivars*: 'Antique Lace' grows 2 feet tall and bears flowers in partial shade. 'Lucida' has dark red flowers, var *suaveolens* (var. *flava*) has canary-yellow flowers on 2 feet tall stems

Chrysanthemum parthenium (feverfew)

'Golden Ball'	Yellow	1–2	Transplant	65–75	20–30	Fresh

> *Other cultivars*: 'Tetra White' has large white flowers on 2 inch stems. 'Fortuna' (1–2 feet) has creamy white centers surrounded by quilled clear white petals. All feverfews have highly aromatic foliage.

> *Other species*: *C. segetum* (corn chrysanthemum) grows 1.5–2 feet tall and usually bears bicolored flowers. 'Eastern Star' has primrose petals with a dark center. 'Paradiso' has gold petals surrounding a dark center. *C. carniatum* (tricolor daisy) is 2 feet tall and available in several colors. *C. coronarium* (crown daisy) 'Beauty of Limmat' bears single creamy yellow flowers with a golden band. *C.* × *spectabile* (shasta daisy) 'Cecilia' has creamy white flowers around a yellow center. Many other cultivars of shasta daisies may also be used

(continued)

Table IV *(continued)*

Genus and species (common name) and common cultivar	Flower color	Height (feet)	Planting method	Germination Temperature (°F)	Time (days)	Uses

Consolida ambigua (rocket larkspur)

'Hyacinth' mix	Various	3–4	Direct sow	50–55	20–30	Fresh, dried

> *Other cultivars*: 'Kelsay' mix is similar but 7–10 days earlier. Chilling seed for 7–14 days at 40°F enhances germination

Consolida orientalis (field larkspur)

'Imperial Giant'	Various	3–4	Direct sow	50–55	20–30	Fresh, dried

> *Other cultivars*: Sunburst series is available in single colors as well as a mix. Plants have been selected for uniformity and stem strength

Cosmos bipinnatus (cosmos)

'Klondyke'	Various	3–4	Direct sow	70–73	7–14	Fresh

> *Other cultivars*: 'Sensation' mix is early flowering and may also be found in single colors; 'Versailles' has lilac-rose flowers with a crimson ring

Craspedia globosa (golden drumsticks)

	Yellow	1.5–2.5	Transplant	70–76	7–14	Fresh, dried

> **Comments**: The flowers are useful for everlasting market. Fresh seed is extremely important for good germination

Daucus carota (Queen Anne's lace)

	White	1.5–3	Direct sow	65–76	10–20	Fresh, dried

> **Comments**: Tolerant of a wide range of soil types, inflorescence is more open, and flowers are more cream-colored than *Ammi majus*

Emilia javanica (florist's paintbrush)

	Gold/red	1–2	Direct sow	60–70	10–14	Fresh

> **Comments**: Useful for fillers in bouquets. Plants produce immense yield, therefore, only a small area is needed. Do not store well, best for local markets. Plants will reseed themselves in southern climates. Separate colors or mix are available

Euphorbia marginata (snow-on-the mountain)

	White	2–4	Transplant	60–68	10–14	Fresh

> **Comments**: Night temperatures should be below 70°F, and if plants are to be produced in the south (zones 7–11), grow in late fall to early spring

Godetia amoena (godetia)

'Grace' series	Various	1–3	Transplant	68–75	10–12	Fresh

> **Comments** : Best for Mediterranean climate such as coastal California. Does not tolerate heat and humidity. *Godetia* and *Clarkia* are used interchangeably although minor botanical differences exist. *Clarkia* is grown similarly to *Godetia*. Plants should be spaced closely together and require early planting for best flowering. See Anderson and Pemberton (1990) for additional details concerning environmental constraints.
>
> *Other cultivars*: 'Sweetheart' has semidouble and double blush-pink flowers; 'Miss Nagasaki' bears salmon-rose blooms. Both are open pollinated. Many other open pollinated cultivars are available

Gomphrena globosa (globe amaranth)

Mixed	Various	1–2	Direct sow	65–78	10–14	Fresh, dried

> **Comments**: Flowers are usually shades of white, purple-violet, or rose. Separate colors are available. 'Lavender Queen' is particularly good

Table IV (continued)

Genus and species (common name) and common cultivar	Flower color	Height (feet)	Planting method	Germination Temperature (°F)	Time (days)	Uses

Gomphrena haageana (globe amaranth)

'Strawberry Fields'	Pink-red	1–2	Transplant	65–78	10–14	Fresh, dried

Comments: Seed of globe amaranth should be clean and scarified for best germination. The extra cost for clean seed is paid for by significantly better germination

Other cultivars: 'Strawberry Fields' ('Woodcreek Red') is fast becoming the standard for *Gomphrena* crops but var *aurea superba*, an orange selection, is also an excellent cut flower

Helianthus annuus (sunflower)

'Taiyo'	Yellow	3–5	Direct sow	64–73	7–14	Fresh

Other cultivars: Many are available; singles include 'Autumn Beauty,' a hybrid of mixed colors, 'Primrose,' sulphur-yellow, and 'Sunbright,' yellow upright flowers with brown centers and no pollen. Double-flowered forms include 'Sungold' with golden yellow flowers on 3–5 feet tall stems

Helianthus debilis (*cucumerfolius*) (cucumber-leaf sunflower)

'Piccolo'	Gold	3–4	Direct sow	64–73	7–14	Fresh

Other cultivars: 'Diadem' bears white to pale yellow flowers on 4 feet tall stems, 'Orion' has pale yellow petals, and 'Stella' produces intense gold flowers with curled petals and maroon centers

Helichrysum bracteatum (strawflower)

'King Size' series	Various	2–2.5	Direct sow	64–73	7–10	Fresh, dried

Other cultivars: Standard series are less expensive than 'King Size' series but less uniform. Flowers are fully double

Other species: *H. cassianum* 'Rose Beauty' has clusters of pink flowers on 2 feet tall stems. *H. subulifolium* 'Golden Sun' has bright green foliage and clear yellow flowers on 1.5–2 feet tall stems

Helipterum roseum (*Acroclinum*) (sunray everlasting)

'Red Goliath'	Red	1–2	Direct sow	64–73	7–10	Fresh, dried

Other cultivars: 'Sensation Giants' have large, fully double flowers on 1–1.5 feet tall stems. 'Red Bonnie' has flowers of rose-pink shades with yellow centers. White shades are also available

Other species: *H. manglesii* ('Rhodanthe'), also known as 'Swan River Everlasting,' bears solitary semidouble flowers on 1–1.5 feet tall stems. Mixes and single colors are available

Iberis amara (rocket candytuft)

'White Pinnacle'	White	1–2	Direct sow	60–68	10–14	Fresh

Comments: Flowers are fragrant and relatively persistent

Lavatera trimestris (mallow flower)

'Loveliness'	White/pink	2–3	Transplant	70–72	7–14	Fresh

Comments: Plants are susceptible to numerous insects and diseases, market yet to be developed. Good potential, particularly in California

Limonium sinuatum (annual statice) (see Fig. 3)

'Fortress'	Various	1–3	Transplant	63–68	10–14	Fresh, dried

(continued)

Table IV *(continued)*

Genus and species (common name) and common cultivar	Flower color	Height (feet)	Planting method	Germination Temperature (°F)	Germination Time (days)	Uses

Limonium sinuatum (cont.)

   Other cultivars: Sunburst series is available as a mix and as separate colors; 'Pacific Strain' is less uniform than 'Sunburst' but seed is less expensive. Plants are available in plugs from many growers. 'Heavenly Blue' and 'Oriental Blue' are popular blue-flowered statice. The yellow forms are more susceptible to root rots (*Pythium, Phytophthora*) and should be treated with a fungicide at planting

Limonium suworowii (rattail statice, also known as *Psylliostachys suworowii*)

| | Rose-pink | 1–2 | Transplant | 61–72 | 14–21 | Fresh, dried |

Comments: Produces long tassel-like flowers. Plants are well established in European cut flower markets. Best for zones 4–7, as tassels are not sufficiently long in warm weather. Flowers do not dry as well as other statice and are best used fresh

Lunaria annua (honesty, money plant)

| | White/purple | 2–3 | Direct sow | 65–70 | 10–14 | Dry pods |

Comments: Plants selfsow in southern areas and should be allowed to do so for continuous production. Flowers are white and purple but pods are similar regardless of flower color. They have a purple-podded form, however, bronze-to-purple pods form in summer.

Matthiola incana (stock) (see Fig. 4)

| Excelsior Giants | Various | 2–3 | Transplant | 68–73 | 7–14 | Fresh |

   Other cultivars: 'Ultra Double' series is uniform and available in a mixture and single colors. Frolic series is $2–2\frac{1}{2}$ feet tall in mixed colors. Some series have been selected for double flowers and claim 80% double flowers. Unfortunately, this is not the case, 50% being more common.

Comments: Many stocks are grown in the field in California without protection. Direct sowing is common in California but reduces the ability to select seedlings for doubleness. In the Midwest and East, they are best grown in the greenhouse. They do not perform well in the South, as the optimum growing temperature is 50°F

Molucella laevis (bells-of-Ireland)

| | Green | 2–3 | Direct sow | 64–73 | 14–30 | Fresh, dried |

Comments: Seed should be chilled at 36–38°F for 2 weeks prior to sowing

Nigella damascena (love-in-a-mist)

| 'Miss Jekyll' | Various | 1–2 | Direct sow | 68–73 | 7–14 | Fresh, dried |

Comments: The seed pods are the most attractive part of the plant and are useful for drying

Oxypetalum caeruleum (tweedia)

| 'Heavenborn' | Light blue | 2–3 | Transplant | 63–73 | 21–35 | Fresh |

Comments: Flowers are easily damaged by wind and rain and should have a moderate covering if grown outdoors. Plants are tolerant of a wide range of temperature and can be produced in the greenhouse successfully (Armitage *et al.*, 1990). Stems exude a milky sap when cut and the foliage is aromatic. Plants are good subjects for the South as they are tolerant of heat and humidity but have potential throughout the country. The color is unique and should be used more in the specialty cut trade

Papaver nudicaule (Iceland poppy)

| 'Wonderland' | Various | 1.5–2.5 | Transplant | 55–68 | 14–21 | Fresh |

Comments: Cut stems bleed excessively and should be plunged in hot water for approximately 10 seconds immediately after cutting, or they will not take up water. An excellent crop for cool weather or a cool greenhouse

Table IV *(continued)*

| Genus and species (common name) and common cultivar | Flower color | Height (feet) | Planting method | Germination | | Uses |
				Temperature (°F)	Time (days)	

Other cultivars: 'Champagne Bubbles' is an F_1 hybrid with large cup-shaped flowers available in a mixture of colors. Seed is more expensive than standard cultivars. 'Matador' bears large scarlet flowers on 1–2 feet tall stems. 'Misato-Carnival' is one of the best series available. 'San Remo' occurs in many colors and is also an excellent cut flower

Reseda odorata (mignonette)

| 'Grandiflora' | White | 1–1.5 | Direct sow | 63–73 | 7–14 | Fresh |

Comments: Flowers are fragrant. Plants are best grown in a cool area

Other cultivars: 'Machet' has rich red flowers on 1 foot tall stems

Rudbeckia hirta (black-eyed susan)

| 'Marmalade' | Gold-orange | 2–2.5 | Transplant | 64–75 | 7–14 | Fresh |

Other cultivars: 'Gloriosa' has yellow and mahogany flowers on 3 feet tall stems. 'Goldilocks' bears semidouble flowers on 2–3 feet tall stems

Salvia leucantha (velvet sage)

| | Lavender-blue | 3–4 | Transplant | Sterile | | Fresh, dried |

Comments: Flowers are sterile and plants can only be raised vegetatively, usually by terminal cuttings. Flowering is a short-day response (Armitage and Laushman, 1989), therefore, flowering does not occur until late summer and fall. Plants may be perennial in zones 8–11. A crop with exciting potential

Other species: *S. farinacea* (mealy cup sage) bears blue or white flowers on 1–1.5 feet tall stems. They are perennial in zones 7–11

Scabiosa atropurpurea (pincushion flower)

| 'Giant Imperial' series | Various | 3–3.5 | Direct sow | 63–71 | 7–14 | Fresh |

Other cultivars: 'Olympia' series is available as a mix and is 3–3.5 feet tall

Other species: *S. stellata* (drumstick flower, ping pong flower) is grown for the bronze dried seedheads

Tithonia rotundifolia (Mexican sunflower)

| 'Goldfinger' | Orange | 2–3 | Transplant | 73 | 14–21 | Fresh |

Other cultivars: 'Torch' bears orange-red flowers on 3–4 feet tall stems

Trachelium caeruleum (throatwort)

| 'Umbrella' series | Blue, white | 2–3 | Transplant | 64–73 | 14–21 | Fresh |

Comments: *Trachelium* is an excellent cut flower for the greenhouse and is occasionally used in the field, particularly in California. Flowers initiate under long-day conditions (Armitage, 1988)

Xeranthemum annuum (immortelle)

| 'Superbissimum' | Violet | 2–2.5 | Direct sow | 63–70 | 10–14 | Dried |

Other cultivars: 'Flore-plena' has semidouble and double flowers available as a mix or in separate colors

Zinnia elegans (zinnia)

| | Various | 2–3 | Transplant | 68–73 | 7–10 | Fresh |

Other cultivars: Both open pollinated and F_1 hybrids are used as cut flowers. Cactus-, dahlia-, double-, and scabiosa-flowered cultivars are available as mixes and as separate colors. Susceptibility to powdery mildew requires constant fungicidal sprays. Staggered planting dates are better than a single plant date for many cultivars

B. Perennial Species

Numerous herbaceous perennials are useful as cut flower crops. Table V outlines some advantages and disadvantages to growing perennials as cut flowers.

In Table VI perennials are listed according to their ability to withstand cold temperatures. Geographic limitations to commercial production are based on the USDA zone hardiness map (Fig. 2). The hardiness map was redrawn in 1990 (USDA, 1990) and reflects average annual minimum temperatures. The hardiness ratings provide guidelines only and are not meant to be absolute. Heat limitations have been provided for all perennial species and are also based on the USDA plant hardiness zone map. Because of the lack of an average warm temperature mapping system, the heat tolerance figures can be quite useful, especially for growers east of the Rocky Mountains.

The hardiness map (Miscellaneous Publication #1475) can be obtained from: Director, United States Arboretum, Agricultural Research Service, USDA, Washington, D.C. 20002.

Table V

Some Advantages and Disadvantages to Growing Perennials as Cut Flowers

Advantages

Crop longevity is usually 3–5 years, thus costs of replanting are minimal compared with annuals

Many "unusual" flowers are available as perennial species. Breeding efforts in perennials are strong; new and better cultivars for cutting are constantly emerging

Perennials are undergoing a "boom" in American floriculture, resulting in strong markets in cut flowers and plantlets

Yield per plant and yield per square foot can be very high

Disadvantages

Often there is little marketable yield the first year if plugs or transplants are used and 2 years if started from seed

Seed germination of some species requires special treatments, such as vernalization or scarification. Germination of some species may take up to 1 year (Pinnel et al., 1985; Armitage, 1989c)

Flowering time of perennials is short, compared with annuals, and extending the season of bloom is difficult, particularly after the first year

Harvesting costs per crop can be high, especially compared with single stem plants, such as bulb crops

Table VI

Perennial Species Useful as Cut Flowers

Genus and species (common name) and cultivar	Flower color	Height (feet)	Spacing (inches)	Planting zone	Harvest season	Uses
Achillea × 'Coronation Gold' (coronation gold yarrow)						
'Coronation Gold'	Yellow	1–3	12 × 12	3–9	Spring, summer	Fresh, dried

Comments: 'Coronation Gold' is a hybrid between A. filipendulina and A. clypeolata and cannot be raised from seed. The foliage is gray-green and fragrant. The flower heads average 2.5 inches in diameter, although 3–4 inches in diameter heads may be produced. Flowers of all yarrow species must be fully open (i.e., stamens just visible) before harvesting

Achillea filipendulina (fern leaf yarrow)						
'Parker's Variety'	Gold/yellow	3–3.5	12 × 12	3–9	Spring, summer	Fresh, dried

Comments: 'Parker's Variety,' also known as 'Cloth of Gold,' is taller and bears larger but fewer flowers than 'Coronation Gold'

Other cultivars: 'Gold Plate' bears large (3–4 inches) diameter flowers on strong upright stems

Achillea millefolium (common yarrow)						
'Cerise Queen'	Red/cerise	1–2	9 × 12	3–9	Spring, summer	Fresh, dried

Comments: Stem strength is better on northern grown plants compared with those in the south. Shelf life (fresh) is only 2–4 days in water. Vase life is the most limiting factor

Other cultivars: 'Fire King' has deep red flowers, 'Rose Beauty,' 'Lilac Beauty,' 'White Beauty' have pink, lilac, and white flowers, respectively. 'Paprika' is an excellent cerise red cultivar

Other species: Galaxy series is a cross between A. millefolium and A. taygetea and has the foliage characteristics of common yarrow but with larger and more colorful flower heads and stronger stems. Some of the available cultivars useful for cut flowers are 'Appleblossom,' light mauve; 'Great Expectations,' pale yellow; 'The Beacon,' red; and 'Salmon Beauty,' salmon. A. ptarmica (sneezeweed) has white flowers. The plants are not as vigorous and flowers not as showy as other yarrow species. 'The Pearl,' which may be grown from seeds, is the most popular double-flowered cultivar

Aconitum napellus (monkshood)						
'Newry Blue'	Blue	3–4	12 × 18	2–6	Summer, fall	Fresh

Comments: An excellent cut flower that can only be grown well in cooler areas of the country. The roots are poisonous, and gloves should be used while harvesting to prevent plant exudate from cut stems entering open cuts. Powdery mildew can be a problem without preventative fungicide application

Other cultivars: 'Bicolor' bears blue and white flowers, 'Spark's Variety' produces dark blue flowers on branched stems

Other species: A. × arendsii has rich, dark blue flowers on strong stout stems. Although not as tall as other species, the stem strength and flower size are excellent. A. carmichaelii has unbranched stems of dark blue flowers. 'Kelmscott' bears violet-blue flowers

Aquilegia × hybrida (hybrid columbine)						
McKana series	Various	1–2.5	8 × 12	3–9	Spring	Fresh

    Other cultivars: 'Crimson Star,' 'Song Bird' series, many others available

    Other species: A. caerulea (Rocky Mountain columbine) bears blue and white flowers on 2 ft stems. Plants persist for up to 5 years. 'Musik' and 'Olympia' series are excellent as cut flowers

Comments: Seeds should be given a 3–6 week cold treatment at 39–41°F. All flowers tend to shatter, so best suited for local markets. Plants generally decline after 3 years of production

(continued)

Table VI (continued)

Genus and species (common name) and cultivar	Flower color	Height (feet)	Spacing (inches)	Planting zone	Harvest season	Uses

Asclepias tuberosa (butterfly weed)

| | Orange | 1.5–3 | 15 × 15 | 3–9 | Spring, summer | Fresh |

Comments: Butterfly weed has a long taproot and is not easily transplanted or divided. Plants are exceptionally long lived and may be harvested for 8–10 years. Susceptibility to spider mites, however, is a serious problem. An excellent cut flower that is particularly suitable for southern and midwest States because of its tolerance to heat and humidity

Aster ericoides (aster)

| 'Monte Casino' | White | 2–4 | 12 × 12 | 3–7 | Fall | Fresh |

Other cultivars: 'Blue Wonder' has blue-pink flowers, 'Constance' and 'White Wonder' bear white flowers, and 'Ester' produces pink blossoms. Hybrids such as Butterfly series and Master and Star series have larger flowers than 'Monte Casino' and are vigorous

Comments: A. ericoides is a short-day plant and will not flower until photoperiod shortens in the fall. Plants may be grown in the greenhouse using chrysanthemum lighting and black cloth facilities for year-round production. Place in water immediately after harvesting. Plants are grown more often in the greenhouse than in the field

Other species: A. novi-belgii (New York aster) and A. novae-angliae (New England aster) may also be used as cut flowers. Many cultivars are available ranging from 1–5 feet in height. Some cultivars flower in the summer ('September Ruby') while others flower in the fall ('Climax'). All require support for best quality stems. A. dumosus and A. cordifolius 'Ideal' also have excellent potential for cut flowers

Astilbe × *arendsii* (astilbe)

| | Various | 2–2.5 | 12 × 15 | 3–7 | Spring | Fresh, dried |

Comments: Astilbes require partial shade and moist conditions to perform well as a cut flower. Flowers are available in white, red, lavender, pink, and magenta. Cut when approximately 60% of the flowers are open

Other species: A. taquetii (late-flowering astilbe) has coarse magenta flower heads on 3–4 feet tall plants. 'Superba' is the main cultivar available. Flowers appear 3–4 weeks later than those of A. × arendsii

Astrantia major (masterwort)

| | Pink/green | 1–3 | 12 × 15 | 4–6 | Spring | Fresh |

Comments: Masterwort should be grown in partial shade with plenty of moisture. The flowers are subtended by greenish bracts, which give them a "shaggy" appearance

Other species: A. carniolica is 1.5–2 feet tall. The cultivar 'Rubra' bears purple flowers. A. maxima has pink or whitish flowers and is larger and more handsome than A. major

Campanula persicifolia (peach-leaf bellflower)

| 'Telham Beauty' | Blue | 1–3 | 12 × 15 | 3–6 | Spring, summer | Fresh |

Comments: This is the best bellflower for cut flowers because it has an excellent vase life and is easy to grow. Plants do best in cool areas, and quality is better north of zone 6 than in southern areas. Bellflowers may also be raised from seed

Other cultivars: 'Alba' has white flowers, 'Moerheimii' bears double blue flowers

Other species: C. glomerata (clustered bellflower) is also useful as a cut flower. Blue and white forms are available. C. pyramidalis is also excellent

Centaurea macrocephala (Armenian basket flower, marco polo)

| | Yellow | 3–4 | 18 × 18 | 2–6 | Summer | Fresh, dried |

Comments: A particularly popular cut flower in Europe, this species is best produced north of zone 7. The bright yellow flowers contrast well with the brown overlapping bracts. Useful fresh or dried

Table VI (continued)

Genus and species (common name) and cultivar	Flower color	Height (feet)	Spacing (inches)	Planting zone	Harvest season	Uses

Other species: C. montana (mountain bluet) has escaped from cultivation and is common on roadsides in zones 2–7. The flowers of the species are blue but pink and white cultivars are available

Cirsium japonicum (Japanese thistle)

'Pink Beauty'	Red	2–2.5	15 × 15	5–7	Spring, summer	Fresh

Comments: The foliage is prickly and not particularly popular with harvesters. The flowers persist for 7–10 days in water

Other cultivars: 'Rose Beauty' has light pink flowers

Delphinium × *hybridum* (delphinium)

Pacific Giant series	Various	2–4	18 × 18	3–6	Spring, summer	Fresh

Comments: Delphinium is produced as an annual in zones 7–10, being planted in early fall and harvested in early spring. As a perennial in the North, plants cut in spring may produce additional flowers in the fall. Treat harvested stems with silver thiosulfate

Other cultivars: Many cultivars for cutting are available including New Century series, 'Belladonna' hybrids and Princess Caroline (peach colored) are presently in high demand. Each series consists of mixes and cultivars of single colors. Plants are often forced in a cool greenhouse over the winter

Other species: D. grandiflorum (Chinese delphinium) is 1–1.5 feet tall and available in various shades of blue

Echinacea purpurea (purple cone flower)

'Bright Star'	Purple	3–4	18 × 18	3–9	Summer	Fresh, dried

Comments: The purple-pink ray flowers are attached around a cylindrical bronze-copper disk. The flowers may be used fresh, or the ray flowers may be removed from the disk and the disk dried. This is a useful technique when plants of the species are used because of their tendency to droop

Other cultivars: 'Alba' and 'White Lustre' have white petals surrounding an orange-bronze disk. 'Magnus' (synonomous with 'Ovation' and 'Bravado') is a seed-propagated purple-flowered cultivar and an exceptional cut flower selection. 'Robert Bloom' bears large cerise flowers

Echinops ritro (globe thistle)

'Taplow Blue'	Blue	3–4	18 × 24	3–7	Summer	Fresh, dried

Comments: The bracts and stems turn a metallic blue color as they mature. The blue is more intense in areas of cool summer nights. Flowers persist for months in water although foliage only persists for 7–10 days

Other species: E. commutatus has white flower heads on 4–5 feet tall plants. E. blancheanus is a similar species and flowers best under long-day conditions (Wallerstein, 1983). E. bannaticus 'Blue Glow' has deep blue flowers on 3 foot tall stems

Eryngium planum (sea holly)

	Blue	2–4	18 × 18	3–8	Spring, summer	Fresh, dried

Comments: The small blue flowers are surrounded by bracts, which turn metallic blue as the flowers mature. Quality is better in areas of cool night temperatures. This is the best sea holly for shipping. Because of the small flowers, little damage is done in transit compared with the larger flowered species (Armitage, 1989b)

Other species: The most popular species in the United States is E. amethystinum. The bracts and flower heads are larger and more colorful. E. alpinum has the largest and showiest flowers but is not summer hardy south of zone 6

Gypsophila paniculata (baby's breath)

'Perfecta'	White	1–3	45 × 60	3–7	Spring, summer	Fresh, dried

(continued)

Table VI *(continued)*

Genus and species (common name) and cultivar	Flower color	Height (feet)	Spacing (inches)	Planting zone	Harvest season	Uses

Gypsophila paniculata (cont.)

Comments: A good deal of research has been conducted on *Gypsophila* (Hicklenton, 1987; Moe, 1988; Shillo and Halevy, 1982). High light and long days are recommended for flowering. A high pH (6.5–7.4) is best for baby's breath. Staggered planting every 14 days helps spread the harvest through the season. Baby's breath can be produced year round in the greenhouse

Other cultivars: 'Bristol Fairy' produces smaller and more petite flowers and is useful for weddings and small containers. 'Flamingo' has double pink flowers, 'Red Seas' produces small, reddish, double flowers

Limonium tataricum (Goniolimon tartaricum) (German statice)

	Pinkish blue	1–3	12 ×15	3–7	Spring, summer	Fresh, dried

Comments: Plants may be raised from seed (62–74°F) and planted in fall or early spring. Flowers are effective fresh but the majority are sold dry. The flowers are often sold by weight, not by the stem. The inflorescence is feathery and highly branched

Other species: *L. latifolium* is similar but not as tall or as easy to grow as *L. tataricum*

Limonium altaica (statice)

	Light blue	1–2	12 × 12	4–7	Spring, summer	Fresh, dried

Comments: An excellent but relatively unknown statice, it has been blooming prolifically in Georgia (zone 7) trials for 5 years

Other species: *L.* × 'Misty Blue' and *L.* × 'Misty Pink' produce beautiful long, wispy flowers of light blue and light pink, respectively. The plants require support. Plants are grown from tissue culture. *L. perezii* (sea foam statice) is grown in large numbers in California, and *L. gmelinii* also has potential as a cut flower, but its worth has not yet been proven throughout the United States (Clemens and Welsh, 1989)

Lysimachia clethroides (gooseneck loosestrife)

	White	2–2.5	12 × 12	3–7	Summer	Fresh

Comments: Plants require cool temperatures for best yield and highest quality. They spread rapidly under all environments, but stem strength is poor when night temperatures remain above 70°F. Do not fertilize heavily. Support netting is useful. Use of a floral preservative is essential

Physostegia virginiana (obedient plant)

'Pink Bouquet'	Pink	2–4	12 × 12	2–9	Summer, fall	Fresh

Comments: Plants spread rapidly and fill all available space within 2–3 years. They require support because the stems exhibit a strong negative geotropic response, which means that if the stems topple, shoot tips bend upward and are of poor quality. *Physostegia* is a long-day plant (Cantino, 1982), and flowers are not produced until mid- to late summer. Yields of 28 stems per square foot have been recorded (Armitage, 1987)

Other cultivars: Variety *alba* is 2–3 feet tall and bears creamy white flowers at 3–5 weeks earlier than 'Pink Bouquet.' 'Summer Snow' bears clear white flowers on 2–3 feet tall stems. 'Vivid' is a common garden cultivar but is too short (1–1.5 feet tall) and compact for the cut flower market

Other species: *P. purpurea* is native to the southeastern United States and flowers in late spring to early summer, allowing a longer cropping time for this group of plants

Platycodon grandiflorus (balloon flower)

	Blue	2–3	12 × 18	3–7	Summer	Fresh

Comments: Plants are long-lived, and 5-year production is not uncommon

Other cultivars: 'Fuji' series has been produced for cut flower production. Pink, white, and blue are available. 'Double Blue' ('Hakone Blue') is a double-flowered cultivar where flowers are fuller and more interesting than the single-flowered forms. 'Komachi' is an interesting selection with flowers that persist as "balloons" rather than opening

Table VI (continued)

Genus and species (common name) and cultivar	Flower color	Height (feet)	Spacing (inches)	Planting zone	Harvest season	Uses

Scabiosa caucasica (scabious, pincushion flower)

| 'Fama' | Blue | 1.5–2.5 | 18 × 18 | 4–6 | Spring, summer | Fresh |

Comments: Plants do not produce well south of zone 7. Plantings can be staggered for extended flowering the first year

Other cultivars: House hybrids are a mixture of blue shades. 'Clive Greaves' is 2 feet tall and bears light blue flowers; 'Compliment,' a seed propagated cultivar, has large, dark lavender flowers; 'Miss Wilmott' has white flowers; and 'Staffa' bears dark blue blossoms

Solidago hybrids (goldenrod)

| 'Lemore' | Yellow | 2–2.5 | 12 × 12 | 3–7 | Summer, fall | Fresh |

Comments: Plants require at least one pinch and should also be supported. Goldenrod is responsive to short days and can be flowered year round in the greenhouse. Many hybrids have been developed by Dutch breeders for the cut flower market and are superior to garden selections

Other cultivars: 'Straehlenkrone' flowers in early summer on 1.5–2 feet tall stems, and 'Super' has lemon yellow flowers on 2–2.5 feet tall stems

Other species: *S. canadensis* grows 3–4 feet tall. 'Baby Gold,' a hybrid between *S. canadensis* and *S. virgaurea*, is popular. *S. odora* (sweet goldenrod) has light yellow flowers and is wonderfully fragrant. *S. virgaurea*, European goldenrod, is up to 2–3 feet tall. 'Golden Rod' and 'Praecox' flower in the fall on 2–3 feet tall stems

X Solidaster luteus (solidaster)

| | Light yellow | 1–3 | 15 × 15 | 3–7 | Summer | Fresh |

Comments: A cross between *Solidago* and *Aster*, this hybrid offers excellent potential for cut flowers. Plants should be pinched and supported. The development of flowers is accelerated by short days and may be produced year round under greenhouse conditions. Both *Solidago* and *Solidaster* are susceptible to rust.

Thalictrum rochebrunianum (meadow-rue)

| 'Hewitt's Double' | Lilac | 2–4 | 18 × 18 | 3–6 | Spring, summer | Fresh, dried |

Comments: The small flowers may be used as fillers and occasionally can substitute for baby's breath. Support is necessary to keep the flowers from becoming tangled and damaged when harvested

Other species: *T. aquilegifolium* (columbine meadow-rue) has fuller purple or white flowers and columbine-like foliage. This is a better species for the south, but flowers are not as persistent. 'White Cloud' is an exceptional white-flowered cultivar. *T. dipterocarpum* (yunnan meadow-rue) is similar to *T. rochebrunianum* and often offered under that name. No difference in field culture occurs

Veronica longifolia (Veronica)

| 'Oxford' | Light blue | 2–2.5 | 12 × 15 | 4–7 | Summer | Fresh |

Comments: Lime should not be added to the soil as plants do better at pH of 5.5–6.0

Other cultivars: 'Blauriesin' has light blue flowers, but pink and white cultivars are also available

Veronicastrum virginicum (culver's root)

| | White, blue | 2.5–3 | 12 × 12 | 3–7 | Summer | Fresh |

Comments: Plants should be fertilized heavily in the spring. Flowers in branched racemes are borne in summer. An excellent cut flower with untold potential. Remove center flower as it begins to expand; lateral flowers will all open at once

Other cultivars: Variety *alba* has white flowers and is more handsome than the blue species

C. Bulbous Species

Table VII provides some advantages and disadvantages to cut flower production from bulbous species.

1. Bulb Longevity

Many growers of cut flowers treat all bulbous plants as annuals, believing that new material every year reduces cultural problems and increases the quality of the cut stem. Yearly replacement is costly in terms of plant material, bed preparation, and labor, but in some species, it is absolutely necessary. Not all crops, however, need to be treated as annuals; some may be left in the ground (assuming non-destructive winter temperatures) for 2 to 5 years. Table VIII shows how several bulb species are affected over time (Armitage, 1989d; Armitage and Laushman, 1990a; De Hertogh, 1989a).

As shown by the data in Table VIII, *Anemone* should be treated as an annual, *Crocosmia* can be left in the ground for 2 to 3 years, while *Liatris* can be produced for at least 3 years (Fig. 5). Inground persistence of common bulb genera is shown in Table IX (based on work in Georgia, zone 7b). Table IX assumes that crops can be wintered over in the field.

2. Extending Flowering Time of Bulb Species

First year flowering of a number of species can be extended by successive plantings in the fall and winter. For example, frozen *Liatris* corms may be planted in the spring and summer to extend the flowering time through the fall. Table X shows how staggered planting in the fall and early winter affect flowering time of various bulb species.

Table VII

Advantages and Disadvantages of Cut Flowers from Bulbous Species

Advantages

Bulb crops can be harvested from late winter until frost

Many bulb species are relatively inexpensive

The shelf life of many, although not all, cut flowers is very good, even in tap water

Flowers are usually borne on a single stalk (scape) and are easy to harvest

Many bulbous species are relatively unknown and fit the need for "different" flowers

A number of species can be field produced and sold prior to Mother's Day in many parts of the country

Disadvantages

Many bulbous crops must be treated as annuals for purposes of commercial production so labor and material costs are repeated annually

Removing bulbs from the field is time consuming and difficult

Yield per bulb is low compared with herbaceous annuals and perennials and necessitates planting many bulbs per square foot, thus increasing material costs

Table VIII

Yield and Stem Quality of Bulb Species over Time[a]

Crop	Year	Survival (%)	Stems per plant	Stem length (inches)
Anemone coronaria	1	90	6.5	10.1
'De Caen'	2	30	7.0	9.4
	3	20	5.4	8.5
Crocosmia × crocosmiiflora	1	100	0.9	22.2
'James Coey'	2	87	1.2	16.0
	3	45	2.0	20.0
Liatris spicata	1	100	2.7	24.6
	2	95	8.0	30.0
	3	95	15.0	39.2

[a]Adapted from Armitage and Laushman (1990a).

Fig. 5. Beds of *Liatris*, 1-year-old planting.

Table IX

In-ground Persistence of Common Bulb Crops Used for Cut Flower Production[a]

Genus	In-ground persistence (years)
Acidanthera	1–2
Allium	3–5
Anemone	1
Brodaea	1–2
Crocosmia	2–3
Freesia	1
Gladiolis	1–2
Iris (Dutch)	1–2
Ixia	1–2
Liatris	3–5
Lilium	1–2[b]
Ornithogalum	1–2
Polianthes	3–5
Zantedeschia	3–5

[a]Adapted from Armitage and Laushman (1990a, b) and De Hertogh (1989a, b).

[b]Lilies are generally treated as an annual because a significant amount of foliage is removed when the flowers are cut which reduces the vigor of the bulb in subsequent years. If approximately one third to one half of the stem remains after harvesting, lilies can be productive in subsequent years.

In the case of both species, an excessive delay in planting significantly decreased the yield. In *Acidanthera*, no differences in flowering time, yield, or stem length occurred until planting was delayed until March. However, with anemones, planting after November resulted in reduced yield although harvest could be extended with delayed planting. Planting after December is not recommended. Research with other bulbous species showed similar species-specific results (Armitage and Laushman, 1990b).

3. Recommended Planting Dates

Based on research at various institutions and commercial recommendations (De Hertogh, 1989a), bulbs for outdoor planting should be planted at the times recommended in Table XI.

Some of the bulb crops useful for cut flower production are outlined in Table XII.

Table X

Effect of Staggered Planting Times on Flowering of Bulb Crops[a]

Crop	Month of planting	Yield (flowers/bulb)	First harvest (date)	Harvest duration (days)	Stem length (inches)
Acidanthera bicolor	11	1.2	July 2	20	28.9
'Muralis'	2	1.3	July 2	26	28.3
	3	0.8	July 11	21	28.1
Anemone coronaria	11	10.2	Feb. 27	65	9.4
'De Caen'	12	4.8	Mar. 30	33	7.6
	1	1.1	Apr. 15	17	5.6
	2	0.2	Apr. 30	5	5.6

[a]Adapted from Armitage and Laushman (1990a).

Table XI

Planting Time of Common Bulb Crops in Northern and Southern United States[a]

Genus	Planting time (based on availability)	
	South (zone 7–11)	North (north of zone 7)
Acidanthera	Nov–Feb	Feb–Mar
Allium	Sept–Dec	Sept–Dec
Anemone	Sept–Dec	Jan–Mar
Brodaea	Sept–Feb	Feb–Mar
Crocosmia	Jan–Mar	Feb–Apr
Iris (Dutch)	Oct–Dec	Jan–Mar
Gladiolus	Jan–Apr	Mar–Apr
Liatris	Nov–Apr	Nov–Apr
Lilium	Oct–Dec	Oct–Dec
Polianthes	Nov–Dec	Mar–Apr
Ornithogalum	Sept–Dec	Feb–Mar
Zantedeschia	Oct–Apr	Apr–May

[a]From De Hertogh, 1989a.

Table XII

Bulb Species Useful for Cut Flower Production

Genus and species (common name) and cultivar	Color	Longevity (years)	Height (feet)	Planting Depth (inches)	Plants per square foot	Flowering time
Acidanthera bicolor (Gladiolus callianthus)						
'Muralis'	White	1	3–3.5	2.5	10	Summer

Comments: The plant and flower habit are similar to a gladiolus. Flowers have a purple throat and a pleasant fragrance

Allium giganteum (giant onion)						
	Lilac-blue	3–5	4–5	6	5	Spring

Comments: One of the most perennial and most expensive of the alliums, the long stout scapes bear 6–7 inch diameter inflorescences consisting of hundreds of small lavender flowers. The bulbs split after the first year and each half can be replanted to increase numbers

Other species: 'Globemaster' is a hybrid between A. christophii, downy onion, and A. elatum. Hundreds of sterile violet flowers are formed on 3 feet tall stems. A. aflatunense is similar to A. giganteum but is only 1–2 feet tall and has smaller flowers. The bulbs, however, are significantly less expensive

Allium sphaerocephalum (drumstick chives)						
	Purple	2–3	2–2.5	3	30	Spring, summer

Comments: The green flower buds turn purple with maturity providing a two-tone effect when harvested. The oval flower heads consist of 50–100 flowers (see Fig. 6). Bulbs are susceptible to *Sclerotium cepivorum*, a fungus that causes bulb rot

Other species: Other species of allium that are useful for cut flowers are A. cowanii, a relative of A. napolitanum (Naples onion), A. triquetrum (three cornered moly), and A. caeruleum (azure onion). A. napolitanum should be grown in a greenhouse. All are best treated as annuals

Anemone coronaria (poppy anemone)						
DeCaen	Various	1	1–1.3	2	10	Spring

Comments: 'De Caen' consists of saucer-shaped single flowers in bright colors. The use of 55% shade results in longer flower stems (Armitage, 1989e). Plants do not tolerate heat, and the advent of warm temperatures inhibits flowering and results in plant dormancy

Other cultivars: 'Mona Lisa' has the longest stems of any cultivar and was developed for the greenhouse cut flower trade. 'St Brigid' has semidouble flowers and is available as a mix or as single colors. 'St Piran' bears single and semidouble flowers

Brodiaea laxa (Triteleia laxa) (brodiaea)						
'Queen Fabiola'	Blue	2–3	1–1.3	2	250	Spring, summer

Comments: Bulbs performed well for 2 years in Georgia (zone 7) trials, but are also eminently suited to western states. Pretreatment of dormant corms with ethylene gas results in more flowers per inflorescence and promoted flowering of small corms (Han et al., 1989)

Crocosmia × crocosmiiflora (crocosmia, monbretia)						
'Emily Mackenzie'	Orange	2–3	2–2.5	3	10	Summer

Comments: Crocosmia is an under-used, relatively unknown genus with enormous potential as cut flowers

Other cultivars: 'A. E. Amos' bears orange-red flowers, 'Citronella' has orange-yellow blossoms, 'James Coey' produces red flowers, while 'Solfatare' bears apricot-yellow flowers

Table XII *(continued)*

Genus and *species* (common name) and cultivar	Color	Longevity (years)	Height (feet)	Planting Depth (inches)	Plants per square foot	Flowering time

Crocosmia × *crocosmiiflora (cont.)*
    Other species: *C. masonorum* has 3 feet long flower stems with orange-red flowers. 'Firebird' is an excellent cultivar. 'Lucifer' bears scarlet-red flowers and is the result of a cross between *Crocosmia* and *Curtonus paniculatus*

Iris (Dutch iris)

| | Various | 1 | 2–2.5 | 2 | 15 | Spring |

Comments: Usually available from October to April. For greenhouse production, precooled bulbs require 8–10 weeks to harvest. For outdoor production, plant noncooled bulbs from October to December or precooled later to extend season (De Hertogh, 1989a). Late frosts may result in severe damage to flower buds and excessive heat during initiation and development of flowers may result in bud blast. Ethylene pretreatment of bulbs results in early flowering in the field (Stuart *et al.*, 1966)
    Other cultivars: 'Blue Ribbon' and 'Ideal' produce dark blue flowers, while 'Telstar' bears light blue flowers. 'White Cloud' and 'White Wedgewood' are excellent white-flowered cultivars. 'Apollo' has yellow and white flowers, and 'Golden Harvest' is a popular dark yellow selection. Many others are available (De Hertogh, 1989a)
    Other species: English iris (*I. xiphiodes*) and Spanish iris (*I. xiphium*) are sometimes used as cut flowers. In general, they are less available, less winter hardy, and more difficult to grow

Liatris spicata (Kansas gayfeather)

| | Purple | 3–5 | 3–4 | 2.5–3.5 | 5 | Summer |

Comments: Flowering season can be extended by using precooled or frozen corms. Plants are only 1–2 feet tall the first year but develop to 3–4 feet in height by second year and require support. This is a well-established specialty flower, and through cold storage treatments, staggered planting times, and manipulation of greenhouse temperatures, they may be produced year round (Geller, 1981; Moe and Berland, 1986; Stimart, 1989)
    Other cultivars: Variety *calliepsis* is synonymous with *L. spicata* but is generally propagated vegetatively resulting in excellent uniformity. 'Floristan White' is a good white-flowered cultivar that persists for 4–5 years and may be produced from seed. 'Kobold' is a dwarf cultivar more useful for the garden than for cut flowers
    Other species: *L. pycnostachya* is taller (up to 5 feet) and coarser in appearance than *L. spicata*. It is not as persistent or as productive

Ornithogalum arabicum (arabicum)

| | White | 1–2 | 1.5–2 | 1–1.5 | 3–5 | Spring, summer |

Comments: Flowers have a black throat and an excellent shelf life. They do not tolerate heat and are most effectively grown on the West Coast
    Other species: *O. thyrsoides* (chincherinchee) also has excellent shelf life, persisting many weeks in water. Plants are best adapted to West Coast climates. *O. umbellatum* (Star-of-Bethlehem) is well adapted to eastern climates and is a prolific producer. Unfortunately, stems are generally less than 1 foot in height, which limits its effectiveness. *O. saundersiae* has been excellent in Georgia trials, producing 3–4-inch tall stems with dozens of white flowers. Early grown from seed

Polianthes tuberosa (tuberose)

| 'Mexican Single' | White | 3–5 | 2–3 | 1–2 | 5 | Summer, fall |

(continued)

Table XII (continued)

Genus and species (common name) and cultivar	Color	Longevity (years)	Height (feet)	Planting Depth (inches)	Plants per square foot	Flowering time

Polianthes tuberosa (cont.)

Comments: Bulbs are not winter hardy north of zone 7. The dried leaves that die back in the fall are an effective winter mulch and need not be removed. The bloom time extends from midsummer until frost

Other cultivars: 'The Pearl' is a pure white double form and equally as productive as 'Mexican Single.' However, the inner row of petals tend to blacken and vase life is reduced when double forms are used

Zantedeschia (calla lilies)						
'Black Magic'	Yellow	3–5	3–3.5	3	5	Spring, summer

Comments: The hybrids ('Black Magic,' 'Pink Perfection,' etc.) appear to be more weather tolerant in eastern climates than *A. aethiopica*, the common white calla. Rhizomes need at least 2 and often 3 years to reach optimum yield and height. Shading (30–55%) results in longer scapes. Much of the breeding and subsequent tissue culturing has occurred in America and New Zealand (Welsh and Baldwin, 1989), and cultivars are now available through specialist suppliers

Other species: *Z. aethiopica* with white flowers and unspotted dark green foliage is popular and easily produced on the West Coast. They are not as productive in the East as the hybrids. *Z. elliottiana*, Elliot's Calla Lily, produces long dark yellow flowers and is a dominant parent of many of the hybrids

Fig. 6. *Allium sphaerocephalum* (drumstick chives). From University of Georgia Cut Flower Research.

D. Woody Plant Species

Many woody species have tremendous potential for cut flowers, and their useful-ness is now being recognized by growers. In the outdoor field situation, woody plants may be used as windbreaks, hedges, or screens and can serve as modifiers of the climate as well as sources of cut flowers. Many are utilized for greens, flowers, fruits, unusual stem shape, and stem color. Little experimental work has been conducted, but many species are potential candidates for cut flowers or greens (Dirr, 1989; Eisel, 1988). Table XIII was constructed by Dr. Michael Dirr at the University of Georgia and presented at the second National Conference on Specialty Cut Flowers (Dirr, 1989).

Table XIII

Woody Species with Potential as Cut Flowers[a]

Genus and *species* and cultivar	Useful properties	Color
Amorpha canescens	Flower/foliage	Purple/gray
Aronia arbutifolia	Fruit	Red, persistent
Aucuba japonica	Foliage	Gold leaves, golden variegated
Buddleia globosa	Flower/fragrance	Yellow ball-shaped flowers
Buddleia × *weyeriana* 'Sungold'	Flower/fragrance	Yellow ball-shaped flowers
Buxus	Foliage	Lustrous dark green leaves
Callicarpa americana	Fruit	Magenta and white, remove leaves
Callicarpa dichotoma	Fruit	Magenta and white fruits, smaller than *C. americana*
Callicarpa bodinieri	Fruit	Magenta and white, smaller than *C. americana*
Callicarpa japonica	Fruit	Lavender, excellent grower
Calluna vulgaris	Flower/foliage	Large range of flower and foliage colors
Calycanthus floridus	Flower/fragrance	Red-brown flowers
'Athens'	Flower/fragrance	Yellow flowers, exceedingly fragrant
Camellia sasanqua	Flower/foliage	Various shades of white, pink, and red
Celastrus orbiculatus	Fruit	Yellow-orange
Celastrus scandens	Fruit	Yellow-orange
Cercis canadensis 'Alba'	Flower	White flowers
Chimonanthus praecox	Flower/fragrance	Waxy yellow flowers, outstanding fragrance
Chionanthus retusus	Flower/foliage	White; lustrous, leathery, dark green leaves
Clethra alnifolia 'Rosea,' 'Pink Spires'	Flower/fragrance	Flowers in July, wonderful fragrance
Cornus alba 'Sibirica'	Stem color	Brilliant red
Cornus florida	Flower	Many colors, single and double
Cornus sericea	Stem color	Red
'Flaviramea'	Stem color	Yellow

(continued)

Table XIII *(continued)*

Genus and *species* and cultivar	Useful properties	Color
Corylopsis species	Flower/fragrance	Soft yellow early spring flowers
Corylus avellana 'Contorta'	Stem shape	Contorted, twisted with handsome catkins
Cotinus coggygria	Flower	Smoky pink to purple panicles in summer
Cytisus	Flower/stem	White, yellow to garnet flowers, green angled stems
Danae racemosa	Foliage	Rich green evergreen "leaves"
Daphne × *burkwoodii*	Flower/fragrance	Pinkish white flower
'Carol Mackie'	Flower/fragrance/foliage	Pinkish white flowers, creamy edge variegated foliage
Daphne odora	Flower/fragrance/foliage	Several forms, all exceedingly fragrant
Erica	Flower/foliage	Winter flowers, many colors
Euonymus alatus	Stem shape	Corky winged stems
× *Fatshedera lizei*	Foliage	Ivy-like evergreen foliage
Forsythia	Flower	Golden yellow flowers
10 new introductions	5 for extreme flower bud hardiness (−25°F)	
Hamamelis × *intermedia*	Flower/fragrance	Yellow, copper, bronze-red
'Arnold Promise,' 'Pallida'		Yellow
'Jelena'		Copper
'Diane,' 'Ruby Glow'		Bronze red
Hydrangea arborescens 'Annabelle'	Flower	Large white, rounded, June, good dried
Hydrangea paniculata 'Grandiflora,' 'Praecox,' 'Tardiva'	Flower	White, summer
Hydranga quercifolia 'Harmony,' 'Roanoke,' 'Snowflake'	Flower/foliage Flower/fruit	Excellent large white panicles in June
Ilex evergreen species		
Ilex deciduous species *decidua, serrata, verticillata, serrata* × *verticillata*	Fruit Flower/fragrance/foliage	Yellow, orange, red fruit
Itea virginica 'Henry's Garnet'		
Jasminum mesnyi	Flower	Double yellow flowers
Kerria japonica 'Picta'	Flower/foliage	Yellow flowers, creamy variegated foliage
Koelreuteria bipinnata	Fruit	Pink-rose capsules, September
Lonicera fragrantissima	Flower/fragrance	Delightful scent
Magnolia grandiflora	Foliage	Lustrous dark green leaves
'Bracken's Brown Beauty,' 'Little Gem,' 'Samual Sommer'	Flower/fragrance/foliage	Yellow tinged purple flowers
Michelia figo		
Michelia doltsopa	Flower/fragrance/foliage	White flowers
Morus bombycis (*australis*) 'Unryu' ('Tortuosa')	Stem shape	Poor man's contorted filbert

(continued)

Table XIII (continued)

Genus and species and cultivar	Useful properties	Color
Myrica cerifera	Fruit/foliage	Gray, waxy fruits, evergreen, scented foliage
Myrica pensylvanica	Fruit/foliage	Gray waxy fruits, deciduous, scented foliage
Nandina domestica	Foliage/fruit/flower	
'Gulfstream,' 'Moonbay,' 'Moyer's Red'	Flower/fragrance/fruit	Delicate, white "blueberry" flowers
Oxydendrum arboreum		
Poncirus trifoliata	Stem color	Bright green thorny stems
Prunus species	Flower	Forced branches, colors vary
Apricot, cherry, peach, plum		
Salix caprea	Flower	Gray, soft furry catkins
Salix discolor	Flower	Gray, soft furry catkins
Salix gracistyla	Flower	Gray-pink, soft furry catkins
Salix melanostachys	Flower	Purple-black, soft furry catkins
Salix 'Golden Curls'	Stem shape/color	Golden, contorted
'Scarlet Curls'	Stem shape/color	Red, contorted
Salix matsudana 'Tortuosa'	Stem shape	Yellow brown, contorted
Salix sachalinense 'Sekka'	Stem shape	Flat, flattened fasciated stems
Sarcococca hookerana	Fragrance/foliage	Early spring fragrance
Skimmia japonica	Flower/fruit/foliage	Superb flowers, red fruits, evergreen foliage
Skimmia reevesiana	Fruit/foliage	Brilliant red fruits
Spiraea × bumalda cultivars	Flower	White, pink, carmine
Spiraea japonica cultivars	Flower	White, pink, rose
Symphoricarpos albus	Fruit	Rose-red
Syringa laciniata	Flower/fragrance	Heat tolerant lilac flowers
Syringa vulgaris	Flower/fragrance	Wide range of colors; singles, doubles
Ulmus alata	Stem shape	Corky wings
Viburnum × burkwoodii, V. × carlcephalum, V. carlesii, V. × juddii	Flower/fragrance	Semisnowball, fragrant flowers
Viburnum macrocephalum	Flower	Large 5 to 6 inch diameter white snowball flowers
Viburnum × pragense	Foliage	Lustrous, large dark green leaves

[a]From Dirr (1989).

E. Ornamental Grass Species

A number of annual and perennial ornamental grasses and grains may also be used as cut flowers. Most are used in dried arrangements, but effective fresh flowers may also be found. Many are not difficult to grow if full sun, well-drained soil, and fertilization are provided. An excellent discussion of grain plants has been done by Godwin (1988). Flowers of grasses must be picked as they expand from the foliage. Allowing flowers to fully expand results in seed heads that quickly shatter. Table XIV is adapted from Hockenberry-Meyer (1988).

Table XIV

Ten Popular Ornamental Grasses for Cut Flowers[a]

Genus and species	Common name	Classification
Agrostis nebulosa	Cloud grass	Annual
Briza maxima	Big quaking grass	Annual
Chasmanthium latifolium	Northern sea oats	Perennial
Eragrostis trichodes	Love grass	Perennial
Erianthus ravennae	Plume grass	Perennial
Lagurus ovatus	Hare's tail grass	Annual
Miscanthus sinensis	Eulalia grass	Perennial
Panicum virgatum	Switch grass	Perennial
Polypogon monspeliensis	Rabbit's tail grass	Annual
Phalaris canariensis	Canary grass	Annual

[a]From Hockenberry-Meyer (1988).

IV. COST OF PRODUCTION

Various studies have shown that production of specialty cut flowers can be profitable if done efficiently. Riall et al. (1988) studied the feasibility of cut flower production in Georgia and concluded that specialty flowers were economical in the field but not in the greenhouse. Brumfield (1989) provided detailed guidelines concerning costs of production of field-grown specialty flowers. As with any business endeavor, the ability to make a profit is dependent on keeping costs of production down. Factors that affect cost include location, size, managerial skill, market channel, time of year, and space utilization (Brumfield, 1989). Overall cost may be broken down into variable costs (costs per crop) and overhead (cost per acre). The cost per crop is relatively unchanged regardless of size, except for quantity discounts, but cost per acre is reduced as acreage increases (Brumfield, 1989). *Variable costs* are allocated to individual crops for materials, such as fertilizer, seed, plants, pots, potting soils, and mulch, and labor costs, including any benefits such as employee compensation, social security, paid holidays and sick leave, and any retirement policies. *Overhead costs* are incurred whether crop is grown or not (i.e., machinery and equipment, depreciation, interest, repairs, taxes and insurance, cost of land, electricity and fuel, and management fee).

In Brumfield's examples (1989), an average return of at least $3.00 per bunch of 10 stems on annual crops was necessary to be profitable. This, of course, will change with changes in yield and as variable and overhead costs change. Many growers average less than $2.00 per bunch and are profitable while others require $3.50 per bunch to remain in business. Although yield must be sufficiently high, quality is, without doubt, the most effective means of obtaining a profitable return. The difference in product price for a profitable operation and one with the same crop but that is not profitable is the difference in the quality of that crop. The ability to secure a market niche is also very important to the profitability of any operation.

V. SUMMARY

Specialty cut flowers have a huge potential in the United States market. At the present time, many specialty flowers are imported. There is no doubt that the market for specialty flowers will continue to rise, but the sites of production are still uncertain. American growers are capable of producing as large a variety of flowers equalling or surpassing the quality of imported production. If the American grower is to compete with international growers, competition must be based on the quality of the product and effective marketing. American growers must have pride in their product, and "Grown in the U.S.A." must mean that the flowers are true to name, color, and quality designation. Postproduction treatments must begin in the field, grading must be honest, and packing must be done with the contents in mind and not to see how many stems can be jammed together. "Grown in the U.S.A." must become a recognizable and proven symbol of quality to wholesalers, florists, and consumers or the specialty cut flower market will be turned over to lawyers and brokers, with no interest in quality, and to overseas producers.

ACKNOWLEDGMENTS

Many thanks to Judy Laushman and Joe Seals for their constructive comments.

REFERENCES

Anderson, R. G., and Pemberton, G. (1990). "Gotta get *Godetia.*" *Greenhouse Grower* **8**(13): 72–74.

Armitage, A. M. (1987). "The influence of spacing on field-grown perennial crops." *HortScience.* **22**(5): 904–907.

Armitage, A. M. (1988). "Effects of photoperiod, light source and growth regulators on the growth and flowering of *Trachelium caeruleum.*" *J. Hort. Sci.* **63**(4): 667–674.

Armitage, A. M. (1989a). "The Georgia report: *Caryopteris* on trial." *Gatherings: Newsletter of the Assoc. of Specialty Cut Flower Growers.* **1**(1): 5.

Armitage, A. M. (1989b). "Notes on Eryngium." *Gatherings: Newsletter of the Assoc. of Specialty Cut Flower Growers.* **1**(2): 10.

Armitage, A. M. (1989c). *Herbaceous Perennial Plants.* Athens, GA: Varsity Press Inc.

Armitage, A. M. (1989d). "Some minor bulbs for greenhouse or field cut flowers." *Proceedings of the Second National Conference on Specialty Cut Flowers.* Athens, GA: Univ. of Georgia, pp. 99–101.

Armitage, A. M. (1989e). "Shading affects field grown cut flowers." *HortScience* (in press).

Armitage, A. M., and Laushman, J. M. (1989). "Photoperiodic control of flowering of *Salvia leucantha.*" *J. Amer. Soc. Hort. Sci.* **114**(5): 755–758.

Armitage, A. M., and Laushman, J. M. (1990a). "Planting date, in-ground time affect flowers of *Acidanthera, Anemone, Allium, Brodiaea* and *Crocosmia.*" *HortScience* **25**(10): 1236–1238.

Armitage, A. M., and Laushman, J. M. (1990b). "Planting date, in-ground time affect flowers of *Liatris, Polianthes* and *Iris.*" *HortScience* **25**(10): 1239–1241.

Armitage, A. M., Seager, N. G., Warrington, I. J., and Greer, D. H. (1990). "Response of *Oxypetalum caeruleum* to light, temperature and photoperiod." *J. Amer. Soc. Hort. Sci.* **115**: 910–915.

Brumfield, R. (1989). "Cost of production." *Proceedings of the Second National Conference on Specialty Cut Flowers.* Athens, GA: Univ. of Georgia, pp. 17–26.

Cantino, P. D. (1982). "A monograph of the genus *Physostegia* (Labiatae)." *Contributors of the Gray Herbarium* **211**: 1–105.

Clemens, J., and Welsh, T. E. (1989). "Statice status." *Proc. Second National Conf. on Specialty Cut Flowers.* Athens, GA: Univ. of Georgia, pp. 123–129.

De Hertogh, A. A. (1989a). *Holland Bulb Forcers Guide,* 4th ed. Hillegom, The Netherlands: Int. Flower Bulb Center.

De Hertogh, A. A. (1989b). "Dutch iris, *Liatris*, and lilies as outdoor cut bulb flowers." *Proceedings of the Second National Conference on Specialty Cut Flowers*. Athens, GA: Univ. of Georgia, pp. 93–98.

Dirr, M. A. (1989). "'Woodies' with cut potential." *Proceedings of the Second National Conference on Specialty Cut Flowers*. Athens, GA: Univ. of Georgia, pp. 168–171.

Eisel, M. C. (1988). "Deciduous woody plants for the florist trade." *Proc. of Commercial Field Production of Cut and Dried Flowers*. St. Paul, MN: Univ. of Minnesota, pp. 57–64.

Evans, R. Y., and Reid, M. S. (1990). "Postharvest care of specialty cut flowers." *Proc. Third National Conf. on Specialty Cut Flowers*. Ventura, CA: pp. 26–44.

Geller, Z. (1981). "Horticultural and physiological aspects in growing of *Liatris spicata*." Master thesis. The Hebrew Univ. of Jerusalem, Rehovot.

Godwin, B. J. (1988). "Grains for the florist trade." *Proc. of Commercial Field Production of Cut and Dried Flowers*. St. Paul, MN: Univ. of Minnesota, pp. 77–86.

Halevy, A. H., and Mayak, S. (1981). "Senescence and post harvest physiology of cut flowers. Part I. " In *Horticulture Reviews*, edited by J. Janick. Westport, CT: Avi.

Han, S. S., Halevy, A. H., Sacks, R. M., and Reid, M. S. (1990). "Enhancement of growth and flowering of *Triteleia laxa* by ethylene." *J. Am. Soc. Hort. Sci.* **115**: 482–486.

Healy, W., and Aker, S. (1988). "Production techniques for fresh cut annuals." *Proc. of Commercial Field Production of Cut and Dried Flowers*. St. Paul, MN: Univ. of Minnesota, pp. 139–146.

Hicklenton, P. R. (1987). "Flowering of *Gypsophila paniculata* cv. 'Bristol Fairy' in relation to irradiance." *Acta Hortic.* **205**: 103–111.

Hockenberry-Meyer, M. (1988). "Everlasting ornamental grasses." *Proc. of Commercial Field Production of Cut and Dried Flowers*. St. Paul, MN: Univ. of Minnesota, pp. 69–75.

Kieft, K. (1989). *Kieft's Growing Manual*, 1st ed. Holland: Blokker.

Laushman, J. M. (1989). "Annual cut flowers." *Proc. of Second National Conference on Specialty Cut Flowers*. Athens, GA: Univ. of Georgia, pp. 120–122.

Moe, R., and Berland, M. (1986). "Effect of various corm treatments on flowering of *Liatris spicata* Willd." *Acta Hortic.* **177**: 197–201.

Moe, R. (1988). "Flowering physiology of *Gypsophila*." *Acta Hortic.* **218**: 153–158.

Nowak, J., and Rudnicki, R. M. (1990). *Postharvest Handling and Storage of Cut Flowers, Florist Greens and Potted Plants*. Portland, OR: Timber Press.

Pinnell, M., Armitage, A. M., and Seaborn, D. (1985). "Germination needs of common perennial seed." *Univ. of Georgia Research Bull.* **331**.

Riall, B. W., Castro, J., and Meeker, M. G. (1988). *Feasibility of Cut Flower Production*. Atlanta, GA: Georgia Tech Institute of Technology.

Sacalis, J. R. (1989). *Fresh (Cut) Flowers for Designs. Post Production Guide*. Columbus, OH: Ohio State Univ.

Seals, J. L. (1989). "Extra special cut flowers from seed." *Proc. of Second National Conference on Specialty Cut Flowers*. Athens, GA: Univ. of Georgia, pp. 59–80.

Seals, J. L. (1990). "Who buys cut flowers?" *The Cut Flower Quarterly* **2**(2): 15–16.

Shillo, R., and Halevy, A. H. (1982). "Interaction of photoperiod in flowering control of *Gypsophila paniculata* L." *Scientia Hortic.* **16**: 84–88.

Stimart, D. (1989). "Liatris." *Gatherings: Newsletter of the Assoc. of Specialty Cut Flower Growers.* **1**(2): 5–7.

Stuart, N. W., Asen, S., and Gould, C. J. (1966). "Accelerated flowering of bulbous iris after exposure to ethylene." *HortScience.* **1**: 19–20.

United States Department of Agriculture. (1989). "Floriculture crops, 1988 summary." *SpCr* **6-1**(89).

United States Department of Agriculture. (1990). *USDA Hardiness Zone Map*. Misc. Publ. 1475.

United States Department of Agriculture. (1991). "Floriculture crops, 1990 summary." *SpCr* **6-1**(91).

Vaughan, M. J. (1988). *The Complete Book of Cut Flower Care*. Portland, OR: Timber Press.

Wallerstein, I. (1983). "The flowering of *Echinops blancheanus* Boiss." *Hassedah.* **63**(5): 992–993, 996.

Warren-Auman, C. (1980). "Minor cut flowers." In *Introduction to Floriculture*, (edited by R. A. Larson). New York: Academic Press, Inc. pp. 183–211.

Welsh, T. E., and Baldwin, S. (1989). "Calla lilies: A New Zealand perspective." *Proc. Second National Conference on Specialty Cut Flowers*. Athens, GA: Univ. of Georgia, pp. 81–90.

Wilkins, H. F. (1988). "Basic considerations for the post harvest care of cut flowers in commercial field production of cut and dried flowers." Minneapolis, MN: Univ. of Minnesota.

II
POTTED PLANTS

8

Bulbous and Tuberous Plants

August De Hertogh

Introduction to Floriculture, Second Edition
Copyright © 1992 by Academic Press, Inc.
All rights of reproduction in any form reserved.

195

I. INTRODUCTION

A. Diversity of Bulbous and Tuberous Plants

Bulbous and tuberous plants are normally forced as cut flowers, flowering pot plants, or both. In addition, some can be fully programmed and sold as pot plants for consumer forcing (growing pot plants). The species described in this chapter (Table I) can be utilized for one or more of these marketable products.

Botanically, bulbous and tuberous species are classified as bulbs, corms, tubers, tuberous roots, and rhizomes (Hartmann *et al.,* 1990). In practice, however, they are referred to as "bulbs" or "Geophytes" (Beauchamp, 1989; Doss *et al.,* 1990). The main storage organs of bulbs and corms are: scales and leaf bases, and stem tissue, respectively. Many of the bulbous species described in Table I have a protective tunic, but each differs greatly in rooting and flowering characteristics as well as precise cultural conditions for forcing. The major storage organs of the tuberous crops are either roots or stems. Because of this diversity, only general forcing information will be presented. For specific forcing information and cultivars for North American conditions, consult the Holland Bulb Forcer's Guide (De Hertogh, 1989) and the Ball Red Book (Ball, 1991). For those interested in research aspects of bulb forcing, several references are available (Beauchamp, 1989; Bergman *et al.,* 1971; Bogers and Bergman, 1985; De Hertogh, 1974; Doss *et al.,* 1990; Hartsema, 1961; Rasmussen, 1980; Rees, 1972; Rees and van der Borg, 1975).

B. Basic Growth Habits and Cultural Requirements

Because of the origins of these species (Table I), many of them have a requirement for an annual warm–cool–warm thermoperiodic cycle for growth and development (Hartsema, 1961). Light, moisture, nutrients, plant growth regulators, and ventilation also are important, and their effects will be covered when essential to the species.

To place the handling of the bulbs for forcing into a logical sequence, a four-phase forcing system has been devised (De Hertogh, 1974). The phases are: (1) *production,* all practices that go into producing marketable sized bulbs; (2) *programming,* all handling procedures from bulb harvest until they are placed under greenhouse conditions; (3) *greenhouse,* all growing conditions in this facility; and (4) *marketing,* the development of the plants until the optimal marketing stage has been reached. For cut flowers or flowering pot plants, all four phases are used. However, when the plants are sold as growing pot plants, the greenhouse phase is either not used, or it is markedly shortened. In view of the low energy requirement, these latter products should become very important to the floriculture industry.

Prior to a description of the production and forcing procedures, mention must be made regarding dormancy in bulbs. The species (Table I) have evolved bulbous or tuberous storage organs as a means of survival under adverse climatic condi-

Table I

Description of Bulbous and Tuberous-Rooted Plants Commonly Forced in Commercial Greenhouses[a]

Taxonomic classification	Common name(s)	Origin of species	Storage organ	Flowering habits
Amaryllidaceae				
Alstroemeria spp.	Inca or Peruvian lily	South America	Rhizome	Multiflowered inflorescences
(*Hippeastrum*)(*Amaryllis*)	Amaryllis	South America	Bulb	Multiflowered inflorescences
Narcissus spp.	Daffodils, jonquils, paperwhites	Central Europe to Mediterranean	Bulb	Single- and multiflowered inflorescences
Araceae				
Caladium × *hortulanum*	Caladiums	Tropical America	Tubers	Single-flowered
Zantedeschia spp.	Calla lily	South Africa	Rhizomes, tubers	Single-flowered
Compositae				
Dahlia spp.	Dahlias	Central America	Tuberous roots	Multiflowered shoots
Iridaceae				
Crocus spp.	Crocus	South Europe to southwest Asia	Corm	Large corms produce 2–3 flowers
Freesia spp.	Freesias	South Africa	Corm	Multiflowered inflorescences
Iris × *hollandia*	Dutch iris	Hybrids of species from Spain and North Africa	Bulb	Single-flowered
Iris reticulata	Dwarf iris, specie iris	Caucasus	Bulb	Single-flowered

Liliaceae				
Allium karataviense	Ornamental onion	Turkestan	Bulb	Multiflowered inflorescences
Convallaria majalis	Lily-of-the-valley	Europe	Rhizome	Multiflowered inflorescences
Hyacinthus orientalis	Hyacinths	Greece to Asia Minor	Bulb	Multiflowered inflorescences
Lilium spp.	Lilies	Northern temperate zones	Bulb	Multiflowered inflorescences
Muscari armeniacum	Grape hyacinths	Mediterranean to southwest Asia	Bulb	Multiflowered inflorescences
Tulipa spp.	Tulips	Mediterranean to China	Bulb	Single- and multiflowered inflorescences
Oxalidaceae				
Oxalis spp.	Wood sorrel	South America and South Africa	Bulbs, tuberous roots, rhizomes	Single- and multiflowered inflorescences
Ranunculaceae				
Anemone coronaria	Windflowers	Northern temperate zones	Tuberous roots	Single-flowered
Ranunculus asiaticus	Buttercups	Europe and Asia	Tuberous roots	Single-flowered

[a]Van Scheepen, 1991.

tions. Thus, from a general exterior examination, many species appear to be dormant during certain developmental periods, but, morphologically and physiologically, this is not the case. For example, Figure 1 illustrates the developmental changes observed in the tulip from July (Fig. 1A) to September (Fig. 1F). During this period, no external changes are evident, but significant cell division and morphological changes have occurred.

C. Factors Affecting Flower Development

In bulbs, floral development is primarily regulated by temperature and specific genetic factors (Table II). It can, however, be modified by all the factors given in Table II, with ventilation and ethylene being two other important ones. Therefore, it is imperative that bulbs be handled with care at all times. While losses of 1 to 5% can occur under normal circumstances, improper handling can lead to severe economic losses.

The major factors influencing flower formation and development are (1) bulb size, (2) leaf formation, (3) temperature and light, and (4) relationship of flower

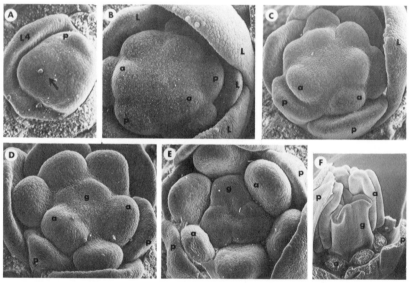

Fig. 1. Scanning electron micrographs of the floral developmental stages of a tulip. Artifacts (*arrow*) on apical surfaces are contaminants from carbon dioxide used in critical point preparation system. (**A**) Stage P_1. First set of perianth (P) and last leaf primordum (L4) are visible. (**B**) Early stage A_1. Leaf primordia (L), two sets of perianth (P), and first set of androecium (A) are visible. (**C**) Late stage A_2. Leaf primordium (L) as well as double sets of perianth (P) and androecium (A) are visible. (**D**) Very early stage G. Double sets of perianth (P) androecium (A) are visible as well as gynoecium (G) being initiated in center of apex. (**E**) Stage G. (**F**) A maturing floral bud. From Shoub and De Hertogh (1975b).

Table II

Factors Influencing Growth and Development of Flower Bulbs

Genetic factors: species, cultivar, bulb size
Temperature integrated over time
Moisture
Nutrients
Light: photoperiod, intensity, or both
Planting medium composition
Plant growth regulators: endogenous and exogenous
Diseases
Insects
Ventilation
Modified atmospheres
Low pressure (hypobaric) storage

formation relative to harvest and specific high, low, or both temperature requirements. These factors have been reviewed by Hartsema (1961), and they will be discussed only briefly.

With the exception of bulbs, like *Narcissus* (daffodils), which are sold on the basis of the number of readily observable noses (double-nose I and II), most bulbs are graded and marketed by their circumference sizes measured in centimeters. Inches are not used. The normal sizes (in centimeters) used for forcing will be cited for each bulb species. By using large sized bulbs, the minimum bulb size requirement is bypassed, and the flower size, the number of flowers produced by the forced bulb, or both is increased.

Closely related to bulb size is the requirement for leaf formation. When a tulip bulb is too small to flower it forms only one leaf. However, when it is a flowering sized bulb, it initially forms the entire complement of leaves, three to five, depending on the cultivar, and subsequently the flower (Fig. 1). In contrast, the number of leaves of the flowering hyacinth can be variable, and it was this characteristic that has permitted the development of the "prepared" hyacinth. Dutch *Iris* must form a minimum of four leaves in order to form a flower. If the bulb forms only three leaves, the plant will not produce a flower. Minimum leaf numbers have not been reported for the other species covered in this chapter.

The most important factor regulating flower initiation and development in bulbous species is temperature, and this factor shall be discussed in detail for each bulb crop. While light is not a controlling factor in tulips, hyacinths, and so on, it does influence the forcing of high quality flowers and pot plants. A minimum of 25 klx is preferred for most bulbs. For Dutch *Iris*, however, light intensity is very important (Fortanier and Zevenbergen, 1973). Flower abortion will occur if sufficient light (>50 klx) is not provided. Light is also critical for lilies (Durieux, 1975) and *Alstroemeria* (Healy and Wilkins, 1985).

A final factor that must be considered is the stage of development of the apical meristem at the time the bulbs are harvested. Some, like tulips, hyacinths, *Crocus, Iris,* and *Muscari,* are vegetative when harvested. Prior to cooling, they must have a full complement of flower parts. Others, like Dutch *Iris* and lilies, do not form a flower until after they have been cooled. The control of this development will be covered in subsequent sections.

II. BULB PRODUCTION PHASE

A. Basic Procedures

For most bulbs, the production phase consists of five steps: (1) harvesting, grading the bulbs into planting stock and marketable bulbs, and preplanting storage; (2) planting, rooting, and low temperature mobilization for flowering, bulbing or both; (3) leaf and flower stalk growth; (4) flowering; and (5) increases in bulb size, numbers, or both. There are two basic systems of reproduction, natural and artificial, that are integrated into this cycle. In addition, some bulbous species have been tissue cultured either for rapid multiplication, to obtain specific pathogen-free clones, or both. Detailed production practices for bulbs can be found in Anonymous (1984), Gould (1957), Langeslag (1989), and Rees (1972). The important procedures related to forcing for the species covered in the chapter (Table I) are summarized in the following sections.

B. General Commercial Bulb Production Practices

Regardless of which country or where the specialized bulb growing location is, the bulb grower is responsible for producing bulbs that are free from serious diseases and insects and that, when handled properly, will produce either marketable potted or growing plants or cut flowers. To do this requires constant attention to planting stock vigor and keeping the bulbs true-to-type. Thus, whether they are grown in the field or in greenhouses, bulbs are routinely inspected and rogued. In many countries, they are given certificates based on the absence of diseases.

As mentioned in Section I,C, bulb growers must produce and subsequently market only flowering-sized bulbs. This is the second most critical production factor. If flowering-sized, pest-free bulbs are produced, the next essential steps are (1) proper time of harvesting and (2) optimal postharvest storage conditions. These latter conditions initiate the programming phase of forcing, and they will be covered in the respective sections of this chapter.

III. TRANSPORTATION FACTORS

Because the majority of bulbs are produced in areas other than the forcing locations, special considerations must be given to transport of the bulbs.

All of the rooting room bulbs in this chapter do best if they are stored under well-ventilated conditions (De Hertogh, 1989). Consequently, they should be shipped and stored in well-ventilated packing materials (e.g., plastic tray cases). Of the nonrooting room bulbs, only Anemones, Dutch *Iris,* Paperwhite *Narcissus,* and *Ranunculus* require ventilated shipping conditions. All the other nonrooting room bulbs (e.g., lilies), require either specialized packing or nonventilated storage conditions. It is also vital that the shipping period should be as short as possible and at the proper temperatures (De Hertogh, 1989). In some cases (e.g., Freesias), air freight is essential. For bulbs shipped by boat from countries such as The Netherlands and Israel, well-ventilated, controlled temperature containers are available. When shipped in large units or when the transport period is in excess of 3 days, automatic temperature recorders should be used to monitor the temperatures en route.

On arrival, forcers should perform three steps (De Hertogh, 1989). First, ventilate those bulbs that require ventilation. If they are in tray cases, this is easily done; however, if cardboard boxes are used for transport, they must be opened. Second, always inspect the bulbs. Check all cultivars for obvious physical damage and for serious diseases (see Section VII). Take a few bulbs of each cultivar and cut or peel them to be certain that all of the organs, and especially the shoot and basal plate, are normal. If tulips are included and they are to be precooled, be certain that the cultivar has reached stage G (Fig. 1E). Consult the Holland Bulb Forcer's Guide (De Hertogh, 1989) for details on this procedure. Third, prior to planting, store the bulbs at the proper temperature.

IV. PROGRAMMING AND GREENHOUSE PHASES FOR ROOTING ROOM BULBS

A. Basic Concepts of Standard Forcing; Cuts and Pots

The objective of rooting room bulb forcing is to produce high quality potted plants or cut flowers. Depending on the species, pot plants should be 6 to 12 inches tall, and cut flowers a minimum of 14 inches and preferably 16 to 20 inches tall. Generally, the greenhouse phase should be less than 30 days for standard forcing (Table III). In addition, the plants should have strong stems and large flowers. Last, the percentage of aborted (blasted) flowers must be less than 5%, with 0% as a goal.

Standard forcing is used for all the rooting room bulbs, and it is widely employed in the United States and Canada. This technique follows the natural bulb production cycles. To utilize this technique, a forcer has to have a controlled temperature rooting room and select those cultivars that are suitable for forcing. For specifics on cultivar utilization, consult the Holland Bulb Forcer's Guide (De Hertogh, 1989) and the Ball Red Book (Ball, 1991).

To avoid difficulties in the forcing of bulbs, forcers should pay attention to the seven planting media factors given in the following list. Specifically, an analysis for

Table III

Concept of Standard Forcing of Rooting-Room Bulb Crops

Forcing phase	Natural seasons	Important forcing processes
Programming	Summer	Harvesting of bulbs; postharvest storage at warm temperatures to control flower development
Programming	Fall	Planting, rooting under cool, moist conditions
Programming	Winter	Low temperature mobilization
Greenhouse	Spring	Leaf growth, flower stalk elongation, and flowering

pH and soluble salt levels is advised. Consult the local county extension office for assistance.

(1) The plant medium must be a *well-drained mix,* and yet it should have a capacity to retain sufficient moisture for growth. It should be as light in weight as possible, but heavy enough to keep the bulbs anchored. The planting medium should not compact readily because bulbs need air as well as moisture. Thus, pure sands, pure peat, and heavy clays are not highly suitable planting media. Loamy soils when mixed with additions such as peat, perlite, vermiculite, or calcined clays make an ideal planting medium for most bulbs (see point 7). Generally, these are mixed in a ratio of one part soil, one part peat, and one part coarse aggregate. When soil is used be certain that it does not contain any residues of harmful pesticides.

(2) The medium must be sterile!

(3) For most bulbs, the planting medium should have a pH between 6.0 and 7.0. Specific pH values are given for each bulb species in the appropriate Sections of the Holland Bulb Forcer's Guide (De Hertogh, 1989).

(4) The soluble salts levels must be low.

(5) At the time of planting, be certain that the medium is moist.

(6) The temperature of the medium should approximate that of the bulbs, somewhere between 50° and 63°F. This means that after sterilization the medium must be cooled before being used.

(7) For some bulbs (e.g., freesias, Dutch *Iris,* and lilies), the media should be fluoride free. Thus, superphosphate and perlite should not be used, and pH should be near 7.

Containers used for bulb forcing must have good drainage. Also, the containers must be clean. The 4 inch pots used for forcing bulbs are normally standard pots. The 5 inch and the 6 inch pots are either standard or $\frac{3}{4}$ azalea pots, while the 8 inch pots are usually $\frac{1}{2}$ size pans. The exact depth depends on the bulb species being forced. They can be made of clay, plastic, or polystyrene. Generally, only new pots are used. When this is done, sanitation is not a problem. If pots have been used previously, they must be cleaned. When plastic pots are used, be certain that all the holes are open. The most suitable ones are those with side drainage holes. When new clay pots are used, it is advisable to either leach them outside or to soak them in water for at least 2 days.

When flats or plastic trays are used for cut flower forcing, be certain to consider the size and the depth of the flat. It has been found that a flat or tray, 14 × 16 × 4 inches, is not only large enough for the bulbs, but also it can be moved with relative ease. This sized container will hold about 50 tulips or 25 large daffodils and will produce approximately four dozen flowers per flat. When flats are assembled, the bottom boards should be separated about 0.25 inch to allow for expansion and to assure adequate drainage. When plastic trays are used, they should not have an excess of drainage holes. If an excess of roots grows out of the tray, this can lead to *Botrytis cinerea* or *Trichoderma* problems (Bergman, 1983).

When large sized bulbs (e.g., tulips, hyacinths, and daffodils) are planted, only the nose of the bulb should be out of the medium. The small-sized bulbs should be planted 0.5 inch deep in the pots.

Because temperature control is paramount, modern forcers use controlled temperature (±2°F) rooting rooms (De Hertogh, 1989). This facility is as essential to flower-bulb forcing as lights and black cloth are to photoperiodic plants like the *Chrysanthemum*. This facility must be capable of permitting not only the rooting of the bulbs but also satisfying the cold-weeks requirement of the bulbs. It should be compartmentalized and capable of exchanging the air every 24 hours. In order to reduce labor costs, many forcers have palletized their rooting room.

The basic operation of the rooting room is to initially place the bulbs at 48°F until the roots of the bulbs protrude out of the holes of planting containers. Then, the temperature is lowered to 41°F until the shoots of the most advanced bulbs are 1 inch out of the bulb. Subsequently, 32°–35°F is used to retard the growth of the shoots until the remainder of the cold requirement is satisfied. As long as the bulbs are in the rooting room, it is essential that they be kept moist at all times. In addition, it is advisable to use preventative fungicidal sprays to control leaf diseases.

In the greenhouse phase, temperatures, watering, light, fertilization, ventilation, sanitation, and pests must be controlled. Many of these factors are interrelated (e.g., watering, sanitation, and ventilation all affect pest control). In the following sections, the greenhouse factors that are important for each bulb type will be provided. Only night temperatures are given. Day temperatures should only be 5°F above night temperatures, when possible.

B. Forcing of Specific Crops

1. Tulips

Standard forced tulips can be utilized either as potted or growing plants or cut flowers (De Hertogh, 1989). The basic differences depend on either the cultivars used or the length of the low temperature treatment. The flowering season extends from late December to early May. Generally, 12 cm and up bulbs are used, but 11 to 12 cm bulbs can be used for some late forcings.

When tulips are harvested, the apical meristem is vegetative, and to force the bulbs requires that they are given a series of warm–cool–warm temperatures (Hartsema, 1961). Flower initiation and organogenesis is controlled by the post-harvest warm temperatures. The precise temperatures used depend on whether the bulbs are forced early, medium, or late in the season.

For early forcing, the bulbs are harvested in mid- to late June and given 1 week at 93°F. The bulbs are then transferred to 63° to 68°F for acceleration of flower initiation and organogenesis (Fig. 1). When the flower reaches stage G (Fig. 1E), in mid- to late August, the bulbs are placed at 44.5° to 48°F for 6 weeks of regular precooling prior to planting. This treatment is given to permit further development of the flower as well as the roots (Shoub and De Hertogh, 1975b). When planting tulips in pots, place the flat side facing the outside. When this is done, the large bottom leaf faces outward. After planting, the bulbs are rooted at 48°F, and then the remainder of the low temperature treatment is given at 41°F. When the cumulative cold treatment equals 15 weeks, bulbs of suitable cultivars can be moved to a 65°F greenhouse. For early forcings, the plants may be stretched in the dark using newspapers, for example, until the lowermost internode becomes visible. Then, the plants must be exposed to light. In the greenhouse, cut and pot tulips should be fertilized with calcium nitrate [$Ca(NO_3)_2$] once a week at 2 pounds per 100 gallons.

For medium forcing (e.g., Valentine's Day), the bulbs are harvested in late June to early July and are placed at 63° to 68°F for flower development. Some cultivars for cut flower usage may have to be precooled at 48°F starting in the first week of September; others can be planted in mid- to late September. For pot plants, only nonprecooled bulbs are needed, and they are planted in early October. The conditions in the rooting room are the same as those described for early forcing. After 16 to 20 cold weeks for cut tulips and 14 to 16 cold weeks for pot tulips, they are placed in a 63°F greenhouse. Again, they should be fertilized, and no dark stretching is required.

For late forcing (e.g., Easter), the bulbs are harvested in July and stored at 73°F to September 1, followed by 68°F to October 1, and then 63°F. Cut tulips are planted in late October or early November and pot tulips in mid-October to mid-November. The exact planting date is dependent on the cultivar, use, and desired flowering date. Rooting room and greenhouse procedures are the same as described for medium forcing.

2. Hyacinths

Hyacinths are primarily used as pot plants (De Hertogh, 1989). However, the flowers can be cut, and, if desired, the individual floret can be used in corsages. In addition, they can be forced on water using special forcing glasses. Bulb sizes from 15 to 16 cm to 18 to 19 cm are used for forcing.

The apical meristem of hyacinths is vegetative when the bulbs are harvested, and to force them requires a series of warm–cool–warm temperatures (Hartsema, 1961). The specific temperature requirements are 10° to 15°F higher than those of the tulip. The flowering season extends from mid-December to April, and to control the development throughout the period, two types of bulbs are available: prepared and regular.

Prepared bulbs, which are used for December and January flowering, are harvested in mid-June and then placed at 86°F for 2 weeks, 78°F for 3 weeks, followed by 73°F until the uppermost floret reaches Stage A_2 (identical to tulip Fig. 1C). The bulbs are then held at 63°F until planted in September. In contrast to tulips and other bulbs, hyacinths are not precooled.

Before planting, hyacinth bulbs should be watered thoroughly. This helps to reduce a dermatitis itch that can occur. After planting, prepared bulbs need only 10 to 12 weeks of rooting and cooling at 48°F before being placed in a 73°F greenhouse. When the florets begin to show color, the temperature should be dropped to 68°F. If desired, some dark stretching can be used for 4 to 5 days immediately after the plants are placed in the greenhouse. There is no data to indicate that hyacinths should be fertilized in the greenhouse.

For medium and late forcings, the bulbs are harvested in late June and early July and stored at 78°F until planted. Planting will be in late September to mid-November depending on the desired flowering date and cultivar. For late forcings (e.g., Easter), it is critical that shoot length should not exceed 4 inches in the rooting room. Therefore, as soon as the plants are rooted at 48°F, the temperature should be lowered to 41°F. When the shoots reach 1 inch, the temperature must be lowered to 32° to 35°F. The minimum cold-week requirement is 13 weeks, and up to 23 cold weeks can be used. These plants can be forced in 59° to 63°F greenhouses and do not require dark stretching. Ethephon (De Hertogh, 1989) can be used to reduce stem topple (Shoub and De Hertogh, 1975a).

3. Daffodils (Narcissus)

Standard forced *Narcissus* are used either as pot plants or cut flowers (De Hertogh, 1989). The season extends from mid-December to April. As with tulips, the major differences are the cultivars and temperatures used in the rooting room and the greenhouse. Generally, double-nosed DN I and DN II bulbs are used for forcing.

At harvest, *Narcissus* have an almost completely formed flower (Hartsema, 1961), the flower being initiated in May shortly after the mother bulb flowers. To force the bulbs, a warm–cool–warm temperature sequence is required.

For early forcing, the bulbs are harvested in July and given 1 week at 93°F. They are then held at 63° to 68°F until precooled in August at 48°F. The handling of cut and pot daffodils after planting in early October is different. Cut daffodils are rooted and cooled continuously at 48°F, which is the optimal temperature for shoot elongation. When a total of 15 to 16 cold weeks has been accumulated, the bulbs are forced in 55° to 59°F greenhouse. If desired, dark stretching can be used. In contrast, pot daffodils are rooted at 48°F, but as soon as they are rooted the temperature is lowered to 41°F. Then, after 15 cold-weeks, they are forced at 60° to 65°F in the greenhouse. These variations in temperature either promote increased length or inhibit growth depending on the product desired. There are no data to indicate that daffodils benefit from fertilization in the greenhouse.

For medium and late forcings, bulbs are harvested in late July or early August and stored at 63° to 68°F until planted. Because time is no longer a major factor, the bulbs are planted in order to provide 17 to 18 cold weeks for cut daffodils and 14 to 16 cold weeks for pot daffodils. They are rooted at 48°F and then cooled at 41°F or 35°F, depending on the growth of the shoots. Shoot growth should not exceed 4 inches in the rooting room. The pot and cut daffodils are forced in the greenhouse at 60° to 63°F or 55° to 59°F, respectively. Ethephon can be used to reduce the flowering height of potted *Narcissus* (De Hertogh, 1989).

4. *Specialty Bulbs* (Allium karataviense, Crocus, Iris reticulata, *and* Muscari armeniacum)

Collectively, these bulbs are referred to as "specialty bulbs" (Langeslag, 1989). Their normal flowering season is from January to mid-March. The exceptions would be *Iris danfordiae,* which can be forced for Christmas, the *Muscari armeniacum,* which can be forced as late as April, and *A. karataviense,* which is forced for late April and May. These plants are used only as pot plants, but the *Muscari* flowers can be cut and used in designs. *Crocus* can also be forced on water using special forcing glasses. Bulb sizes used for forcing are *A. karataviense* 12 and up cm, *Crocus* 10 and up cm., Dwarf *Iris* 6 and up cm, and *Muscari* 9 to 10 cm.

When harvested in June or July, the apical meristem of these bulbs are vegetative. They require a warm–cool–warm sequence.

For early forcing, *Crocus* and *Muscari* are given 1 week at 93°F followed by 63° to 68°F until precooled at 48°F. *Iris reticulata* bulbs are stored at 73°F to July 1, and then at 63° to 68°F until precooled in late August. The bulbs are planted at 48°F in early October. As soon as they are fully rooted, the temperature must be dropped to 41°F and later to 32° to 35°F. They require 15 to 16 cold weeks. It is important to keep the shoots of these species as short as possible. These plants can flower in the rooting room without light if not properly handled. The plants should be forced in a 58° to 60°F greenhouse without fertilization.

For medium forcing, *Crocus* and *Muscari* are stored at 63° to 68°F until planting. *Iris reticulata* bulbs are again stored at 73°F to August 1, and then at 63° to 68°F. After planting, they are rooted at 48°F, and then the temperatures held at 41°F followed by 32° to 35°F; 15 to 16 cold weeks are optimal. For these forcings, a 55° to 59°F greenhouse is preferred.

A. karataviense bulbs are harvested in August and stored at 68° to 73°F until planted. They are subsequently rooted and cooled at 48°F, 41°F, and 32° to 35°F. It requires 21 to 22 cold weeks and is forced at 55° to 60°F (Dosser, 1980). Biweekly applications of 20–20–20 at 200 ppm nitrogen are advised.

V. PROGRAMMING AND GREENHOUSE PHASES FOR NONROOTING ROOM BULBS

A. Basic Concepts for Forcing; Cuts and Pots

In the past few years, the forcing of nonrooting room bulbs as potted plants and cut flowers has markedly increased. The concept for the forcing of these bulbs is outlined in Table IV.

It is important to lift these bulbs on time, use flowering-sized bulbs, and proper planting media as described under the rooting room bulb section of this chapter. Also, optimal greenhouse conditions are needed. For details on the forcing of these bulbs, consult the Holland Bulb Forcer's Guide (De Hertogh, 1989). Only the salient factors will be described for the genera described in this section.

B. Forcing of Specific Crops

1. Alstroemeria

The physiology of flowering of this crop has been reviewed by Healy and Wilkins (1985). Normally, No. 1 sized, rooted rhizomes are planted in late summer to early fall. At present, they are used only as cut flowers and will need support netting. This crop is flowered year-round. Light is critical, and they need a 13 to 16 hour photoperiod and a light intensity of at least 50 klx. Plants grow best at 50° to 60°F. They are normally grown for 2 or more years and require a $Ca(NO_3)_2$ fertilizer program. Thinning of weak and blind shoots is necessary.

2. Amaryllis (Hippeastrum)

The physiology of flowering of this crop has been reviewed by Rees (1985a). In principle, it is one of the easiest crops to force and is primarily used as a pot plant. Bulbs are produced in South Africa for fall and early winter forcings and The Netherlands and Israel for later winter and spring forcings. Bulbs larger than 22 to 24 cm need to be used. They should be forced at 70° to 80°F and require no specific light intensities and no fertilization.

Table IV

Concept of Nonrooting Room Bulb Forcing

Forcing phase	Natural seasons	Important forcing processes
Programming	Summer, fall/winter, or both	Harvesting of bulbs; grading and proper storage for either early, late, or retarded forcings
Greenhouse	Winter, spring, or both	Planting, rooting, leaf growth, and flowering

3. Anemones

Horovitz (1985a) and Ohkawa (1987) have reviewed the growth and flowering of *A. coronaria.* Generally, tubers 3 to 5, 5 to 6, or 6 cm and up are used (De Hertogh, 1989). This crop must be forced at 45° to 50°F; temperatures over 60°F must be avoided. It requires a *very well-drained* planting medium and a good fertilizer program. Anemones do best at a light intensity of 25 klx.

4. Caladiums

Although the flowering of caladiums has been reviewed by Wilkins (1985a), the principle use of the forced plant is its foliage. It comes in a wide range of colors and shapes (Harbaugh and Tjia, 1985). A critical factor to observe is that the tubers must be stored at 70° to 75°F and the plants forced at 75° to 80°F. Caladiums do not tolerate temperatures below 70°F for any period of time. The quality of some cultivars can be improved by de-eyeing (Harbaugh and Tjia, 1985). A light intensity of 30 to 40 klx is preferred.

5. Calla Lilies (Zantedeschia)

Wilkins (1985d) has reviewed the flowering of *Zantedeschia.* There are several species and interspecific hybrids available for forcing either as potted plants or cut flowers. The major producing areas are the United States, New Zealand, Israel, and The Netherlands. Tuber or rhizome size and gibberillic acid (GA_3) preplant dips (Corr and Widmer, 1987) can affect the number of flowers produced, and enhancing the number of flowers is commercially important. Callas are very susceptible to *Erwinia* soft rot. Thus, they should always be carefully inspected on arrival. After planting, they should be kept moist, but not overwatered, and forced at 60° to 65°F. They require only a light fertilizer program.

6. Dahlias

Dahlias are forced only as potted plants (De Hertogh, 1989). Their growth and development has been studied by Barrett and De Hertogh (1978) and reviewed by Runger and Cockshull (1985). It is critical that only specially selected cultivars be forced. Normally, No. 1 tuberous roots are used. They should be forced at a light intensity of 50 klx and at 63° to 65°F. Once they begin to grow, they require frequent waterings and a moderately heavy fertilizer program. To dwarf some cultivars, A-Rest (ancymidol) must be used as a soil drench 10 to 14 days after planting.

7. Dutch Iris

These bulbs are only used as cut flowers (Ball, 1991; Gould, 1957). In the United States and Canada, the flowering season extends from December to May. The principle source of bulbs for forcing in both the United States and Canada is the state of Washington. Also, under normal circumstances the bulb suppliers carry out the programming of Dutch *Iris,* and forcers plant them immediately on arrival.

The physiology of flowering of *Iris* has been reviewed by Rees (1985b). When harvested, the apical meristem is vegetative. To flower, this species requires a sequence of high–cool–warm temperatures (Stuart *et al.,* 1955). 'Wedgwood' and 'Ideal' are the cultivars most commonly used for early flowering. They are given either 10 days at 89.5°F or exposed to 100 to 500 ppm ethylene for 24 hours at 68°F, then held at 89.5°F for 3 days, followed by 63°F for 2 to 4 weeks, and then 6 weeks at 48° to 50°F. After this, they are planted in a 55°F greenhouse. The greenhouse must provide high light intensities, and the planting medium must be kept moist. In the greenhouse, the bulbs should be fertilized with 200 ppm nitrogen of 20–20–20 and $Ca(NO_3)_2$ on a 3 to 4 day alternate basis. This should start about 2 weeks after planting in order to allow the roots to develop.

For later forcings, the bulbs are held (retarded) at 86°F until 8 weeks prior to the desired planting date. At this time, they receive 2 weeks at 63°F followed by 6 weeks at 35° to 48°F. It is not desirable to have sprouts emerge from the bulbs, and if this is observed, the lower temperatures are used. The greenhouse conditions are the same as described above for early forcing.

8. Freesias

The physiology of flowering has been reviewed by Gilbertson-Ferriss (1985). Freesias are used both as cut flowers and potted plants (De Hertogh, 1989). Normally, 5 to 7 cm corms are used. Because of their rapid response to temperature changes and the precise need to control the growth and development of this crop in order to ensure that high quality plants are produced, close coordination must be carried out between the bulb supplier and forcer. Also, for long distance transport, air freight is preferred. The cut flowers will require support and tall potted plants may, too. Freesias are very sensitive to fluoride injury. They should be forced at 50° to 55°F as cut flowers and 55° to 60°F as pot plants and at light intensities over 25 klx. Paclobutrazol (Bonzi) can be used as a preplant corm dip to reduce the height of potted plants (De Hertogh, 1989).

9. Lilies (Asiatic and Oriental Hybrids)

The physiology of flowering in *Lilium* has been reviewed by Rees (1985c). The forcing of Asiatic and Oriental hybrids as cut flowers and potted plants has markedly increased in the past decade. In general, 10 to 12 up to 18 to 20 cm bulbs are used. The exact size is highly cultivar dependent (De Hertogh, 1989). It is very important that for the earliest forcings that Asiatic and Oriental hybrids be precooled at 35°F for a minimum of 6 and 8 weeks, respectively. For later forcings the bulbs must be frozen in at 30°F. Lilies need a light intensity of at least 25 klx to minimize flower abortion and abscission. They should be forced at 55° to 63°F and need a moderate fertilizer program using $Ca(NO_3)_2$ and potassium nitrate (KNO_3). Root-rot diseases can be a serious problem, and preventive fungicidal preplant dips and postplanting soil drenches should be used.

10. Lily of the Valley (Convallaria)

The physiology of flowering has been reviewed by Wilkins (1985b). This is one of the easiest crops to force. Only 3-year-old pips should be used. All pip storage is at 25° to 28°F. Once they have been frozen-in for at least 8 weeks, they can be thawed, planted, and forced. They require a low light intensity greenhouse and should be forced at 63° to 74°F.

11. Oxalis *Species*

The physiology of some species has been reviewed by Wilkins (1985c). In general, the most widely forced species are *O. bowiei, O. deppei, O. hirta, O. martiana, O. purpurea, O. regnellii,* and *O. versicolor.* The forcing requirements for these species are in the Holland Bulb Forcer's Guide (De Hertogh, 1989). Depending on the species, they are used in standard pots, hanging baskets, or both. They force best at 50° to 60°F and at medium light intensities (50 klx). Most of them require a moderate to high fertilization program with 20–20–20.

12. *Paperwhite* Narcissus

The flowering of *Narcissus tazetta* has been studied by Yahel and Sandler (1985). Most of the paperwhite *Narcissus* forced in the United States are produced in Israel. The bulb sizes used range from 13 to 14 cm to 16 cm and up. Flower initiation and subsequent retarding is carried out at 75° to 86°F. After November 1 or whenever shoot growth is observed, they should be stored at 35°F. Regardless of which temperature is used for storage, bulbs for forcing should receive 2 to 4 weeks at 63°F before planting. Paperwhites should be forced at 60° to 63°F and at medium light intensities (50 klx). They do not require fertilization when forced.

13. Ranunculus

The physiology of flowering has been reviewed by Horovitz (1985b). Tuberous roots of 3 to 5, 5 to 6, and 6 cm and up are used for forcing either as cut flowers or potted plants (De Hertogh, 1989). *Ranunculus* require a *very well-drained* planting medium and a good fertilizer program. The crop must be forced at 45° to 50°F. Temperatures over 60°F must be avoided. *Ranunculus* force best at a light intensity of 25 klx or higher.

VI. PLANT GROWTH REGULATORS FOR BULBS

At present, three plant growth regulators are approved for use on certain forced bulbs. They are A-Rest (ancymidol), Bonzi (paclobutrazol), and Florel (ethephon). Ancymidol is used to control the marketable plant heights of potted tulips, potted dahlias, and potted lilies. Ethephon is used to control the marketable plant heights and stem strength of potted hyacinths and potted daffodils. Paclobutrazol is used for potted freesias. The specific concentrations and precautions for using these plant growth regulators are discussed in the Holland Bulb Forcer's Guide (De Hertogh, 1989).

VII. DISEASES, INSECTS, AND PHYSIOLOGICAL DISORDERS

A. General Aspects

There are many diseases, insects and disorders of bulbous and tuberous plants (Bergman, 1978, 1983; Gould and Byther, 1979a,b,c; Lane, 1984; Moore *et al.,* 1979). If proper precautions are taken, very few are encountered in forcing (De Hertogh, 1989). Only the most prevalent ones will be mentioned. Some of these originate in the production fields and carry over to forcing. Others occur as a result of improper programming, improper greenhouse phase practices, or both.

B. Diseases

1. Alstroemeria

In general *Alstroemeria* has few serious disease problems. However, *Botrytis, Pythium,* and *Rhizoctonia* can occur.

2. Amaryllis (Hippeastrum)

The major disease is *Stagonospora curtisii.* It is characterized by red blotches on the bulb scales, floral stalk, or floral bud. If observed on the bulb, dip in mancozeb before planting. If observed while forcing, use a mancozeb spray.

3. Anemones

The major disease of anemones is *Botrytis.*

4. Calla Lily (Zantedeschia)

Outside of viruses, the major diseases that can be encountered are Phytophthora root rot, a Phytophthora leaf blight, and a bacterial (*Erwinia*) soft rot. If fair amounts of *Erwinia* are observed, all rhizomes or tubers of that stock should be discarded.

5. Daffodils and Paperwhite Narcissus

The primary disease of daffodils that affects forcing is Fusarium basal rot. This is a disease that originated in the production fields and continued developing after harvest. Forcers should inspect the basal plate of the bulbs for the presence of soft or dark brown tissue. Any infected bulbs should be discarded. Normally, paperwhite *Narcissus* are not infected by *Fusarium. Botrytis* can occasionally be a problem during forcing and in outdoor growing.

6. Dahlias

The main diseases encountered with tuberous rooted dahlias are viruses. These are normally carried by the tuberous roots, and forcers are advised to *purchase only special-grown stocks* of the cultivars used for forcing. Plants with severe symptoms should be discarded.

7. Dutch Iris

The two principle greenhouse diseases generally encountered are *Fusarium* and *Penicillium.* In both instances, the bulbs should be checked carefully on arrival and then dipped in thiabendazole prior to planting. In addition, a sterilized planting medium should be used.

8. Freesias

The major corm disease is *Fusarium. On arrival,* corms should be checked carefully and, when necessary, dipped in thiabendazole. The major foliar and flower disease is *Botrytis.*

9. Hyacinths

The primary preplanting disease that is encountered is *Penicillium.* The disease is identified by the presence of a blue mold. It usually becomes a problem when the relative humidity is high. This can occur during transport or after long periods of dry storage. Unless the disease attacks the basal plate or the flowers, it is usually not a serious problem. The disease can be controlled either by drying the bulbs under well-ventilated conditions or by dipping them prior to planting in approved fungicides. *Botrytis* can occasionally be a problem during the greenhouse phase.

10. Iris reticulata

Iris reticulata can be infected by *Mystrosporium adustum,* a fungal disease commonly known as "ink spot." It starts in the production fields and can be identified by a black spot on the bulb scale. Bulbs should be inspected on arrival and infected bulbs should be discarded.

11. Lilies

The most important disease affecting lilies is the *root rot complex.* It must be assumed that the fungi are present when the bulbs arrive. Thus, bulbs should be given a preplant bulb dip for 10 to 30 seconds, and then they need to be drenched every 3 to 5 weeks after planting. In addition, a sterilized planting medium should be used.

Botrytis can also affect lilies. There are a large number of fungicides and an exotherm smoke available to control the disease.

12. Lily of the Valley (Convallaria)

Botrytis can be encountered.

13. Ranunculus

The two major diseases that can be encountered are *Rhizoctonia* and viruses.

14. Tulips

a. **Fusarium.** This bulb rot can occur as either a carry-over disease from the production fields, or healthy bulbs can be infected by planting them in infested soils. The latter can be a problem in the forcing of special precooling of tulips. *On arrival,* the bulbs should always be checked. The fungus can be identified in several ways. First, a sour smell is a good indicator of the presence of *Fusarium.* Second, look for a *white mold* growing on the tunic. This is usually a sign that the bulb has *Fusarium.* Do not confuse this with the blue-gray mold of *Penicillium!* If removal of the tunic reveals soft tissue, the bulb should be discarded. Third, check to see if the bulb is soft to the touch. Last, look for very lightweight bulbs. These are bulbs that have been totally consumed by the fungus. In the event that a sample lot of a given cultivar contains more than 10% *Fusarium* infected bulbs on arrival, serious consideration should be given towards discarding the entire lot. Experience has shown that the amount of hidden damage caused by ethylene produced by *Fusarium*-infected bulbs can be quite extensive.

If bulbs are healthy, infections by *Fusarium* can be prevented by using a sterilized planting medium. Also, when tulips are special precooled, always use a preventative fungicide dip and soil drench program.

b. **Penicillium.** A second disease that is commonly observed on tulips is blue mold or *Pencillium.* Generally, this is not a highly serious disease. These bulbs should be either dusted with a suitable fungicide if placed into dry storage for an extended period of time or dipped in a fungicide immediately prior to planting.

c. ***Soil-Borne Diseases.*** There are several soil-borne diseases (e.g., August virus and *Rhizoctonia*), that can attack tulips after planting. Thus, it is essential that the planting medium be sterile. This will eliminate these causal agents.

d. **Botrytis.** In both the rooting room and in the greenhouse, the most prevalent disease is *Botrytis tulipae* (Fire) on the shoots. Forcers should always use a preventative program with *Botrytis tulipae.* There are several effective fungicides that can be used to spray the plants during these stages of development. When large amounts of roots are exposed in the rooting rooms, *Botrytis cinerea* can attack the roots. This is primarily a cut flower forcing problem. The best way to prevent this disease is to use forcing trays with a limited amount of holes and to maintain a high relative humidity.

e. ***Viruses.*** In some instances, the flower petals will show breaking symptoms (e.g. yellow streaks in a red cultivar). When this occurs, the virus was carried by the bulbs, and the forcer has no control over the disease. Instances of large amounts of broken tulips should be reported to the bulb supplier.

f. **Pythium.** It is essential that steps be taken to prevent *Pythium* from becoming a problem. A well-drained sterile planting medium must be used.

C. Insects

The greenhouse insects that can be encountered during forcing are aphids, bulb mites, whiteflies, mealybugs, spider mites, sweet potato whitefly, and thrips (Lane, 1984). Most can be readily controlled by a number of approved insecticides (De Hertogh, 1989).

D. Physiological Disorders

There are only a few serious physiological disorders of bulbs during forcing (De Hertogh, 1989). Most of these can be avoided or minimized by using the procedures outlined in earlier sections of this chapter. The major ones encountered are given in Table V.

VIII. HANDLING OF MARKETABLE PRODUCTS

A. General Aspects

Specifics on handling procedures for the species discussed in this chapter have been described in the Holland Bulb Forcer's Guide (De Hertogh, 1989).

Table V

Major Physiological Disorders of Forced Bulbs

Genera	Type of physiological disorder
Alstroemeria	Flower abortion, leaf scorch
Anemones	Short flowers, yellowing of foliage
Crocus	Flower abortion
Daffodils (*Narcissus*)	Bull-nosing (flower abortion)
Dahlia	No shoot production, sterile clump
Dutch *Iris*	Three leaves (blindness), flower abortion or blasting, stem topple
Dwarf *Iris*	Flower abortion or blasting
Freesias	Flower abortion or blasting, leaf scorch
Hyacinths	Spitting (casting of the bud)
Lilies	Flower abortion or blasting, flower abscission, leaf scorch, stem pull injury
Ranunculus	Short flowers, yellowing of foliage
Tulips	Flower abortion or blasting, flower blindness (antholyse), stem topple
Zantedeschia (calla lilies)	Flower and leaf abnormalities, chalking

B. Growing Pot Plants

When rooting room bulbs are marketed as growing plants they should be marketed directly out of the rooting room. It is essential, however, that the bulbs have been given a proper programming treatment. Before leaving the greenhouse, tulips should be treated with ancymidol (De Hertogh, 1989). Consumers only need to place these plants in a well-lighted area and water daily. It is also possible to market *Amaryllis (Hippeastrum)*, Caladiums, and Paperwhite *Narcissus* as growing pot plants. They can be sold as soon as they are rooted and begin to show root growth.

Table VI

Optimal Stages for Marketing Bulb Plants and Flowers

Genera	Basic usage	Optimal marketing stage
Allium karataviense	Pot	When flower head is expanding out of the sheath
Alstroemeria	Cut	When first flowers are fully colored
Amaryllis (Hippeastrum)	Pot	When floral spike and leaves are 20–30 cm (8–12 inches) long
Anemones	Cut/pot	When flower shows color, but is closed
Caladiums	Pot	When leaves are fully expanded
Calla lily (*Zantedeschia*)	Cut/pot	When flower is fully opened, but before pollen is shed
Crocus	Pot	Sprout
Daffodils (*Narcissus*)	Pot	Pencil stage
	Cut	Goose-neck (see Fig. 2)
Dahlias	Pot	When first flower is fully opened
Dutch *Iris*	Cut	Tight bud
(blue ribbon/prof. blaauw)	Cut	Falls are just opening
Freesias	Pot/cut	When first floret is fully opened
Hyacinths	Pot	Green-bud (see Fig. 3)
Iris reticulata and *I. danfordiae*	Pot	Sprout
Lilies	Pot/cut	When first flower is puffy and showing color
Lily of the valley	Pot	When first bell shows color
(*Convallaria*)	Cut	When terminal bell has lost green color
Muscari (grape hyacinths)	Pot	When florets begin to show some color
Oxalis spp.	Pot	When leaves are fully expanded and flower buds are visible
Paperwhite *Narcissus*	Pot	When floral stalk and leaves are 20–30 cm (8–12 inches) long
Ranunculus	Pot	When flowers begin to color
	Cut	When flower is fully colored, but not open
Tulips	Pot	Green-bud
	Cut	50% colored

C. Flowering Pot Plants

The optimal flower "bud" stages for potted plants are given in Table VI (see also Fig. 3). Some of these plants can be cold stored prior to marketing, but others cannot. Consult the Holland Bulb Forcer's Guide (De Hertogh, 1989) for details.

D. Cut Flowers

The optimal flower "bud" stages for cut bulb flowers is presented in Table VI (see also Fig. 2). Some of these flowers can be stored for limited periods. Consult the Holland Bulb Forcer's Guide (De Hertogh, 1989) for details.

Fig. 2.   The gooseneck stage of flower development of the daffodil. This is the proper stage to cut this flower.

Fig. 3.   The green-bud stage of flower development for marketing potted hyacinths.

REFERENCES

Anonymous. (1984). "Bulb and Corm Production." London: Her Majesty's Stationery Office.

Ball, V., ed. (1991). *The Ball Red Book,* 15th ed. Reston, Virginia: Prentice-Hall Co.

Barrett, J. E., and De Hertogh, A. A. (1978). Growth and development of forced tuberous-rooted Dahlias. *J. Am. Soc. Hortic. Sci.* **103:** 772–775.

Beauchamp, R. M., ed. (1989). "Proceedings—International symposium on bulbous and cormous plants." *Herbertia* **45(1/2):** 179.

Bergman, B. H. H. (Chairman). (1978). *Ziekten en Afwijkingen bij Bolgewassen. Deel II: Amaryllidaceae, Araceae, Begoniaceae, Compositae, Iridaceae, Oxalidaceae, Ranunculaceae.* Lisse, The Netherlands: Laboratorium voor Bloembollenonderzoek.

Bergman, B. H. H. (Chairman). (1983). *Ziekten en Afwijkingen bif Bolgewassen. Deel I: Liliaceae.* Lisse, The Netherlands: Laboratorium voor Bloembollenonderzoek.

Bergman, B. H. H., Eijkman, A. J., van Slogteren, D. H. M., and Timmer, M. J. G., eds. (1971). "First international symposium on flowerbulbs." *Acta Hortic.* **23:** 1–440.

Bogers, R. J., and Bergman, B. H. H., eds. (1985). "Fourth international symposium on flowerbulbs." *Acta Hortic.* **177:** 1–692.

Corr, B. E., and Widmer, R. E. (1987). "Gibberellic acid increases flower number in *Zantedeschia elliottiana* and *Z. rehmannii.*" *HortScience.* **22:** 605–607.

De Hertogh, A. A. (1974). "Principles for forcing tulips, hyacinths, daffodils, Easter lilies and irises." *Scientia Hortic.* **2:** 313–355.

De Hertogh, A. A. (1989). *Holland Bulb Forcer's Guide.* Hillegom, The Netherlands: International Flower Bulb Centre.

Doss, R. P., Byther, R. S., and Chastagner, G. S., ed. (1990). "Fifth international symposium on flower bulbs." *Acta Horticulturae* **260:** 582.

Dosser, A. (1980). "*Allium karataviense*—A Lonely Little Onion in a Petunia Patch." *N.C. Flower Growers Bull.* **24:** 11–12.

Durieux, A. J. B. (1975). "Additional lighting of lilies (cv. 'Enchantment') in winter to prevent flower bud abscission." *Acta Hortic.* **47:** 237–240.

Fortanier, E. J., and Zevenbergen, A. (1973). "Analysis of the effects of temperature and light after planting on bud blasting in *Iris hollandica.*" *Neth. J. Agric. Sci.* **21;** 145–162.

Gilbertson-Ferriss, T. L. (1985). "*Freesia* × *Hybrida.*" In *Handbook of Flowering.* edited by A. Halevy, vol. III. Boca Raton, Florida: CRC Press, p. 34–37.

Gould, C. J., ed. (1957). *Handbook on Bulb Growing and Forcing.* Mt. Vernon, Washington: Northwest Bulb Growers Assoc.

Gould, C. J., and Byther, R. S. (1979a). "Diseases of Bulbous Iris." *Washington State Univ. Ext. Bull.* **710.**

Gould, C. J., and Byther, R. S. (1979b). "Diseases of Narcissus." *Washington State Univ. Ext. Bull.* **709.**

Gould, C. J., and Byther, R. S. (1979c). "Diseases of Tulips." *Washington State Univ. Ext. Bull.* **711.**

Harbaugh, B. K., and Tjia, B. O. (1985). "Commercial forcing of Caladiums." *Fl. Coop. Ext. Ser. Cir.* **621.**

Hartmann, H. T., Kester, D. E., and Davies, F. T., Jr. (1990). *Plant Propagation, Principles and Practices,* 5th ed. Englewood Cliffs, New Jersey: Prentice-Hall.

Hartsema, A. M. (1961). "Influence of temperatures on flower formation and flowering of bulbous and tuberous plants." In *Handbuch der Pflanzenphysiologie,* edited by W. Ruhland, vol. XVI. Berlin, and New York: Springer-Verlag, pp. 123–167.

Healy, W. E., and Wilkins, H. F. (1985). "*Alstroemeria.*" In *Handbook of Flowering,* edited by A. Halevy, vol. I. Boca Raton, Florida: CRC Press, pp. 419–424.

Horovitz, A. (1985a). "*Anemone coronaria* and related species." In *Handbook of Flowering,* edited by A. Halevy, vol. I. Boca Raton, Florida: CRC Press, pp. 455–464.

Horovitz, A. (1985b). "*Ranunculus.*" In *Handbook of Flowering.* edited by A. Halevy, vol. IV. Boca Raton, Florida: CRC Press, pp. 155–161.

Langeslag, J. J. J. (Chairman).(1989). *Teelt en Gebruiksmogelijkheden Van Bijgoedgewassen.* Lisse, The Netherlands: Tweede Druk. Ministerie Landbouw Visserij en Consulentschap Algemene Dienst Bloembollenteelt.

Lane, A. (1984). *Bulb Pests.* London: Her Majesty's Stationery Office.

Moore, W. C., Brunt, A. A., Price, D., Rees, A. R., and Dickens, J. S. W. (1979). "*Diseases of Bulbs.*" London: Her Majesty's Stationery Office.

Ohkawa, K. (1987). "Growth and flowering of *Anemone coronaria* L. 'de Caen.'" *Acta Hortic.* **205:** 159–167.

Rasmussen, E. (1980). "Third international symposium on flowerbulbs." *Acta Hortic.* **109:** 1–533.

Rees, A. R. (1972). *The Growth of Bulbs.* New York: Academic Press.

Rees, A. R. (1985a). "*Hippeastrum.*" In *Handbook of Flowering,* edited by A. Halevy, vol. I. Boca Raton, Florida: CRC Press, pp. 294–296.

Rees, A. R. (1985b). "*Iris.*" In *Handbook of Flowering.* edited by A. Halevy, vol. I. Boca Raton, Florida: CRC Press, pp. 282–287.

Rees, A. R. (1985c). "*Lilium.*" In *Handbook of Flowering,* edited by A. Halevy, vol. I. Boca Raton, Florida: CRC Press, pp. 288–293.

Rees, A. R., and van der Borg, H. H., eds. (1975). "Second international symposium on flowerbulbs." *Acta Hortic.* **47:** 1–446.

Runger, W., and Cockshull, K. E. (1985). "*Dahlia.*" In *Handbook of Flowering.* edited by A. Halevy, vol. II. Boca Raton, Florida: CRC Press, pp. 414–418.

Shoub, J., and De Hertogh, A. A. (1975a). "Floral stalk topple: A disorder of *Hyacinthus orientalis* L. and its control." *HortScience.* **10:** 26–28.

Shoub, J., and De Hertogh, A. A. (1975b). "Growth and development of the shoot, roots, and central bulblet of *Tulipa gesneriana* L. cv. 'Paul Richter' during standard forcing." *J. Am. Soc. Hortic. Sci.* **100:** 32–37.

Stuart, N. W., Gould, C. J., and Gill, D. L. (1955). "Effect of temperature and other storage conditions on forcing behavior of Easter lilies, bulbous iris and tulips." *Rep. 14th Int. Hortic. Cong.* **1:** 173–187.

Van Scheepen, J., ed. (1991). "International checklist for Hyacinths and miscellaneous bulbs." Royal General Bulb Growers Association (KAVB), Hillegom, The Netherlands.

Wilkins, H. F. (1985a). "*Caladium × hortulanum.*" In *Handbook of Flowering,* edited by A. Halevy, vol. II. Boca Raton, Florida: CRC Press, pp. 101–104.

Wilkins, H. F. (1985b). "*Convallaria majalis.*" In *Handbook of Flowering,* edited by A. Halevy, vol. II. Boca Raton, Florida: CRC Press, pp. 321–323.

Wilkins, H. F. (1985c). "*Oxalis.*" In *Handbook of Flowering,* edited by A. Halevy, vol. III. Boca Raton, Florida: CRC Press, pp. 442–444.

Wilkins, H. F. (1985d). "*Zantedeschia.*" In *Handbook of Flowering,* edited by A. Halevy, vol. IV. Boca Raton, Florida: CRC Press, pp. 523–524.

Yahel, H., and Sandler, D. (1985). "Retarding the flowering of *Narcissus tazetta* cv. 'Ziva.'" *Acta Hortic.* **177:** 189–195.

9

Azaleas

Roy A. Larson

I. INTRODUCTION

A. Economic Development

Commercial production of azaleas for forcing in the United States was about equal to poinsettias in 1950, and surpassed potted chrysanthemums and Easter lily production (Table I). By 1950, flowering potted chrysanthemums were available throughout the year, and 3.5 million more pot mums were sold than azaleas. Poinsettia and azalea sales remained about equal until 1970. Longer-lasting poinsettia cultivars, better growth control methods, and mass-market retailing increased consumer demand for poinsettias, while azalea production remained static. Azalea production was not even tabulated for 15 years because statistics would have revealed individual earnings in many states, but in 1986, azalea statistics were inserted in the census once again.

There are only about half as many establishments forcing azaleas now as there were 20 years ago, but the quantity sold has increased by approximately 4% (Table II).

Several factors probably were responsible for the decline in popularity of forcing azaleas among commercial growers. Disease problems caused by fungi and nematodes resulted in serious crop losses until better control measures were achieved. The production time for an azalea crop, from propagation to flowering, could be as long as 3 years, so the price required to realize a profit is a handicap. The cost has delayed the entry of azaleas into the mass-market outlets, where low retail prices are widely advertised. The need to break flower bud dormancy before flowering can occur is an obstacle and expense not encountered with many other flowering potted plants, adding another disadvantage to azalea production.

These handicaps have not yet been satisfactorily resolved, but the beauty of a well-grown azalea plant in full flower does attract customers (Fig. 1). There is a good range of flower forms, sizes, and colors. In some regions, the florist azalea can also be used as landscape material, increasing its versatility. Azalea forcers

Table I

Comparison of United States Production of Four Floriculture Potted Crops from 1950 to 1989 [a]

	Quantity sold ($\times$ 1,000)						
Crop	1950	1959	1970	1986	1987	1988	1989
Azaleas	3,410	6,128	9,750	7,085	8,619	8,584	10,134
Chrysanthemums	—	9,750	21,542	25,480	33,064	32,133	32,302
Easter lilies	2,825	4,119	5,359	6,921	8,097	7,633	8,296
Poinsettias	3,707	6,819	8,951	36,674	44,633	41,154	46,245

[a]Data obtained from Fossum (1973) and the Bureau of the Census and the Agricultural Statistics Board, USDA (Anonymous, 1990).

Table II

Statistics on National Production of Potted Finished Florist Azaleas[a]

Year	Number of establishments	Quantity sold (× 1,000)	Wholesale value (× $1,000)
1950	1,928	3,410	4,315
1959	1,735	6,128	8,253
1970	1,745	9,750	16,770
1986	694	7,085	31,411
1987	987	8,619	37,515
1988	969	8,584	37, 542
1989	944	10,134	44,126

[a]Data compiled by Bureau of Census and the Agricultural Statistics Board, USDA (Anonymous, 1989).

Fig. 1. A well-grown azalea plant in full flower.

have reduced production time and space by letting azalea specialists in warmer areas do the propagation and early culture, while the forcers only complete the final phase of the crop. One can now purchase azalea plants in all stages of development from small liners to budded-up plants ready for forcing.

A promising sign for the future of forcing azaleas was the announcement in 1989 that azaleas had surpassed kalanchoes as the top flowering potted plant crop sold through the auction houses in the Netherlands. Almost $20,000,000 worth of flowering azaleas were sold in 1988, a 15% increase over 1987. The commercial production of flowering azaleas in 2 inch and 4 inch pots also is increasing. These smaller units are available at price levels not possible with older, larger plants and place azaleas into new market outlets. They also can be used as "color spots" in dish gardens sold in retail florist shops.

Florida, Alabama, and Oregon are the leading states in the production of azaleas. The climate in the southeast is conducive to rapid vegetative growth, which enables producers to get plants up to forcing size more quickly than would occur in most other regions. Structures that can be covered with polyethylene film in the winter are often used (Fig. 2). Oregon perhaps has supplanted California as the chief area for production of liners on the West Coast. High land values have prompted several prominent California firms to sell their property.

B. Characteristics of Azaleas for Forcing

The azalea most commonly used for forcing is *Rhododendron obtusum.* There are five other azalea subseries in the genus (Leiser, 1975). Rhododendrons are in the Ericaceae (heath) family, which is composed of 70 genera and approximately 1900 species. Linnaeus named the genus *Azalea* in 1753, but it was reclassified in 1834. Nine species of *R. obtusum* have usually been used by plant breeders in the

Fig. 2. Outdoor production in the summer, but the structure can be covered with polyethylene film for winter protection.

development of new cultivars. The genealogy of azaleas is somewhat confusing, however, and the parentage of some cultivars is not always known.

Flower types, as well as color, make the cultivars very diverse. Plants can have single, single hose-in-hose, semidouble, semidouble hose-in-hose, double, and double hose-in-hose flowers (Lee, 1958). Some cultivars do have fragrant blossoms, but fragrance is not a major criterion in the evaluation of cultivars. Foliage color, texture, and shape also can be quite variable.

There are several qualities that should be possessed by a cultivar if it is to be acceptable in commercial floriculture. Traits considered by a grower to be important might not be appreciated by the consumer, such as rate of growth, but the traits desired by azalea specialists are (1) ease of propagation; (2) rapid growth; (3) relative tolerance of insects and disease organisms; (4) uniform initiation and development of flower buds; and (5) attractive, long-lasting flowers. In some regions the ability of a cultivar to withstand outdoor conditions could increase its versatility, but that characteristic is not considered to be as important as those listed above.

Cultivars with red flowers are most popular now, but they do not dominate the market as they once did. One prominent azalea propagator reported that color percentages for their sales are 40% red, 50% pink or variegated, and 10% white for much of the year, but 80% red for Christmas and Valentine's Day. Red has declined in popularity because the color would clash with the colors advocated by interior decorators. Growers have to be aware of such trends, even though they cannot readjust their color assortments as quickly as decorators can change their color schemes.

A rather comprehensive list of cultivars was prepared by Stadtherr (1975) but several cultivars prominent now should be added to the list. It would be difficult to select accurately the most popular cultivars grown because cultivar popularity differs within the United States. For example, the cultivar 'Nancy Marie' is popular on the West Coast but is seldom seen in the eastern United States. European customers have a wider assortment of azalea cultivars available to them, and some top cultivars there have not yet been widely distributed in North America. An attempt has been made to acquaint the reader with some prominent United States cultivars, listed in Table III. Their years of introduction to the industry are also shown, when such information is known.

Many of the cultivars are sports of established cultivars, such as the white-flowered 'Dogwood' cultivar mutating to 'Coral Dogwood' (1975), 'Variegated Dogwood' (1979), and 'Nancy Marie' (1985). 'Gloria,' a very popular cultivar in the United States, was a sport of 'Dorothy Gish' in 1969. The cultivar 'Hellmut Vogel' has given rise to at least 17 sports, most of which are grown in Europe.

II. PROPAGATION

Most azalea plants that are used for forcing are now propagated by specialists who are located in areas of the country where climate is favorable for efficient and

Table III

**Some Popular Cultivars Used for
Forcing in the United States in 1990**[a]

Cultivar	Year
Red flowered	
Chimes	1934
Hellmut Vogel	1967
Hershey's Red	
Hexe	1878
Knut Erwen	1934
Mission Bells	1967
Nordlicht	1981
Prize	1968
Red Wing	1952
Roadrunner	
Salmon-to-Orange flowered	
Dorothy Gish	1935
Marianne Haerig	1977
Pink flowered	
Coral Bells	
Pink Dream	
Solitaire	
Variegated flowered	
California Sunset	1944
Gloria	1969
Inga	1933
Leopold Astrid	1933
Nancy Marie	1985
Nicolette Keessen	
Variegated Dogwood	1979
White flowered	
Alaska	1935
Dogwood	1972
Marcel	
Paloma (actually very light pink)	
Snow	About 1920
White Gish	

[a]List compiled after personal communication with William Aulenbach, Yoder Brothers, Barberton, OH, and Freddie Blackwell, Blackwell Nurseries, Semmes, AL.

economical production of stock plants and liners. Florida, Alabama, Oregon, and California exemplify such areas.

Terminal cuttings are removed from stock plants that have been frequently monitored and treated to assure pest-free material, as most of the diseases can attack azaleas at any stage of growth. Most cuttings are removed from stock plants in June or July, though cutting removal is not confined to these months. Shoots that are immature and succulent do not root as readily as those that are firm and almost brittle. Terminal cuttings are 3 to 4 inches long, and only leaves that would interfere with placement in the propagation medium are removed. A rooting hormone generally is used to promote or hasten rooting, as an excessive duration in the propagation benches increases exposure to disease problems. Leaching of nutrients from the foliage also will be excessive if cuttings are kept under mist too long. Mist propagation is most often used.

Cuttings are "stuck" in beds or flats filled primarily with acid peat moss. Other ingredients such as perlite can be mixed with the peat moss to improve aeration and drainage, in a ratio of 4:1 (peat moss:perlite) on a volume basis. The propagation medium should be free of pests, such as disease organisms (primarily fungi and nematodes), insects and insect eggs, and weed seed.

Six to eight weeks often are required for rooting. The rooted cuttings should then be transplanted, as efficient production is not realized if transplanting is delayed. Propagators use different types of small containers or flats, but peat moss is the most popular potting medium. Care must be exercised when rooted cuttings are planted in peat moss in plastic pots, flats, or trays, as the medium can stay too wet. Root systems will not be as vigorous in an excessively wet medium, and some of the most devastating disease organisms affecting azaleas are favored by wet conditions.

Azaleas also have been propagated through tissue culture. One commercial nursery in Washington annually sells 400,000 micropropagated azaleas (Edick-Caton, 1990). More research seems to have been done on other rhododendron species, however (Anderson, 1978; Blazich and Acedo, 1988; Chee, 1985; Meyer, 1982; McCulloch, 1984). Advantages of tissue culture over more traditional methods of vegetative propagation are increased production of new cultivars or established cultivars that are difficult to root, maintenance of pathogen-free material, year-round propagation, and reduction of stock plant costs (McCulloch, 1984). Low average survival rates and lack of uniformity are two disadvantages (Anderson, 1978). Micropropagation from flower buds has been advocated (Meyer, 1982) because flower buds are easier to disinfect than are hairy, fairly large shoot tips.

Budding and grafting have been used in propagating azaleas, but high costs of such laborious procedures have lessened their use.

III. PLANT CULTURE

It often takes at least 2 years to go from the propagation phase to anthesis. Most of that time is used to increase plant size, going from the rooted cutting to a plant with multiple shoots that is large enough to bring premium prices.

A. Vegetative Development

Progressive growers attempt to provide optimum conditions for vegetative growth so plants will grow vigorously and reach marketing size as quickly as possible. Those optimum conditions include temperature, daylength, light intensity, potting medium, water, and nutrients.

1. Temperature

A major reason for the prominence of the southeastern region of the United States in azalea production is warm temperatures throughout much of the year. Vegetative growth will be vigorous at night temperatures of 65° to 70°F with day temperatures 10° to 15°F higher (Larson and Biamonte, 1972). Plants grown outdoors on the West Coast will not grow as rapidly because of the cooler night temperatures that usually prevail.

Media and water temperatures should also be monitored occasionally, to make certain that growth is not retarded because of adverse conditions. Water and nutrient absorption and translocation will occur more readily if the medium and water temperatures are at least 60° to 65°F, than if the temperatures are lower. Chlorosis of the foliage can be evident, even with an adequate fertilizer program, if the water is too cool. Nutrient deficiencies can be corrected if a hot water heater is used to temper the water, without any increase in nutrient concentration in the fertilizer program. Some disease problems also could be intensified when cool media or water temperatures inhibit development of the root system, and the medium remains wetter than is desired.

2. Light

Light intensity and duration are known to affect vegetative growth, while the effects of light quality are not as well known. High leaf temperatures usually accompany high light intensities, so the relationship between temperature and light intensity cannot be ignored. Water loss through transpiration can be excessive at high light levels, and foliar necrosis or abscission can occur. Lateral shoots can be shorter and firmer under high light conditions than if the plants are placed under reduced light intensities. Light intensities of 2500 to 4000 foot-candles are considered optimum during the vegetative stage. Rooted cuttings just removed from mist propagation require a greater reduction in light intensity than do well-established plants. Removal of shoot tips to increase lateral branching (pinching) will expose the upper canopy of leaves to high light intensity levels, and "sunburn" can occur on such tender foliage. Syringing or shading to reduce leaf temperatures would be helpful. Reduction of light intensity outdoors usually is achieved with saran cloth or a similar shading material. The same material also can be placed over a greenhouse structure if the plants are grown inside or a greenhouse shading compound could be applied. A light intensity reduction of 50% is often the goal of azalea producers.

Azaleas once were considered to be unaffected by daylength (day neutral), but studies conducted over 50 years ago (Skinner, 1939) indicated that azalea plants had a tendency to remain vegetative under long days but would initiate flower buds under short days. These responses have been used to advantage in year-round flowering programs (Larson, 1975). Long days (16 to 18 hours or interrupted nights) could be used following the pinch to stimulate vegetative growth.

Light quality, affected by shading material or crowding, can also affect shoot growth and quality, but research is lacking on this light factor.

3. Water

Acid peat moss is almost universally used in azalea production. Acidic pH, good water and nutrient holding capacity, and high organic content are characteristics of peat moss, and these are prime requisites for vigorous growth of azaleas. Irrigation practices must be closely monitored when peat moss is used, however, as overwatering can cause the medium to remain too wet, subjecting root systems to increased chances for infection by water molds, such as *Phytophthora cinnamomi*. A delay in the application of water can cause the peat moss to dry out to such an extent that rewetting is difficult. Root damage will usually be followed by leaf burn and defoliation.

Air temperature, light intensity, relative humidity, and plant and container sizes will affect the frequency and amount of water that must be applied. Plants that have just been pinched should not require as much water as the plants needed just prior to pinching, as considerable leaf area is removed and transpiration is decreased.

Water quality is gaining increased national attention but azalea growers have been aware of water quality problems for many years. In some areas, expensive water treatment facilities have been installed to make acceptable water available. The expense has actually caused some growers to move elsewhere. Root growth of azaleas can be hampered because of a poor source of water, and minor element deficiencies seem to occur quickly under such conditions.

Water can be applied in several different ways. Overhead watering with sprinklers is often practiced, but azalea plants are subject to some foliar diseases, and almost all of them are more damaging when foliage is wet. Splashing water droplets can disperse the disease organisms. Report on the use of capillary mats and ebb and flow watering systems to irrigate azaleas are lacking.

Growers and researchers have evaluated different frequencies on time of watering. Keever and Cobb (1985) found that the application of water 2 to 4 hours prior to the hottest part of the day was beneficial as a cooling mechanism. Split applications at 10 A.M. and 3 P.M. produced similar results, while intermittent irrigation during the day had little effect.

Excessively cool water temperatures (perhaps less than 50°F) can cause poor water and nutrient absorption and translocation. Nutrient deficiencies will occur, though fertilizer rates could be adequate. The use of tempered water can correct the problem.

Relative humidity is a factor that cannot always be readily controlled in field production, but the vigorous, fast growth of azaleas in southern Florida and the Gulf region of Alabama might be at least be partially attributed to high relative humidities. Syringing of the foliage will raise the humidity momentarily, but the increased likelihood of increased foliar disease problems with syringing has already been mentioned.

4. Nutrition

Mastalerz (1977) and Nelson (1985) have discussed nutrition of floricultural crops, and only some specific azalea nutrient requirements will be presented here. Use of a suitable growing medium can simplify fertilization programs. Elements, such as iron and manganese, are available in acidic media, but deficiency symptoms quickly appear as the pH rises. Application of the appropriate minor elements can lessen the chlorosis, but the cause is not corrected until a lower pH is once again attained. The optimum range of pH for azaleas is 4.5 to 5.5.

Twigg and Link (1951) and Oertli (1964a,b,c) have described nutrient deficiency symptoms on azaleas (Table IV). Critical elemental levels, as reported by Mastalerz (1977), are shown in Table V. An excellent summarization of azalea nutrition was written by Kofranek and Lunt (1975).

At North Carolina State University, we have tried to keep the florists' azalea fertilization program simple, and we seldom encounter difficulties with it. We use an acid peat moss medium with a pH of 5.2 to 5.5. We use a 21–7–7 (neutral) soluble fertilizer at 20 ounces per 100 gallons of water. Fertilizer is applied weekly

Table IV

Some Nutrient Deficiency Symptoms on Azaleas[a]

Nutrient	Deficiency Symptoms	Reference
Nitrogen	Older leaves turn yellow, chlorosis is uniform over entire leaf	Oertli (1964a)
Phosphorus	Reddish-purple blotches in the middle area of the leaf; blotches eventually turn brown; leaf drop occurs first at bottom of shoot	Oertli (1964a); Twigg and Link (1951)
Potassium	Interveinal chlorosis of young leaves; lesions eventually develop, particularly near leaf tips	Twigg and Link (1951)
Calcium	Cessation of growth; young leaves are very small; leaf tips become burned or distorted	Oertli (1964b)
Magnesium	Older leaves become chlorotic, beginning at the tips of the leaves; leaf drop can be severe	Oertli (1964b); Twigg and Link (1951)
Iron	Interveinal chlorosis in young leaves; midrib and side veins remain green while remainder of leaf can be almost white	Oertli (1964b); Twigg and Link (1951)
Copper	Browning of shoot tip; chlorosis, dwarfing and cupping of young leaves; shortened internodes	Twigg and Link (1951)
Boron	Cessation of growth; distorted new growth; death of shoot tips	Twigg and Link (1951); Oertli (1964c)
Sulfur	Chlorosis of young foliage; small area at leaf tip can remain green	Oertli (1964b)

[a]Twigg and Link (1951) used the cultivar 'Coral Bells,' and Oertli (1964 a,b,c) used the cultivar 'Sweetheart Supreme.'

Table V

Critical Leaf Analysis Values for Azaleas[a]

Element	Deficiency range[b]	Normal range[b]	Excess range[b]
Nitrogen (N)	<1.8 %	2.0–3.0 %	3.0 or greater
Phosphorus (P)	<0.20 %	0.29–0.50 %	0.65 or greater
Potassium (K)	<0.75 %	0.80–1.60 %	
Calcium (Ca)	<0.20 %	0.22–1.60 %	
Magnesium (Mg)	<0.16 %	0.17–0.50 %	
Manganese (Mn)	<30 ppm	30–300 ppm	400 ppm or greater
Iron (Fe)	<50 ppm	50–150 ppm	
Copper (Cu)	< 5 ppm	6–15 ppm	
Boron (B)	<16 ppm	17–100 ppm	200 ppm or greater
Aluminum (Al)			Very tolerant
Zinc (Zn)	<15 ppm	5–60 ppm	
Sodium (Na)		4500 ppm	1500 ppm or greater

[a]From Mastalerz (1977).
[b]Percent of dry weight.

in the summer, but only every 2 weeks in cooler, darker times of the year. Urea formaldehyde (such as Milorganite or Borden's 38) is applied as a top-dressing after the roots are well established, at a rate of one teaspoon per 6 ½ inch pot. A second application will be made about 3 months later if the crop is still in the vegetative stage.

Nutrient requirements are reduced immediately after pinching, as water frequency is reduced, but once new growth begins, the appropriate fertilization program should be resumed.

5. Pinching

Azaleas are pinched to increase shoot numbers, plant size, floriferousness, and also as a mechanism for timing flowering. The first mentioned reasons will be discussed in this section on vegetative development, while the use of pinching to schedule flowering will be considered in the section on flowering.

The final size of azalea plants will be largely determined by the number of times plants are pinched, if growing conditions are satisfactory. In many places, azaleas are only pinched once each year, but the plants could be pinched every 3 to 4 months if faster increases in size were desired. This can only be done under protected conditions or in climates where low temperatures are not encountered. The expenses encountered in indoor culture must be considered, but new vegetative growth could always be occurring under the proper environmental conditions. A night temperature of 65°F and long days will enhance vegetative growth. Fertilization programs would have to be more precise than under conditions where plants are only pinched once annually. Carbon dioxide injection has also been suggested for maximum growth.

Pinching can be done manually or chemically, but most plants are pinched with pruning shears or electric clippers. Some propagators use the pinch as a way to get cuttings so the plants serve dual roles as stock plants and eventually as flowering plants. If such a practice is followed then the pinch involves the removal of shoots about 3 to 4 inches long. If cutting production is not an objective of pinching, then only the tips of the shoots need to be removed. More leaf axils then remain, so one might expect more lateral shoots than when a harder pinch is made.

There are different chemicals that have been used to pinch azaleas. The fact that azaleas are multibranched plants makes chemical pinching worthwhile. Fields of azaleas that might require weeks to be pinched can be chemically pinched in hours, so labor costs are significantly reduced. The crop will be more uniform in development as well, as all plants are pinched at the same time.

Off-Shoot-O was the first chemical pinching agent of economic importance (Stuart, 1967, 1975) but its use has declined. Effectiveness of Off-Shoot-O is influenced by temperature, relative humidity, stage of apex development, and cultivar. The chemical works by physically damaging the apex, and the material has to come in contact with the apex for pinching to occur. One can tell within about 24 hours if shoot tip damage has occurred.

Dikegulac (Atrimmec) was the second prominent chemical pinching agent. Its mode of action is biochemical, so the chemical does not have to come in direct contact with the apex. The material is translocated through the phloem, and DNA synthesis is affected (Bocion et al., 1975; de Silva et al., 1976). It is not affected as much by the factors that influence the effectiveness of Off-Shoot-O (Larson, 1978). The effectiveness of Atrimmec cannot be determined until at least 2 weeks after its application. Lateral shoot initiation and development are delayed compared to those on plants that are manually pinched, and new leaves are often very narrow. Some azalea growers do not use Atrimmec alone, but prune the large, long shoots to get the desired plant shape, break apical dominance, and then apply Atrimmec 2 days later to stimulate lateral branching.

Other new chemicals are being tried, but EPA label clearance is lacking at this time.

B. Flower Initiation and Early Development

Flower initiation can be promoted by temperature, daylength, and chemicals (Criley, 1975, 1985; Pettersen, 1972). Under natural conditions, initiation will occur in late summer, when temperatures are warm and the days are getting shorter (nights are getting longer). Flowers buds continue to develop until they reach the stage of dormancy, which is overcome by cold temperature.

Flower initiation can be stimulated by the use of photoperiod control or growth regulators. This control is needed in the year-round production of flowering azaleas (Anonymous, 1967). Immediately following the final or "timed" pinch, plants should be under natural or simulated long days, to promote vegetative growth. When lateral shoots are sufficiently long (about 6 weeks after the pinch) plants can be either placed under a 9- or 10-hour daylength, achieved by pulling black cloth or other opaque material over the plants, or treating the plants with a growth regulator (Stuart, 1964, 1965). Two of the most frequently used chemicals are B-Nine SP or Cycocel. Both are applied as foliar sprays. B-Nine SP can be applied once at a rate of 2500 ppm or twice at 1500 pm with the applications 1 week apart. Cycocel is applied at a concentration of 2500 ppm.

Heursel (1985) has investigated the effectiveness of Bonzi (paclobutrazol) in promoting flower bud initiation (Heursel, 1985). A concentration as low as 25 ppm has promoted flower bud initiation. Bonzi is quickly effective so it has been suggested that this chemical be applied 1 week later than when B-Nine SP or Cycocel are used.

Plants treated with growth regulators will be somewhat smaller than untreated plants because stem elongation is affected. The pinching chemical, Atrimmec, also exhibits growth retardation, so shoot length could be excessively affected when chemicals are used to regulate both lateral shoot development and flower initiation.

The optimum temperature for flower initiation is about 65°F. The influence of daylength has already been discussed. Light intensity also can affect flower initiation. Azalea plants generally will initiate flower buds more readily and in greater number under fairly high light intensities as long as temperatures are not excessively high. Multiple flower buds will be produced more often under high light intensities (Criley, 1985). Plants that are too crowded often will have flower buds confined to the uppermost, exposed canopy of shoot apexes, while few flower buds will be formed on the heavily shaded shoots. Criley (1985) advocated some shading during the later stages of development, to attain more uniform forcing later. It also was suspected that high light intensity might intensify bud dormancy.

C. Flower Bud Dormancy

Flower buds will develop to a certain stage but not proceed any further until plants have been exposed to cool temperatures to break dormancy. The flower buds become dormant so flowering does not occur in late fall or winter, when freezing temperatures would kill the flowers. Azaleas grown outdoors receive the cold temperature treatment by being subjected to winter conditions, and bloom in the spring when dormancy has been broken and warm temperatures occur. Growers wishing to force azaleas out-of-season must provide cool temperatures with a controlled procedure. Growers in the cooler regions of the country can expose plants to cool temperatures until late fall or early winter. Protection from frost is required in the holding house, which is usually shaded so leaf temperatures will not be high while the root systems are subjected to cool temperatures. Plants can be moved to warmer houses for forcing, or the temperature raised in the same greenhouse where the cold storage environment was provided.

In warm climates, or at certain times of the year in cooler climates, dormancy can only be broken if the plants are placed in refrigerated storage. 'Kurume' azalea cultivars only require 4 weeks at 35° to 50°F, while most other cultivars require 6 weeks for dormancy to be broken. Lights will not be needed at 35° to 40°F, but about 12 hours of light daily will be required at temperatures ranging from 41° to 50°F. Leaf drop will occur if lights are not provided at the warmer temperatures. At one time light intensity was considered to be unimportant in the cooler, and very low light intensities were provided. Research has shown, however, that plants placed in coolers with light intensities of 125 foot-candles for 12 hours daily will flower faster and be of better quality than those exposed to lower light intensities. Plants precooled at 45° to 50°F also will force faster than those kept at 35° to 40°F. Not all cultivars are as responsive, however.

Plants in the warmer, lighted coolers must be watered regularly, or leaf abscission will occur. A relative humidity of approximately 90% is suggested. Air exchanges in the cooler are necessary. One exchange every 4 hours has been suggested (Anonymous, 1988) and can be accomplished with an exhaust fan and a time clock.

One survey conducted among azalea forcers revealed that very few growers provided optimum cool storage conditions (Anonymous, 1988).

Gibberellic acid (GA) has been found to be a satisfactory substitute for cool temperature to break dormancy (Boodley and Mastalerz, 1959; Larson, 1983; Larson and Thorne, 1985; Martin et al., 1960). Unfortunately, GA does not yet have EPA clearance for this purpose. Several applications of GA_3 and GA_4 and $_7$ at 1000 ppm can be as effective as 4 to 6 weeks in the cooler, and the plants do not have to be moved from the greenhouse to the cooler and back to the greenhouse. Flowers often are larger on GA treated plants. Caution must be exercised so that GA is not applied too often or at excessive concentrations, or flower quality will be impaired. Flower buds must be at a later stage of development when GA is used, as the dormancy-breaking activity begins very quickly, while the response is slower when cool temperatures are used.

A combination of 3 weeks of cool temperature and three applications of GA at 250 to 400 ppm can also be effective. Labor still is required to move plants, a cool storage facility is necessary, but twice as many plants can be moved through the cool storage than when only cool temperatures are used.

Plants treated with GA should be in bloom about 5 to 6 weeks after the initial treatment. Plants placed at cool temperatures will require 4 to 6 weeks in the cooler, and another 4 to 6 weeks for forcing, so the production schedule can be reduced by about 1 month when GA is used.

D. Flowering

When plants are first moved from cool temperatures, to be forced at 60° to 65°F night temperatures in the greenhouse, they usually require shading or syringing to avoid high leaf temperatures and possible sun scald. The temperature of the potting medium will remain cool for several hours, while leaf and air temperatures could be quite high.

Light intensity between 2500 and 4000 foot-candles will be sufficient during the forcing period. Flower buds will not open uniformly at lower light intensities, while leaf and petal burn could occur at light intensities much higher than 4000 foot-candles, particularly if the medium dries out. Water should be applied to medium rather than the foliage, as diseases such as *Cylindrocladium,* powdery mildew, and *Botrytis* can be accentuated with wet foliage. Ebb and flow watering has not been practiced with azaleas. The delicate root systems and susceptibility to root-infecting pathogens such as *Phytophthora cinnamomi,* make growers hesitant to use subirrigation.

Crowding of plants during forcing is not a good practice and could nullify months or even years of optimum cultural conditions. Flowers on shoots in the upper canopy will develop more quickly than those on the crowded, lower shoots. Disease problems also can be worse under crowded conditions.

Plants that were pinched well in advance of the dormancy-breaking treatment can have an excessive number of bypass shoots that surround the flower buds (Fig. 3). The flower buds might even abort if bypassing is excessive. Manual removal is one way those shoots can be removed once they have been produced, but that chore is tedious, costly, and can be damaging if care is not exercised and the flower buds are accidentally removed. Perhaps the best way to control bypassing is to break dormancy as soon as the buds are sufficiently developed, so

Fig. 3. Vegetative bypass shoots surrounding a flower bud.

the bypass shoots will not be initiated. Efforts have been made to eliminate or control bypassing with chemicals (Whealy *et al.*, 1988). Paclobutrazol (Bonzi) was most effective. It did not delay flowering or adversely affect flower diameter. The treatments were applied 7 weeks before the plants were placed in cool storage, so the chemical did not prevent initiation of bypass shoots but did inhibit bypass shoot elongation. Bypass shoots were less than $\frac{1}{2}$ inch long on treated 'Gloria' plants, while bypass shoots on untreated plants averaged 3.5 inches in length. Whealy *et al.*, (1988), used Bonzi at 200 ppm, but some growers have been satisfied with the results obtained with 100 ppm.

Temperature, cultivar, and dormancy-breaking treatment can affect the time required for forcing. Some cultivars will be salable in 3 to 5 weeks at 60° to 65°F, while others might require 6 weeks or more. Forcing time is usually regulated by temperature. Higher temperatures will hasten flowering, while lower temperatures will slow it down. Quality will not be impaired if cool temperatures are needed to delay flowering, but night temperatures exceeding 70°F could cause poor pigment development, "soft" flowers, and reduced plant quality.

IV. CONTROL OF PESTS

Azaleas are subject to some insect and disease problems that can make successful production very difficult. The duration of an azalea crop does prolong exposure to attack by pests. A portion of the production cycle might occur outside, and also requires exposure to cool temperatures to break flower bud dormancy. Such conditions can increase the vulnerability of azaleas.

A. Insects, Mites, and Nematodes

There are some insect and insect-like pests that can be very damaging to azaleas. Azalea lace bug (*Stephanitis pyrioides*) will cause upper leaf surfaces to become mottled, while the underside of the foliage will turn bronze in color. This pest is particularly troublesome in field production, particularly in dry weather. Leaf miner (*Gracillaria azaleella*) will cause the leaf tip to roll back and turn brown (Fig. 4). Severe infestations will give an azalea crop a brownish hue. Eggs and hatching larvae are protected from insecticides because they are encased in the leaf tips. Elimination or at least control of the adults, or elimination of the larvae as they leave the protection of the rolled leaf tips, will require several applications of an effective insecticide.

Spider mites can cause conspicuous damage, both in the field and greenhouse. The most commonly found spider mite in the field is the southern red mite (*Oligonychus ilicis*), while the two-spotted mite (*Tetranychus urticae*) is most common on azaleas in the greenhouse (Streu, 1975). Mottling and bronzing of leaves occur because mites feed on the underside of the foliage. The signs of mite infestation

Fig. 4. Azalea leaf miner damages. Note curled leaf tip, which shields the eggs and turns brown. (Photograph courtesy of James R. Baker.)

(webs, dust, debris) are much more noticeable than the minute mites, which usually can only be identified with the help of at least a hand lens. Mites were among the first pests to show resistance to pesticides, and that trait has continued to be a barrier to control. Different types of miticides must be used so resistance is less likely to occur. Dry conditions will intensify mite populations and damage, so syringing can be practiced as one method for at least partial control. Such a practice can increase the incidence of foliar diseases, however, so it should be used with caution.

Other pests reported in azalea references are whiteflies, aphids, and thrips, but they are not as troublesome as those already discussed. Azalea roots can be damaged and plants killed because of root knot nematode (*Meloidogyne incognita*). The stylet nematode (*Tylenchorhynchus claytoni*) and stunt nematode (*Trichodorus christiei*) also can affect azaleas. Nematocides and medium pasteurization have been used to reduce the likelihood of nematodes, and the problem is not as severe as it was years ago.

B. Diseases

There are several very serious azalea diseases (Coyier and Roane, 1986; Strider *et al.*, 1985). Some disease organisms cause azalea plants to die rather quickly, while at least one major disease results in a rather slow decline of the plants. Azaleas are subject to disease throughout the duration of the crop (Aycock and Daughtry, 1975).

It would be difficult to rate diseases in order of importance, but one of the most prevalent is Phytophthora root rot (also called azalea decline or littleleaf disease) cause by *Phytophthora cinnamomi.* The first name does indicate that root systems are attacked and eventually destroyed, and its other common names are indicative of the often rather slow progress of the disease (Fig. 5). The disease symptoms will be hastened under stress conditions, such as can occur in the greenhouse or during the cold storage treatment. The source of the inoculum has often been traced to the irrigation water or the propagation or potting media. Benson (1986) studied the relationship of media, temperature, moisture, and light intensity. Plants grown in the shade had greatest disease severity, as the medium remained moist and cool, favoring the pathogen.

There are some fungicides that can be used periodically to lessen the severity of the disease, but the best control is sanitation throughout the life of the crop (Hoitink and Schmitthenner, 1972).

Cylindrocladium blight and wilt, caused by *Cylindrocladium scoparium,* progresses much more quickly than does Phytophthora root rot. It can affect plants at any stage of development. Though referred to as a blight or wilt, it could also be recognized as a stem canker, a leaf spot, or a root rot. Such a diversity in symptoms requires culturing and identification by experts. Spores can be spread by air or water and can gain entrance through leaves or roots. Overhead irrigation and syringing of foliage to reduce mite infestations are common practices but can cause spread of *Cylindrocladium* as spores are splashed from infected to healthy plants.

Fig. 5.   Plant infected by *Phytophthora cinnamomi.* (Photograph courtesy of R. K. Jones.)

Infection can occur early on stock plants or in the propagation stage but may not be evident until plants are under some form of stress. This delayed reaction has often led growers to believe that plants were healthy, only to learn differently at a later date, such as during forcing.

Foliar sprays or media drenches with appropriate fungicides can help control the disease. The examination of cuttings for leaf spots, media pasteurization, sanitary cultural practices, and the avoidance of splashing water also are effective control measures.

Botrytis is a pathogen that infects many floricultural crops, but it is not a major problem for greenhouse forcing azaleas. It occasionally does occur in poorly ventilated storage units during the breaking of flower bud dormancy or when azaleas are shipped to markets in cardboard boxes. Powdery mildew can also be troublesome at high humidities and with susceptible cultivars. Once established, it can be difficult to control, even with frequent applications of fungicides.

Some disease organisms are not serious on greenhouse forcing azaleas when they are in the production phase but can be troublesome when the plants are placed in the landscape. Leaf and flower gall (*Exobasidium*) and Ovulinia petal blight are examples.

Effective fungicides are available for control of most azalea diseases, but label clearance varies too much from state to state to recommend specific chemicals. Growers in many states have access to insect and plant disease clinics, and such experts should be consulted for the solution of pest problems.

V. PHYSIOLOGICAL DISORDERS

There are several problems that can be encountered in azalea production, but no pest can be detected or blamed for its occurrence.

A. Failure to Flower

Failure of azaleas to flower uniformly, if at all, is a very serious problem, but one not usually encountered by an alert, experienced grower. Late pinching, improper temperatures, and excessively low light intensity during the flower initiation and early developmental stages can be reasons. Vegetative shoots surrounding the flower buds frequently result when too much time elapses from time of pinching to initiation of the flower dormancy-breaking treatments as mentioned previously by (Whealy et al., 1988; Fig. 3).

B. Defoliation

Defoliation can be a serious problem, and there are several reasons why evergreen azalea plants could lose their leaves. Almost any type of stress could be responsible, such as lack of water, fertilizer injury, disease, and ethylene pollution. Leaf abscission caused by ethylene is most likely to occur in a cold storage unit that might have some ethylene-producing material in it, such as apples or leafy vegetables. Inadequate lighting (intensity or duration, or both) in a cooler when the temperature is over 40°F will almost always cause leaf abscission. Low relative humidity, lack of water in the medium, and excessive air movement can also cause the foliage to abscise. Leaf burn will precede abscission in some instances, such as excessive fertilizer or diseases, but green leaves can abscise if the causes are ethylene, low light, or lack of moisture.

C. Sparse or Irregular Flowering

Adequate time from the final pinch to the dormancy-breaking treatments must be allowed to enable shoot apexes to complete the reactions that are necessary for uniform flowering to occur. Vegetative growth follows the removal of the shoot apex, and flower initiation and development then occur. Flower bud initiation will be delayed, and further develoment will be inhibited, if plants are pinched too late in the season. Uneven flowering can also occur if shoot apexes on a plant are of different ages. Shoots that were present but too small to pinch will often form flower buds sooner than lateral shoots that evolved from the pinch. Those flower buds could also force faster after the cold storage treatment. Excessive crowding of plants during the vegetative and early reproductive stages could also result in the flowers in the upper part of the canopy opening sooner than those on the sides of the plant, where light was reduced.

D. Herbicide Injury

Many azalea liners are produced outdoors, and the use of herbicides, either by the grower or a neighbor, could cause difficulties. Twisted leaf petioles, leaf burn or spotting, and many other symptoms could reveal contamination by an herbicide. Herbicide use in the greenhouse is really restricted, and only chemicals labeled for greenhouse use should be applied. Frequently encountered disorders have been listed and described by Hasek and Kofranek (1975).

VI. MARKETING OF PLANTS

Greenhouse-forced azaleas are often referred to as "florist azaleas" and retail florist shops still account for the sale of most of the plants, particularly those in 6 inch pots or larger containers. Mass-market sales of such items are increasing, however. The long production cycle and other costs generally require a pricing structure that has limited its distribution through mass-market outlets. The production of smaller plants with a reduced production duration has improved the opportunities for marketing in larger volume at lower prices. There is a demand for azaleas grown in 4 inch pots (Fig. 6), and an innovation has been the production and sale of azaleas in 2 or 2½ inch pots.

Other handicaps in azalea marketing have been the lack of standards of quality, marked variability from one cultivar to another with regard to growing habit and vigor, and regional differences in desired plant shape.

A. Grading

Growers generally buy liners based on diameter, such as 8 × 10 or 10 × 12 inches. It is assumed that there will be more shoots on a plant with a large diameter than on a smaller one, and shoot number eventually will be reflected in flower number. Some propagators have tried to correlate plant diameter and shoot number in their grading systems, while other propagators will indicate how many pinches the plants have received.

Florists and others selling flowering azalea plants are faced with the same dilemma. There is a need for a generally accepted rating system, and such an attempt has been made by the joint efforts of the Produce Marketing Association and the Society of American Florists (Anonymous, 1990). Their grading system is shown in Table VI. Plant size, number of flowers and stage of development, and plant appearance are the important characteristics, while container size has a minor role.

B. Packing and Handling

Azalea liners once were dug in the field, placed in boxes, and shipped to the azalea forcer. Now almost all azaleas are grown in containers, causing less of a

Fig. 6.   Sales of azaleas in 4 inch pots are increasing.

Table VI

Azalea Grades Established by the Produce Marketing Association and SAF[a]

Grade	Product dimension[b] (inches)	Number of flower buds	Stage of openness	Characteristics	Care tags
Small	Height, 8–10 Top of plant width, 6 Pots, $4-4\frac{1}{2}$	8–10	Minimum 50% of flowers in candle bud stage	Dark green foliage, strong stems, healthy root system, soil must be damp	Yes
Large	Height, 12–16 Top of plant width, 9 Pots, $6-6\frac{1}{2}$	15–20	Minimum 50% of flowers in candle bud stage	Dark green foliage, strong stems, healthy root system, soil must be damp	Yes
Extra large	Height, 16–20 Top of plant width, 13 Pots, 7–8	30–35	Minimum of 50% of flowers in candle bud stage	Dark green foliage, strong stems, healthy root system, soil must be damp	Yes

[a]From Anonymous (1990).
[b]Top of plant width dimensions given are "minimum."

set-back in handling and shipping. Efforts are usually made to decrease the heat of the plants prior to boxing and shipping. Propagators are reluctant to ship plants when the potting medium is wet, as shipping costs are increased, but plants in dry media could be desiccated by the time they arrive at the greenhouse. Compromises have been made, though shipping damage caused by drying has not been eliminated completely.

Forcers should unpack the boxes immediately on arrival, apply water if necessary, and place the plants in a protected environment (Anonymous, 1988). Even more attention must be given to plants that are going from the forcing area to the retail outlet, as open flowers and well-developed flower buds must be considered. Improper handling can cause damage to the flowers, negating months of careful production. Usually only 25 to 30% of the flowers should be fully open at the time of marketing, to increase the longevity of the beauty of the plants once they have been purchased. Plants should be unpacked or removed from sleeves as quickly as possible, and the medium examined for moisture content.

C. Retailing

Most retail outlets have limited storage for potted plants, and all might have to be exposed to room temperatures. It would be better if some azaleas could be held at a cooler temperature to delay increased flower opening, or the plants might be overly mature when they are finally sold.

Azalea cultivars do differ in the lasting ability of the flowers. Nell *et al.,* have listed some of the cultivars, based on flower longevity (Table VII). Not all cultivars have been evaluated but more information should continue to be released. Black et al., (1990) also have shown that plants that have been subjected to 6 weeks of cold storage to break dormancy had greater flower longevity than when GA was used.

Proper display of plants in the retail outlet can influence sales. Red-flowered varieties often do not show to their best advantage in an area where only fluorescent lights are used. More balanced lighting can show the customer what the plants will look like under other lighting conditions. All plants should have care tags attached to them, as most customers will expect the plants to last at least 2 to 3 weeks, but they might not know how to achieve that goal. Purchasers should surely be made aware of the importance of watering the plants properly, as the prominence of acid peat moss as the potting medium makes rewetting of a dry medium very difficult. Inadequate water was considered by Solit and Solit ( ) to be the most frequent mistake made by azalea purchasers.

Table VII

Azalea Cultivars That Are Considered to Be "Long Lasting" as Flowering Potted Plants[a]

Cultivars
Red flowered
Ambrosia
Hershey's Red
Mission Bells
Prize
Roadrunner
Salmon-to-orange flowered
Dorothy Gish
Pink flowered
Coral Bells
Solitaire
Variegated or bicolor
California Sunset
Gloria
Leopold Astrid
White flowered
Alaska
Dogwood
White Gish
Whitewater

[a]From Nell et al., (1986).

The potential use of the forced azalea as a landscape plant when its use as an attractive house plant is over should not be ignored in areas of the country where the climate is sufficiently mild to justify such a practice. Time of year, cultivar hardiness, and method of outdoor planting will have major impacts on its success or failure as a landscape plant.

REFERENCES

Anderson, W. C. (1978). "Rooting of tissue cultured rhododendrons." *Proc. Int. Plant Propagators' Soc.* **28:** 135–139.

Anonymous. (1967). *Schedule and Growing Procedure for Year Around Azalea Production.* West Chicago, IL: Geo. J. Ball, Inc.

Anonymous. (1988). *Yoder Dormant Azaleas. Recommendations for Successful Cooling and Forcing.* Barberton, OH: Yoder Brother, Inc.

Anonymous. (1990). *Recommended Grades and Standards for Potted Plants.* Alexandria, Va.: Produce Mktg. Assoc. (PMA) and Society of American Florists (SAF).

Aycock, R., and Daughtry, B. (1975). "Major diseases." In *Growing Azaleas Commercially,* edited by A. M. Kofranek and R. A. Larson, Sale Pub. No. 4058, pp. 78–88. Univ. of California, Berkeley.

Benson, M. (1986). "Relationship of soil temperature and moisture to development of Phytophthora root rot of azalea." *J. Environ. Hortic.* **4**(4): 112–115.

Black, L. A., Nell, T. A., and Barrett, J. E. (1990). "Dormancy-breaking method effects on azalea longevity." *HortScience.* **25**(7): 810.

Blazich, F. A., and Acedo, J. R. (1988). "Micropropagation of 'Flame' azalea." *J. Environ. Hortic.* **6**(2): 45–47.

Bocion, P. F., Huppi, G. A., deSilva, W. H., and Szkrybalo, W. (1975). "A group of new chemicals with plant regulatory activity." *Nature (London)* **258:** 142–144.

Boodley, J. W., and Mastalerz, J. W. (1959). "The use of gibberellic acid to force azaleas without a cold temperature treatment." *Proc. Am. Soc. Hortic. Sci.* **74:** 681–685.

Chee, R. (1985). "The history of rhododendron micropropagation provides some tips for success." *Am. Nurseryman.* **161**(10): 42–47.

Coyier, D. L., and Roane, M. K., eds. (1986). *Compendium of Rhododendron and Azalea Diseases.* St. Paul, MN: APS Press.

Criley, R. A. (1975). "Effects of light and temperature on flower initiation and development." In *Growing Azaleas Commercially,* edited by A. M. Kofranek and R. A. Larson, Sales Publ. No. 4058, pp. 52–61. Univ. of California, Berkeley.

Criley, R. A. (1985). "Rhododendrons and azaleas." In *CRC Handbook of Flowering,* edited by A. H Halevy. Boca Raton, FL: CRC Press, pp. 180–197.

deSilva, W. H., Bocion, P. F., and Walther, H. R. (1976). "Chemical pinching of azaleas with dikegulac." *HortScience.* **11**(6): 569–570.

Edick-Caton, L. (1990). Personal communication. Briggs Nursery, Inc. Olympia, WA.

Fossum, M. T. (1973). *Trends in Commercial Floriculture Crop Production and Distribution: A Statistical Compendium for the United States 1945–1970.* Alexandria, VA: Marketing Facts for Floriculture (SAFE).

Hasek, R. F., and Kofranek, A. M. (1975). "Problems of evergreen azaleas." In *Growing Azaleas Commercially,* edited by A. M. Kofranek and R. A. Larson, Sale Publ. No. 4058, pp. 97–99. Univ. of California, Berkeley.

Heursel, J. (1985). "Bonzi-ein neuer wachstums regulator fur immergrune azaleun." *Deutscher Gartenbau.* **39**(37): 1742, 1744–1746.

Heursel, J. (1986). "Bonzi makes the culture of azaleas more certain." *Gartenbau and Gartenbau-wissenschaft.* **86**(33): 1216–1218.

Hoitink, H. A., and Schmitthenner, A. F. (1972). "Control of Phytophthora root rot (wilt) of Rhododendron." *Am. Hortic.* **51:** 42–45.

Keever, G. J., and Cobb, G. S. (1985). "Irrigation scheduling effects on container media and canopy temperatures and growth of 'Hershey Red' azalea." *HortScience.* **20**(5): 921–923.

Kofranek, A. M., and Lunt, O. R. (1975). "Mineral nutrition." In *Growing Azaleas Commercially* edited by A. M. Kofranek and R. A. Larson, Sale Publ. No. 4058, pp. 36–46. Univ. of California, Berkeley.

Larson, R. A. (1975). "Continuous production of flowering azaleas." in *Growing Azaleas Commercially,* edited by A. M. Kofranek and R. A. Larson, Sale Publ. No. 4058, pp. 72–77. Univ. of California, Berkeley.

Larson, R. A. (1978). "Stimulation of lateral branching of azaleas with dikegulac sodium (Atrinal)." *J. Hortic. Sci.* **53**(1): 57–62.

Larson, R. A. (1983). "Enhancement of azalea production with plant growth regulators." *Sci. Hortic.* **34:**108–109.

Larson, R. A., and Biamonte, R. L. (1972). "Response of azaleas to precisely controlled temperatures." *J. Am. Soc. Hortic. Sci.* **97**(4); 491–493.

Larson, R. A., and Thorne, C. B. (1985). "Environmental versus chemical manipulation of azalea flowering." *N.C. Flower Growers' Bull.* **29**(2): 1–9.

Lee, F. P. (1958). "The Azalea Book." Princeton, New Jersey. Van Nostrand.

Leiser, A. T. (1975). "Taxonomy and origin of azaleas used for forcing." In *Growing Azaleas Commercially,* edited by A. M. Kofranek and R. A. Larson, Sale Publ. No. 4058, pp. 9–14. Univ. of California, Berkeley.

McCulloch, S. (1984). "Micropropagation of rhododendrons." *Am. Rhododendron Soc.* **38**(2): 72–73.

Martin, L. W., Wiggans, S. C., and Payne, R. N. (1960). "The use of gibberellic acid to break flower bud dormancy in azaleas." *Proc. Am. Soc. Hortic. Sci.* **76:** 590–593.

Mastalerz, J. W. (1977). *The Greenhouse Environment. The Effect of Environmental Factors on Flower Crops.* New York: Wiley.

Meyer, M. M., Jr. (1982). "*In vitro* propagation of *Rhododendron catawbiense* from flower buds." *HortScience.* **17**(6): 891–892.

Nell, T. A., and Barrett, J. E. (1986). Chrysanthemum and azalea variaties for the interiorscape. Univ. of Florida, Staff Report Series, *ORH* **86:** 1.

Nelson, P. V. (1985). *Greenhouse Operation and Management,* 3rd ed. Reston, VA: Reston Publ. Co., Inc.

Oertli, J. J. (1964a). "Azalea nutrition disorders. 1. Nitrogen, phosphorus and potassium deficiencies." *Florists' Rev.* **134**(3482): 20, 62.

Oertli, J. J. (1964b). "Azalea nutrition disorders. 2. Calcium, magnesium and sulphur deficiencies." *Florists' Rev.* **134**(3483): 21, 62.

Oertli, J. J. (1964c). "Azalea nutrition disorders. 3. Chlorosis, tipburn, result of iron and boron deficiencies." *Florists' Rev.* **134**(3484): 31, 80.

Pettersen, H. (1972). "The effect of temperature and daylength on shoot growth and bud formation in azaleas." *J. Am. Soc. Hortic. Sci.* **97:** 17–24.

Skinner, H. T. (1939). "Factors affecting shoot growth and flower bud initiation in rhododendrons and azaleas." *Proc. Am. Soc. Hortic. Sci.* **37:** 1007–1011.

Solit, K., and Solit, J. (1987). "Keep Your Gift Plants Thriving." Pownal, Vermont. Storey Communications, Inc.

Stadtherr, R. J. (1975). "Commercial cultivars." In *Growing Azaleas Commercially,* edited by A. M. Kofranek and R. A. Larson. Sales Publ. No. 4058, pp. 17–29. Univ. of California, Berkeley.

Streu, H. T. (1975). "Insect, mite and nematode control on azaleas." In *Growing Azaleas Commercially,* edited by A. M. Kofranek and R. A. Larson. Sale Publ. No. 4058, pp. 89–96. Univ. of California, Berkeley.

Strider, D. L., Benson, D. M., Jones, R. K., and Aycock, R. (1985). "Major infectious diseases of hybrid evergreen azaleas." *N.C. Flower Growers' Bull.* **29**(2): 9–12.

Stuart, N. W. (1964). Report of cooperative trial on controlling flowering of greenhouse azaleas with growth retardants. *Florists' Rev.* **133**(3477): 37–39, 74–76.

Stuart, N. W. (1965). "Controlling the flowering of greenhouse azaleas." *Florist Nursery Exch.* **144**(11): 22–23.

Stuart, N. W. (1967). "Chemical pruning of greenhouse azaleas with fatty acid esters." *Florists' Rev.* **140**(3631): 26–27, 68.

Stuart, N. W. (1975). "Chemical control of growth and flowering." In *Growing Azaleas Commercially*, edited by A. M. Kofranek and R. A. Larson. Sales Publ. No. 4058, pp. 62–72. Univ. of California, Berkeley.

Twigg, M. C., and Link, C. B. (1951). Nutrient deficiency symptoms and leaf analysis of azaleas grown in sand culture." *Proc. Am. Soc. Hortic. Sci.* **57:** 369–375.

Whealy, C. A., Nell, T. A., and Barrett, J. E. (1988). "Plant growth regulator reduction of bypass shoot development in azaleas." *HortScience* **23**(1): 166–167.

10

Potted Chrysanthemums

G. Douglas Crater

Introduction to Floriculture, Second Edition
Copyright © 1992 by Academic Press, Inc.
All rights of reproduction in any form reserved.

I. INTRODUCTION

A. History

Year-round potted chrysanthemums continue to be one of the top flowering potted plants produced internationally. Numerous cultivars of chrysanthemums were grown in Europe prior to 1800 as garden flowers. Since that time, the chrysanthemum has been refined and developed to the point where it is now a year-round crop with no strong holiday association and the primary pot plant grown in much of the world. The real growth in the pot mum industry has occurred since the 1940s, and new innovations are being developed. The pot mum ranked number one in pot plant production in the 1970s and early 1980s. In the 1990s it is still a major pot crop that ranks second only to poinsettias in the United States.

B. Botanical Information

The chrysanthemum is a member of the Composite or Aster family. There are between 100 and 200 species of this aromatic annual or perennial herb. It is native to the Northern Hemisphere, chiefly Europe and Asia, with a few from other areas. Many authorities claim that the cultivated chrysanthemum originated in China more than 2000 years ago. Chrysanthemums are primarily grown as ornamentals, but one species is cultivated as a source of the important insecticide, pyrethrum. The cultivated chrysanthemum that is used for pot culture is a hardy or semihardy aromatic plant with flowers that exhibit a wide range of colors. The chrysanthemum has been developed into several groups by flower type, such as single, cascade, anemone (with center cushion)(Fig. 1), pompon (globular), decorative (Fig. 2), spider, quilled, incurved (Fig. 3), daisy and large exhibition. The species of chrysanthemums that is used for pot culture is *Dendranthema grandiflora* Tzvelev., commonly known as *Chrysanthemum morifolium* Ramat. Today there are at least 100 chrysanthemum cultivars available to growers. Chrysanthemum cultivars that are suitable for pot mum culture must exhibit the following characteristics: (1) form a well-shaped plant, (2) branch readily, (3) produce flowers quickly on relatively short stems, and (4) have flowers of the desired color, shape, and size.

Chrysanthemums bloom under short days and long nights; therefore, they are termed photoperiodic. Cultivars are classified according to their response group. Response groups are determined by the time from the initiation of short days to flowering (length of time required to bloom). Cultivars used in most areas are classified as 7- to 12-week response groups. For example, a 10-week response group takes 10 weeks from initiation of short days to flowering. The time from planting until short days are begun is not included in the response time. To grow a crop of pot mums takes about 1 to 3 weeks longer than the response group that is used. For potting to flowering, the pot mum is about a 3-month crop.

Fig. 1. Anemone-type flower.

Fig. 2. Decorative-type flower.

Fig. 3.   Quilled-type flower.

C. Prominent Cultivars

With regard to specific cultivars that should be used or are recommended, every supplier or propagator has many good cultivars available. As plant breeders and hybridizers develop new cultivars, many of the older cultivars are replaced. Usually sales personnel from supply firms and propagators can recommend good cultivars. Each grower will find that certain cultivars will do better than others.

Many shapes, sizes, and colors of chrysanthemums are available and are divided into three major types: standards, disbuds, and sprays. The types or cultivars of potted chrysanthemums that are most widely sold are shown in Table I. All of these cultivars come in many colors, with most emphasis on whites, yellows, lavenders, pinks, and bronzes.

II. PROPAGATION

A good quality cutting is essential for success in pot mum production. Quality cuttings are disease- and insect-free, and they are uniform in size. Most propagators who supply cuttings do an excellent job of providing the grower with high quality cuttings. Some growers propagate their own cuttings, but it is difficult for such a grower to produce uniform, disease-free cuttings regularly. Stock plants and propagation benches also utilize bench space that might be used more profitably for growing other crops or flowering pot mums.

Table I

Type of Cultivars

Cultivar	Comments
Standards	Largest flowers. Lateral flower buds are removed to produce one large flower per stem.
Spider	Narrow and lacy-petaled flowers
Daisy	Wider petals, resembling the ordinary daisy
Spoon	Flat-petaled flowers that are spoon-shaped
Decorative	Disbudded, single flower with flat, as opposed to incurved, petals
Disbuds	These are also grown with a single flower per stem. The plant is usually pinched to encourage several stems to grow. Each stem is disbudded to produce a single flower per stem.
Sprays	These are not disbudded, all side shoots are allowed to flower. The plant is pinched as in the "disbuds." In most cultivars the terminal bud is removed allowing all side breaks to develop. The pompon, daisies, anemones, and decoratives are popular spray types.

A. Stock Plant Culture

Stock plant production requires adherence to sound practices during culture. Disease and insect problems will affect crop quality and can completely ruin a crop. Pest-free cuttings are essential each time stock plants are replanted. Before the stock bed is replanted it should be fumigated or pasteurized to reduce or prevent diseases, insects, nematodes, and weed problems.

Rooted cuttings are usually spaced 6 × 8 inches in beds similar to cut flower beds. Watering, fertilizing, and insect and disease control will be similar to any crop of chrysanthemums. Stock plants should be grown under full sun (especially during late fall, winter, and early spring). Daylength should always be above the critical daylength for flower bud initiation (14½ hours) or supplemented with an additional 4 hours of light per night at 10 foot-candles (fc) between 10:00 or 11:00 P.M. and 2:00 or 3:00 A.M.

When should stock plants be replaced? Too many growers try to get one extra crop of cuttings from their stock plants and wait too long to replant. Usually stock plants should be removed and new plants set in their place after five or six flushes of cuttings. After the fifth or sixth flush, new cuttings form on old breaks and are more likely to set crown buds. Chances of maintaining disease-free stock also diminish with each month in the greenhouse.

Patented varieties cannot legally be propagated unless the royalty fee is paid on each cutting. Many of the better cultivars are patented to prevent everyone from rooting them without compensation to the originator of the cultivar.

B. Propagation of Plants

Cuttings should be taken from disease-free stock and should be removed from stock plants by breaking or snapping them just above a node. It is best *not* to use a knife or a cutting instrument, because these transmit disease organisms readily. Cuttings are easily rooted from terminal, vegetative shoots. The cutting should be approximately 2 inches long, depending on cultivar and time of year. Unrooted cuttings can be stored for up to 4 weeks. This is sometimes done so cuttings can be accumulated for a specific planting date. Cuttings are usually stored in polyethylene bags with cool air circulating around the bags.

In many cases, unrooted cuttings are stuck directly into the pot in which they will be finished. This procedure will save time and labor, but can also result in a pot where one or more cuttings might not root, or root more slowly thereby creating an uneven mature plant. Direct sticking is becoming more popular.

Propagators of chrysanthemums sell two or three types of cuttings. Unrooted, callused, or rooted cuttings can be obtained. When rooting a cutting, most propagators use a rooting hormone to increase the number of roots, ensure more uniform rooting, and hasten rooting. If a rooting hormone is applied, the cutting should not be dipped into powder or liquid because this is an easy way to spread diseases. The best method of applying the rooting hormone is to dust it over the basal end of the stems with a small duster.

Improved rooting can be obtained with indole-3-butyric acid (IBA). It has been reported that rooting hormone in talc may not produce uniform rooting because of the amount of talc that would adhere to each cutting.

Many different media can be used in propagation. A well-drained medium is desired. Cuttings root best when the rooting medium stays between 70° and 80°F. Bottom heat to maintain this temperature accelerates rooting. Cuttings should be misted intermittently when first stuck. Mist schedules will vary because of specific conditions.

Cuttings should be spaced approximately 1 inch apart in rows, and the rows should be 1 to 2 inches apart. Some propagators apply a dilute solution of nutrients through the mist system to enhance plant growth. Some sources suggest not fertilizing until day 12 in propagation. This then requires a foliar fertilization formula [potassium nitrate (KNO_3) is most common]. Cuttings should be removed from the propagation bed when new roots are between ¼ and ½ inches long. These rooted cuttings can also be stored at 32°F for 1 or 2 weeks if necessary.

Growth and development of young plants will be best if the night temperature is kept at 65°F, and the atmosphere is kept moist. Cuttings should be prevented from wilting after they are planted. One of the best methods is to place cuttings under intermittent mist. The mist should keep cuttings turgid but not overly wet and should be turned on 1 hour after sunrise and turned off 1 hour before sunset. On clear, sunny days the mist should run 6 seconds every 15 minutes in the summer and 6 seconds every 30 minutes during the winter. This cycle should be adjusted to local conditions.

If a mist system is not used, then a white polyethylene tent over the cuttings will help maintain a high humidity. Usually cuttings are kept under the polyethylene tent for 5 to 10 days or until some root growth can be seen.

Many growers have a separate propagation area. This area has bottom heat under the benches, and the temperature is usually several degrees higher than the rest of the greenhouse area. It also must have lighting facilities to provide long days (short nights) for vegetative growth when the natural daylength is less than 14½ hours long.

An alternative to propagation of mums by rooted cuttings is to produce plants via tissue culture methods. Chrysanthemums have been successfully propagated in culture by organogenesis since the late 1960s. New shoots are produced from either preexisting meristems, or more rarely, from differentiated tissue, such as leaves or flower petals. Shoots (monopolar) produced using this method must be rooted, and the plants acclimatized. The primary advantages conferred by this method are the elimination of viruses, other pathogens, or both from the plants and the recovery of induced mutations. Plants produced by bud proliferation or adventitious bud formation, however, may not be uniform and may have characteristics that deviate from normal. Although culture-induced changes or variability are desirable for the development of new cultivars, practical propagation demands clonal trueness to type.

Somatic embryogenesis (May and Trigiano, 1989) is another *in vitro* method for propagating plants and involves the formation of bipolar structures that possess both root and shoot apices. Unlike their zygotic counterparts, somatic embryos are not formed from the fusion of gametes and are not associated with either endosperm, seed coats, or fruit formation. Somatic embryos originate from single cells and, therefore, should circumvent the possibility of chimeric plants. Somatic embryogenesis is a future tool that could be used in the near future to develop artificial seed that can be used for genetic or propagation purposes.

III. PLANT CULTURE

A. Vegetative Stage

The vegetative stage in pot mum production is the time when long days (short nights) are provided to give the plant time to become established and to grow enough before flower buds are initiated. This stage is used to manipulate the final height of the plant.

1. Growing Medium

a. Mixes Containing Soil. It is difficult to suggest a growing medium that will be acceptable to every grower or for every condition. Water, nutrients, and soil are interrelated, and there is no one mix that will fit every grower's situation. A good growing medium will be loose and well drained. Most topsoils do not have the qualities required to produce optimum root and plant growth. Without good root growth the plant cannot absorb an adequate amount of water and fertilizer, and a plant will be produced that will not be as vigorous or produce as many breaks as will a well-rooted plant. Mixes containing soil usually are pasteurized with a chemical or steam before use. Before a soil is used in a mix, a sample should be sent to a soil testing laboratory for anaylsis.

For areas where topsoils are high in clay, the following proportions are suggested on a volume basis: one part soil, one part organic matter, and one part inert material. If a soil is a loam or does not have an extremely high clay content, then two parts of soil should be used. Organic matter and inert material could be any of the ones previously mentioned. For areas where topsoils are high in sand, the following proportions are suggested on a volume basis: one part sandy, and one part organic matter. If the soil is very loamy or has some clay, then some inert material should be added to loosen up the mix. A mix composed of two parts sandy soil, two parts organic matter and one part inert material should be used in this type of situation. If the soil has a very fine sand, then the mix should be one part sandy soil, one part organic matter, and one part inert material. Some growers use a very small amount of soil, approximately 10% in their mix. This is done primarily for the buffering capacity it provides.

Many growers forget about soil pH, but this should be adjusted and corrected before cuttings are planted. Optimum soil pH for pot mums is between 5.7 and 6.7. A soil test should be made to determine what adjustments must be made to the growing medium. Methods used to raise or lower pH can be found in greenhouse management texts (Ball, 1985; Nelson, 1985).

Many growers carry mixing too far and run the soil and peat through a soil shredder. The procedure shreds coarse peat moss into a powder form or pulverizes a soil with good structure into powder form. The end result of using a soil shredder is a fine soil mix that will provide very poor aeration for plant growth. A concrete mixer or similar equipment is much better for mixing soil. Five minutes or less is long enough to do a thorough job of blending components.

b. Artificial (Soilless) Mixes. Many growers are using "artificial" media that they mix themselves or purchase premixed. There are several reasons for using artificial mixes. First, it is increasingly difficult to find good topsoil that is uniform in physical and chemical properties. Second, handling and mixing soil requires time and labor, and labor costs are constantly increasing. Third, an artificial mix is usually free of disease organisms, insects, and weed seed.

Many growers have found that purchasing soilless media is more economical than mixing their own media. When they compared cost of obtaining, mixing, sterilizing, and handling soil to costs of buying an artificial mix already prepared and ready to use, they found that the artificial mix was not excessively expensive. There are several excellent prepared artificial mixes available, and growers have found them to be very satisfactory.

The volume of mix required to fill various sized pots is shown in Table II.

Table II

Volume of Mix Required to Fill Pots

Size of standard pot (inches)	Number of pots	
	Per cubic foot	Per bushel
3	90	112
4	40	50
5	24	30
6	19	24
7	11	14
8	7	9

2. Potting

It is advisable to grade cuttings according to size before potting. A grower who plants both tall and short cuttings in the same pot can get an uneven product at flowering. Even if the grower tries to correct the problem by pinching tall cuttings hard and short cut cuttings soft, this approach will still result in an unbalanced plant because more breaks will develop from the soft pinch. Before planting, the cuttings should be grouped into categories to insure the production of uniform crops. These categories could be (1) average (medium stem diameter and fairly well rooted), (2) short (thin stem diameter and not well rooted), and (3) tall (medium to thick stem diameter and fairly well rooted).

Cuttings should be planted shallowly, and the roots should just barely be covered by the mix. They should also be planted at the same depth they were in the propagation medium. Cuttings planted too deeply are susceptible to root or stem rot. Growers using peat-lite mixes tend to plant cuttings deeper without problems. If cuttings are planted at a 45° angle, so they lean out over the rim of the pot instead of straight up, a better-shaped plant will result. Usually more breaks and better flowers will be obtained because more light can reach the center of the plant where there is usually less light. Planting the cuttings as close to the rim of the pot and equal distances apart helps to ensure a more uniformly shaped finished plant.

Avoid planting rooted cuttings in a dry medium. Immediately after cuttings have been planted they should be watered-in thoroughly. They should be watered-in twice, and the second watering should contain a soluble fertilizer. With a 20% nitrogen fertilizer, a rate of 1 ounce of fertilizer to 3 gallons of water is recommended. Research has shown the most critical period of pot mum fertilization is during the first half of the growing time. Therefore, the fertility level of the soil mix should be in the optimum range as soon after planting as possible. Early application of fertilizer will get nutrient levels to the optimum range.

3. Spacing

Many growers believe that the amount of space given to a pot mum is almost directly proportional to the final quality. Space management must go hand in hand, however, with what price a pot mum will bring. Spacing too closely during early stages of growth must be avoided, however, to ensure maximum development of lateral shoots.

When cuttings are first planted the pots are often not spaced but are placed pot to pot. Usually, after plants are removed from the mist area or from under the polyethylene tent, they are placed at their final spacing. Spacing pots at their final spacing immediately after pinching is very important to ensure that plants are not shaded by each other and receive the maximum amount of sunlight (Fig. 4). The most popular spacing of 6-inch pots is 14 × 14 inches to provide 1.4 cubic feet per pot. Table III lists some spacings that are used successfully by many pot mum growers.

Fig. 4. Good spacing. Tubes for watering can be seen.

Table III

Number of Cuttings and Spacing According to Pot Size

Pot size (inches)	Cuttings/Pot	Spacing
3	1	pot to pot or 5 inch centers
4	1	7 inch centers
5	2 or 3	10 inch centers
6	3 to 5	12 inch centers
7	5	14 inch centers

It is possible to space the pots several times, gradually giving them more space as the plants grow. Gradual spacing may conserve some space, but it requires more labor, planning, and discipline. Closely spaced plants cast shade on adjacent plants and do not develop as well as plants spaced more generously.

In some cases, one may want to grow pot mums at a close spacing because of market demands and economics. However, better quality plants should be produced when more spacing is provided. Also, a grower has to temper recommendations with economic considerations. Information on other size containers, hanging baskets, and seasonal suggestions are found in Section V,G of this chapter.

4. Number of Cuttings per Pot

Pot mums in 6-inch pots should average 20 to 30 flowering shoots per pot. Four or five cuttings should be used during the spring, summer, and fall months, and five cuttings should be used per pot for the winter months. Only one cutting is required for a 4-inch pot.

5. Photoperiod

Lighting of cuttings is done to prevent flower buds from forming too early during periods of long nights (short days). To prevent plants from initiating flower buds, the dark period must not be any longer than 7 continuous hours of darkness. Lighting is done between approximately September 1 and March 31, depending on latitude, to keep plants vegetative. The number of weeks that pot mums are lighted is usually determined by how fast the plants grow and whether they are tall, medium, or short growing cultivars. Lighting, shading, and planting schedules are usually furnished by major chrysanthemum propagators or suppliers. The following recommendations should be observed to supply correct lighting for the cuttings:

1. For benches 48 inches wide, 60-watt bulbs with reflectors should be placed 3 feet apart and 24 to 36 inches above the tops of the plants, while 100-watt bulbs can be spaced 6 feet apart.
2. Approximately 1¼ watt of light per 3 square feet of bench area should be provided.

3. Lights should not be turned on before 10:00 P.M. and should be left on for 3 to 4 hours, depending on the time of year (intermittent or cyclic lighting can be substituted here).

There are two types of lighting used, continuous lighting or cyclic lighting. With continuous lighting, at least 10 fc of light intensity is required to prevent premature budding. When lighting, precautions must be taken to prevent light pollution to plants in a flowering program.

Many growers use flash or cyclic lighting to prevent premature flower initiation of the potted chrysanthemum. Artificial light may be supplied for 20% of the normal lighting period or 6 minutes for each 30 minutes, using a minimum of 10 fc of light. Instead of lighting consecutively for 4 hours, a total of 48 minutes of light spread over a 4-hour time period will provide comparable results. With increased electric rates many growers are changing to intermittent lighting. Compared with continuous lighting systems, only a fraction of the normal electrical input is necessary with intermittent lighting (approximately 20%). These recommendations should be followed

1. Make sure plants are getting a minimum of 10 fc at the top of the plants.
2. Run intermittent lighting for the same 4 hours one would normally use for continuous lighting.
3. With intermittent lighting the lights will need to be on 6 minutes out of every 30 minutes (on 6 minutes, off 24, on 6 minutes, off 24, etc.) until 4 hours are completed.

Chrysanthemums are lighted because they form leaves and increase in stem length under long days whereas they form flower buds and the stems terminate with flowers under short days. Chrysanthemums should have less than 13½ hours of light for the development of flower buds. Flower bud initiation will occur at 14½ hours, but shorter daylength (longer nights) are required for further development.

When sticking unrooted cuttings, it usually takes approximately 5 more days of lighting than when rooted cuttings are planted. Research has shown that plants must have some root initials before they are responsive to photoperiods.

In northern states, (north of the 42nd parallel), the use of high intensity discharge (HID) lights from August 15 to February 28 is recommended the first 7 to 21 days following the planting of rooted cuttings. Plants are lighted for 16 hours each day, using 400 watt lamps spaced on 10 foot centers.

6. Temperature

Temperature requirements vary with the stage of development. For rooting cuttings, temperature plays a very important role in the success of root formation. Plants should have temperatues of at least 70°F during the day, and bottom heat is best. A rooting medium should not be below 70°F until after roots are formed.

The most favorable temperature for growth of young pot mum plants is higher than for older plants. For this reason, it is suggested that pot mums be grown at a minimum night temperature of 64°F for the first 4 weeks after potting. The minimum night temperature for the next 4 to 5 weeks should be 60° to 62°F. The day temperature can be 10° to 15°F higher than night temperatures (see growth regulation regarding the research on temperature and height control). The maximum temperature should be 90°F. Flowering of many cultivars can be delayed and pigments of dark colored petals do not develop properly. The minimum temperature for growing pot mums is 50°F, but only after flowers are developed. The cooler temperature intensifies color in several cultivars, but it can also cause pink coloration in white petals and bronzing in pink, undesirable traits. Because many pot mum growers have plants of all stages in one house, usually only one temperature can be used.

7. Light Intensity

Pot mum growers have found that it is usually best to grow pot mums under fairly high light intensities. Vegetative growth, quality, and production of pot mums will be improved under high light. Sometimes in midsummer, growers, especially in the southern United States, apply a light shade to their greenhouses to reduce heat and to prevent sunburn on newly opened flowers. This shading is usually 20 to 35% shade and never over 50%. Shading helps cool the greenhouse and reduces watering requirements. Excessive shade (over 50°F) can reduce the quality on certain cultivars, as measured by quality and number of blooms.

8. Fertilization

A proper fertilization program is essential in the production of pot mum crops. It will improve flower keeping quality, reduce disease problems, and produce higher yields. There are many kinds of fertilizers that can be used, such as soluble inorganic, organic, and slow-release fertilizers. Soluble inorganic sources of fertilizers are most commonly used but some changes in fertilization programs can be anticipated because of potential ground water contamination.

Pot mums use large quantities of nitrogen and potassium during the vegetative growth stage. During the last third of the growing cycle the nitrogen rate should be cut in half. It is very important that a pot mum be fertilized immediately after planting. Research has shown that a pot mum's requirements for fertilizer are most critical during the first half of the growing time. They should be fertilized when the cuttings are first watered-in and no later than the second or third watering, depending on time of year.

After the initial fertilization of 1 ounce to 3 gallons of water, one can fertilize with an injector, with 200 ppm nitrogen and 200 ppm potassium every time the plants are watered. A higher rate should be used if fertilizer is applied weekly. With a soilless medium 250 to 300 ppm each of nitrogen and potassium is recommended.

Improved growth has been observed where a slow-release fertilizer has been used in addition to constant liquid fertilization. Slow-release fertilizer can be applied immediately after planting and prior to watering-in at the rate of 1 level teaspoon per 6-inch pot or one might wait a few days until the plants are established. This rate varies with different types of slow-release fertilizers so recommendations on the bag should be noted. Best results occur by applying the fertilizer as a top dressing. When mixed in the soil, a slow-release fertilizer will continue to release nutrients as long as the medium is warm and moist.

Soluble salts in the growing medium primarily come from the applied fertilizer and are composed primarily of ammonium, nitrate, calcium, magnesium, potassium, sodium, bicarbonate, chloride, and sulfate ions. When the soluble salts level becomes too high roots are damaged, which reduces their ability to absorb nutrients and water. Many times plants injured by soluble salts show the same symptoms as plants that are not getting enough fertilizer or water. Actually, plants that are burned by soluble salts levels may be deficient in some nutrients, not because these nutrients are not in the medium, but because plants cannot absorb the nutrients because of root damage from the toxic soluble salts levels. Some other symptoms that may be observed from high soluble salt levels are marginal leaf burn, leaf chlorosis, stunting, yellowing of new growth, excessive wilting, small flowers, and, in some cases, a reduction in growth.

Some growers water very lightly and do not apply enough water for it to run out the bottom of the container; this procedure results in a buildup of salts in the medium. If the medium then becomes dry, soluble salts become more concentrated. All pot mum crops should be watered thoroughly at least once every 2 weeks so that water removes excess salts out the bottom of the container and prevents a buildup of a high level of soluble salts.

One should be alert for a problem with soluble salts. It is a good idea to have a soil analysis run periodically so one can observe the salts and nutrient levels. A foliar test is also recommended because this will supply more information on the nutrient status in the plants.

During the months of October through March one should avoid using fertilizers high in ammonium (40% or higher). Research has shown that nitrate fertilizers during these months improve stem strength, flower quality, and keeping quality.

9. Watering

Hand watering is still the most common practice for small production areas. Many large growers even use hand watering because they feel they can more closely supply the correct amount of water at the proper time. This practice, however, is time-consuming, and more growers are switching to some automated type of watering. Some growers feel that automatic watering is a necessity. Watering crops automatically saves labor, results in more thorough watering, almost eliminates soil compaction, and keeps foliage dry, reducing disease problems.

Good results have been obtained with the automated watering system that uses plastic tubes running to each pot (Fig. 4). Frequency of watering may be controlled by either a weight scale or an electric timer. Regardless of the method used, 1 pint of water should be applied each time 6-inch pots are irrigated. With automatic watering, excellent drainage is imperative.

One should always apply enough water when irrigating pot mums. Applying an insufficient amount of irrigation water may lead to two problems: (1) excessive soluble salts and (2) water stress. Both problems, of course, result in poor crop quality.

Mat watering is also being used very successfully with pot mums. Several types of mats are available. Water is applied to mats and is soaked up through the bottom of the pot. Algae are sometimes a problem but can be controlled chemically or by covering the mat with a perforated black plastic cover.

"Ebb and flow" is one of the latest watering systems to be put into use by commercial pot plant growers. The trays often are movable benches that provide 85% growing space, compared to the 65 to 70% that stationary benches would utilize for growing. The system costs approximaely $4 to $6 per square foot, depending on the size of the facility. The crop is watered and fertilized from the bottom, so soluble salts levels must be monitored carefully.

10. Pinching

Pot mums are pinched to produce multistemmed plants. "Pinching" is the removal of the shoot apex so that the maximum number of lateral shoots will develop. Before the plant is pinched it should have made enough growth so the pinch can be made in new growth.

a. "Hard" Pinch. "Hard" pinching is when the stem is pinched back to allow fewer than 6 leaves to remain on the stem. This usually severely limits the number of breaks that may develop. One reason hard pinching is used is to even up the height of pot mums. At pinching the short cuttings are pinched soft, whereas the taller ones are pinched hard. This can still result in uneven finished plants (plants should be graded before planting).

Pinching is usually done between the 10th and 14th day after planting. Plants should be actively growing with 10 to 14 leaves (counting those just unfolding around the bud) on the stem. It should be noted that the pinch date does not affect time of flowering.

b. "Soft" Pinch. A "soft" pinch is when a small amount of the stem tip is removed (rolled out). When a plant has had a soft pinch and was grown properly, it should have approximately 10 leaves below the pinch. New growth of 1½ to 2 inches should occur on the newly planted cuttings before the pinch is made. This normally occurs from 10 to 14 days after planting, or when the roots reach the bottom of the pot. It is important that root growth is adequate to support the emerging lateral shoots after the pinch is made. One should remove ½ to ¾ inch of new growth. Pinching too soft may contribute to after-pinch stretch and a top-heavy plant. Excessive stem removal results in fewer breaks from the mature tissue, which then results in fewer blooms. Many growers allow six to eight leaves after the pinch.

Table IV gives a pinching schedule that may be used as a guide. If the crop is ready for pinching several days earlier than the guide indicates, the plants should be pinched. If the crop is not ready, the pinch should be delayed until growth requirements are met. If this delay is extensive, cultural practices should be reviewed.

c. Delayed Pinch. A delayed pinch is one made after the start of short days. It was initially designed to control height before growth regulators were available. Floral initiation occurs quickly on lateral shoots, and fewer leaves are produced than if long days are continued. The plant also appears to be loose and open. The practice is commonly used on tall growing cultivars, however.

Some general pointers for pinching pot mums include the following:

1. Before pinching establish a good root system and 1½ to 1¾feet of vigorous new growth.
2. Remove ¼ to ¾ inch.
3. Pinching should be done before the start of short-day treatment for best quality plants.
4. Pinching after short-day treatment is started is called *delayed pinching* and should be avoided except for tall cultivars.
5. If the plant is not well enough established at the scheduled pinching time, cultural practices should be evaluated and corrected.

Table IV

Pinching Schedule

	Average crop ready for pinching	
Season and Treatment	North (days)	South (days)
Late spring through early fall plantings		
Tall treatment	7–10	7–10
Medium treatment	8–11	8–11
Short treatment	10–14	10–14
Early spring and late fall plantings		
Tall treatment	10–14	10–14
Medium treatment	11–15	11–15
Short treatment	14–17	14–17
Midwinter plantings		
Tall treatment	14–17	10–14
Medium treatment	15–18	11–15
Short treatment	17–21	14–17

11. Growth Regulators

An ideal pot mum is approximately 2 to $2\frac{1}{2}$ times as high as its pot. Many mass market outlets prefer shorter mums—maybe $1\frac{3}{4}$ times as high as the pot. Therefore, height control is of utmost importance to pot mum growers. Under long-day conditions chrysanthemums form leaves and increase in stem length; under short-day conditions flower buds form and the stem terminates. Therefore, daylength, and especially the number of long days, can be used to influence the final height of the plant. Genetically short-growing cultivars are provided with more long days to increase stem length than are tall growing cultivars.

The number of long days controls height to a certain extent, but additional control of height is sometimes needed and can be done with growth regulators. Chemical growth retardants not only retard stem elongation but also can result in dark green foliage and stronger stems. Growth regulators are used for the following reasons:

1. Controlling the height of the plant
2. More compact growth
3. Darker green foliage
4. Stronger stem strength
5. Improved resistance to wilting
6. Improved keeping quality of flowers

B-Nine (daminozide) and A-Rest (ancymidol) are the most commonly used growth regulators, with B-Nine predominating. Two new growth regulators (Bonzi and Sumagic) are being studied, and if approved for chrysanthemums would be very effective growth regulators. B-Nine is the most widely used because it is easy to apply and plant response is predictable. It is applied as a foliar spray to newly emerging shoots or to visibly budded plants to control plant height. Plants should be sprayed with a 0.25% (2500 ppm) solution (1 tablespoon per gallon of water) when the lateral shoots are $1\frac{1}{2}$ to 2 inches long. Pot mums are usually sprayed about 2 weeks after pinching, and, in some instances, a second spraying is needed for tall cultivars. With some fast-growing cultivars and especially during summer months, a rate of 0.50% (5000 ppm) (2 tablespoons per gallon) may be necessary. Growers are now spraying after potting when cuttings are not wilted. This helps prevent the elongation of internodes prior to pinching. One gallon of 2500 to 5000 ppm solution (depending on the time of year) should be srayed per 200 square feet of bench area. Table V should help simplify mixing B-Nine, however, mixing recommendations by the manufacturer should be followed.

When B-Nine is applied too late, pink flowers have a tendency to fade, while some white-flowered cultivars have cream-colored petals. B-Nine and Alar contain the same active ingredients.

A-Rest is a plant growth regulator that is very effective in reducing internode elongation of pot mums. It is usually applied when new shoots are 2 to 3 weeks old

Table V

B-Nine Mixing Rates

Solution (%)	Amount of B-Nine mixed in gallon water		Parts per million (ppm)
	Ounces	Grams	
0.15	0.20	6.7	1,500
0.25	0.40	11.1	2,500
0.40	0.65	18.1	4,000
0.50	0.80	22.3	5,000
1.00	1.60	44.5	10,000

and not over 6 inches long. It is effective either as a spray or a soil drench. When bark is used as a potting medium the effectiveness of A-Rest is reduced when applied as a drench.

Research by Heins (1985) has produced a new method to control the height of chrysanthemums without using growth regulators. Heins found that the warmer the day temperature the taller the plants. The tallest plants occurred when a high day temperature was combined with a low night temperature. To control height growers could grow their crops at a slightly cooler day temperature than night temperature (negative DIF).

12. Carbon Dioxide

Carbon dioxide enrichment of the greenhouse atmosphere is recommended in some regions on bright days during the winter months. Injection of 750 to 1000 ppm should begin approximately 1 hour prior to sunup and conclude approximately 1 hour prior to sundown or until ventilation occurs. Higher rates provide no appreciable benefits.

Carbon dioxide levels of approximately 1000 ppm can provide the following benefits: (1) earlier flowering (4 to 7 days decrease in production time), (2) increased stem strength, (3) larger flowers, (4) increased flower number and more lateral branching, (5) overall quality improvement, and (6) increase in rate of photosynthesis.

The major sources of carbon dioxide for greenhouse use include (1) pure carbon dioxide from compressed or cryogenic liquid, (2) burning hydrocarbon fuels, and (3) decomposition of organic matter. Each of these sources has its advantages and disadvantages. The organic material is erratic and unpredictable. Burning of hydrocarbon fuels is fairly inexpensive, but if the fuel is not pure or if incomplete combustion results, then toxic gases can be produced. Compressed or cryogenic liquids or solids ("dry ice") appear to work best but are also the most expensive.

B. Flowering Stage

1. Photoperiod

Dendranthema grandiflora (*Chrysanthemum morifolium*) is classified as a short-day plant, although flowering is controlled by the length of the dark period. The plant remains vegetative under long days (daylength of 14.5 hours or longer, 9.5 hours of night or darkness). When days are less than 14.5 hours, flower buds are initiated, but flower buds do not develop until the daylength is less than 13.5 hours (10.5 hours of night or darkness). Some cultivars will vary slightly from this critical photoperiod, but these durations are valid for most commercial cultivars in the United States.

A grower can have complete control over flowering by manipulating the hours of darkness a plant receives. This can be done using natural daylengths, by lighting, by shading, or by a combination of these manipulations. Lighting is used to keep plants vegetative when natural daylengths are short, and shading with a blackcloth is used to get plants to flower when daylengths are too long.

a. Lighting. See Section III,A,5.

b. Shading. During spring and summer months, days are longer than the critical photoperiod of 13.5 hours. To develop flower buds on pot mums, at least 10.5 hours of darkness are required for each 24-hour period. A good grade of black sateen cloth or black polypropylene is required to prevent penetration of more than 2 fc of light. Shading should be practiced daily until flower color develops. Failure to shade each day results in delayed flowering and taller plants, but some growers do skip pulling black cloth on Saturday nights. This usually will delay flowering by 1 day for every skipped day.

A no-light or dark period could be 13.5 hours long, but this creates two problems in greenhouse operations. The first is a time problem for getting the covering and uncovering done by employees, whose normal working hours are from 8:00 A.M. to 4:30 or 5:00 P.M. The second is a problem of heat buildup under black cloth or plastic, particularly in the summer. The first problem can be worked out by either extending the dark period or having someone to pull the black cloth or plastic at the proper times. The second problem, heat buildup, can be avoided by covering after the temperature has decreased. This would mean delaying shading until after 7:00 P.M. An alternative would be to provide positive air movement (fan and pad cooling) through the plant area under the shade material. It is not very easy to do this alternative and still maintain the desired level of darkness, but it can be done and is strongly recommended for southern growers.

Plants can be shaded by manually pulling the shade cloth over the plants or by mechanizing the procedure. A time clock can activate the drive mechanism to open or close the shade at preset times. During the summer months the mechanized shading is ideal because shade cloths can be pulled at 6:00 or 6:30 P.M. and reopened after dark to reduce plant temperature. The automated system can recover the plants around 5:00 A.M. and then reopen at 8:00 A.M. This reduces high temperature and the resultant delay in flowering.

c. After Lighting. Research indicates that many cultivars flower successfully when shaded for 35 days, followed by long days. Because many cultivars do not respond to such treatment, growers are cautioned to try "after lighting" only on a trial basis.

2. Scheduling

There are various schedules designed to grow plants of a certain height. The height desired will depend on the market. As mentioned previously, the standard height is for the plant to be 2 to 2½ times higher than the pot. Propagators and distributors have suggested schedules that growers can follow, but these schedules are for average conditions.

A sample schedule for 10-week pot mums is shown in Table VI. Nine-week mums would flower 1 week sooner; 11-week cultivars would flower 1 week later.

C. Response Groups

Commercial varieties are divided into response groups that refer to the number of weeks from start of short days until flowering. A 9-week cultivar takes 9 weeks to flower from the start of short days, whereas an 11-week cultivar takes 11 weeks. The most common response groups are the 8-, 9-, and 10-week cultivars for use in a year-round schedule. In the South, some response groups will flower in less time than in the North.

D. Temperature Effects

Temperature can really affect the growing schedule of a pot mum crop. When the night temperature exceeds 80° to 85°F, many cultivars do not flower in the allotted time. There are at least three ways to combat heat delay. The first method is to grow cultivars that have less tendency for heat delay. The second is to cover plants later in the evening and uncover later in the morning. During the hottest months it would be best to cover at 9:00 P.M. and uncover at 9:00 A.M. In addition,

Table VI

Schedules for Tall, Medium, and Short Cultivars

Plant height desired	Potting date	Pinching date	Lighting dates (long dates)	Flowering date
Tall	Jan. 28	Feb. 11	Jan. 28–Feb. 3	April 15
Medium	Jan. 28	Feb. 11	Jan. 28–Feb. 10	April 22
Short	Jan. 28	Feb. 11	Jan. 28–Feb. 17	April 29

it would be advisable to use cloth instead of plastic, because cloth will allow for some air exchange. The third is to air-condition the greenhouse under the covering, using a method similar to the fan and pad system of cooling.

A minimum night temperature of 62°F should be maintained to ensure proper flower development. Lower temperatures will result in uneven flowering, if flower initiation and development occur at all.

A low night temperature of 55° to 60°F provided during the final 2 to 3 weeks results in pink shades for many white-flowered cultivars and bronzing of yellow cultivars. A low temperature finish for cultivars, however, intensifies flower color.

E. Cropping Tips

In scheduling, enough time should be allowed before pinching so that enough new growth will be made to result in about 10 leaves below the pinch. Short-growing cultivars usually require at least 1 week of long days after the pinch in order to provide for the additional stem length. Short days for the tall-growing varieties are started about 1 week before the pinch to reduce the height at which the plants are finished. With good growth regulators available this really is not best. This is called a "delayed" pinch, as it occurs after the start of short days. It must be remembered that chrysanthemums do not grow as rapidly in cool, dark weather as in sunny, warm weather. Therefore, it usually takes more long days during the cool, dark winter months.

F. Garden Mums

Certain cultivars can be flowered as spring pot plants without shade or lights. Garden mums respond to photoperiod the same as any pot mum does. Garden cultivars make good Easter and Mother's Day items because the plants can be cut back and planted outside after the flowers fade. These plants will usually flower again in the fall and for several years thereafter.

For growing spring flowering garden mums, rooted cuttings should be potted around the first of March. A soft pinch is made about 2 weeks after potting. The plants will not require any supplemental lights or shade. The temperature must be kept at 60°F until flower buds are initiated. Then the temperature can be dropped to 50°F for the remainder of the growing season. Some cultivars require pinching of center buds although others do not. They can be grown in 4- to 6-inch containers, but are usually grown in the smaller containers with a close bench spacing. Better quality plants will result if some space is provided between containers.

G. Single-Stem Pot Mums

Some growers are planting large-flowered mum varieties and growing them single-stem in 6-inch or larger pots. These plants are not pinched. This type of pot mum produces fewer but much larger flowers and makes an entirely different quality plant for special sales. Market demand will vary, but these plants usually sell best during winter and spring holidays. Some growers even grow cut mums in this manner.

H. Year-Round Program

In a year round program, a specific quantity of pots is produced throughout the year. The potential production is about four pots per square foot of bench area per year. The space required for a year-round program will vary according to the number of pots produced. A medium-size grower will produce approximately 1000 6-inch pots each week. When plants are moved to their final spacing of 1.4 square feet per pot, this will require approximately 1400 square feet of bench space for short days until the plants flower. If top quality is not deemed essential the pots can be spaced to 1 square foot each or approximately 1000 square feet for this operation. This is minimum spacing, and cultivar selection is of utmost importance.

A small grower may want to produce 50 pot mums every other week and will need about 70 square feet of bench space for each crop. This is a minimum program for a small retail grower. Allowing 12 weeks for each crop, six crops would be in the greenhouse at any one time, thus 500 square feet would be more practical.

I. Disbudding

Disbudding is the process of removing immature flower buds, which are not desired on the chrysanthemum stem, to provide for either a small number of large flowers or a larger number of small flowers. Most pot mums are disbudded to remove lateral flower buds so the plants will be more attractive and uniform and have larger flowers. Usually with the daisy types and some other cultivars, the terminal bud is removed so the lateral buds will develop and provide for numerous, small flowers. With certain cultivars that are grown as pot plants, no disbudding is required.

Disbudding should be done as soon as the buds can be easily rolled out between the thumb and forefinger without injuring the terminal. If disbudding is done too early the terminal buds or stems may be damaged. If done too late, however, it will result in later flowering, small flowers, and longer stems (a dramatic reduction in the quality of the finished product). The bud and pedicel should be completely removed, with no stubs remaining on the stem.

There are three distinct types of disbudding used to produce the desired appearance of the pot mum.

1. Disbudding is the removal of all lateral buds, leaving one terminal bud per stem. Disbudding is best done when the lateral buds can be removed without injury to the terminal bud. Large decorative varieties are most often disbudded.

2. Center bud removal (CBR) is the removal of the center bud (terminal bud) allowing all side buds (laterals) to develop into *spray pots*. Center bud removal is used for daisies, anemones (Fig. 5), and small flowering decorative cultivars.

3. Multiple bud removal (MBR) is sometimes used to reduce clubbiness of the flower formation. Multiple bud removal is done much like a soft pinch when flower buds can be felt, but not seen. This occurs approximately 3 to 4 weeks after the pinch, when there are between four and six leaves on the new shoots.

IV. CONTROL OF INSECTS, DISEASES, AND PHYSIOLOGICAL DISORDERS

A. Insects

1. Insect Types

The following are insect pests on pot mums:

Aphids are sucking insects less than 3 mm long. There are various kinds and colors. Aphids disfigure young growth and can be difficult to eliminate.

Corn earworms feed on flower buds and petals of the open flower and also consume large amounts of young, tender foliage.

Fig. 5. Anemone-type flowers on plant given the center bud removal treatment.

Cutworms usually cut off plants at the soil line; however, some species eat the foliage or flowers.

Cyclamen mites are usually found at the very tip of the stem, but they are too small to be seen with the naked eye. These pests are generally recognized by the damage they do. They cause malformation of the new growth and sometimes cause a bronzing color on young foliage.

Foliar nematode is a microscopic eelworm that must have a film of moisture on the leaf to move from one plant to another. It enters the leaf and begins feeding through damaged areas and stomata. After it enters the leaf the foliage dies, leaving a wedge-shaped brown area between leaf veins. Keeping foliage dry prevents foliar nematodes from spreading.

Fungus gnat larvae do much damage the roots and enter the vascular system in the stem, severely injuring or killing the plant.

Grasshoppers eat leaves, stems, and flowers and do much damage in a short time.

Larvae of *leaf miners* tunnel through the foliage between the upper and lower epidermis, making irregular, light-colored patterns in the leaf. Injury to foliage causes it to be unsightly and decreases market value. With a serious infestation, some leaf drop can occur.

Leaf rollers are small caterpillars that chew the underside of the foliage. The upper epidermis becomes desiccated, causing a parchment-like appearance. They cause leaves to roll. The caterpillar forms a web in the rolled leaf, and later emerges as a moth.

Loopers, commonly known as cabbage loopers, damage pot mums by eating holes in young leaves, destroying shoot tips, and eating the tender growing flower petals.

Mealybugs are soft-bodied, sucking insects with white cottony masses that cover the gray pests. They are usually found in leaf axils and feed on the sap in leaves and stem.

The *chrysanthemum midge* makes galls on foliage and stems. Inside the galls are larvae that feed for about 28 days before they emerge to mate and repeat the cycle.

Slugs and snails eat foliage, making conspicuous damage and occasionally attack the flowers. They feed at night but their slime trails will often reveal their presence.

Sowbugs usually feed on decaying organic matter. They often become so plentiful that they eat the bases of stems on newly planted cuttings.

Spider mites are serious pests on pot mums because they suck the juices from the leaves and cause a light mottling of the leaves. Mites are very small and usually cannot be seen with the naked eye. Reproduction is rapid, and a single female can have over one million descendants in a month at temperatures above 70°F. They also acquire resistance to organic phosphate insecticides.

Spittle bugs are rarely a pest on pot mums, but they may sting the young growth and cause minor disfiguration. A frothy white mass reveals that this pest is present.

Symphylids tend to stunt plant growth when they injure root systems.

Tarnished plant bugs damage the plant by puncturing the stem. This can cause excessive branching and sometimes prevent flowering. They also feed on foliage, causing small, grayish dead areas in the leaf and often wilting the leaf tips.

Termites can sometimes be a problem in wooden benches with wooden legs. They can bore into stems of plants, causing wilting.

Thrips are slender sucking insects, just visible to the eye, and they infest foliage, developing buds, and open flowers, primarily during the spring and sometimes in midfall. They rasp and the foliage and flowers and cause the light-colored streaks on the foliage and brown or light streaks on flowers. Western flower thrips transmit tomato spotted wilt virus.

Whitefly can be a pest on pot mums. The nymphs suck the juices of the leaves, and when found in large quantities, can damage the plant.

2. Insect Control

Control of insect pests on pot mums is a major problem. In addition to using insecticides and miticides, growers should control weeds around their greenhouses. These weeds may harbor insects that will then move into the growing area. Pesticides should be applied before insect populations build up on the crop. A grower should inspect the crop daily so that insects will not become a problem. Early detection and identification is a key to control. Most insects feed on the lower sides of the leaves, so the grower should especially examine these areas and be certain the undersides of leaves get sprayed with the pesticide and that no insects remain.

There are several different methods of insect control: (1) chemical control, (2) biological control, and (3) integrated pest management.

a. Chemical Control. Chemical control is the most widely used method for insect control. There are several chemicals registered for insect control on chrysanthemums. If improperly used, registered chemicals can also damage plants. Many chemicals have been overused, resulting in pesticide resistance by certain insects. Label recommendations must be followed.

b. Biological Control. Biological control with predators and parasites is becoming more commonly used for control of insects. Larvae of cabbage loopers and similar pests have been controlled successfully with a Bacillus and was one of the first successful examples of the biological control of insects in flower crop production.

c. Integrated Pest Management (IPM). The IPM is the area experiencing the widest acceptance and implementation by growers. This is a relatively new idea of pest control. One of its major principles is to not spray until a pest reaches its economic threshold level. The economic threshold level is that insect population at which the loss from damage is equal to the cost of control. Integrated pest management uses all suitable techniques such as cultural, biological, and chemical control. In order to implement this program some type of insect scouting practices must be used. This could involve regular observation walks through the growing area or it could be the use of yellow sticky strips to monitor types and numbers of insects within the growing area.

This new IPM method also encourages the use of preventive measures relating to insect control. Some of these are

1. Grow resistant cultivars.
2. Maintain sanitary conditions in growing areas.
3. Destroy heavily infested plants.
4. Remove all weeds and tall grass near the growing area.
5. Screen all vents and fan openings so insects cannot enter.
6. Do not allow personal plants in the production area, and examine all plants coming into the production area.
7. Be aware that insects are attracted to certain colors, such as yellow and to a lesser extent white and light blue. Personnel should be discouraged from wearing clothing with such colors in the production area.
8. Do not use excessive amounts of fertilizer, as there seems to be a correlation between excessively lush growth and faster development of some pests.

The use of yellow sticky traps was recommended earlier. These are used to help monitor the insect population. The traps should be placed in the canopy of the crops. They can be obtained from suppliers or built by the grower. Most are painted yellow. Blue is used to detect the western flower thrips. One must continually check and record what is found on the boards. When a critical number is found in an area, then it is time to take appropriate control measures.

B. Diseases

There are two basic ways that diseases can be controlled. First is prevention, which is the most effective and economical method. By knowing what factors cause or contribute to the disease, one can remove or prevent them. The second method is use of an eradicant, usually in the form of a chemical, to control, kill, or prevent the disease organism from reoccurring or attacking a healthy plant.

There are several procedures that can prevent or control diseases besides the use of chemicals:

1. Keep foliage dry, because most fungi and bacteria require several hours of free moisture for spores to germinate.
2. Provide good air circulation. This will reduce free moisture on foliage.
3. Remove and destroy severely affected plants.
4. Remove weeds growing around the greenhouse.

5. Select cultivars that have more resistance to disease, although presently there is not much difference among cultivars.
6. Provide for good drainage and do not keep the growing medium too wet.
7. Make sure the growing medium has been properly pasteurized.
8. Prevent recontamination by cleaning all equipment, containers, and supplies that are taken into the growing area. Disinfectants such as sodium hypochlorite or LF 10 can be used.
9. Avoid overwatering. Wait until the medium is dry before rewatering.
10. Thoroughly clean the growing area when a crop is removed and before another crop is brought into this area.
11. Purchase disease-free cuttings from cultured stock.

Chrysanthemum diseases can be divided into four general groups according to causal agents: bacteria, fungi, viruses, and root knot nematodes. The following are some of the major diseases found on pot mums.

1. Bacterial Diseases or Stem Rot

Plants in all stages of growth may exhibit the symptoms of these diseases. In newly planted cuttings it first appears as early wilting. Examination of cuttings will show the stems to be softened and internally rotted with red streaking associated with the vascular system. In most cases, plants will wilt at some stage of the disease's progress, retarding growth. Breakdown of the infected older plants may start at the location of a pinch and work down and out into the lateral shoots. Complete loss of the plants is not uncommon. It is possible, but not common, for wilting to occur as late as anthesis. Plants with this disease should be destroyed.

2. Fungal Diseases

a. Basal Stem and Root Rot. Basal stem and root rots account for the loss of plants during the high humidity, high temperature periods of growing. One or two fungi are often involved.

Rhizoctonia stem rot usually occurs when cuttings are planted too deeply or are constantly overwatered. *Rhizoctonia* may be distinguished by the following symptoms:

1. The decay originates near the growing medium surface rather than in the root system. It commonly attacks recently planted cuttings.
2. The diseased area has a brown coloration and may result in a "stringy" dry rot.
3. Strands of brownish mycelium may be seen with a hand lens on the growing medium surface or on affected plant tissue.
4. Where plants are crowded, such as in propagating beds or containers, the fungus may form a brown-colored web over the plants.
5. The expression of the disease is often rapid and may rot mature plants upward from an infection at the growing medium surface.

Pythium is a root and stem rot that affects some cultivars more than others. *Pythium may be distinguished by the following symptoms:*

1. Rotting may start at the root tips.
2. The fungus does not produce a characteristic color; roots will be brown or black depending on the stage of actively growing plants.
3. The fungus might only prune or partially injure the root system, causing a slow retardation in growth.
4. In almost all cases the root system may have been injured by overwatering or high soluble salts just prior to infection.
5. Good drainage in a pasteurized medium can reduce losses.

Field observations often reveal both *Pythium* and *Rhizoctonia* attacking the plant at the same time. Spread of the organisms occurs via plants, splashing water, or via dirty hose nozzles, pots, tools, and growing media. Poor growing medium aeration, inadequate drainage, high salt concentrations, or all of these are known to predispose the plants to infection and increase the severity of expression of the diseases.

Cottony Stem Blight. The organism is easily recognized by the dense, white, wet-appearing, cottony mass-like growth on the stem. Associated with this growth, often within the stem cavity, are hard black seed-like sclerotia or fungal masses sometimes up to ¼-inch across. In poorly ventilated, damp houses, the disease spreads rapidly. However, expression of the disease is rarely seen on greenhouse chrysanthemums and then only when poor regulation of temperature and humidity prevail. Usually proper ventilation prevents this organism from being a problem.

b. Foliar and Flower Fungal Diseases. *Ray blight.* The ray blight fungus, *Ascochyta chrysanthemi,* may attack any part of the chrysanthemum plant during its normal growing cycle. The name "ray blight" is derived from the fact that symptoms often appear on the ray florets of a developing flower.

1. Petals on one side of the flower turn dark brown or black from their base outward and tend to stick together.
2. Rot may progress down the flower stem.
3. Blossoms fall apart.
4. Leaves and unopened buds also may be attacked, producing a mushy, black brown spotting and rotting under continued moist, moderately warm conditions. The rot may extend from the leaves into the stem, causing a brown to black lesion that may eventually girdle the stem and cause death at the top of the plant. In severe cases, rot will weaken the stems sufficiently to permit flopping over of the flower heads. This sort of blight is also seen when previously infected cuttings are being rooted under mist.
5. The infected leaf tissue may resemble Septoria leaf spot symptoms, but it is differentiated by the lack of the circular spot pattern common to *Septoria* and the infection lesion of *Ascochyta.*

Botrytis. The *Botrytis* fungus also may attack any part of a chrysanthemum plant during its normal growing cycle and frequently causes the greatest losses on mature flowers in transit. *Botrytis* is sometimes called gray mold because of its appearance. Expression of the disease varies considerably from Ascochyta ray blight.

1. It does not attack the flower bud from inside and is usually found on a more mature or fully opened flower.
2. Foliage infections are characteritically soft brown and on drying do not have any distinguishing features.
3. Foliar *Botrytis* is found most often on the lower leaves shaded in the center of the pot.
4. Mature infections will produce visible threads of mycelium that may produce tufts or growths of brown, spore-laden mold.
5. The organism most readily attacks injured tissue, such as that burned from excess fertilizer or sunburn.
6. It is most troublesome when the air is very moist. Dry air, adequate spacing, and removal of infected plant parts can reduce the severity of *Botrytis.*

Powdery mildew. Whitish, powdery growth occurs on the upper surface of leaves. The growth begins as distinct spots, which later enlarge and merge together. Some cultivars are more susceptible than others. It is prevalent under high humidity conditions, especially in dark weather.

Septoria leaf spot or leaf blight. Distinct circular irregular blotches or spots appear on the foliage. These are grayish-brown in color, become brittle in the center and sometimes fall out. Affected leaves may turn yellow and die. Some cultivars are more susceptible than others.

Rust. Chocolate brown powdery pustules first appear on lower leaves. As the disease progresses up the plant, leaves will turn yellow and fall. Plants are weakened and may eventually die.

Verticillium wilt. Sometimes known as Siedeivitz disease, it is a vascular fungus that affects most cultivars. Margins of leaves turn yellow and eventually wilt or dry up. This begins at the base of the plant and works up the stem. It may affect one side of a stem or plant, more than the other. No distinct spotting occurs. Examination of the stem tissue of infected plants may reveal brown streaking in the vascular bundles. *Verticillium* seldom kills plants. Cuttings taken from stock plants that have *Verticillium* generally have the fungus within them. Cultivars vary in resistance.

Fusarium wilt. This disease is becoming more common in potted greenhouse crops. *Fusarium* is a fungal wilt disease, causing symptoms similar to those described above for Verticillium wilt and is often mistaken for Verticillium wilt. *Fusarium* is generally more prevalent under warm temperature conditions, and affects stems, which are usually decayed, with brown streaks extending upward. The plants wilt at time of flowering.

3. Virus or Virus-like Disease.

Virus particles cannot be seen by light microscopes. They multiply within the cells of affected plants without killing those cells. They are present in the sap of diseased plants and many can be transmitted to a healthy plant through wounds or by mechanical means. The chrysanthemum is known to be affected with at least seven viruses: *yellows, spotted wilt, flower distortion, aspermy, rosette, mosaic,* and *stunt.* Stunt is a slow moving viroid disease that severely dwarfs plants and causes fading of pink, red, and bronze flowered cultivars. Spotted wilt is one that has caused great concern. Western flower thrips will spread spotted wilt virus from plant to plant.

Yellows is a mycoplasma that affects the flower. A portion of the flower is yellow-green rather than its normal color, and flowers are usually much smaller than normal. This mycoplasma is spread by aphids and leaf hoppers that feed on infected plants and transfer it to healthy plants.

Chrysanthemum phloem necrosis. Chrysanthemum phloem necrosis is a disease caused by a mycoplasma-like organism (MLO). These are not new to floriculture as aster yellows is a MLO disease. MLOs are transmitted by insects like leaf hoppers, which feed upon the sap-conducting vessels. MLOs are not generally transmitted mechanically, and they do not live outside of the plant host or their insect vector.

Symptoms of phloem necrosis include early leaf deterioration, leaf flecking and necrotic sectoring, veinal necrosis, closed stomata, a general decline in vigor, and some changes in flower shape. Partial control might be achieved through the use of lower temperatures and higher nitrogen and potassium levels. Chemicals and antibiotics are either ineffective or not safe for general use, based on experience with other MLO diseases.

4. Root Knot Nematodes

These cause gall-like knots on roots and usually stunt the plant severely. Clean stock and pasteurized media are the best preventative measures for root knot nematodes.

Control procedures will, in most cases, include several cultural procedures as well as chemical recommendations. One must check on the chemical registration of any pesticides used as control measures.

C. Physiological Disorders

Pot mums can have many problems that are related to cultural practices and are not disease or insect-related. Physiological disorders can be divided into a number of categories. The following problems will be discussed as to cause and possible solutions.

1. Plants Too Short

Plants that are too short can be caused by several factors, such as poor root growth, insufficient nitrogen in the early growing stages, failure to provide enough long days, or excessive use of growth retardants.

2. Plants Too Tall

Excessive height can be caused by too many long days, crowded conditions, inadequate light intensity, and temperatures that are too high. A high day temperature combined with a low night temperature can cause plants to grow much taller.

3. Uneven Flowering

Uneven flowering may be a problem, especially in winter, because of cool nights, as flower buds fail to form at night temperatures below 60°F. Stray lights striking plants during desired dark periods can also cause this problem, as can heat delay.

4. Not Enough Shoots

The development of too few shoots usually occurs during the first 2 weeks after plants are pinched. Some problems that may develop during this time that could cause fewer shoots to develop are poor root growth, cool night temperatures, very dry air, too hard a pinch, and insufficient nitrogen.

5. Malformed Flowers

Malformed flowers can be caused by diseases or insects, but poor control of daylength, temperature, or both can also affect flower form.

6. Failure to Flower

Ethylene pollution from faulty heaters, incorrect day length manipulation, light leaks when shading, and extreme temperatures are some factors that can cause this disorder.

7. Poor Growth

Poor growth can be caused by not enough sunlight in very dark weather, too much or too little fertilizer, improper pH, overwatering, and poor drainage. Growers need to keep records of their cultural practices and also some notes on extreme weather conditions. These conditions will help determine the causes of some poor growth problems.

8. Wilting

Plants will often wilt if the weather has been overcast for a few days followed by bright sunny days. This results from the plant not being able to absorb water quickly enough. This condition should correct itself after a few bright days.

9. Crown Buds

The major distinction between crown buds and terminal buds is that leaves beneath a crown bud are strap-shaped, while leaves beneath a normal terminal bud are lobed. Vegetative shoots can emerge around a crown bud if the causative factor is not corrected. Crown bud formation results from failure to apply the blackcloth consistently, lighting time clock failure, insufficient light intensity during the lighting period, or application of the blackcloth too late after pinching, as well as products of incomplete combustion from heaters.

10. Bract Buds

During the warm summer months it is not unusual to find overdeveloped individual floret bracts on some cultivars. Bract buds may develop when day or night temperatures exceed 81°F, the blackcloth is removed too soon, the blackcloth is torn, or incandescent lighting is not uniform over the plants.

11. Early Flower Initiation

When long days are not provided immediately after planting, flowers will initiate too early, causing undesirable finished products. In spring, growers have assumed daylengths were long enough to maintain vegetative growth when they were not, causing early flower initiation.

12. Sun Scald

Sun scald develops when high temperatures and high light intensities cause rapid evaporation of moisture from flower petals and is often confused with Botrytis blight. Sun scald and Botrytis blight are easy to distinguish because the former occurs on younger petals at the center of the flower, and the latter infests the tips of older petals. *Botrytis* will often affect petals damaged by sun scald, however. Sun scald can be prevented by supplying a light shade over the plants as color starts to show on the flower buds and by proper selection of cultivars for certain areas or time of the year.

13. Heat Delay

To avoid heat delay, maintain temperatures below 86°F under short days (when plants are under opaque shade). During hot days, delay covering plants with cloth or plastic until early evening. Thirteen hours of continuous darkness must be provided. Therefore, if the plants are covered at 8:00 P.M., they should not be uncovered until 9:00 A.M. When it is possible, ventilate under the cloth or plastic. Where conditions for heat delay are common, cultivars should be selected that are less susceptible to this disorder.

14. Cold Delay

Maintaining sensitive cultivars at temperatures below 60°F at night can result in delayed and uneven flowering.

15. Faded Flowers

Plants exposed to cool temperatures (below 60°F) or flowers that are old will have a pinkish hue on white flowers. High temperatures, above 75°F, will often cause fading in dark colored flowers or failure of the floral pigments to develop.

16. Herbicide Injury

Sometimes growers will use volatile herbicides in or near the greenhouse. These fumes often damage the plants. If these chemicals are used in the greenhouse, several cropping periods can be damaged by the residual effect of the chemical. Very few herbicides have label clearance for greenhouse use.

17. Air Pollutants

Leaf discoloration is sometimes caused by air pollutants. Stunted growth, malformed flowers, and delayed flowering have often been attributed to air pollutants.

V. HANDLING OF THE FINISHED PRODUCT

A. Harvesting

Pot mums are ready to go to market when flowers are about one-half to fully opened. Many growers send pot mums to market as soon as the flower petals begin to unfold but before the flowers are fully open. These earlier harvested plants do not usually obtain as large a flower size as those that are allowed to mature before shipping. For local sales, mature flowers are much more desirable.

In order to ensure that all flower buds open properly under an indoor (low light) environment, potted chrysanthemums should not be sold to the consumer until 50% of the flowers are open.

A grading system has been established by the Produce Marketing Association (PMA) and the Society of American Florists (SAF) for potted chrysanthemums. (Anonymous, 1990). There are six grades based on size and ranging from miniature (9 to 12 inches tall with at least four flower buds) to jumbo (15 to 20 inches tall with at least 18 single and 45 multiple flowers). Miniature grade plants are usually in 4- to 4½-inch pots, small and medium grade plants are in 5- to 5½-inch pots, large grade plants are in 6-inch pots, extra large jumbo grade plants are in 7- to 8-inch pots. A new series of potted chrysanthemums, grown singly in 2- or2½-inch pots, has been introduced, but grades have not been established. Five stages of floral development also are described. The stages are

1. Color showing, petals upright.
2. Plant has ¼ to ⅓ open flowers.
3. Plant has ½ open flowers.
4. Plant has to ¾ open flowers.
5. Mature, all flowers open.

The same stages are used for potted chrysanthemums grown as spray (center bud removal) or decorative (disbudded) plants.

Some tips to ensure better quality pot mums at sale time are

1. Terminate fertilization levels when flowers begin to show color. Reducing fertilizer levels can extend pot mum longevity by 10 to 14 days. Finish pot mums with potassium nitrate the last half of the growing season to improve stem strength and keeping quality. It is important to avoid high fertilizer rates in order to maximize longevity of the flowers.
2. Decreasing finishing temperatures to 56°F for the last few weeks of the schedule can improve flower color, plant strength, and keeping quality. Such a practice is usually not possible, however, because plants at different stages are in the same greenhouse.
3. Maintain good light levels as the crop matures, to maximize carbohydrate production and increase longevity.
4. Do not let flowering plants wilt. This will decrease quality and longevity.
5. Ship chrysanthemums cool (under 65°F) to help extend their keeping quality.

B. Packing

Pot mums are packaged by inserting them in paper or polyethylene sleeves to protect them during shipment. Pots are often placed upright in cardboard boxes in groups of six. For retail sales, pots are sometimes wrapped in colored foil to improve their appearance, and care tags are inserted.

C. Storage

Pot mums can be stored up to 2 weeks without reducing quality. The cooler the storage temperature, the longer the pot mums will keep. When storing pot mums for 2 weeks, the temperature should be approximately 40°F. Pot mums will not store very long if the temperature is above 56°F. If the growing medium becomes dry a few times, the longevity of the plant is decreased.

If pot mums are stored in light, they will retain their quality longer than when stored in the dark. A minimum of 50 fc should be provided. Light prevents foliage from becoming depleted of its food materials.

D. Shipping

Pot mums are often grown within 100 miles of the market. Shipping long distances is usually very expensive. This is a reason why the potted chrysanthemum is a good crop for a local grower. Pot mums will hold up well in shipment, if handled properly.

E. Keeping Quality

The most important factor in ensuring long life for the pot mum is related to maintenance of an active growing root system during the latter part of the growing season. Major causes of root losses are overwatering, high soluble salts, and root rot organisms. All of these can be controlled and will provide pot mums that will last much longer and better satisfy the customer.

Nutrition also affects the keeping quality of pot mums. Too much nitrogen fertilization (especially ammonium or urea nitrogen) at the end of the growing season will decrease the keeping life (Crater, 1973). Many growers completely withdraw fertilization during the last 2 to 4 weeks of the growing season. If a nitrogen fertilizer must be applied late in the growing season, research has shown that nitrate nitrogen is better than an ammonium nitrogen fertilizer. Therefore, for optimum longevity, nitrogen fertilization must be carefully controlled.

F. Consumer Care of Product

A pot mum can be held at the retail outlet for up to 2 weeks and still be attractive and salable if it is cared for properly. The retailer should do the following to ensure adequate durability:

1. Remove plants from the shipping box and sleeves as soon as they arrive, and water thoroughly if needed.
2. Remove broken or bruised blossoms and leaves.
3. Properly display the attractive plants.
4. Do not overcrowd pot mums in the display or storage area.
5. Water plants as they need it. Do not let them wilt, but also do not keep the medium too wet. Let the medium dry out slightly between waterings.
6. Keep plants out of direct sunlight.
7. Do not set plants in front of an air conditioner or heat ventilator.

To increase sales of pot mums, the customer should be taught how to take care of the plant. Proper watering procedures must be explained. The plant should be placed in a bright location in the home, but not in direct sunlight, and should be kept out of drafts. Broken or crushed leaves or flowers should be removed. Most pot mums will not require fertilizing in the home. Also most pot mum cultivars are not suitable as garden plants. Specific cultivars have been selected as garden mums for outdoor use.

G. Special Marketing Ideas

Higgins (1989) suggests that there are four new product forms of chrysanthemums (Anonymous, 1989). They are described as follows:

1. Hanging Basket Mums

a. Grow just like a potted mum.

b. Both potted mum and garden mum cultivars are well suited for this crop.

c. Potted mum schedules work. The basic schedule is to plant the rooted cuttings, provide 2 weeks of long days, pinch, and go to short days for flowering.

d. Diversity in flower color and flower forms is available in the cultivars suited for hanging baskets.

e. Hanging basket mums can be scheduled weekly, biweekly, monthly, or for holidays, just as for regular potted mum crops.

f. Hanging basket mums may utilize overhead space or empty bench space and help to increase the profitability of a business.

g. A well-grown hanging basket mum may be used in homes, restaurants, and other interiorscape locations.

2. Bulb Pan Mums

a. Bulb pan mums offer more symmetry and grace than a 6-inch potted mum. The plants are shorter and develop flowers on a lower plant that creates a mounded effect in the finished product.

b. Follow regular potted mum schedules. Bulb pan mums adapt well to normal schedules with very little modification. Growth regulator treatments will be needed.

c. 6-inch pan:4 plants per pot, 7-inch pan:5 plants per pot, 8-inch pan: 6 plants per pot, and 10-inch pan:8 plants per pot.

d. Bulb pan mums are useful and attractive on end tables, coffee tables, and receptionist's desks. In offices and interior plantscaping locations, a planting of bulb pan mums can provide a more pleasing and natural appearance than taller plants that are in 6-inch pots.

3. Dish Garden Mums

a. A small flowering chrysanthemum can provide a "color spot" in a dish garden.

b. There are diverse flower forms and colors available.

c. Production is in 18, 24, and 32 cell pack units.

4. Super Mini Potted Mums—4½ Inches

a. This is the same as the European potted mum.

b. It has more symmetry and balance than a typical 4-inch potted mum as it has three cuttings per pot, while the typical 4-inch pot mums consists of one pinched plant.

c. It is an excellent product for home, office, or gift use.

    d. The super mini mum is an efficient use of space. Light for only 7 to 10
       days. Most culitvars are in the 8-or 9-week response groups. The finishing
       area may be turned over at least six times a year.
    e. The extra plants per pot reduce grower shrinkage. If a stem snaps off, it
       is rarely missed.
    f. Most importantly, there is a tremendous labor savings. After spacing out
       from the propagation area, the pots are not touched until they are boxed.

The Fleurette series creates a very novel product in the miniplant series.

H. Cost of Production

Many growers do not know precisely what it costs to grow their crops of pot
mums. They determine their selling price by what other growers are getting.
Growers should know the costs of production. To determine production costs, five
major cost areas must be considered: (1) overhead, (2) labor, (3) materials, (4)
marketing costs, and (5) profit expected.

REFERENCES

Anonymous. (1989–90). Gloeckner. 1989–90 Chrysanthemum Manual. New York: Fred C. Gloeckner
       Co., Inc.
Anonymous. (1989). Pot Mum Culture. Barberton, OH.: Yoder Brothers.
Ball, G. (1985). All about pot mums. In Ball Red Book, edited by G. Victor Ball. Reston, VA: Reston Publ.
       Co., pp. 406–419.
Boodley, J. W., Kumpf, J., and Pollinger, B. (1983). "An evaluation of soil vs peat-lite media on
       post-production life of selected mums." Conn. Greenhouse Newsletter. 117: 11–12.
Cathey, H. M. (1954). "Chrysanthemum temperature study. A. Thermal-induction of stock plants of
       Chrysanthemum morifolium." Proc. Amer. Soc. Hort. Sci. 64: 483–491.
Crater, G. D. (1973). "The effect of different forms of nitrogen on growth, nitrogen content, stem
       strength, keeping quality of the standard chrysanthemum." PhD. Diss. Columbus, Ohio.
Crater, G. D., Tilt, K. M., and Vetanovetz, R. (1987). "Conversion chart for greenhouse and nursery
       operations." Agric. Ext. Serv. Univ. of Tenn. PB1249. Knoxville, TN.
Elliott, G. C., and Nelson, P. V. (1983). "Relationships among nitrogen accumulation, nitrogen assimila-
       tion, and plant growth in chrysanthemums." Physiologia Plantarum. 57(2): 50–259.
Hanan, J. J., Holley, W. D., and Goldsberry, K. (1978). Greenhouse Management. Advanced Series in
       Agricultural Science 5. Berlin: Springer-Verlag.
Hanzel, B., and Carlson, W. H. (1986). "The pot mum pinch—How to do it." BP News. 17(4): 1–2.
Heins, R. D. (1985). "Growing green." GrowerTalks. 49(2): 18.
Higgins, E. (1989). Four New Product Forms of Chrysanthemums, Yoder Grower Guides 1989. Bar-
       berton, OH: Yoder Brothers.
Ishida, A., Masui, M., Nukaya, A., and Shigeoka, H. (1983). "Effect of macro- and microelements and
       boron on growth, keeping quality and leaf marginal burn in chrysanthemums." J. Japanese Soc.
       for Hort. Sci. 52(3): 302–307.
Laurie, A., Kiplinger, D. C., and Nelson, K. S. (1969). Commercial Flower Forcing, 7th ed. New York:
       McGraw-Hill.
Mastalerz, J. W. (1977). The Greenhouse Environment. The Effect of Environmental Factors on Green-
       house Crops. New York: Wiley.

Matteoni, J. A. (1988). Mum Decline—What Is It? *BP News.* **12.**

May, R. A., and Trigiano, R. N. (1989). "Somatic embryogenesis in *Chrysanthemum morifolium.*" *HortScience.* (Abstr. 597), 139. 86th Annual Mtg. Tulsa, OK.

Miller, R. (1988). "Turning the table on pests: A new strategy for pest control on chrysanthemums." *GrowerTalks.* **51**(9): 76–82.

Nelson, K. S. (1978). *Greenhouse Management for Flower and Plant Production.* Danville, IL: The Interstate Printers and Publishers, Inc.

Nelson, P. V. (1985). *Greenhouse Operation and Management,* 3rd ed. Reston, VA: Reston Publishing Company.

Pound, W. E., and Tayama, H. K. (1984). "The effect of the plant growth regulator Accel in increasing axillary shoots of mums." *Ohio Florists' Assoc. Bull.* **656:** 5–7.

Tayama, H. K., and Roll, T. J., eds. (1989). *Tips on Growing Potted Chrysanthemums.* Bulletin FP-767. Ohio Coop. Ext. Serv. The Ohio State University, 11–1989. Columbus, OH.

Waters, W. E., and Conover, C. A. (1967). *Chrysanthemum Production in Florida.* Bull. No. 730. Agric. Exp. Stat., Univ. of Florida Gainesville.

Wodecki, M. J., and Holcomb, E. J. (1989). "Varying concentration of IBA affects rooting of chrysanthemum cuttings cv. 'Bright Golden Anne.'" *Penn. Flower Growers Bull.* **391.**

Yang, K., and Zhao, Y. (1983). "An investigation into the procedure for growing outside chrysanthemum plants." *J. of Nanjing Tech. College of Forest Products.* **1:** 31–38.

11

Gloxinias, African Violets, and Other Gesneriads

R. Kent Kimmins

Introduction to Floriculture, Second Edition
Copyright © 1992 by Academic Press, Inc.
All rights of reproduction in any form reserved.

I. INTRODUCTION

There are approximately 125 genera and over 2000 species in the family Gesneriaceae. Of this number about 300 species have been cultivated. These species have a worldwide distribution, including tropical America, Spain, Asia, and Africa. They are found on limestone cliffs, in rain forests, on the forest floor, and on mountains over 15,000 feet high (Burtt, 1967).

Individual species range in size from plants less than 2 inches tall (*Sinningia pusilla*) to small trees (*Cytandras* species). The flowers range from only slightly bell-shaped to those with a long tubular or cylindrical shape.

For the commercial grower the two most important members of the Gesneriaceae are *Sinningia speciosa* (florists' gloxinia) and *Saintpaulia ionantha* (African violet). Other gesneriads that are increasing in popularity are the *Episcia* hybrids, hybrid *Streptocarpus,* and *Achimenes* hybrids. Specific details pertaining to the culture of several Gesneriad genera are mentioned in the text and are summarized in Table I.

II. *Sinningia speciosa*—FLORISTS' GLOXINIA

Sinningia speciosa was first named *Gloxinia speciosa* in 1817 by Conrad Loddiges, an English nurseryman, after he had studied the new plant from Brazil. Gloxinia is the common name used in commercial floriculture.

Sinningia speciosa has one or more stems with paired leaves. The blade of the leaf is a large and very pubescent. The bell-shaped flowers are approximately 2 inches wide and usually pale lavender. They are produced on plants 4 to 6 inches tall, although the size may be variable. The type species came from Brazil. Most

Table I

Environmental Conditions Suitable for Growth and Flowering of Major Commercial Gesneriads

Plant	Watering of medium	Temperature (°F)	Light (klx)	Humidity (%)
		Conditions for culture		
Gloxinia	Moist	70	25.8	50–70
African violet	Moist	70	10.7–11.8	70
Episcia	Moist	64	11.8	75
Miniature Sinningia	Moist	70	10.7–11.8	70
Streptocarpus	Dry slightly between waterings	60	12.9	70
Achimenes	Moist	64	11.8	70

florists' gloxinias are of the convariety *fyfiana,* the name originating for *Gloxinia* ×
fyfiana (Moore, 1957). The red cultivars are commercially most popular (Fig. 1).

Through hybridization and selection, gloxinias may be large-growing cultivars
for 5- to 6-inch pots or compact types grown in 4-inch pots. Depending on the type,
they may be single- or double-flowered, with colors ranging from pure white
through pink, lavender, and red, to dark purple. Popular single cultivars include the
'Velvet' series, the 'Bridget's Best' series, and Small's 'Super Compact' series for
4-inch pots.

Plants are grown from seed for commercial production. A plant with a large
single head of flowers can be produced using this method in approximately 6 to 7
months in a 5-inch pot or 5 to 6 months in a 4-inch pot.

There are some specialists who produce and sell seedlings. The seeds are sown
on a soilless medium and placed under intermittent mist in a shaded house at 70°F
night temperature for rapid germination. When the seedlings are large enough to
handle, they are transplanted into $2\frac{1}{2}$-inch pots. Either plastic or clay pots may be
used successfully. A soilless medium including peat moss, sand, perlite, and
vermiculite can be used. When the seedlings are approximately 3 months old, they
are shipped to the grower, where they are transferred into the final pot.

The seedling production procedure can also be used by the retail or wholesale
grower if warm greenhouse conditions are available and space is not limited. When
the seedlings are transplanted by the grower, it may be easier to transplant the
seedlings into flats as soon as they can be handled. The plants should be spaced
1 to 2 inches apart. When the foliage of the plants begins to touch, the plants

Fig. 1. *Gloxinia* greenhouse production.

should be transferred to a 4-, 5-, or 6-inch azalea pot, depending on the finished size of the cultivar being grown. Plastic or clay pots may be used with success.

The media for growing the transplanted seedlings can be soilless mix or a soil mixture such as 1:1:1 (light organic soil, peat moss, and coarse sand or perlite, on a volume basis). A pH of 6.0 should be obtained by adding dolomitic limestone to the medium. Plants may be potted with the bottom two leaves buried in the pot for a more sturdy finished plant, should be spaced on 10-, 12-, or 16-inch spacing, and watered immediately. It is very important that water not be allowed to remain on the foliage. A fungicide can be added with the water to prevent disease from injuring roots or leaves, especially if the plants are shipped.

The best production of gloxinias is achieved at a night temperature of 70°F. Optimum growth can be obtained with a radiant flux of 19.4 to 26.9 klx. The growing medium should never be allowed to become dry. Tube watering and mat watering methods have been used with excellent results. The fertilization program should begin immediately after transfer to the finishing pot. A complete fertilizer (such as a 15–16–17 peat-lite special) can be alternated with calcium nitrate at the rate of 2 pounds per 100 gallons of water. Fertilization is recommended every 10 days and is increased as the plants grow larger. Some growers do use a 200 ppm nitrogen, phosphorus, and potassium constant fertigation program using 15–15–15 or 20–20–20 at 2 to 3 pounds per 100 gallons. There have been some problems in the winter using the 20–20–20, which is high in ammoniacal nitrogen. Osmocote at $\frac{1}{4}$ to $\frac{1}{2}$ teaspoon per 6-inch pot can be used effectively. With sufficient natural light and temperature, additional lighting should not be necessary. If, however, the night temperature is 60°F, additional lighting may be beneficial to reduce flower delay. Good results are obtained with 100 watt incandescent bulbs spaced 4 feet above the plants and turned on 4 to 5 hours each night. The additional lighting seems most effective immediately after the seeds have germinated. When plants are well budded, more uniform flowering may be obtained by removing the first two flower buds as soon as color is evident.

The use of chemicals to control plant height might be necessary, particularly in the summer when heavy shade to reduce temperature causes stretching of the internodes. Sweet (1983) noted that B-Nine, at a diluted strength of 1000 ppm in solution, worked well when applied 1 to 2 weeks after transfer into the finished pot. He noted that late application would not shorten the plant enough to be a quality plant. Sydnor *et al.* (1972) found that B-Nine applications resulted in compact plants and that flower color was intensified on pink cultivars used in the study.

III. *Sinningia* × *hybrida*—COMPACT GLOXINIAS

Sinningia × *hybrida* is a comparatively new compact gloxinia hybrid. Seed can be sown in January for June flowering. The seeds should be sown in a fine medium at 72° to 77°F for good germination. The fine seed should not be covered. In 2 to 4 weeks after sowing, seedlings can be transplanted into flats or 2 $\frac{1}{4}$-inch pots. If the seed are sown in plugs, they can be transplanted into the finishing pots, usually $3\frac{1}{2}$- to $4\frac{1}{2}$ inch pots. As with most gloxinias, the seedlings should be planted slightly deeper than they had been in the seed flats (Aimone, 1987).

When the seedlings in the flats are 6 to 8 weeks old, they can be transplanted into the finishing pots. They should be planted in a well-drained medium that is high in organic matter, with a pH of 5.5 to 6.0. Fertilization programs are similar to those used for other gloxinias. The plants can be grown at 64° to 68°F until the buds are seen, when the temperature can be dropped to 60°F. Light shade should be provided to produce light levels of 25.8 klx.

The cropping time is less than for other gloxinias. More flowers are produced, but they are smaller.

IV. *Saintpaulia ionantha*—AFRICAN VIOLET

The African violet is considered one of the most popular flowering potted plants in the United States (Fig. 2). Hybridizers have greatly improved the selection of cultivars that are available to the consumer.

The African violet was discovered in 1892 in Tanga, East Africa, by Baron Walter Von Saint Paul, governor of German East Africa. Herrmann Wendlan, a prominent German botanist, named the genus *Saintpaulia* in honor of its discoverer.

Fig. 2. African violets on mat watering system.

The African violet was first introduced into the United States about 1894. George Stumpp, a New York florist, purchased two plants in Germany and brought them to the United States. In 1927 Walter L. Armacost of the Armacost and Royston Nursery in Los Angeles imported the first seed into the country from England and Germany, thus starting the first commercial production of the African violet in the United States. The first plants named and offered for sale in this country were grown by Armacost and Royston in 1936. Several of the original cultivars are still available today.

African violet plants have very short stems with leaves arranged in a rosette. The blades of the leaves vary from ovate–elliptic to round, and they are usually hairy. The margins may be entire or may be undulate. Flowers are produced in a cyme with five to ten flowers per cyme. The calyx is five-lobed. The corollas are usually blue, blue-violet, bicolored, or nearly white.

Most of the hybrid African violets developed in the United States are progeny of the original 10 named cultivars introduced in 1936. Some of the original cultivars are still available, such as 'Blue Boy' and 'Neptune.' The cultivars now being grown include 'Rhapsody,' 'Ballet,' and 'Optimara.' Many of these cultivars originated in Europe. The 'Nortex' series was developed in the United States. These cultivars have been bred to produce large, long-lasting flowers that stand well above the foliage.

New forms of African violets are causing increased interest among some breeders. The miniature African violets and the trailing African violets are becoming popular. The 'Optimara' series 'Little Jewel' and 'Little Indian' are in great demand. The various sizes of the African violets are adding additional variety to the market.

African violets can be grown from seed, but only a few cultivars will come true from seed. A flowering plant can be produced from seed in about 10 months. Seed should be sown on a screened soilless medium or milled sphagnum peat moss. A 70°F night temperature with high relative humidity should be maintained. When the seedlings are large enough to handle, they should be transplanted to 4-inch pots and filled with a highly organic, well-drained medium. Many of the soilless media provide excellent conditions for active growth. If a soil medium is used it should be steam-pasteurized. The seedlings should be grown under 10.7 to 11.8 klx of light.

The principal commercial method for propagating African violets is by leaf petiole cuttings. Mature leaves that are firm and a good green color are selected. The petiole is cut at a slant to about 1 to $1\frac{1}{2}$ inches long. A rooting hormone may be used to promote more rapid root formation. Flats are filled with a pasteurized rooting medium. Many different materials are used by growers. A few of the media include peat moss and sand; peat moss, sand, and vermiculite; and vermiculite alone. The leaves are stuck into the prepared flats so that leaves do not touch each other. The flats are then placed in a shaded 70°F greenhouse. In approximately 8 to 12 weeks the plantlets will be ready to be transplanted. They are separated from the leaf when they are about 1 inch tall and are potted into a $2\frac{1}{2}$-inch pot, and then later shifted to a 4-inch pot. It takes 8 to 10 months from propagation to flowering, when plants are grown in 4-inch pots.

There has been much interest in producing African violets through tissue culture. Both petiole and leaf tissue micropropagation have been successful. In some micropropagation experiences the cultivars have not always remain stable, but Larson (1985) reported that the cultivar 'Valencia,' a pinwheel-flowering chimera, produced plants 96% true to the cultivar. Tissue cultured plantlets take about 1 year to attain a flowering, salable stage.

Many media are used for finishing seedlings or rooted plantlets. Soil mixtures that are loose, well-drained, and high in organic matter are often suggested for African violets. Soil mixtures should be pasteurized before use. Many new soilless mixtures containing peat moss, vermiculite, perlite, sand, and bark are now available for growing violets, and excellent results may be obtained if the mixture can stay uniformly moist (Gurdian and Nell, 1984).

Plants should be potted at the same depth as they were growing in the propagation medium. Clay or plastic azalea pots may be used for African violets. Plastic pots are preferred by some growers because the pots are clean, lightweight for shipping, and retain moisture better. Leaf petioles also do not bruise, as they do when they rub on the rims of clay pots.

African violets grow most rapidly at a night temperature of 70°F. Bench heating has resulted in shortened production time of 2 weeks when the soil temperatures were held at 64°F. Plant quality and flowering were also improved (Fig. 3). Several

Fig. 3.   Bench heating in African violet production.

growers have found that a slightly lower night temperature intensifies the flower color.

Light intensity for best growth and flowering is 10.8 to 11.8 klx. If the irradiance is above 13.9 klx, chlorophyll destruction will occur, causing the leaves to turn yellow. African violets can be grown very satisfactorily under artificial light. Stinson and Laurie (1954) found that excellent plants were grown under daylight fluorescent lamps at 6.5 klx for 12 to 18 hours daily.

Media in which violets are grown should be kept moist, but not soaked, at all times. If plants dry to the wilting stage there can be root damage, particularly if the soluble salts level is high. This is one reason that most media contain large amounts of organic matter. The method of watering is very important in preventing damage to the foliage. Damage called ring spot occurs when the water temperature is about 15°F colder than the air temperature. Most growers use tube or mat watering systems to avoid foliage damage, or they apply water during the cooler times of the day.

The use of supplemental carbon dioxide at the rate of 800 to 1000 ppm may be beneficial, especially if the greenhouse ventilators are closed for a long period of time. Increased carbon dioxide may reduce the amount of light required to obtain a high quality plant (Stromme, 1985).

V. *Episcia* SPECIES—FLAMING VIOLET

Native to Central and South America and the West Indies, 10 species of *Episcia* are in cultivation. Three of the species, *Episcia cupreata, Episcia lilacina,* and *Episcia reptans,* have been hybridized for commercial use.

Most of the hybrids originated from *Episcia cupreata,* a native of Colombia and Venezuela. The copper-colored leaves are 3 to 4 inches long and 2 to 3 inches wide. Hybrids of *E. cupreata* have foliage from 1 to 8 inches long. Flowers of the hybrids are orange-red with a yellow throat, and are sometimes spotted.

Episcia lilacina originates from Costa Rica, Nicaragua, and Panama. The flowers are large, lilac-blue, often with a lemon-yellow throat. The foliage is a very dark bronze with a lighter green veination, and approximaely the same size as that of *Episcia cupreata.*

The third *Episcia* species used in hybridization is *Episcia reptans,* which originated in Colombia, Brazil, and Guyana. The foliage is dark green with pale green and silver markings. The flowers are deep red with a pink throat.

Some of the most colorful and popular cultivars are 'Chocolate Soldier,' 'Painted Warrior,' 'Tri-Color,' 'Pinkiscia,' 'Chocolate and Cherries,' 'Mrs. Fanna Haaga,' and 'Filigree.' Another hybrid from two different species is *Episcia cygnes (E. dianthiflora* × *E. puntata)* (Fig. 4).

Most *Episcia* hybrids are propagated by stolons. The most desirable plants are obtained when young, medium-sized, 2- to 3-inch stolons are used for cuttings. All but the top four leaves should be removed. If the leaves remaining are too large,

Fig. 4. *Episcia* in flower.

the bottom two leaves may be reduced in size. The stem should be cut $\frac{1}{2}$ inch below a node. An indolebutyric acid (IBA) rooting hormone should be applied. Cuttings can be stuck in pots or flats filled with any well-drained pasteurized medium. Various media have been used, including the peat-lite mixes. Cuttings should be maintained at a relative humidity of 80 to 90% if possible or under intermittent mist. Cuttings should be shaded until they are rooted, which should take 2 to 3 weeks. After rooting, a dilute fertilizer solution of a complete fertilizer should be applied.

Cuttings grown in flats should be potted into $2\frac{1}{2}$- to 3-inch pots. Other methods of propagation, such as leaf petiole cuttings as described for African violets, have the disadvantage of not producing plants true to the parent leaf variegation.

Plants should be shifted to larger pots as they grow. Some growers make hanging baskets, and others sell the plants in small pots or shift them to 4-inch pots for later sale. Regardless of the final use, the plants should be spaced so the leaves on adjacent plants do not touch each other.

The best hybrid *Episcia* plants are produced at 64°F. Some hybrids will die if the temperature is below 60°F. The optimum temperature in the greenhouse is 64°F at night and up to 90°F during the day, with a 75% relative humidity. At higher temperatures and humidity it is very important to have good air circulation to prevent mildew.

Good flower production is achieved at a light intensity of 11.8 klx. Hybrids with light colored leaves will flower at a lower intensity, but the dark-leaved hybrids need an intensity of at least 11 klx to flower.

The growing medium should remain moist, and no cold water should be allowed to touch the leaves. Tube or mat watering systems are recommended for production of these plants. The fertilization program recommended for African violets can be used on episcias.

VI. *Sinningia pusilla* AND THE MINIATURE HYBRIDS

The smallest of the gesneriads is *Sinningia pusilla*. It was introduced into the United States in the late 1950s, although it was collected and described in the early nineteenth century. The leaves are about 1 inch long. Their very short stems are almost not apparent. The 1-inch tubular, lavender-to-lilac colored flowers are displayed above the crown of the foliage. *Sinningia pusilla* has been used for germ plasm for many of the miniature hybrids. Some of the most popular cultivars are *S. pusilla*, 'Dollbaby,' 'Little Imp,' 'Bright Eyes,' 'Snow Flake,' and 'Coral Baby.'

Propagation of these plants can be accomplished by seed or tubers. The species can be obtained from seed but the hybrids must be started from tubers to obtain the true cultivars. A well-drained medium similar to that used for African violets can be used. Plants should be spaced so leaves do not touch each other on adjacent plants. The plants can be grown to blooming size in 2- to 3-inch pots and should not be crowded in pots that are too small.

Good plant production is obtained if the night temperature does not exceed 70°F. Light intensity of 10.7 klx is adequate for good foliage and flower production. Excellent flowers can be produced under fluorescent lights, with 12 to 16 hours of illumination daily.

The miniature hybrids should be grown at a relative humidity of 80%. The growing medium should remain moist because if it becomes dry, the plants will go dormant. The dormant period may be only for a few weeks, but it could be several months. Plants should be fertilized at the recommended rates for African violets every 7 to 10 days.

VII. STEMLESS *Streptocarpus* HYBRIDS

Most of the *Streptocarpus* plants available to growers and consumers today are hybrids of many different species. In England in 1947 the cultivar 'Merton Blue' was crossed with *Streptocarpus johannis*. The result was a cultivar with excellent characteristics called 'Constant Nymph.' In 1952, the Weismoore hybrids were introduced in Germany. Radiation treatment was applied in the 1960s to obtain additional 'Nymphs.' Addtional breeding efforts have provided a broader color range of Streptocarpus plants within the two groups.

Most of the hybrids produce leaves up to 12 inches long in an asymmetric clump, similar to a rosette. Flower stalks carrying many flowers arise from the center of the clump, producing a well-balanced potted plant (Fig. 5).

Fig. 5. *Streptocarpus* hybrid.

Most commercial propagation is by leaf sections to maintain the cultivars. A mature leaf with the midvein removed is one of the methods recommended. The two pieces are placed lengthwise in propagation media (peat-lite type). As many as 20 plants may be obtained in this manner from one original leaf. The medium should be kept moist at 70°F. Plantlets appear in about 2 months and are large enough to transplant in about 3 months. A blooming plant can be obtained in about 6 months.

Streptocarpus plants should be grown in a well-drained medium. Unlike the medium for the African violet, this medium should not be acidic. Some growers suggest no peat moss at all, but others use a mixture such as (1:1:1) peat moss, vermiculite, and perlite or (1:1:1) peat moss, soil, and perlite. Azalea pots should be used because of the shallow root systems of these plants. Plastic pots 2½ or 4 inches in size are usually used to grow *Streptocarpus*. Pots should be spaced so the leaves on adjacent plants will not touch as the plants grow.

Temperature is extremely important in *Streptocarpus* culture. Most of the hybrids will do poorly if the day temperature exceeds 80°F. Night temperatures of 60°F are satisfactory. A very high humidity is not necessary; 50% humidity is sufficient.

Streptocarpus plants will produce flowers if they have a light intensity of about 12.9 klx, slightly higher than that for African violets. With sufficient light, hybrids bloom throughout the year.

The growing medium should be allowed to dry slightly between waterings. These plants are easily overwatered, and the root system can be damaged by overfertilization. Balanced fertilizer should be used at one-half the manufacturers recommended rate every 2 weeks for vigorous blooming plants.

VIII. *Achimenes*

Achimenes are more popular now because of the new cultivars that are being offered to the grower and the consumer. A wide range of colors is available from recent breeding efforts.

Plants can be grown from seed, cuttings, or rhizomes. Seed sown between December and March produce flowering plants in 5 to 7 months. Seed should be sown on a well-drained medium, such as a peat-lite mix. The seed, which are extremely small, should not be covered, but the medium must be kept moist. Germination should take 14 to 21 days when kept at 75° to 80°F.

Six to eight weeks after sowing, the seedlings should be ready to transplant into cell-packs. A peat-lite mix should be used for the medium. After 4 to 6 weeks in the cell-packs, the seedlings can be planted into their final pots or baskets. Plants should be adequately spaced to avoid stem elongation. Pinching may be necessary to prevent excessive shoot elongation and to promote lateral branching. Good ventilation is necessary to prevent disease.

Achimenes should be grown in bright, filtered sunlight up to 5000 foot-candles (fc). Direct sun can damage leaves and flowers. Plants can be grown at temperatures ranging from 75° to 90°F during the day, and 67° to 70°F at night.

A 20–20–20 fertilizer at 200 ppm nitrogen should be applied to the plants at each watering. A 15–30–15 fertilizer can be used to promote additional flowering (Wiles, 1988). The fertilization should continue until the plants start to go dormant.

Warm water (70° to 75°F) will produce better quality plants than cold water. Plants should be watered in the morning, and the medium kept evenly moist. If the plants are allowed to dry too much they will go dormant.

Growth regulators may be effective on some cultivars. B-Nine (5000 to 7500 ppm) has been applied as a single spray application when the shoots are 2 to 3 inches long, but label clearance is lacking.

Dormancy usually begins in early fall. The leaves will begin to turn yellow, and flowering will stop. At this time watering frequency should be reduced, and fertilization should stop. When the stems have turned brown, the plants should be cut back and stored at 50°F. Dormancy lasts 2 to 5 months, depending on the cultivar (Wiles, 1988).

It takes 2 to 4 months for plants to flower when they are started from rhizomes. They are usually started between February and April. The rhizomes are laid on a peat-lite medium and covered with one-half inch of medium. They can be started in pots or cell-packs and watered well initially, then they are watered sparingly until the rhizomes have started to grow. Sprouting should begin in 3 to 6 weeks when the rhizomes are held at 60° to 75°F nights. After sprouting, plants can be grown at 60°F and kept evenly moist. After growth has started, the plants can be moved to their final pots or hanging baskets (Fig. 6).

Cuttings can be taken, and flowering plants can be produced in 3 months. Depending on the time the cuttings are obtained, plants may be weaker than the plants grown from seed or rhizomes.

IX. PESTS AND DISEASES

There are a few pests that can cause problems with the Gesneriads. Most of the pests can be controlled with chemical sprays, but some of the genera are very sensitive to certain chemicals, and serious damage can result. Only recommended pesticides should be used. The most common insect pest is the mealybug. African violets, episcias, *Streptocarpus,* and *Achimenes* can be infested if the pest is not controlled early. Usually mealybugs will be found in the axils of the leaves or on the underside of the foliage.

Another pest found on Gesneriads is cyclamen mite. It is particularly bothersome on African violets and gloxinias. Cyclamen mites cause distorted growth in the crown. The leaves become very brittle and grow in a dwarfed, curled manner.

Spider mites may be found on gloxinias. In very large infestations, the undersides of the leaves may be covered with webs.

Fig. 6. *Achimenes* basket in full flower.

An insect that has become more widespread is the western flower thrip. Damage on the flowers of most *Gesneriads* appears as areas of broken color along the edges of the flowers. On dark colored flowers, the damage appears as light or almost white areas along the flower edge. On light colored flowers, the damage is often brown streaks. Foliage is affected by distortion of new, expanding leaves. Often the foliage appears worm eaten.

At certain times loopers or army worms can be problems on gloxinias. Plants should be monitored frequently, because such pests can severely damage a crop very quickly.

Root knot and foliar nematodes can be problems on African violets. The best cure is to discard infected plants. Pasteurization of the growing media, pots, benches, and other equipment is the primary preventative measure. Several of the diseases associated with Gesneriads can be prevented or controlled by proper spacing, good air circulation, proper sanitation, and proper watering practices. Crown rot, root rot, stem rot, *Botrytis,* and mildew can all be controlled with proper care.

Viruses, especially the tomato spotted wilt virus, may be found on Gesneriads. This virus is transmitted by the western flower thrip. If plants are found to have a virus, they should be destroyed. It is very important that plants used for propagation be virus-free. If cuttings are made, the knife should be flamed between each plant, to prevent spreading the virus to other plants. Obviously a major method of control of tomato spotted wilt virus is control of the vector. Very fine-meshed screens can be placed over ventilating inlets, to prevent entrance of the thrips. Yellow sticky traps also can be used to detect the presence of thrips and other insect pests.

X. HANDLING

All members of the Gesneriads can be handled in a similar manner. Many of the plants will be shipped in paper or plastic sleeves. Care should be taken when shipping large leaved plants, such as gloxinias, that the leaves are not broken because they are so brittle as well as large. Many growers are producing smaller plants with more flexible leaf petioles to prevent this problem.

There has been interest in the use of silver thiosulfate (STS) on plants that drop their flowers during shipment. Agnew *et al.* (1985) used the STS on *Streptocarpus* plants. The longevity of other genera may be enhanced with STS, but the chemical does not have EPA approval.

Plants should be shipped by carriers who can control the temperatures in their trucks. Many of the Gesneriads can be damaged or killed with temperatures below 50°F or above 90°F.

When the plants are received, they should be unpacked immediately. They should be given light, placed in a warm temperature, and watered with warm water. The consumer should be informed concerning the proper care of the plants. This is best done by care tags that can be placed on the plants before they are sold by the retailer.

REFERENCES

Agnew, N. H., Albrecht, M. L., and Kimmins, R. K. (1985). "Reducing corolla abscission of *Streptocarpus* × *hybridus* under simulated shipping conditions with silver thiosulfate." *HortScience.* **20**(1): 118–119.

Aimone, T. (1985). "Culture notes—*Streptocarpus.*" *GrowerTalks.* **48**(11): 22–26.

Aimone, T. (1987). "Culture notes—*Sinningia hybrida.*" *GrowerTalks.* **51**(4): 23.

Bilkey, P. C., and McCown, B. H. (1979). "*In vitro* culture and propagation of *Episcia Sp.* (flame violets)." *Proc. Am. Soc. Hortic. Sci.* **104**(1): 109–114.

Burtt, B. L. (1967). "Gesneriads as a family." *Plants and Gard.* **23:** 54–57.

Gurdian, F. A., and Nell, T. A. (1984). "Fertilization of African violet in various media." *Benchmarks.* **3**(2): 1–3.

Johnson, B. (1978). "*In vitro* propagation of gloxinia from leaf explants." *HortScience.* **13**(20): 149–150.

Koch, G. M., and Holcomb, E. J. (1982). "Recommendations for producing Improved Red Velvet gloxinias." *Florists' Review.* **171**(4430): 32–33.

Langefeld, J. (1984). "Culture notes—*Sinningia speciosa.*" *GrowerTalks.* **48**(7): 16–20.

Larson, R. A. (1985). "African violet chimeras: A practical use of micropropagation." *N.C. Flower Grower's Bull.* **29**(6): 9–10.

Moore, H. E., Jr. (1957). *African Violets, Gloxinias, and Their Relatives.* New York: Macmilan.

Neal, K. (1987). "African violets." *Greenhouse Manager.* **6**(7): 36–42.

Sanderson, K. C., Martin, W. C., Shu, L., Evans, C. E., and Patterson, R. M. (1985). "Fertilizer and irrigation effects on medium leachate and African violet growth." *HortScience.* **20**(6): 1062–1065.

Stinson, R. F., and Laurie, A. (1954). "The effect of light intensity on the initiation and development of flower buds in *Saintpaulia ionantha.*" *Proc. Am. Soc. Hortic. Sci.* **75**: 730–738.

Stromme, E. (1985). "Gesneriaceae." In *CRC Handbook of Flowering,* edited by A. H. Halevy, vol. III. New York: CRC Press, pp. 48–52.

Sweet, J. (1983). "Cultural techniques for gloxinias, exacums, streptocarpuses and begonias." *Florists' Review.* **171**(4441): 23, 25.

Sydnor, T. D., Kimmins, R. K., and Larson, R. A. (1972). "The effects of light intensity and growth regulators on gloxinias." *HortScience.* **7,** 407.

Wiles, L. S. (1987). "Culture notes—*Streptocarpus.*" *GrowerTalks.* **51**(5): 16–18.

Wiles, L. S. (1988). "Culture notes—*Achimenes.*" *GrowerTalks.* **51**(12): 12.

12

Poinsettias

David E. Hartley

Introduction to Floriculture, Second Edition
Copyright © 1992 by Academic Press, Inc.
All rights of reproduction in any form reserved.

I. INTRODUCTION

A. History

The poinsettia, *Euphorbia pulcherrima* (Willd. ex Klotzch), is native to Mexico, and one of a large number of plants belonging to the family Euphorbiaceae. The genus *Euphorbia* alone contains some 700 to 1000 species. The poinsettia was cultivated by the Aztecs in an area near the present-day Taxco before Christianity came to the Western Hemisphere. During the seventeenth century, a group of Franciscan priests near Taxco began using the flower in nativity processions because of the poinsettia's brilliant red color and holiday blooming time (Ecke *et al.,* 1990). Poinsettias were first introduced into the United States in 1825 by Joel Robert Poinsett, from whom the plant gets its common name. Poinsett was serving as the first United States ambassador to Mexico when he found the flowers growing near Taxco. Poinsett, also a botanist, had some plants sent to his home in Greenville, South Carolina, from there he distributed plants to some botanical gardens and friends. The poinsettia was grown commercially and sold for Christmas during the late 1800s and early in the 1900s.

During the early 1900s, the Ecke family began cultivating field-grown poinsettias in southern California for the local fresh cut-flower market. During the 1920s cultivars were selected for both cut flower and pot plant production. As improved sports appeared, they were sold to Eastern greenhouse operators for flowering pot plant production. The modern era of poinsettia culture began with the introduction of the seedling cultivar 'Oak Leaf' (Ecke *et al.,* 1990). From 1923 until the early 1960s, all of the principal cultivars of any commercial importance were selections or sports from this original 'Oak Leaf' seedling, most of them having been selected and developed by Paul Ecke of Encinitas, California.

The poinsettia has become a symbol of Christmas in many parts of the world, especially in the United States, Western Europe, and Canada. Its popularity has steadily grown until it is now the most important flowering pot plant in the United States. More than 41 million pots with a wholesale value of more than $147 million were reported to have been sold in 1988 from 28 of the largest floriculture production states (Anonymous, 1989). By comparison, about 20 million pots with a wholesale value of $47 million were sold in 1977. The leading states for poinsettia production are California, Pennsylvania, Ohio, and Florida.

Part of the popularity of poinsettias can be attributed to the genetic improvements that have been made in the cultivars since the 1960s. The poinsettias of today are longer lasting, which adds value for the consumer, and they are easier to grow, which makes their production less challenging for the grower. New colors, a variety of sizes, and interesting forms have also been developed to enhance the interest and value of poinsettias.

B. Botanical Information

The poinsettia is a shrubby perennial plant that does not survive freezing temperatures. Like many other species in the Euphorbiaceae family, poinsettias have specialized cells that produce latex. When plant tissues are wounded or cut, this white, milky fluid flows freely from the wound. The latex sticks to the hands and can become a nuisance when vegetative cuttings are being harvested.

Poinsettias are photoperiodic and flower in response to a change in daylength (Grueber, 1985b). They are classified as short-day (more correctly, long-night) plants that initiate and develop flowers as the nights become longer than the days. Under natural daylength conditions in the Northern Hemisphere, poinsettias initiate flowers in early fall. Full flower development occurs about 8 to 10 weeks later under greenhouse temperatures, or just in time for the Christmas holiday season.

Flowers of the genus *Euphorbia,* including the poinsettia, are characterized by a single female flower, without petals and usually without sepals, surrounded by individual male flowers all enclosed in a cup-shaped structure called a cyathium. On the side of the cyathium, one to four or more glands are borne. On the poinsettia these glands secret a sweet honey. The showy red, pink, or white portions of the plant, popularly referred to as the flower, are modified leaves, called bracts. One bract is attached near the base of each cyathium. There may be 20 to 30 cyathia and bracts in each "flower." In addition, there may be one to five colored leaves between the green leaves and the bracts, often referred to as transitional bracts.

C. Development of Cultivars

The poinsettia cultivars grown commercially in greenhouses before 1960 were mutations of a few seedling selections and their characteristics, responses, and cultural requirements were quite similar. Cultivars introduced after 1960 have been seedling selections of diverse parentage that makes generalization of specific responses much more difficult.

With the introduction of the 'Paul Mikkelsen' cultivar in 1963, poinsettias entered a new era. This cultivar, with stiff stems and foliage retention characteristics, provided the trade with the first long-lasting cultivar of any commercial importance. In 1968, an attractive, long-lasting red cultivar, 'Eckespoint C-1' was introduced. This cultivar made an excellent branched plant and provided the grower with a predictable multiflowered plant of high quality. European growers began showing more interest in poinsettias in the early 1960s. In 1964, the first 'Annette Hegg Red' was grown in Norway. In 1968, it was grown by all major producers in Europe and was introduced into the United States and Canada. The original 'Annette Hegg Red' sported into several other red, pink, white, and bicolored cultivars. The Hegg cultivars provided an entirely new type of multiflowered plant because of their ability to produce five to eight branches from a pinch and their ease of growing.

During the early 1970s, Gregor Gutbier of Linz, West Germany, began poinsettia breeding and ultimately introduced several new cultivars. 'Gutbier V-10 Amy,' introduced in 1976, and its white, pink, and marbled sports have a short growing habit. These cultivars are grown extensively in southern Florida, primarily because they will initate flowers under the high temperature conditions of the fall and because they stay relatively short. Generally, they are not recommended as major cultivars for any geographical area except southern Florida. 'Gutbier V-14 Glory' was introduced in 1979. This medium-tall cultivar has wide bracts and branches well. It has become the major poinsettia cultivar in the southern part of the United States. *Botrytis* during the later stages of flowering can be a problem with this heavy, bushy plant. There are pink, white, and bicolored sports of 'Gutbier V-14 Glory.' Eduard Gross of Blanzac, France, has been breeding and selecting poinsettias for the French market, where they sell flowering poinsettias year round. In 1988, 'Gross SUPJIBI' was introduced in North America. This cultivar is short with thick, wide tetraploid-type bracts, and it branches quite well.

In 1988, 'Eckespoint Lilo' was also introduced (Fig. 1). Its bright ruby red bracts contrast with dark green foliage and makes an interesting presentation. This cultivar requires certain cultural techniques to ensure good branching. Positive features of this cultivar are early blooming, fast recovery of drooping bracts after unsleeving, and excellent foliage retention for the consumer. 'Eckespoint Celebrate,' introduced in 1988, brought a "new look" to the market. The big bright-red bracts stay perfectly erect after sleeving and through its life with the consumer. It is particularly

Fig. 1.   Pinched, multibloom plant of 'Eckespoint Lilo,' a poinsettia cultivar with excellent keeping qualities.

suitable for a fancy single stem plant. A free branching sport, 'Eckespoint Celebrate 2' was available in 1990 for pinched, multibloom plants. In 1989, new novelty colors, a golden yellow 'Eckespoint Lemon Drop' and a bronzed pink 'Eckespoint Pink Peppermint,' were introduced. Several new forms are being developed, particularly some incurved bract and leaf types.

II. GROWTH REQUIREMENTS AND PLANT RESPONSES

A. Flowering

The poinsettia flowers in response to a change in daylength and is classified as a short-day plant. Not unlike the chrysanthemum, another important florist crop, flowers are initiated in the poinsettia as the days become shorter. Under natural daylength conditions in the Northern Hemisphere, flowers are initiated in early fall. Research at North American universities has established that flower initiation occurs in late September when the daylength becomes 12 hours and 20 minutes or less (Grueber, 1985a). Considering the fact that twilight extends the effective daylength beyond sunrise and sunset, flower initiation occurs approximately September 23 to 27 under normal conditions. The actual development of flower parts, determined by microscopic examination, does not occur until 7 to 10 days later. For all practical purposes, flower buds are initiated approximately October 1 in the Northern Hemisphere.

The development of a poinsettia flower from a bud to a fully developed and salable flower takes about 8 to 10 weeks under optimum greenhouse conditions. Two principal conditions must be met to ensure the normal development of a poinsettia flower. First, the daylength must continue to shorten, and second, the temperatures must be within an optimum range. If, for example, daylength were controlled by lighting or the use of a black cloth so that the poinsettia received 12 hours of light and 12 hours of darkness, flower bud initiation would occur but further development of the flower would be delayed. Under natural daylength conditions in the Northern Hemisphere, the days do become shorter and the flowers develop normally. If it is desirable to flower poinsettias outside of their natural flowering season, it is advisable to reduce the daylength to a 10-hour period by covering the plants with black cloth for a 14-hour period. The rate of flower development is proportional to the average greenhouse temperature (Berghage et al., 1988). The optimum night temperature range for most poinsettia cultivars is 60° to 70°F. Within this temperature range, normal flower initiation and development occur; however, flower development is more rapid at 70° than at 60°F.

Because the natural flowering of poinsettias occurs in late November and early December in the Northern Hemisphere, the poinsettia has become a part of the Christmas holiday tradition (Fig. 2). It has not been necessary to manipulate the flowering time of the poinsettia except that the popularity of this crop has created a demand for plants that are in full bloom early in November. With most of today's

Fig. 2. A commercial greenhouse of flowering poinsettias for the Christmas holiday season.

important cultivars, it has become necessary to initiate flowering in part of the crop earlier than would occur under natural daylength conditions. This necessitates the use of a black cloth to shorten the daylength, usually beginning in early to mid-September and continuing to mid-October when the natural daylength is short enough to ensure the continued development of the flowers.

B. Light

Light intensity, quality and duration must all be considered for the culture of poinsettias. All three factors are manipulated to ensure high quality flowering plants. The energy requirements, or light intensity requirements, for the poinsettia are not as great as for some other important flowering crops. However, they do respond to high light intensities during most of their growing period. In the vegetative stages, light intensities of 4000 to 6000 foot-candles (fc) are adequate assuming, the plants receive these light levels on a daily basis. Under most circumstances, a greenhouse grower would encounter both sunny days and cloudy days so that the light intensities may vary from 10,000 fc on a sunny day to 2000 or less on a cloudy day. Therefore, poinsettias benefit from being grown with all of the light available in a growing structure, except during times of the year when light intensity and temperatures are unusually high. For example, it may be advisable to shade greenhouses during the summer months, not only to reduce the light intensity but to reduce heat stress.

Light duration or daylength control is important in the control of flowering. This becomes particularly important for the specialist propagator of poinsettias who must maintain vegetative stock plants throughout the year as a source of cuttings. Natural daylengths during the winter months in the Northern Hemisphere are short enough to induce flowering. Therefore, the daylength must be extended by artificial light to keep the plants vegetative. This is normally done by lighting in the middle of the night to break up the long, dark period from approximately September 15 to April 15.

Light quality is important for both flowering and stem elongation in poinsettias. Red light is more effective and efficient than blue light in preventing flower initiation in vegetative plants. Incandescent lamps, therefore, are usually chosen over fluorescent lamps in maintaining vegetative stock plants. Stem elongation, however, is enhanced by red light. One of the reasons that plants stretch when they are grown close together is that the natural light is filtered by leaves before it reaches the stems. Therefore, by providing more space between plants, the unfiltered light that reaches the stems is less conducive to stem elongation.

C. Temperature

Poinsettias originated from semitropical areas and will not withstand freezing temperatures. Even temperatures above freezing and below 45°F can induce chilling injury in the poinsettias. Optimum temperatures for poinsettia growth are between 60° and 80°F. Growth and development of poinsettias below 60°F is very slow. Temperatures above 80°F may be detrimental to optimum growth of the vegetative plant. It is also well established that night temperatures higher than 72° to 75°F may be detrimental to flower initiation and development in the poinsettia (Larson et al., 1988). A traditional temperature regime for poinsettias might have been 65°F during the night and 75° to 80°F during the day for most of the life of the poinsettia. However, studies have shown that these traditional temperatures may be modified to change the growth response of the poinsettias (Heins et al., 1988). For example, stem elongation is greater as the difference between day and night temperatures becomes greater. Therefore, it is possible to control the height of flowering poinsettias by reducing day temperatures and increasing night temperatures. This can be done without changing the average daily temperature so that flower development continues at the same rate, while at the same time reducing the plant height.

D. Media

The most important features of the growing media for poinsettias is that they must be clean, reproducible, and provide the right physical properties for the root environment. The term "soil mix" is no longer descriptive of the media used to grow poinsettias because they usually do not contain any native soil. Most growing media today are described as peat-lite mixes, generally containing a combination of sphagnum peat moss plus one or more other ingredients like perlite, vermiculite, bark, wood residuals, expanded clay products, Styrofoam beads, or sand (Fonteno *et al.*, 1981). These formulated media are not unique to poinsettia culture and often are the same media used for many other flowering pot plants. Although it is still recommended to steam pasteurize growing media for poinsettias, many of the peat-lite mixes are successfully used without pasteurization. This is possible because all of the ingredients in a peat-lite mix are relatively free of weed seed, disease organisms, or toxic materials. Most of the ingredients like peat moss, perlite, or vermiculite are relatively uniform from source to source, thus making each batch consistently uniform. In general, the peat-lite mixes also have excellent physical properties including good aeration and good water holding capacities. Many poinsettia growers prepare their own growing media, and several commercial enterprises offer ready-to-use growing media. A good peat-lite mix containing at least 50% sphagnum peat moss and a combination of perlite, vermiculite, processed wood bark, or all of these can be used to grow both stock plants and flowering poinsettia plants. In many cases, these same media are used for the propagation of poinsettia cuttings. However, specialized media, such as preformed foam products or rockwool, are more frequently used as media for rooting cuttings. These preformed rooting media offer the advantages of being clean, easy to handle, and possessing the proper physical properties, such as good aeration and water holding capacities for rapid rooting of poinsettia cuttings.

E. Water

Poinsettia culture is more successful if good quality water is available. When available, water with less than 1 mmho/cm of soluble salts and a pH of 6 to 7 would be ideal. Irrigation waters that contain high levels of sodium, bicarbonates, and other soluble salts interfere with the nutrition and growth of poinsettias (Pruitt *et al.*, 1987). The availability of good quality water is essential to the proper growth and development of poinsettias at all stages of their growth. Water is not only applied to the growing media for uptake by the roots, but it is also applied to leaf surfaces during propagation to reduce the loss of water from cuttings. Excessive use of water on cuttings during propagation leaches nutrients from the leaves and reduces the quality of the cuttings. For this reason intermittent mist systems are used that only apply enough water to form a fine film on the leaf surface and that turn on only frequently enough to maintain this film. As cuttings begin to root, the amount and frequency of the water applied through the mist system is reduced.

Judicious applications of water to the growing medium is one of the most important cultural practices in growing poinsettias. Water is often the vehicle for applying plant nutrients. Soluble fertilizers are often dissolved in irrigation water and applied at each irrigation, or with every second or third watering. While some poinsettia crops may still be manually watered with a water hose, most poinsettia crops are irrigated with a water system to distribute water to each plant or container. The system most frequently used is a drip tube system with small spaghetti sized tubes going to each pot off a water main.

As water conservation and ground water pollution become important issues, the use of recirculating water systems are becoming more common for poinsettia culture. These include ebb and flow systems as well as flood irrigation systems, both of which irrigate pots from the bottom by capillary action. Once the pots have absorbed the irrigation water, the water surrounding the pots is returned to a reservoir and recirculated at the time of the next irrigation. These subirrigation systems provide maximum efficiency of water use as well as a closed system with no water run-off. They must be managed, however, to avoid a buildup of soluble salts in the growing medium. They do not cause leaching of soluble salts as do systems that apply water to the top of the pot.

Irrigation frequency depends on the water holding capacity of the growing medium as well as the size and growth rate of the poinsettia plant. Small, recently potted plants may not need to be watered more frequently than once a week. However, once the plants have established an active root system and are growing rapidly, it is not uncommon to water poinsettias every other day. There are no fixed formulas on how often a poinsettia should be watered. This is strictly up to the judgment of the grower who must water frequently enough to avoid wilting and water stress to the plants and, at the same time, not over water to exclude good aeration of the growing media.

F. Nutrition

Soil mixes of the past, which often contained native soils, provided many of the essential nutrient elements for poinsettias. Today's peat-lite mixes contain few or none of these nutrient elements. Therefore, it is essential to fertilize poinsettias with a nutrient solution that contains most of the essential nutrient elements. It is possible that irrigation water will supply a partial requirement of a few of the plant nutrients. Also, peat-lite mixes are often amended with fertilizers to provide part of the plant nutrients. It is no longer adequate to supply only nitrogen, phosphorus, and potassium to grow a good quality poinsettia plant. Fortunately, some commercial fertilizer formulations for poinsettia production contain some, if not all, essential nutrient elements. Growing media are often amended with limestone or superphosphate to provide part of the calcium, magnesium, and phosphorus requirements of poinsettia plants. It is not uncommon, however, to see symptoms of nutrient element deficiencies in poinsettia plants that have not been properly fertilized. Calcium (McDaniel *et al.,* 1986; Woltz and Harbaugh, 1986), magnesium (Cox and Seeley, 1980), manganese, molybdenum (Cox, 1988; Hammer and Bailey, 1987), nitrogen, and zinc deficiencies frequently occur among poinsettia crops.

Fertilizer rates for poinsettias are usually based on the amount of nitrogen applied in the irrigation water. A recommendation of 250 to 300 ppm refers to the level of nitrogen in a fertilizer solution. Hopefully, the other essential nutrients will also be provided in a ratio to supply adequate nutrition for the plant. Most complete fertilizers now used do supply adequate levels of nitrogen, phosphorus, and potassium, and also all of the essential minor nutrient elements. For example; fertilizers with the formulation of 20–10–20 or 15–5–25 often contain magnesium, iron, manganese, boron, copper, molybdenum, zinc, and sulfur, in addtion to nitrogen, phosphorus, and potassium.

Fertilizer rates are usually higher at the beginning of a flowering poinsettia crop than near the end. It is often said that the most critical time in the poinsettia crop is during the first month of its culture. Fertilizer rates of approximately 400 ppm nitrogen are often used in the first month to provide the nutrition the plant needs to get off to a good start. During the period of time when the plant is making its most growth in October and early November, fertilizer rates are often recommended at the level of 250 to 300 ppm nitrogen with each irrigation. To tone-up or improve the postharvest keeping quality of a flowering poinsettia, fertilizer rates are decreased, or fertilizer is sometimes eliminated all together after the flower bracts have formed. After flower bract formation, poinsettias require small amounts of nutrient elements, and continued high rates of nutrition at this time keep the plants in a "soft" state of growth. Continued fertilization predisposes flowering poinsettias to bract edge burn, *Botrytis,* and poor keeping quality for the consumer (Nell and Barrett, 1985).

G. Ventilation

Ventilation of greenhouses is important for poinsettia culture, not only to help maintain the optimum temperatures, but to create a more suitable environment for growth and disease control. In warmer climates of the United States, ventilation alone may not be adequate to maintain desired temperatures. In many cases, fan and wet pad evaporative cooling systems may be required to avoid excessive daytime temperatures. Fortunately, most poinsettia crops in the United States are flowered during the fall season when temperature control is easier. Generally, it is not desirable for greenhouse temperatures to exceed 85° to 90°F for good quality growth. For those growers who have poinsettia stock plants during the summer, good ventilation and adequate cooling systems may be of more importance.

If temperatures are within the desired range for optimum growth, the free movement of air around plants and across leaf surfaces usually creates a more desirable growing environment. For one thing, air movement helps in the cooling of the plant tissues as well as resupplying carbon dioxide to the plant for photosynthesis. Air moving across plant surfaces can lower the relative humidity around plants, creating a less desirable environment for diseases like *Botrytis.* With adequate ventilation, it is possible to produce higher quality plants. Cooling of the plants allows the grower to increase the light intensity without raising the temperatures above their optimum level. With more light, the plants carry on more photosynthesis and transpire more water. This may increase the water requirement for the plants, but it also provides an opportunity to supply more nutrients with more frequent irrigations.

Horizontal air flow (HAF) systems have proven to be exceptionally beneficial for poinsettia production. Horizontal air flow uses low volume fans to circulate air in a horizontal pattern within the greenhouse, and HAF is particularly effective at night or during periods when ventilators are closed. The slow but constantly moving circular air patterns even out temperatures within the greenhouse by eliminating hot and cold pockets of air. It also helps reduce the relative humidity around the plants, thus reducing the probability of *Botrytis* infestation.

H. Growth Regulation

1. Environmental Control

Height control is of great importance in the production of high quality poinsettias. Without cultural practices to regulate growth, most cultivars would not be within the prescribed height or size ranges required by the market. Adjustments of temperatures, light intensity, light quality, relative humidity, and irrigation practices all have an ultimate bearing on plant growth. Work done at Michigan State University has demonstrated the importance of temperature control in regulating the growth of poinsettias (Heins *et al.,* 1988). Their work suggests that stem elongation, and thus the final height of a poinsettia, is influenced by the difference between day and night temperatures. Assuming that day temperatures are greater than night temperatures, the greater the difference between day and night temperature, the greater the stem elongation. For example, poinsettias grown at temperatures of 65°F (night) and 75°F (day) (10°F difference) would be taller than poinsettias grown at 70°F day and night, zero degrees difference. By temperature manipulation, the grower has more control over the final height of the poinsettia. It should be pointed out that those growers in northern climates who can more precisely control greenhouse temperatures would be able to take advantage of this method of height control more than a southern grower who may not have as much control over high day temperatures.

2. Pinching

Most poinsettias in today's market are grown as pinched plants. Not only do pinched plants have more flowers per plant, but pinching also provides a tool for controlling the ultimate height of a poinsettia. All other factors being equal, the nearer one pinches a poinsettia to the flower initiation date, the shorter the plant. Therefore, by timing of the pinch date, the grower has some control over the height of the poinsettia crop. By pinching too near the flower initiation date, poinsettias may not make enough growth before flower initiation, either to be physically strong or tall enough for the quality of plant desired. However, pinching poinsettias too far in advance of the flower initiation date may result in plants that are either too tall or more difficult to control by other cultural means. Growing conditions in different parts of the United States influence the date of the pinch. In northern climates with more cloudy weather, poinsettias may be pinched before the first of September, while in southern climates, poinsettias may be pinched well after the first of September to avoid tall plants.

3. Chemical Growth Regulation

Several chemical growth regulators are available to help retard the growth of poinsettias (Fig. 3). It should be pointed out that some successful poinsettia growers produce high quality poinsettia crops without the use of chemical growth retardants. However, growth retardants are more often used than not. Growth retarding chemicals are applied either as a drench to the growing medium or as a spray application to the foliage. These applications are usually made just before the time the poinsettia would make its most rapid growth. Timely applications of chemical growth retardants slow the rate of stem elongation without causing visible toxic effects to the plants.

Some of the several chemicals labeled for use on poinsettias are chlormequat (Cycocel), ancymidol (A-Rest), SADH (B-Nine) and paclobutrazol (Bonzi). Cycocel is perhaps the most widely used. Rates for Cycocel application are often in the range of 1000 to 3000 ppm. Generally, the recommended time for the first application of Cycocel is when the lateral branches on a pinched plant are 1 to 2 inches in length. This is usually in late September (Wilfret, 1985). Repeated applications of Cycocel may be necessary in early October to attain the desired plant height control. It is not recommended to apply Cycocel much after the middle of October. Late applications not only reduce plant height but also reduce bract size and may delay bract development.

Fig. 3.   Chemical growth retardants are used to reduce stem elongation. A single 45 ppm foliar spray of Bonzi, *left,* was more effective than three foliar applications of Cycocel at 1500 ppm, *center.* Untreated plant, *right.*

A-Rest is another effective growth retarding chemical for poinsettias. It is applied as a drench to the growing medium. A-Rest is not recommended for foliar applications to poinsettias, and drench applications are time consuming and costly. Therefore, the use of A-Rest is not as popular as the use of Cycocel. Also, the results from the use of A-Rest may not be as predictable as with the use of Cycocel. A-Rest is often used at the rate of 0.25–0.50 mg per 6-inch pot.

B-Nine is often used in combination with Cycocel as a tank mix of the two chemicals. The combination of these two growth retarding chemicals seems to be more effective than either chemical used alone. In fact, care should be taken not to use this combination too frequently or at too high a rate (Berghage and Heins, 1986). B-Nine used at a relatively late date often slows the development of flower bracts and delays the flowering of poinsettias. It is therefore recommended that B-Nine used in combination with Cycocel not be applied to poinsettias after October 1.

Bonzi is used in relatively low rates of 15 to 60 ppm as a foliar spray or as a growing media drench. Bonzi has proven to be an effective growth retarding chemical, but its use requires strict guidelines to achieve the desired results (Barrett, 1987). Sumagic (uniconizole) is very effective, both as a drench or as a foliar application, at very low rates.

4. Breeding

Recently introduced cultivars have been bred and selected for shorter growth habits. By the proper selection of cultivars, the poinsettia grower now has the option of growing high quality poinsettias with less reliance on the use of chemical growth retardants. Modern poinsettia cultivars generally produce more lateral branches following pinching. This greater branching potential also seems to result in plants with shorter overall height. Breeding and selection of improved poinsettias have made height control a more manageable practice.

III. COMMERCIAL PRODUCTION

A. Stock Plants

Vegetative cuttings can be obtained from specialist propagators either for flowering plants or for stock plants. Some growers choose to propagate their own cuttings from stock plants obtained as cuttings from March through June. Cuttings received in March or April are usually pinched twice before cuttings are harvested. Those received in late May or June may only receive one pinch. Stock plants can be grown either in greenhouse benches or in containers, but most stock plants are grown in various sized containers from 6- to 12- inch pots. The earlier the cuttings are received, the larger the stock plant container needs to be. A common practice is to pot up rooted cuttings in 4-inch pots on arrival, and later shift them to larger pots as the plants grow. Many growers find their plants grow fastest and produce the most cuttings if the cuttings are directly planted into larger pots from the beginning. Any clean growing medium suitable for growing flowering poinsettias is usually suitable for growing stock plants. Steam-pasteurized soil mix with good aeration and good water holding properties or clean peat-lite mixes produce good quality stock plants.

A typical schedule for growing stock plants would be to receive rooted cuttings from a specialist propagator between Easter and Mother's Day. The cuttings would be planted into an 8- to 10-inch pot and pinched approximately 3 weeks later when 8 to 9 fully expanded leaves can be left on each plant (Wilkins and Grueber, 1983). Approximately 5 weeks later, each branch that results from the first pinch can be pinched again, leaving two to three mature leaves per branch. This would usually occur in early June. Cuttings that result from the second pinch would then be ready for harvest approximately 6 weeks later in mid-July. Most cuttings for 6- 7- or 8-inch pots are propagated beginning sometime in July. Cuttings that return from the July harvest can be taken in mid- to late August for 4- or 5-inch pots or for unpinched poinsettias.

The cultural and environmental requirements for stock plants are similar to those for the early stages of flowering plants. Monthly growing medium drenches with fungicides are recommended to protect against the root and stem rotting diseases common to poinsettias. Stock plants can be fed a constant liquid feed of approximately 250 to 300 ppm nitrogen along with a good balance of the other major and minor nutrient elements (Bierman, 1988). Night temperatures of 65°F or higher should be maintained for good active growth of stock plants. Day temperatures should not exceed 85°F, if possible. When temperatures exceed 85°F during the day, applying shade to the greenhouse or over the stock plants is usually beneficial.

Whiteflies can be serious pests on poinsettia stock plants. It is very important to begin a whitefly control program from the beginning to avoid a buildup of large populations.

The amount of space required for a poinsettia stock plant depends on how early the cuttings are received, how many times the plants are pinched, and the size of the container in which they are grown. Cuttings received in March or April and pinched twice may require as much as 4 square feet per plant by July when the cuttings are harvested. Plants received in June and pinched once may only need 1 to 2 square feet per plant. Poinsettia stock plants will produce more cuttings of better quality if they are not crowded and are given adequate space. It is very important to pinch stock plants on a regular schedule and not allow the branches to become too long or overly mature. If cuttings are not taken from branches during the first cutting harvest, the branches should be pinched so that they will produce fresh new cuttings for the second harvest.

B. Propagation

Several important factors are essential for the successful propagation of poinsettia cuttings; cleanliness, a mist system, and optimum temperatures. Cuttings must be kept as clean as possible at all times to avoid disease problems. All tools for taking cuttings should be sterile, and persons harvesting cuttings should wash their hands with soap and water and dip their hands in a disinfectant. All containers for transporting cuttings from the stock plants to the propagation area should be clean and sterilized. Benches, rooting media, mist systems, and containers for propagation should also be clean and disinfected.

An intermittent mist system or a fog system is generally required for the successful propagation of poinsettias. Any system that maintains a film of water on the leaves until roots are formed is usually adequate. The use of bottom heat to keep the rooting media warm is often required, even during the warm summer months. Optimum rooting of cuttings occurs if the rooting medium is 72° to 74°F. However, the air temperature around the cuttings need not necessarily be that high. If day temperatures can be kept below 85°F, there is less stress on the cuttings, and they root more rapidly.

Terminal (tip) cuttings 2.5 to 4 inches in length are harvested from stock plants. A sharp knife or other cutting tool that can be easily and frequently disinfected can be used to remove cuttings from the stock plant. It makes no difference whether the cutting is taken at a node or between nodes. Remove leaves from cuttings only if lower leaves will interfere with placing the cuttings in the rooting medium. Cuttings are often taken early in the morning or late in the evening when the cuttings are turgid and not when they are stressed from the heat of the day. Cuttings may be gathered into any suitable, clean container and should be covered while being transported to the propagation area. Rooting hormones of approximately 1000 to 2000 ppm indole-3-butyric acid (IBA) often aid in the rooting of poinsettia cuttings. The rooting hormone may be applied either as a powder dusted on the base of the cuttings or dipped in a liquid hormone solution before they are stuck in the rooting medium.

Cuttings can be rooted in a variety of media. There are specialized propagation media of rockwool or foam that are used for rooting. Poinsettias, however, can be rooted in the growing medium used for flowering plants. This is done by rooting the cuttings in small containers filled with growing medium and later transplanting them into the final containers. Some poinsettia growers prefer to root their cuttings directly in the finishing pot. Regardless of the medium, it is recommended to have prepunched holes in the rooting medium so that the cuttings will not be bruised during sticking. The cuttings should be placed in the preformed holes to a depth of approximately 1 inch. The mist or fog system should begin immediately on sticking. During extremely warm weather, it may be necessary to operate the intermittent mist during the night for the first two nights. There is no set formula for the percent of time the intermittent mist system must be on. It is necessary to maintain a thin film of water on the leaf surface without applying excessive amounts of water until a new root system begins to form on the cuttings. This is normally 10 to 14 days. Then the length of time the mist system is on can be reduced gradually. It is beneficial to reduce the light intensity on unrooted cuttings to 1500 to 2500 foot-candles (fc) during the day. Once roots are established on the cuttings, the light intensity can be increased. Most poinsettia cultivars will root under ideal conditions in 14 to 21 days. They may better withstand transplanting, however, if the roots are allowed to develop in the propagation medium for 21 to 28 days.

C. Scheduling Production

Cuttings for larger containers are propagated earlier than cuttings for smaller containers. For the midsection of the United States, cuttings for 7- and 8-inch pots are normally taken in early July and planted in the finishing pots in early August. Cuttings for 6-inch containers may be taken in mid-July and potted in mid-August. Cuttings for 5-, 4-inch, or smaller pots are often taken in early August and potted in early September (Table I).

Most poinsettias are grown as pinched, multibranched plants. The scheduling of a flowering poinsettia crop usually centers around the time the plants are pinched. If a poinsettia is pinched too early, it may be difficult to control the height of the plant. If it is pinched too late, there may not be enough time for the plant to develop into a strong, good quality plant. Conditions vary throughout the United States. A 6 inch poinsettia plant may be pinched in late August or early September in the northern part of the United States, but may not be pinched until mid-September in southern regions.

Poinsettia plants should be spaced to their final spacing soon after pinching and before the plants become crowded (Table II). This increases the amount of light and air around each plant and reduces the need for chemical growth retardants. Chemical growth retardants are usually necessary to obtain strong, good quality plants. It is also necessary to provide adequate nutrients to get plants off to a good start. Poinsettias are often fertilized with 400 ppm of nitrogen during the first month after transplanting and reduced to 250 to 300 ppm during October and November. Monthly fungicide drenches beginning at transplanting time are good insurance

Table I

Suggested Schedule for Production of Pinched, Multibloom Poinsettias[a]

		Dates of propagation				
Pot size (inches)	Number of cuttings per pot	Stick cuttings in propagation medium	Direct root in finishing pot	Transplant rooted cuttings	Pinch date	Number of leaves after pinch
8	3	July 20	Aug. 1	Aug. 15	Aug. 25	6
7	2	July 25	Aug. 5	Aug. 20	Sept. 1	6
6	2	Aug. 1	Aug. 10	Aug. 25	Sept. 5	5
6	1	Aug. 1	Aug. 10	Aug. 25	Sept. 5	6
5	1	Aug. 5	Aug. 15	Aug. 30	Sept. 10	5
4	1		Aug. 20		Sept. 15	4

[a]Based on midwestern conditions in the United States. Northern areas may require earlier scheduling; southern areas may require later scheduling.

Table II

Suggested Final Spacing for Pinched, Multibloom Poinsettias

Pot size (inches)	Number of cuttings per pot	Spacing on center (inches)	Bench area per pot (square feet)
8	3	19 × 19	2.5
7	2	17 × 17	2.0
6	2	15 × 15	1.5
6	1	14 × 14	1.3
5	1	12 × 12	1.0
4	1	8 × 9	0.5

against stem and root rotting diseases. Greenhouse temperatures should be maintained at 68° to 70°F during the night for the first month and then reduced to 64° to 65°F. If possible, it is desirable to maintain day temperatures below 85°F.

Most of today's poinsettia cultivars initiate flower buds approximately October 1 and flower 8 to 10 weeks later. By selection of cultivars, it may be possible to have flowering poinsettias from mid-November to mid-December without daylength manipulation. However, poinsettias for early to mid-November sales may need to be covered with a black cloth for earlier flower initiation.

IV. INSECTS, DISEASES, AND PHYSIOLOGICAL PROBLEMS

A. Insects and Other Pests

Poinsettias are subject to attack by various insect pests under greenhouse conditions. Whiteflies, fungus gnats, mealy bugs, and spider mites are the most prevalent pests of poinsettias, although other pests may occasionally cause problems.

1. Whitefly

Whitefly has been the most serious insect pest of poinsettias for several years. Greenhouse whitefly (GWF) was the predominate species of whitefly attacking poinsettias prior to 1986. The sweet potato whitefly (SPWF) has also become a serious pest of poinsettias (Price et al., 1987). The SPWF is considered by many to be more difficult to control than the GWF. There are subtle differences in the appearance of the two species, but their biology and control are similar. Whiteflies can be observed anytime of the year in the greenhouse. Yellow sticky traps are used to determine and monitor the presence of whiteflies in greenhouses. Control measures should be taken with the first sign of infestation. When spraying plants, care should be taken to thoroughly cover the lower surfaces of the leaves.

2. Fungus Gnats

Fungus gnats are small, dark-colored flies that are about $\frac{1}{8}$ inch long. Adults move rapidly over plants and usually congregate in dark corners or under benches. Prior to 1966, these flies were considered only "nuisance pests" that did not warrant any control measures. The larvae, however, have been shown to be capable of invading live plant tissue, feeding on tender roots, and causing cuttings and plants to wither and die. Larvae may be especially active during propagation of poinsettia cuttings. This rooting environment is ideal for fungus gnats, and the rooting medium should be checked frequently for larvae activity. Several insecticide sprays and drenches provide effective control.

3. Spider Mites

Mites are not insects. They have eight legs instead of the six legs that characterize insects. Spider mites do not always infest poinsettia crops, although they can become a serious pest. Mites are very small; therefore, plant damage may occur before they are detected. Several species of mites have been reported to infest poinsettias. Mite infestations may be found on many kinds of plants in or outside the greenhouses throughout the entire year. Most spider mite control is achieved by pesticide applications. There are, however, several cultural practices that can help reduce mite problems. Weeds and infested poinsettia plants should be eradicated. Mites are often carried into the greenhouses on employees' clothes, so it is important not to handle infested plants before working with plants that are unaffected.

4. Mealybugs

Mealybugs are among the most serious pests of greenhouse plants. Infested plants look unsightly because of the whitish cottony mass of insects and the black sooty mold fungus that develops on honeydew excretions. Mealybugs are present in the greenhouse throughout the year. The presence of ants should alert the grower to check ant trails to determine if ants are carrying mealybugs into the greenhouse. Pesticide spraying appears to be the most economical way to control mealybugs. However, the waxy coating of the mealybug makes it difficult for pesticide sprays to penetrate and kill the insect. Repeated spray applications may be necessary.

B. Diseases

Pathogens of primary importance include fungi and bacteria. For disease to occur, the organism and the host plant must be in close proximity. Fungi infect plants through wounds, natural openings such as stomates, and intact epidermal surfaces. Bacteria infect primarily through wounds or natural openings. Disease control can only be attained by using clean plants, clean growing media, complete sanitation, and by providing the appropriate environment.

1. Pythium Root and Stem Rot

Pythium usually causes a brown rot of root tips and cortex that may progress up the stem. The root cortex will often slough off. Infected plants are stunted (Fig. 4), and lower leaves may turn yellow, curl, and fall off. Under severe conditions, entire plants are killed. The disease is most common in poorly drained growing media and may be carried over in growing media or on infected plants. It is easily spread in water. There are no airborne spores, but the fungus may be blown around on dust particles. Pythium root and stem rot is effectively suppressed with periodic fungicide drenches.

2. Rhizoctonia Root and Stem Rot

Rhizoctonia solani causes a brown rot of stems beginning at the soil line, and roots may have brown lesions. Leaves often become infected where they touch the rooting medium under mist propagation. Infected plants are stunted, and lower leaves may turn yellow and fall off. Complete plant collapse occurs under severe conditions. This fungus may be carried over in growing media or on infected plants and is easily spread by water. There are no airborne spores, but the fungus may be blown around on dust particles. Fungicide drenches are effective in suppressing this disease.

Fig. 4. Healthy root system, *left.* Plant infected with root rotting pathogens, *right.* Note the reduced growth of the diseased plant.

3. Botrytis Blight/Gray Mold

Botrytis cinerea is probably the most prevalent disease of poinsettias. It can occur during all stages of development and attacks all plant parts. The rot begins as a water soaking and tan to brown lesions, regardless of the tissue affected (Fig. 5). Under humid conditions, a gray, fuzzy mold composed of mycelia and spores form on this rotted tissue. Sometimes young plants rot near the soil line. On more mature stems, a tan canker may develop, girdling the stem and resulting in leaf wilt above the lesion. Red bracts develop a purplish color in infected areas.

Gray mold is caused by a fungus whose spores are airborne and can be assumed to be present everywhere at all times. The fungus grows well on dead organic material. Infection is favored on injured, aging, or succulent tissue and moderately cool temperatures, although the fungus can grow at temperatures ranging from 32° to 96°F. Free moisture must be present on the plant surface for spore germination and high relative humidity (90 to 100%) is necessary for the gray mold sporulation stage. The first line of defense is control of the environment. Avoid physical injury to plants and maintain air circulation, especially at night. Horizontal air flow is an effective means of achieving good air circulation. Space plants to allow for air movement through the plant canopy. Do not water late in the day in greenhouses covered with double layers of film plastic. Use night heat plus ventilation to

Fig. 5. *Botrytis* infected poinsettia leaves, with water soaked, tan to brown lesion symptoms.

lower humidity and avoid splashing water on the foliage and bracts. Keep temperatures above 60°F if at all possible and remove all dead plant material. The dense habit of multiflowered type poinsettias presents a special problem of leaf and bract overlap. Fungicides may be applied for control of Botrytis blight. Thorough coverage is important, and frequent treatment is necessary to provide protection to newly developing plant tissue.

4. Rhizopus Blight

Although *Rhizopus stolonifera* is a relatively uncommon disease of poinsettias, it can be very destructive under favorable conditions in propagation. In Florida, Rhizopus blight has also affected blooming poinsettias, attacking flowers, leaves, and stems. The symptoms consist of a soft, mushy brown rot, with white mycelia and black fruiting bodies (sporangia). The abundant mycelia give the disease a "bearded" appearance. The spores are airborne and may be spread by water. The fungus can survive in plant debris and requries high temperatures (70° to 90°F), high relative humidity (minimum 75%), and wounded or weakened host tissue to attack. The mode of attack is similar to bacterial soft rot with an enzyme being released to cause cell deterioration. When wounds callus or heal rapidly, the fungus is not able to become established. It could be referred to as a "high temperature *Botrytis*." Suppression is best attained by sanitation and careful handling of cuttings to avoid injury. There are no fungicides currently registered to control Rhizopus blight on poinsettias.

5. Bacterial Soft Rot

Erwinia carotovora soft rot occurs primarily in propagation. Cuttings develop a soft, mushy rot beginning at the basal end within 3 to 5 days of "sticking." The bacterium is prevalent on dead plant material and can be carried on windblown dust, nonsterilized tools, and the hands of workers. It spreads readily in water and may be found in pond water. Wounded tissue, waterlogging of rooting medium, high temperatures, and other factors that stress the cuttings favor this organism. There is no effective chemical to control this disease. Grow stock plants under cover or in other controlled environments, and use good sanitation practices throughout the harvest and propagation of cuttings.

C. Physiological Disorders

1. Leaf Distortion

For many years, leaf deformity has been seen in some stock plants, and often on pot plants in greenhouses. The symptoms are extremely variable. In some cases, damage has occurred only at the tip of the immature leaf, and it will give the appearance of having been chopped off at a later stage of development. Where the entire margin of the leaf has been affected in earlier stages, later growth of all except the margin causes a "puckered" appearance, as if a drawstring around the leaf margin had been pulled up tight.

On Christmas season plants, leaf distortion frequently occurs in late September and early October after the plants have been moved from the propagation to the finishing area. Branches that develop after pinching may have two to three misshapen and distorted leaves. In most instances, these leaves remain distorted, but green, throughout the forcing period. Leaves that expand later are usually normal and hide the damaged leaves by market time.

The causes of leaf distortion are not well understood. It seems that when cells in very young leaf tissues are ruptured or killed, the leaf becomes misshapen as it expands. Drying of tissue, burn from fertilizer, nutrient deficiency (particularly molybdenum),and chemical burn have all been suspected of damaging young leaf cells. Plants under stress from bright light, extremely warm temperatures, or moving air often have more leaf distortion.

It is helpful to provide shade and syringe the foliage until roots are well established and the side branches begin to develop.

2. Bract Edge Burn

Although *Botrytis* causes an injury that is typically observed as a burn, all such injury is not necessarily caused by the fungal disease. Severe bract burn has been encountered where extreme rates of fertilizer have been used. Under these conditions, the leaves may show no damage. One theory is that during growth, there is a diluting effect of plant-absorbed fertilizer, but at flowering, new tissue development has virtually ceased, and the fertilizer salts accumulate in the youngest mature and most sensitive tissue—the bract. This accumulation causes cell damage, usually starting on the bract edges.

Where slow-release fertilizers are used, the rate should be modest, and the application should be early enough to ensure almost complete depletion at time of flowering. It may also help to increase the intensity of irrigation in the finishing stages to ensure adequate leaching and removal of accumulated salts.

Bract burn may also be triggered by environmental factors. Research has demonstrated the effects of water stress, temperature, light intensity, and humidity on bract burn, which typically occurs on transitional bracts (those structures that are usually colored but are between the green leaves and true bracts). Changes in the environment, such as large temperature fluctuations, may be conducive to this malady.

Cultivar differences are usually noted. Those cultivars with large bracts, like 'Gutbier V-14 Glory,' are more susceptible to bract edge burn than the 'Annette Hegg' cultivars, for example. Investigations at the University of Florida (Nell and Barrett, 1986a), have led to a better understanding of bract edge burn, also named "bract necrosis." In essence, cultural and environmental factors that keep poinsettias in a soft, actively growing condition late into the production period promote bract necrosis. Some of these factors are high fertilizer rates and heavy watering during the final 4 weeks of production; fertilizers containing nitrogen mostly in the ammonium form; and high relative humidity within the greenhouse environment.

3. Premature Cyathia Drop

During some Christmas flowering seasons the true flowers, or cyathia, may drop from the center of the bract presentation before the flowers reach maturity. This may occur before the plants are ready for market, particularly in northern climates with low light conditions. Because this detracts from the appearance of the poinsettia and makes them appear to be overly mature, it also reduces their economic value.

Miller (1984) determined that premature cyathia drop is caused by low light levels, high forcing temperatures, or both. Water stress aggravates the problem. These conditions allow the food reserves of the plant to become depleted. As the food reserves become low, the plant reacts by dropping the cyathia. It is not uncommon for skies to become overcast during late October and November in northern states. Lower light levels and cooler temperatures accompany the cloudy weather. These are not ideal conditions for high quality poinsettias. With lower light levels plants do not make as much food. Also, as poinsettia bracts develop, they shade the green leaves below and further restrict the amount of light available for photosynthesis. If poinsettias are grown at lower than optimum temperatures during the early part of the production period, flower development may be delayed when the cloudy weather begins. To speed-up the rate of flower development it then becomes necessary to raise greenhouse temperatures. As temperatures increase the plant's food reserves are used at a faster rate, increasing the possibility of cyathia drop.

4. Latex Eruption

Plants belonging to the Euphorbiaceae family contain latex, which is exuded when cells are injured. This became a problem in poinsettia production when the cultivar 'Paul Mikkelsen' and its sports first became popular. The malady is sometimes termed "crud." The mechanism is one of bursting cells resulting from high turgor pressure with latex spilling over the tissue and, on drying, the creation of a growth-restricting layer. When this occurs on developing stem tips, distortion or stunting of growth results. The exuding of latex has also been observed on fully expanded leaves, sometimes giving the appearance of mealy bug infestation because of the white splotches scattered over the leaf surface.

All of the contributing factors have not been clearly defined, but several obvious ones include high moisture availability and high humidity, both of which result in high fluid pressure within the cells. Low temperature is an important contributing factor. Mechanical injury from rough handling or from excessively vigorous air movement may also increase injury to cells. High rates of photosynthesis may contribute by building up a high osmotic pressure in cells from carbohydrate accumulation. Control is best attained by using growing media that dry out in a reasonable length of time and by avoidiing extremes of high humidity, particularly during the night. Moderate shading in extremely bright weather might also be helpful. Sudden lowering of temperature can trigger the reaction. Fortunately, most cultivars are not highly sensitive to this problem.

5. Stem Splitting

Under certain conditions, poinsettias that do not normally branch unless pinched will suddenly produce stem branches at the growing tip. Careful examination will reveal that the true stem tip has stopped growth or aborted. Splitting is actually a first step in flower initiation. The stimulus to flower increases with the age of the stem and with the lengthening of night. Even with short nights and normal 60° to 70°F temperatures, splitting can be expected if the stem is permitted to grow until 20 to 30 leaves are present. Stem tips that are continuously propagated have an increasing tendency to flower. To ensure against this, lights should be supplied to stock plants until May 15. Propagations prior to August 1 should be grown as multiflowered, branched plants with tips discarded. It is always good insurance to discard early pinched shoots instead of trying to propagate them. Also, stems that are heavily shaded by a canopy of higher foliage may be subjected to enough reduced light to cause them to split even in periods when daylength would be considered adequately long to keep apexes vegetative.

6. Bilateral Bract Spots ("Rabbit Tracks")

This condition is characterized by breakdown of the tissues between the veins located on either side of the bract midrib. It occurs in late November and early December during the flowering process. Although the plant is not killed, the condition can effectively lower the quality of the plant, rendering it unsalable. Bilateral bract spots seem to occur under a wide temperature range, in various types of greenhouses, with both gas and oil heat, and on different cultivars. The condition might be confined to a few plants or be found throughout the greenhouse. A study was conducted in Germany to consider certain factors that may predispose poinsettias to bilateral bract spots (Seeley, 1980). The most obvious observation was that cultivars of the 'Annette Hegg' family are more susceptible than most other commercial cultivars. Also, cultural and environmental factors had their greatest influence during the bract development stage. In this study, incidences of bilateral bract spots were associated with high relative humidity or changing humidity levels. High temperatures, above 70°F, and especially high night temperatures during the bract development phase, caused a greater incidence of bilateral bract spots. High levels of nitrogen fertilization near the end of the crop and high nitrogen content in the plants also caused a higher frequency of bilateral bract spots.

V. MARKETING OF PLANTS

A. Packing

Poinsettia cultivars have been genetically improved to last longer and give more value to the consumer. Leaf and bract retention may be remarkably long in most consumer's homes or offices. Yet, the colorful bracts remain somewhat fragile and must be properly protected during handling and shipping. Bracts may be broken or bruised by rough handling or by rubbing against other surfaces. Poinsettia tissues are instantly frozen and killed when exposed to freezing temperatures.

Individual plant sleeves and boxes protect plants as they are transported from the greenhouse to the retail outlet. Tapered sleeves made of paper, plastic, or polyester fiber work equally well. Sleeves are used for all sizes and shapes of poinsettias, from the smallest minipots to the largest tree forms. They should be longer than the plants are tall, so that all of the bracts are below the top of the sleeve (Hartley, 1984). Placing sleeved plants in corrugated paper boxes not only gives additional protection during transport, but also makes handling and shipping more efficient. Boxes must be taller than the plant sleeves to prevent bracts from rubbing against the top of the box.

B. Shipping

Data collected in 1988 indicate that 89% of all flowering poinsettias produced in the United States were sold at wholesale (Anonymous, 1989). The obvious conclusion is that most poinsettias are transported to market. Shipping, although necessary, seldom improves the quality of flowering plants. During shipping, poinsettias are sensitive to water stress, extreme temperatures, mechanical damage, and exposure to ethylene.

The growing medium in the pots should be moist, but not overly wet during shipping. Excess moisture may cause conditions within the shipping container that are ideal for the development of *Botrytis* (gray mold) on the bracts. Optimum shipping temperatures appear to be in the 50° to 55°F range (Scott *et al.*, 1983). Poinsettias are killed by freezing temperatures, and chilling injury was reported by Nell and Barrett (1986b) during 40°F simulated shipping conditions. Temperatures above 60°F may increase the severity of epinasty (droopiness) and reduce the keeping quality for the consumer.

Although boxes and sleeves are necessary to reduce mechanical damage during shipping, prolonged storage of poinsettias in any type of sleeve may result in a condition known as epinasty, a droopy appearance of the plant. The mechanical upward bending of leaves and bracts during sleeving causes them to release much more ethylene than unsleeved plants (Saltveit *et al.*, 1979). When the plants are unsleeved, their bracts and leaves hang downward. The longer the plants are left in their sleeves, the worse the epinasty. Ideally, plants should not be sleeved for more than 24 hours from the greenhouse to the retail sales area. Generally, recovery from moderate epinasty is possible after 48 hours of proper care in a well-lighted environment of 65° to 75°F. Some cultivars are more resistant ot epinasty than others. 'Eckespoint Lilo' and 'Eckespoint Celebrate' are very resistant and the 'Gutbier' family of cultivars shows good resistance. The cultivars of 'Annette Hegg' origin are generally very susceptible to epinasty.

C. Keeping Quality

Cultivar selection, greenhouse culture, environmental conditions, handling and shipping procedures, and consumer care all contribute to the keeping quality of a poinsettia. 'Eckespoint Lilo' and the 'Annette Hegg' cultivars are examples of poinsettias with excellent keeping quality because of their long-lasting leaf and bract retention traits. 'Gutbier V-10 Amy,' on the other hand, is susceptible to premature leaf drop and has only fair keeping qualities. Another popular cultivar, 'Gutbier V-14 Glory,' has very good keeping qualities.

Several cultural practices can be used to condition poinsettias to improve their keeping quality as well as their ability to withstand handling and shipping conditions. Reducing temperatures, light intensity, and fertilizer rates late in the crop production cycle have all been used to improve keeping quality. Nell and Barrett (1985) showed that plants supplied with lower rates of fertilizer and lower ratios of ammonium nitrogen late in the crop also had lower incidences of bract necrosis. Miller *et al.* (1984) were able to reduce premature cyathia drop by lowering temperatures near the end of crop production. Nell and Barrett (1986c) demonstrated that plants lost fewer leaves in interior conditions when plants were shifted to lower light prior to anthesis. These results indicated that poinsettias could be acclimatized during production. Maintaining healthy plants with active root systems is also essential for good keeping quality.

D. Consumer Care

Keeping quality can be extended by proper care at the retail floral department and in the consumer's home or office. Plants must be unpacked and unsleeved immediately on arrival. If plants are left in unopened containers too long, the leaves will turn yellow and fall off. Place poinsettias in a cool (60° to 65°F) bright area, but not in direct sunlight. Check plants daily, and water the growing medium when it feels dry to the touch. Poinsettias must be kept moist at all times. When watering, it is best to water thoroughly until water seeps through the drain holes, but do not allow poinsettias to sit in water. Poinsettias should not be placed where they will be exposed to cold drafts or where hot air will blow directly on them. If possible high humidity is much better than low humidity.

REFERENCES

Anonymous. (1989). *Floriculture Crops 1988 Summary.* USDA Agri. Stat. Board, Washington, D.C.

Barrett, J. E. (1987). "Controlling poinsettia height with Bonzi." *GrowerTalks.* **51**(6): 51–52.

Berghage, R., and Heins, R. (1986). "B-Nine and Cycocel tank mix can be too effective." *Greenhouse Grower.* **4**(8): 96, 98, 100.

Berghage, R., Heins, R., Erwin, J., Carlson, W., and Biernbaum, J. (1988). "Increase bract size with temperature and light." *Greenhouse Grower.* **6**(10): 40, 43–44.

Bierman, P. M. (1988). "Nutritional factors affecting leaf edge burn on poinsettia stock plants." Master's thesis, University of Minnesota, St. Paul.

Cox, D. A. (1988). "Lime, molybdenum and cultivar effects on molybdenum deficiency of poinsettias." *J. Plant Nutr.* **11:** 589–603.

Cox, D. A., and Seeley, J. G. (1980). "Magnesium nutrition of poinsettias." *HortScience.* **15:** 822–823.

Ecke, P. Jr., Matkin, O. A., and Hartley, D. E. (1990). *The Poinsettia Manual.* Encinitas, California: *Paul Ecke Poinsettias.*

Fonteno, W. C., Cassel, D. K., and Larson, R. A. (1981). "Physical properties of three container media and their effect on poinsettia growth." *J. Am. Soc. Hort. Sci.* **106:** 736–741.

Grueber, K. L. (1985a). "Control of lateral branching and reproductive development in *Euphorbia pulcherrima* Willd. ex. Klotzch." Ph.D. Diss., University of Minnesota, St. Paul.

Grueber, K. L. (1985b). "Euphorbia pulcherrima." In *Handbook of Flowering* edited by A. H. Halevy, vol. II. Boca Raton, FL: CRC Press, pp. 488–495.

Hammer, P. A., and Bailey, D. A. (1987). "Poinsettia tolerance of molybdenum." *HortScience.* **22:** 1284–1285.

Hartley, D. E. (1984). "Poinsettias need special care and handling." *Greenhouse Grower.* **2**(11): 78–79.

Heins, R., Erwin, J., Berghage, R., Karlsson, M., Biernbaum, J., and Carlson, W. (1988). "Use temperature to control plant height." *Greenhouse Grower.* **6**(9): 32, 34–37.

Larson, R. A., Hartley, D. E., and Thorne, C. B. (1988). "High temperatures cause late, low quality poinsettias." *GrowerTalks.* **52**(6): 88, 90, 92, 94.

McDaniel, G. L., Graham, E. T., and Lawton, K. A. (1986). *The Role of Nitrogen and Calcium on Stem Strength of Poinsettia.* Res. Report No. 86–15, Tennessee Agric. Exp. Sta.

Miller, S. H. (1984). "Environmental and physiological factors influencing premature cyathia abscission in *Euphorbia pulcherrmia* Willd." Master's thesis, Michigan State Univ., East Lansing.

Miller, S. H., Heins, R. D., and Carlson, W. (1984). "How to prevent center drop in poinsettias." *Greenhouse Grower.* **2**(9): 41–42.

Nell, T. A., and Barrett, J. E. (1985). "Nitrate-ammonium nitrogen ratio and fertilizer application method influence bract necrosis and growth of poinsettia." *HortScience.* **20:** 1130–1131.

Nell, T. A., and Barrett, J. E. (1986a). "Growth and incidence of bract necrosis in 'Gutbier V-14 Glory' poinsettia." *J. Am. Soc. Hort. Sci.* **111:** 266–269.

Nell, T. A., and Barrett, J. E. (1986b). "Influence of simulated shipping on the interior performance of poinsettias." *HortScience.* **21:** 310–312.

Nell, T. A., and Barrett, J. E. (1986c). "Production light level effects on light compensation point, carbon exchange rate and post-production longevity of poinsettias." *Acta Hort.* **181:** 257–262.

Price, J., Schuster, D., and Short, D. (1987). "Managing sweetpotato whitefly." *Greenhouse Grower.* **5**(12): 56–58.

Pruitt, J., Payne, R., and Schwartz, M. A. (1987). "Maintain safe soluble salts levels." *Greenhouse Grower.* **5**(7): 58, 60–61.

Saltveit, M. E., Jr., Pharr, D. M., and Larson, R. A. (1979). "Mechanical stress induces ethylene production and epinasty in poinsettia cultivars." *J. Am. Soc. Hort. Sci.* **104:** 452–455.

Scott, L. F., Blessington, T. M., and Price, J. A. (1983). "Postharvest effects of temperature, dark storage duration and sleeving on qualtiy retention of 'Gutbier V-14 Glory' poinsettia." *HortScience.* **18:** 749–750.

Seeley, J. G. (1980). "Rabbit tracks of poinsettia." *New York State Flower Ind. Bull.* **121:** 1–2.

Wilfret, G. J. (1985). "Effect of planting date and growth regulators on poinsettia height." *Proc. Florida State Hort. Soc.* **97:** 289–291.

Wilkins, H. F., and Grueber, K. L. (1983). "Thoughts on successful poinsettia stock plant cutting production." *Minnesota State Flor. Bull.* **32:** 8–9.

Woltz, S. S., and Harbaugh, B. K. (1986). "Calcium deficiency as the basic cause of marginal bract necrosis of 'Gutbier V-14 Glory' poinsettia." *HortScience.* **21:** 1403–1404.

13

Easter Lilies

Thomas C. Weiler

I. INTRODUCTION

Interior flowering plants are traditionally associated with the Easter holiday, but the plant species emphasized varies with country and throughout history. In northern Europe the Easter plants are *Narcissus,* usually *N. pseudonarcissus; Primula;* and perhaps other garden plants that bloom in early spring. Only in North America is there a strong tradition of marketing *Lilium* species for the holiday. This tradition, which began before 1930 (Griffith, 1930; Lumsden, 1929), is unusual because lilies bloom in the garden in mid- to late summer, not at Easter time. It is a tradition dependent on greenhouse forcing technology.

How *Lilium* became attached to the Easter holiday is uncertain, but it may be related to the common name of the first species grown for the holiday, *Lilium candidum* or the Madonna Lily. Also, while today we think of the Easter lily as a potted plant, the original *Lilium* product was a cut flower, and *Lilium* cut flowers are still produced for Easter and other market periods.

Bulbs of the original Easter *Lilium, L. candidum,* were imported from France (White, 1940). By 1930, *Lilium longiforum* Thunb. was the principal commercial *Lilium* species grown (Griffiths, 1930), and these bulbs were shipped from Bermuda and Japan. Some lilies were sold in pots when Bahr (1922) wrote his text, but the use of lilies as flowering potted plants must have developed after the introduction of compact *L. longiflorum* cultivars from nurseries in the Pacific northwestern United States. For white flowered lilies, preference for *L. longiflorum* cultivars has persisted until the present time (Roberts and Blaney, 1967a,b). However, demand at Easter is now increasing for colored species and hybrids of many other *Lilium* species (DeGraaf, 1969). These crops are discussed in Chapter 8 by De Hertogh and in the Holland Bulb Forcer's Manual (De Hertogh, 1988).

Lilium longiflorum, indigenous to the southern islands of the Liu-Chiu (Liukiu, Ryukyu) Archipelago of southern Japan, was first collected by Carl Peter Thunberg in 1777. In 1819, bulbs of the species were transported to London, and *L. longiflorum eximium* was brought from Japan to the Botanical Garden in Ghent, Belgium, by von Siebold in 1830. By 1856, the variety was introduced into the gardens of Bermuda by sailors and missionaries (Holmes, 1956).

Wilson (1925) described his encounter with the species in Japan as follows:

> On Okinawa, the main island of the Liukiu Archipelago, I saw this Lily growing wild in pockets in the coral rock by the sea. It appears to be a maritime species and, unlike most Lilies, a limestone plant. . . . About its being truly indigenous on . . . (Oshima) island and the others which make up the northern and central groups of the Liukiu Archipelago (there is no question). . . . The most vigorous plants were not more than a metre tall, and the bulbs divided up in such a manner as to produce nests or groups of plants.

Nuttonson (1952) described the Liu-Chiu Archipelago as consisting of islands, islets, rocks, and exposed coral reefs. The terrain is hilly to mountainous, and the islands are commonly fringed by coral reefs (leading to an alkaline soil pH). The vegetation is lush because of high rainfall, high relative humidity, mild winters, and warm summers. Extensive forests are common there. Based on temperature, precipitation, cloud cover, relative humidity, and soil characteristics, he calculated climatic analogs in the United States: Brooksville, Okechobee, Orlando, St. Augustine, and Tampa, Florida and Burrwood, Louisiana.

Common to all lilies are a perennating storage structure and an imbricate (scaly) bulb. Lily bulbs increase in size by yearly increments of growth and contain scale leaves 1 or more years old. A new growing point forms in the bulb each year, and from this, new scale leaves and a new shoot are generated. Flower formation terminates the stem. Bulbs consist of a basal plate from which arise alternate layers of fleshy, scale-like leaves that are superimposed and overlap each other. Roots arise from the under surface of the basal plate.

A large bulb is an accumulation of several years' growth (Pfeiffer, 1935). Industry terminology was developed to describe increments of perennial growth that are added to the bulb each year. The growing or recently matured element of the bulb is termed the "daughter bulb" (Woodcock and Stearn, 1950). The older portion of the bulb is termed the "mother bulb." When a new element forms and grows, the mature daughter bulb of the previous season becomes a portion of the mother bulb, and the new element becomes the daughter bulb. In large bulbs, more than one meristem may arise, and the structure splits into two or more smaller bulbs (Okada, 1952).

Easter lilies pass through a juvenile phase before they can be stimulated to flower (Blaney and Roberts, 1966a,b; Brierley and Curtis, 1942; Stuart, 1944, 1945; Thornton and Imle, 1940), but this stage is not well described. Overwintering (cold, vernalization) treatment causes rapid shoot emergence (Stuart, 1945) and induces flower formation. The equivalent of 6 weeks of 35° to 45°F is required (Brierley, 1941; Hartley et al., 1964; Miller and Kiplinger, 1966a; Smith, 1963; Stuart, 1945, 1967; Weiler and Langhans, 1968). Treatment must be longer below or above this temperature range.

Temperatures higher than 70°F are not cold inductive, and if applied after cold treatment and before formation of flower buds erase previous cold treatment (Lattimer, 1954; Miller and Kiplinger, 1966b; Weiler and Langhans, 1968, 1972a). This erasure is beneficial to remove unknown amounts of cold induction received during fall harvest and shed storage. Erasure is detrimental if it occurs at the end of the cold treatment because the crop will not force within a normal scheduled period.

The transition from bulb formation to shoot formation is not easily predicted, but it is influenced by one portion of a bulb on other portions of the bulb, the bulb size, growth rate, and environmental conditions (Roberts and Tomasovic, 1975; Roberts et al., 1985; Roh and Wilkins, 1977f). Mature bulbs contain about one-half of the leaf primordia that will develop on the stem during forcing (Roberts, 1980). Thus, during forcing, the remaining one-half of the leaves initiate and all the leaves unfold. Bulbs harvested late respond more readily to storage treatments than bulbs harvested early (Lin and Wilkins, 1975; Roberts et al., 1985).

Easter lily flowering also can be photoinduced (Gill and Stuart, 1967; Waters, 1966; Waters and Wilkins, 1967; Weiler and Langhans, 1972b; Wilkins et al., 1968a; Wilkins et al., 1970). A number of light sources and daily intervals can be used (Roh and Wilkins, 1977b,c,d,e; Wilkins and Roh, 1972). The stimulus is usually perceived by stem leaves, but exposed bulb scale leaves are also perceptive (Pertuit and Kelly, 1987; Roh and Wilkins, 1974b).

After induction by cold treatment and long photoperiods, flower buds initiate on the shoot arising from the bulb (Shoushan, 1950). Kosugi (1952) and DeHertogh et al. (1976) described the stages of inflorescence and flower formation in Easter lily. The inflorescence terminates the stem and is not formed until the final one-half of the leaves initiate. The production of flower buds terminates the life of the shoot, and a new axis arises in the axil of the last-formed scale leaf of the bulb (Pfeiffer, 1935). Meristematic activity of the vegetative bud of the new daughter bulb and root formation at buried stem axils of the stem are observed near the time of flower initiation at the elongating shoot tip.

Physiological studies of Easter lilies show that RNA content of shoot apex cells is related to growth rate (Yasui, 1973). The growing shoot and flower buds are supplied metabolites from the older ("mother") elements of the bulb, and under low light conditions even by the younger ("daughter") element of the bulb (Miller and Langhans, 1989a,b). After bloom, the photosynthesizing shoot supplies metabolites to the new developing ("daughter") element of the bulb as well as to bulblets on buried stem nodes (Wang and Breen, 1987). Phytochrome has been found in Easter lilies, including in bulb scales. As in other plants, the pigment responds to red and far red light (Roh and Wilkins, 1974a,b). Hormonal involvement in the Easter lily life cycle was studied as early as 1942 (Stewart and Stuart, 1942). Studies in this area have been continued by others and are too numerous to summarize in this chapter. So far, a clear explanation of plant behavior based on hormonal patterns has not emerged. Gibberellic acid (GA) applications to plants promote, while abscisic acid and ethylene applications slow, stem elongation and plant growth rate (DeHertogh and Blakely, 1972; Hiraki and Ota, 1975; Lin et al., 1975a,b). Plants are especially responsive to GA during the temperature treatments that suppress stem elongation discussed later in this chapter (Zieslin and Tsujita, 1988).

II. PRODUCTION

Easter lily culture was comprehensively reviewed in the manuals by Gould (1957), Kiplinger and Langhans (1967), and Roberts et al. (1985). They document the pioneering accomplishments of the United States Department of Agriculture scientists, such as Oliver, Griffiths, Brierley, Emsweller, and Stuart, and others, such as Blaney and Roberts. Summaries have been made by Wilkins (1980), Miller (1985), and DeHertogh (1988). Terminology associated with Easter lily culture was summarized by DeHertogh et al. (1971).

A. Statistics

Lilium longiflorum is an important holiday potted crop in the North America. In 1970, the most recent year comparison can be made (Bureau of the Census,

1970), 7.5 million bulbs were produced in the United States and 4.1 million potted plants and 3.7 million cut flowers were produced. In 1987, the 28 leading producing states reported that 7.6 million Easter lily pots were sold at firm gate at wholesale value for $28 million (Anonymous, 1988). The recent large numbers of imported *Lilium* bulbs [35.8 million in 1986 (Trade Statistics and Analysis Branch, 1986)] and imported *Lilium* cut flowers—(32.7 million (Anonymous, 1986)—are primarily other species of *Lilium.*

In one study, unit costs as a percent of total cost of production for plants sold were approximately 37 to 53% for direct materials, 9 to 18% for direct labor, and 33 to 52% for overhead, depending on efficiencies of scale and mass or florist market segment (Brumfield *et al.,* 1981).

B. Breeding and Selection

A number of plant breeding and selection objectives and techniques have been employed to improve Easter lilies (Emsweller, 1950; Roberts and Blaney, 1967b; van Tuyl, 1985; van Tuyl *et al.,* 1986; Wolf and van Tuyl, 1984).

Developments from many laboratories have led to progress in compatibility of self- and interspecific crosses, embryo rescue, haploids from anther culture, and somatic hybridization. These techniques promise to revolutionize hybridizing opportunities; however, their summary is beyond the scope of this chapter.

Today's primary *L. longiflorum* pot cultivar, 'Nellie White,' was introduced in 1955 from the Pacific Northwest. 'Ace' is often grown, and a number of other cultivars are known (Table I). Improved cultivars have been released with greater frequency from, among several sources, the Pacific Flower Bulb Growers' Association breeding program, Brookings, Oregon.

Table I

Current and Past Easter Lily Cultivars for Pot Culture and Approximate Greenhouse Forcing Time for Case-Cooled Bulbs[a]

Cultivar	Origin/year	Approximate greenhouse forcing time (days)
'Slocum's Ace' (Ace)	USA/1935	110–120
'Croft'	USA/1928	120
'Harbor'	USA/1981	90–100
'Harson'	USA/1964	85–90
'Nellie White'	USA/1955	100–110
'White American'	The Netherlands/1987	80–102

[a]From Roberts and Blaney (1967), Box and Payne (1967), Laiche (1982), Bonthuis and van Tuyl (1984), DeHertogh (1988), Anonymous (1981).

C. Propagation and Propagule Handling

1. Bulb Production and Handling

Harris, who was acquainted with *L. longiflorum* in Bermuda, introduced the plant into the United States in Philadelphia about 1876 as *L. Harrisii* (Holmes, 1956). This was the beginning of a commercial bulb supply to North America. However, by 1887, disease limited Bermuda bulb production, and bulbs were increasingly supplied by Japan. Bermuda and Japan exported 3 million bulbs to the United States in 1908 (Oliver, 1908); thereafter, Japan became the principal exporter of bulbs. By the early 1940s, the United States imported 25 million bulbs from Japan annually (Anonymous, 1946). Bulbs were not received from Japan during and immediately after World War II, and Mexico, the Netherlands, and Bermuda then became important bulb sources. A total of 3 million bulbs were imported from these countries in 1945 (Butterfield, 1947). This interruption of the bulb supply provided an opportunity for the establishment of a North American bulb production industry.

Original sites of domestic bulb production were the Pacific Northwest (Smith River, California to Harbor, Oregon), Florida, and Louisiana (Fig. 1). Four million bulbs were produced in the United States in 1946, of which 2.7 million were grown for Easter (Anonymous, 1946). Today nurseries in the Pacific Northwest produce most of the bulbs grown for pot culture (Miller, 1985), but some bulbs are imported from Japan.

Fig. 1.   Easter lily bulb production field, Smith River, California.

Bulbs 7 to 10 inches or more in circumference are sold to greenhouse potted plant producers (Blaney and Roberts, 1967; Stuart, 1967). Although natural division of the bulb into two or more bulbs takes place in greater frequency as the bulb size increases (Okada, 1952) and is a method of propagation, commercial propagators rely especially on bulblet formation, either from leaf axils of that portion of the stem submerged in the soil or from scale leaves removed from bulbs and incubated (Blaney and Roberts, 1967; Griffiths, 1930). Bulb production from bulblets requires 2 to 3 years, while production from scales requires 3 to 4 years (Miller, 1985). While *L. longiflorum* rarely forms bulbil propagules in above ground leaf axils, cytokinin applications starting when shoots were 6 inches long promoted their formation (Nightingale, 1979).

In vivo propagation from bulb scale segments and *in vitro* propagation from bulb scales and scale pieces (cotyledons, flower pedicel segments; flower parts including stamen filaments, petal, and ovary) have been extensively studied, but summary of these findings is beyond the scope of this chapter. Because the development of stem bulblets competes with the daughter bulb for metabolites from the photosynthesizing stem, selection for cultivars that do not form bulblets may speed bulb production. To select a bulblet-free genotype would require reliance on tissue culture for propagules (Wang, 1988).

Crop growth and development related to bulb production was extensively reviewed by Roberts *et al.* (1985). Leaf collection and analysis to monitor crop mineral nutrient status is best performed early in the season as soon as matured leaves can be collected from the crop (Chaplin and Roberts, 1981). Infection of roots with mycorrhizal fungi may benefit seedlings and bulbs lacking nutrient reserves when grown in infertile soils (Ames and Linderman, 1978). In one study, optimum bulb production was found when smaller bulbs were planted 4 inches deep and spaced 8 inches apart, and when larger bulbs were planted 6 inches deep and spaced 16 inches apart (Kamel *et al.,* 1978). Predicting bulb growth rate and maturation is based on environmental conditions, especially bulb temperature, both in the field and during handling (Roberts *et al.,* 1983; Wang and Roberts, 1983).

Plants normally bloom in the fields in July, but flower buds are removed by application of a disbudding chemical. This conserves photosynthate for the developing daughter bulb (Roberts and Fuchigami, 1972; Wang and Breen, 1984). Too early disbudding limits the photosynthetic area by reducing stem elongation and leaf expansion, and does not enhance daughter bulb weight. Removing buds when they are 1 to 1.5 inches long increases bulb weight (Wang, 1988; Wang and Breen, 1986).

Harvest dates for bulbs occur consistently in mid-September. Large grade bulbs are sold to potted plant producers, while smaller grade bulbs ["yearlings" and "scalets" (Miller, 1985)] are replanted in November for 1 or 2 more years growth. Cases of commercial-grade bulbs are packed by alternating layers of bulbs and packing medium (Fig. 2), usually peat moss at 30 to 50% moisture content (Stuart, 1954). A varying interval of "shed (common) storage" near ambient outdoor temperature typically occurs between bulb harvest and shipment of graded bulbs

Fig. 2.   Lily bulbs packed in peat moss in opened case.

to regional cooling facilities of bulb suppliers or to local producers. Polyethylene lining of cases to conserve bulb moisture may be advantageous to efficient cold induction of flowering and finished plant growth and quality (Prince and Cunningham, 1990; Stuart, 1954).

Bulbs are graded based on bulb circumference and sold in cases of a standard size and similar packed weight. Thus, the number of bulbs in a case varies with bulb grade: 300 bulbs of 6.5- to 7-inch circumference, 250 bulbs of 7- to 8-inch circumference, 200 bulbs of 8- to 9-inch circumference, 150 bulbs of 9- to 10-inch circumference, and 100 bulbs of 10-inch or greater circumference (Blaney and Roberts, 1967; Miller, 1985). In addition, two stems from one bulb can be obtained from "double-nosed" bulbs (i.e., large bulbs with two growing points).

2. Production from Leaf Cuttings

A finished plant production scheme was proposed that bypasses the field bulb production phase completely (Tammen *et al.,* 1986). It is based on several principles: (1) pathogen-free plants can be developed and used as stock plants; (2) stem leaves readily root and form approximately 2 bulblets per leaf (Roh, 1982), thus leaves provide a source for large numbers of propagules; (3) 0.2-inch diameter bulblets grow continuously without cold treatment; and (4) growth is vegetative under short photoperiods, but 2 to 3 weeks of long photoperiods when plants reach 25 to 35 leaves will induce flowering. Plants flower about 75 days after long photoperiod treatment begins. At this time, however, the Easter lily crop is finished by traditional methods.

D. Pot Plant Production

1. Product Design

Choices of bulb grade, single- or double-nosed bulbs, and number of bulbs in a standard pot are based on market demand for the number of flowers per plant and the fullness of pot. Single-nosed, 6- to 10-inch circumference, bulbs are common for 5- to 6-inch pots, whereas, double-nosed bulbs 9 inches or greater in circumference are common for 7-inch pots. Multiple bulbs may be planted into 8- to 10-inch pots.

2. Scheduling

Easter is an unusual holiday from a crop scheduling point of view. Its date varies yearly from 26 March to 25 April. Scheduling the crop becomes a challenge from three perspectives: (1) pretreatment of the crop varies yearly, (2) the crop often is grown in different months and environmental conditions from year to year, and (3) spring greenhouse space use and crop rotation vary yearly. (For example, little time remains after a late Easter to produce bedding plants for April and May markets, so fewer Easter lilies and bedding plants may be produced that year.)

On October 1 or later, bulbs arrive in the forcing areas and may be cold-treated in the shipping cases by the bulb supplier or by the potted plant producer. After flower-inductive treatments, typical greenhouse forcing time for an Easter lily crop is 90 to 110 days, depending on cultivar and method of cold treatment (Tables I and II). To be marketed, plants must flower approximately 1 week before the holiday.

3. Cold Induction of Flower Formation

Potted plant forcing involves a series of procedures. Three approaches are used to achieve the required cooling of Easter lily bulbs (Table III). Bulbs case-cooled may be received by a greenhouse crop producer either before cold storage or from the bulb supplier after cold storage. Home case-cooled bulbs are not precooled by the bulb supplier. On receipt, cases are placed in a 35 to 45°F refrigerator for 6 weeks before potting and placement in the forcing greenhouse. Supplier case-cooled bulbs are precooled by the bulb supplier. On arrival, they are potted and placed in the forcing greenhouse (Fig. 3). Controlled temperature forcing (CTF) begins with non-precooled bulbs. On arrival, bulbs are potted, watered, and placed at 63° to 65°F for 3 weeks for rooting and leaf formation on the shoot within the bulb. The high flower-bud count and improved plant form typical of this treatment are ascribed to this rooting and leaf forming period. After 3 weeks, plants are placed where bulb temperature will remain 35° to 45°F for 6 weeks. Forcing begins after storage (Carlson and DeHertogh 1967; DeHertogh and Einert, 1969). Natural cooling begins with nonprecooled bulbs. On arrival bulbs are potted, watered, and

Table II

Typical Forcing Schedule for Easter Lilies[a]

Weeks before Easter	Date	Procedure or plant stage		
		Case-cooled	CTF[b]	Natural cooling
26	———	Bulb digging completed		
25	———	—		
24	———	Bulb shipments arrive in forcing areas		
23	———	Start 40°F storage[c]	Pot.[d] Put at 60°–65°F	Pot.[d] Put at 32°–70°F
22	———		↓	
21	———		Start 40°F storage	
20	———			
19	———			
18	———			
17	———	Pot. Put in greenhouse		
16	———			
15	———	↓		
14	———		Put in greenhouse	Put in greenhouse
13	———	Shoots emerge from soil		
12	———	Shoots 1–3 inches		
11	———	Shoots 3–5 inches		
10	———	Shoots 5–9 inches; approx. 47 leaves to be unfolded; flower buds initiating; grow slow plants at warmer average temperature and advanced plants cooler		
9	———	Shoots 9–12 inches; approx. 35 leaves to be unfolded		
8	———	Shoots 9–12 inches; approx. 25 leaves to be unfolded; feel flower buds		
7	———	Shoots 9–15 inches; approx. 12 leaves to be unfolded		
6	———	Buds visible		
5	———	Buds 1/2–1 inch		
4	———	Buds 1–2 inches and bending		
3	———	Buds 2–3 inches		
2	———	Buds 3–5 inches		
	———	Start cold-storing advanced plants		
1	———	Plants ready to sell		
0	———	Easter		

[a]Modified from Weiler *et al.* (1988).

[b]CTF, Controlled temperature forcing.

[c]If long-day treatment is used for 2 weeks after shoot emergence, cold treatment is reduced by 2 weeks (i.e., begins 2 weeks later).

[d]If long-day treatment is used for 2 weeks after shoot emergence, planting is 2 weeks later and cold treatment is reduced by 2 weeks.

placed in a protective structure kept near outdoor temperatures. To avoid bulb freezing, a source for occasional heating may be needed. Bulbs require 1000 hours (6 weeks) of temperatures 35° to 45°F (2° to 7°C). If unusually warm weather prevents the accumulation of 1000 hours of cool temperature, long photoperiod, "insurance" lighting can be used to substitute for deficient cold treatment as discussed in the next section.

Table III

Number of Leaves at Bloom for Easter Lily Cultivars[a]

Cultivar	Pretreatment at 60°–65°F (weeks)	Cold storage (weeks)	Total leaves at bloom
'Nellie White' (CTF)	3	6	81 ± 8
'Nellie White' (Case-cooled)	—	6	72 ± 9
'Ace' (CTF)	3	6	92 ± 11
'Ace' (Case-cooled)	—	6	85 ± 10

[a] ± standard deviation for one bulb grade 8–9 inches circumference of two cultivars averaged over 16 years (1969 to 1984) for controlled temperature forced (CTF) and case-cooled crops (Wilkins, 1985).

Both CTF and natural cooling result in better shaped plants with more leaves and flowers. These methods are also more labor intensive and the crop requires extensive space from October until December for root development and cooling. In locations where late fall may be warm, large refrigerated facilities may be needed.

Because storage temperature is critical to success with Easter lilies, insertion of thermometers into cases or growing mix during prestorage and storage treatments assists management. With the above treatments crop forcing time is normal (i.e.,

Fig. 3. Temporary forcing of potted Easter lily bulbs, until Christmas crops are removed from the greenhouse and lilies can be placed on the benches.

90 to 110 days depending on cultivar, cold treatment method, and forcing conditions). More cold than required speeds the date of bloom further and reduces the number of leaves and flowers.

Holding bulbs for late planting is difficult if at the same time the deleterious consequences of too much cold storage are to be avoided. However, such holding is required for a late Easter, Greek Easter, or other purposes. Storage above 40°F promotes shoot growth and above 70°F erases cold induction. Holding bulbs near 32°F is best (Stuart, 1954). Van Nes (1981) suggests storage at 33°F before potting and holding at 28°F. Temperatures below 28°F may irreversibly damage the bulb and kill the growing point (Miller, 1989).

4. Photoinduction of Flower Formation

Following cold treatment and shoot emergence, long photoperiods continue flower inductive treatment, as discussed earlier. Long photoperiods applied during early growth, before flower formation, can be beneficial to crops that are incompletely cold treated, but detrimental to crops that are adequately cold treated. For inadequately cold treated plants, application of long photoperiods after shoot emergence completes the induction of flower formation. When details of cold treatment are uncertain, this treatment is called "insurance lighting" (Wilkins *et al.*, 1970). Long photoperiods usually are created with 2 to 3 μmol/sec/m^2 [10 to 20 foot-candles (fc)] incandescent light from 10 p.m. to 2 a.m. for 2 weeks, starting when shoots emerge from the growing mix. This treatment is especially useful to force for early Easter dates when time is minimal for cold treatment. Because emergence is rarely uniform between plants, growers often sort pots into two to three groups, based on stage of shoot development, to precisely light and force the crop. Because long photoperiods can produce the same effects as overcooling (fewer leaves and flowers, and rapid forcing time), lighting is applied only if insufficient cooling is suspected. Overapplication of insurance lighting or long photoperiods applied after flower formation elongate expanding internodes (Smith and Langhans, 1962b), which may lead to excessive plant height.

For early Easter dates and short forcing times, a commercial treatment of 4 weeks of cold and 2 weeks of long photoperiods starting when shoots emerge from the soil is useful. This allows producers to gain up to 2 weeks extra forcing time when compared to traditional methods.

5. Predicting Date of Bloom during Crop Growth

Bringing the Easter lily crop into bloom 5 to 10 days before Easter is an art as well as a science. The Easter date varies annually, while the date of bulb harvest remains fairly constant. Thus, duration of storage and forcing time varies yearly. Also the forcing time from plant to plant is variable, so periodic sorting of the crop from those "advanced" and placed at a cooler temperature to those "behind" and placed at a warmer temperature generally is required. Because of these several factors, up to 10% of the crop is unmarketable (Brumfield *et al.*, 1981).

Beginning about 1950, prediction of bloom date improved when flower bud size was correlated to date of bloom at known forcing temperatures (Erickson, 1948).

By frequently measuring the size (length) of the first bud to open, plants could be sorted and moved to various temperatures to bring each plant into bloom near the marketing date. This method becomes useful after flower buds become visible, 4 to 6 weeks before bloom. For example, Roh and Wilkins (1973) found that when grown at 60°, 70°, 80°, and 90°F, time from visible bud to bloom was 50, 28, 25 and 24 days, respectively. For 'Nellie White,' the temperature and bud length relationship was two-phased (Healy and Wilkins, 1984). Buds 2 inches or less responded more dramatically to temperature differences than buds greater than 2 inches. Precise temperature adjustments can be calculated based on current bud length using two regression equations. For example, at 50°, 60°, 65°, 70°, and 75°F, time for buds 2 inches to bloom was 56, 26, 24, 20, and 20 days, respectively.

About 1967, scientists began to correlate rate of leaf unfolding early in the forcing period to date of bloom (Blaney et al., 1967; Grueber et al., 1987; Wilkins and Roberts, 1969). This enhanced the prediction of harvest date before flower buds were visible and permitted earlier sorting of the crop and environmental adjustment to speed or slow growth. The method works as described in the following paragraphs.

(1) Each grade (size) of bulb, when cooled for a specified duration, will flower after forming and "unfolding" a predictable number of leaves (Table III). The actual number of leaves varies somewhat from year to year, and larger bulbs unfold leaves at a faster rate (Lange and Heins, 1990).

(2) Leaf unfolding occurs between shoot emergence and the date flower buds are visible, 4 to 6 weeks before bloom.

(3) At any stage before flower buds are visible, a plant consists of unfolded leaves and leaves yet to unfold. The number of leaves yet to unfold is found by sacrificing a few (e.g., 10) representative plants and counting, with use of a needle and hand lens, the number of unfolded leaves clasping the growing point. The number of leaves yet to unfold divided by the remaining number of days before the desired date of visible bud is the rate of leaf unfolding that must occur if the plants are to grow and be marketed as planned. For example, if Easter occurs on 15 April and 51 leaves remain to be unfolded on 27 January, the following calculation can be made (A) To permit marketing, bloom should occur 7 to 10 days before Easter (e.g., 5 April); (B) Under normal forcing conditions, flower buds are visible about 30 days (4 to 6 weeks) before bloom (Fig. 4). Thus, the required visible bud date is 6 March (Fig. 5). (C) There are 40 days from 25 January to 6 March. (D) Thus, for timely bloom, the rate of leaf unfolding from 25 January to 6 March must be 1.28 leaves per day (51 leaves per 40 days). (E) The unfolding rate, 1.28 leaves per day, can be monitored on several representative plants by discreetly marking the last unfolded leaf, and every few days counting the number of new leaves unfolded since the last observation. Temperature can be raised or lowered as necessary to adjust the leaf unfolding rate (Karlsson et al., 1988).

6. Greenhouse Environmental Management

Optimum environmental conditions for Easter lily forcing are summarized in Table IV.

Fig. 4. Flower bud initiation occurs at approximately this stage.

Fig. 5. Flower buds readily visible.

Table IV

Optimum Environmental Levels for Easter Lilies[a]

Environmental factor	Production optimum
Temperature (°F)	
Bulb growth	55°–65°
Planting to shoot emergence	65°–70°
Emergence to flower initiation (shoots 4–6 inches)	60°–65°
Flower initiation to visible flower buds	61° (night), 65° (day)
Visible flower buds to bloom	60°–81°
Light (μmol/sec/m^2)	Full
pH	7 in soil mix, 6.2 in peat-lite
Fertilization rate (ppm nitrogen)	Perhaps 200 ppm nitrogen each watering
Water solute potential (dS/m at 1:2)	Less than 1.5
Water matric potential (cm tension)	Less than 30–100
Relative humidity (%)	—
Carbon dioxide (ppm)	1000

[a]From Smith and Langhans (1962a); Wilkins *et al.* (1968b); DeHertogh and Einert (1969); Roh and Wilkins (1977c); Wang and Roberts (1983); Healy and Wilkins (1984); Weiler *et al.*, (1988).

a. Managing the Root Zone Environment. Easter lily growing mix, irrigation water, and fertilizer must be carefully integrated because they influence optimization of air, water, and mineral nutrients; avoidance of toxins; and minimization of disease development in the root zone. Many growing mixes have been successfully tried on Easter lilies, including soil-based mixes, peat-lite (soilless) mixes, and rockwool (Goldsberry and Maffei, 1986). Because Easter lilies in bloom tend to be "top-heavy," a relatively high bulk density is preferred, and calcined clay, pasteurized field soil, or pasteurized sand are often chosen as a mix component. However, plants in soil mixes may grow and flower more slowly (Goldsberry and Maffei, 1986) if the mix is poorly aerated or holds too much water. Aeration is maximized with large particles of peat moss, perlite, and vermiculite. For most mixes, water storage is limited. Storage can be maximized by selecting water retaining components such as peat moss or other organic matter and vermiculite. Water absorbing polymers are sometimes added to improve water storage. Within the volume of a pot, few components of contemporary mixes provide significant nutrient storage capacity, and a slow release fertilizer is sometimes added for this purpose (Hasek and Sciaroni, 1970).

Prevention of leaf tip burn and leaf scorch through avoidance of fluoride in the root zone is a major goal in Easter lily production. The growing mix must be carefully formulated to minimize fluorine availability (Marousky and Woltz, 1977; Marousky, 1981; Seeley and Monroe, 1976). Fluorophosphates, such as superphosphate, are excluded from the mix or only added in small amounts, and the mix is limed to obtain a pH of near 7 for mixes containing 25% or more soil and 6.2 to

6.5 for peat-lite and other organic mixes. Fluoride is less available at these high pH levels. Avoidance of ammonium-based nitrogen fertilizers helps retain high pH (Marousky and Woltz, 1977; Tizio and Seeley, 1976). Some sources of phosphorus are refined and relatively free from fluoride. Incorporating a fluoride-free slow release fertilizer into the growing mix is one option for safely providing phosphorus to a crop, while applying soluble fertilizer during crop growth is another. One or both of these practices is common. Superphosphate contains gypsum (calcium sulfate), which is a major source of calcium and sulfur for crops. Because little superphosphate is incorporated for Easter lilies, gypsum itself may be added to the growing medium. Mineral nutrient deficiencies of Easter lilies are described by Seeley (1950).

"Standard" pots (those as deep as wide) are suggested for Easter lilies to permit both basal and stem roots to develop and to maximize drainage. Bulbs are planted so the nose (top) of the bulb is 1 to 2 inches below the growing mix surface. This is accomplished by placing about 1 to 2 inches of growing mix at the pot bottom, putting the bulb on it, and filling in the remaining pot volume with growing mix. Deep bulb placement encourages both early growth of the root system attached to the bulb basal plate and development of stem roots that form when plants are about half grown. If root diseases overtake the basal root system, the stem root system usually can be kept healthy to finish and market the crop.

Selection of deep, standard pots also deters root disease development by maximizing drainage of gravitational water from the pot after irrigation. Capillary length is shortened by adding gravel or other coarse materials to the bottom of the pot. Rather than promoting drainage, such materials diminish drainage.

High quality irrigation water and reliance on calcium nitrate as a major component of a liquid fertilization program are important to successful production. Compared to 40° to 55°F irrigation water, root systems may be more extensive and plants may grow faster if irrigated with 70° to 85°F water (Goldsberry and Maffei, 1986; McMurray, 1978). Finished plants, however, differ little because of irrigation water temperature, unless the water is especially cold.

b. Managing the Aboveground Environment. Optimum plant height, many buds and flowers, little flower bud abortion, and green leaves all the way down to the base of the plant are long-standing goals of Easter lily crop forcers. Flower bud numbers usually are derived by formation of secondary buds from the pedicel node of primary buds and tertiary buds from the pedicel node of secondary buds (Roh and Wilkins, 1977c). Factors that maximize formation and retention of flower buds are large bulb size, optimum cold- and photoinductive treatments, high light flux (Heins et al., 1982; Miller and Langhans, 1989a,b), and moderate forcing temperatures.

Light promotes compact plants and development of all flower buds. Transmission of full natural light through the glazing material is essential for greenhouse forcing of Easter lilies. Overhead obstructions to light, such as pipes, shelves, and hanging plants, make production of high quality finished plants difficult.

Plant spacing greatly affects light penetration into the crop canopy. Cost analysis models for pot spacing on benches, when tested over a wide range of space and labor costs, demonstrate that pot crops are least expensively grown with two moves. Easter lilies are started pot-to-pot. More plants will fit on a bench if pot positions are alternated between rows, that is, pots are placed in an equidistant triangular pattern rather than a square pattern. Leaves of one plant begin to touch another's 4 to 6 weeks after forcing starts. At that time pots are spaced to provide 2 inches of space between pots. At this move plants are sorted based on growth differences. Advanced plants are placed in a cooler area, average plants compose another group, and small, slow plants are placed in a slightly warmer area. Final spacing occurs at about the visible bud stage, 4 to 6 weeks before flowering. Space between plants is expanded to 4 inches. The above is suggested for 6-inch pots. Actual spacing dates and spacings vary with each operation and reflect differences in bulb grades, pot sizes, greenhouse light levels, desired crop quality, and management styles. Generous spacing minimizes the number of plants produced but maximizes plant quality. It may also reduce income. Close spacing causes stems to elongate because plants receive little lateral light. Growth regulation helps to prevent excessive plant stretch from interplant light competition. At extremely close spacings or when respacing is delayed, lower leaves may turn yellow and die because of excessive shading.

Newly potted bulbs are grown for the first 3 weeks at 63° to 70°F nights to ensure good rooting, then temperature usually is lowered to 60°F night, 70°F day. Temperature is especially critical before and during flower bud initiation, which is mid-January for an early Easter and mid-February for a late Easter. Temperatures higher than 70°F delay flower formation by erasing the effects of cold treatment. After these dates, night temperatures, and to a lesser extent, day temperatures, are controlled by judgments related to date of bloom. Buds become visible 4 to 6 weeks before bloom.

Forcing temperature regimes have been proposed to further increase flower bud number. Lowering forcing temperatures during flower bud formation (DeHertogh et al., 1976; Roh and Wilkins, 1977a,c) and split night temperature schemes to conserve fuel (Gent et al., 1979; Laiche, 1982) have been proposed. These are sometimes effective, but considerations of the total environmental relationship and profitability are necessary. Ultimate success depends on bulb history and treatment; environmental factors during forcing, such as light flux, forcing temperature, and carbon dioxide enrichment; and market requirements.

7. Integrated Plant Height and Form Management

Market requirements for final plant height are met through the integration of several aspects of crop management. Light intensity, photoperiod, temperature, ethylene generation from mechanical stroking of leaves (Hiraki and Ota, 1975), and the relationship of night temperature to day temperature are major influences on total stem elongation. Mechanical treatment of the crop is not a current practice.

Assessment and management of plant height varies with bulb treatment; for example, plants from case-cooled bulbs grew to half their final height in 6 to 8 weeks, while those from controlled temperature forced bulbs reached half their final height after 9 to 11 weeks (Hammer and Hopper, 1989).

Desirable plant height and form are primary market requirements. Depending on market requirements, preferred height (including pot) ranges from 2.62 to 3.5 times the pot diameter (e.g. 16 to 21 inches for a 6-inch pot). The top four internodes of Easter lilies contribute as much to plant height as the bottom 30 nodes, and the greater the number of flowers on the plant the greater is the length of the four top internodes (Gianfagna et al., 1986). Nevertheless, for many production situations contemporary L. longiflorum cultivars do not require additional height management if grown at full light, adequately spaced, and forced at normal temperatures. Occasionally, the following height management measures must be undertaken.

a. Height Management with Light Adjustment. Lilies are a high light requiring crop. They also respond to length of day (photoperiod). Year to year variation in crop height is partly caused by differences in light each year. Stray light from other greenhouse cropping areas and from street lamps also affects the crop. Plants remain compact at full light flux. Even insulated greenhouse coverings that attenuate light slightly promote stem elongation. Providing full light necessitates clean, highly transmissive coverings and the placement of the pots in the brightest greenhouse locations. Plants typically "stretch" when placed on benches near walls, under overhead black shade cloth storage, or under hanging baskets or shelves. Close spacing of plants also stretches Easter lilies. Respacing the crop twice during production so leaves between plants barely touch is both economical and an effective plant height regulant.

If plants are too tall or too short, height can be adjusted by controlling photoperiod. As already mentioned, Easter lilies elongate late in production as the last four internodes elongate. This usually happens from 3 weeks before flower buds are visible to 2 weeks after flower buds are visible. New growth of Easter lilies estimated to be too short is stretched by long photoperiods starting when plants are 4 to 6 inches tall. Two to three $\mu mol/sec/m^2$ (10 to 20 fc) incandescent light nightly from 10 p.m. to 2 a.m. are effective. New growth of Easter lilies estimated to be too tall is kept compact by daily cover with blackout shading (e.g., from 5 p.m. to 8 a.m.) to create longer nights, starting when plants are 4 to 6 inches tall. This is especially useful during the longer natural photoperiods of March and April with late Easters.

b. Height Management with Temperature Adjustment. Past crop growers and scientists noticed that warm temperatures stretch lily crops (Kohl, 1958; Roh and Wilkins, 1973). However, Heins et al. (1987) developed a system to manage crop height based on the relationship between day and night temperature. This system is based on three principles. First, crop forcing time is primarily related to air temperature (Wang and Roberts, 1983) and heat accumulation rather than light accumulation (Heins et al., 1982). Second, crop scheduling (e.g., days from plant-ing to flower) is controlled by average daily temperature (i.e., the number of heat units accrued). While details of the following discussion are approximate, the logic is as follows. Forcing temperatures of 60°F night, 70°F day are often suggested for a lily crop. The crop grows slowly between 30° and 55°F (Hartley, 1968; O'Rourke and Branch, 1987; Wilkins, 1986). Using a base temperature of 55°F, accumulation of heat units can be calculated. Assume that the 60°F night temperature is set from 5 p.m. until 8 a.m. (15 hours), and the 70°F day temperature is set from 8 a.m. until 5 p.m. (9 hours). The number of heat units accrued per day are 75 degree-hours during the night [(60−55) × 15] plus 135 degree-hours during the day [(70−55) × 9], or a total of 210 degree-hours. Similar heat unit accumulation could be achieved by other day/night temperature configurations. Examples are provided in Table V. Third, crop height is controlled by the difference between day and night temperature. Night temperatures higher than day temperatures (a negative differ-ential) encourage compact growth while night temperatures lower than day tempe-ratures (a positive differential) encourage elongated growth (Table V). Temperature differential (DIF) is carefully controlled to achieve a predetermined finished crop height.

Crops growing in higher night temperatures than day temperatures may dis-color, and leaves may downturn. When treatment is prolonged, these character-istics may become irreversible.

Studies have demonstrated that after a warm night, lowering day temperature for only 2 hours early in the day is nearly as effective for limiting height as lowering the temperature for the entire day (Erwin et al., 1989).

c. Height Management with Growth Regulators.
Chemical growth retard-ants are also used to manage plant height. Chemical height control began in the

Table V

Forcing Temperature Regimes Providing Similar Daily Heat Unit Totals[a]

Greenhouse temperatures (°F)		Total heat units (degree-hours)	Differential (DIF)
Night	Day		
60	70	210[b]	10
64	64	216[c]	0
69	55	210[d]	−14

[a]Data are calculated for a 15-hour night and 9-hour day temper-atures. The heat unit accumulation base temperature is 55°F. Temper-atures are rounded to the nearest degree.
[b]$(60-55) \times 15 + (70-55) \times 9 = 210$
[c]$(64 - 55) \times 24 = 216$
[d]$(69-55) \times 15 + (55-55) \times 9 = 210$

1960s with Phosfon-D incorporated into the growing mix (Stuart *et al.,* 1961; Wiggans *et al.,* 1960). Subsequently some treatments with Phosfon-L were found effective (Holmes *et al.,* 1975). Ancymidol (A-Rest, Reducymol), introduced in the 1970s, was reliably effective at low doses when applied as a foliar spray or growing mix drench (Hasek *et al.,* 1971; Larson and Kimmins, 1971), and remained the preferred chemical throughout the 1980s. Paclobutrazol (Bonzi), introduced in the mid-1980s, was found effective as a growing mix drench at high doses (Gianfagna and Wulster, 1986; Jiao *et al.,* 1986), compared to drench or spray applications of uniconazole (Sumagic) (Bailey and Miller, 1989; Hammer, 1987). Triazole fungicides, chemically related to some growth retarding chemicals, also reduce stem elongation (Wulster *et al.,* 1987).

The amount of growth retardant to be applied is governed by several factors: (1) the chemical used; (2) the amount of height reduction desired; (3) the lily cultivar; (4) bulb size (plants from larger bulbs grow taller); (5) method of application; (6) growing mix (bark-containing mixes may render ineffective a portion of a drench application); and (7) cultural conditions. For shorter growing types such as 'Ace' and 'Nellie White,' the amount of height reduction needed usually is small, compared to taller growing types, such as 'Georgia,' a cultivar primarily grown for cut flowers. Product formulation affects the degree of retardation, and the addition of a surfactant and application as a foam sometimes increase activity (Holmes *et al.,* 1975).

Time of application and method are interrelated. Obviously, early application provides early moderation of elongation. Early application is feasible either by dipping bulbs into growth retardant solution before planting or by drenching the growing mix after planting. Bulb dips also have been studied because growing mix components may reduce drench application effectiveness (Larson *et al.,* 1987; Lewis and Lewis, 1980, 1981, 1982). Many growing mixes are suitable for drench applications; the chemical is present and effective for an extended period. Treatment before or at shoot emergence results in broad leaves closer to the pot rim that improves plant form (Weiler, 1978).

While early application is made before the need for growth retardation is visually apparent, application after growth begins is popular because the need for retardant application is obvious before treatment. Either drench application to the growing mix or spray application to the foliage after sufficient absorbing leaf area is present is made when plants are 3 to 7 inches tall. Multiple, low dose spray applications may be made. This approach increases labor costs associated with application, but minimizes unattractive clustering of nodes midway up the finished plant.

8. Forcing under Artificial Light

An alternative to greenhouse forcing may be feasible: Easter lilies forced in an insulated building under artificial lighting (Wilkins *et al.,* 1986). Components of a successful system are 8 hours of 400 μmol/sec/m^2 fluorescent light, retardation of stem elongation, and inhibition of ethylene-caused flower bud abortion under low light conditions (Mastalerz, 1965) with silver thiosulfate (STS) (Prince *et al.,* 1987; van Meeteren and de Proft, 1982).

9. Management with Crop Models

The advent of reliable environmental sensors and computerized models of crop behavior over a range of environmental conditions permits producers to optimize environment toward specific operational and market goals. Because the Easter lily growth rate is more closely correlated to temperature than to light, Karlsson *et al.* (1988) developed a temperature-based model between 57° and 86°F. They found equal effects of day temperature and night temperature on growth. For Easter lilies, computerized models will permit the simultaneous and constant management of growth rate, crop height, and perhaps management of other components of the production system, such as pest populations.

10. Pests

A number of pests limit crop growth and quality and are summarized in Tables VI and VII. In addition, numerous weeds compete directly with the crop, or they are hosts for other pests. Pest management is important both during field bulb production and during finished plant forcing. High qualtiy propagules, sanitation, careful monitoring of pests, and other aspects of integrated pest management are essential to production success.

11. Physiological Disorders

Other adversities occur without a known causal organism. These are referred to as abiotic or physiological disorders.

Table VI

Major Insect and Mite Pests of Easter Lily[a]

Pests	Genera
Insects	
Aphids	*Aphis, Myzus, Acyathosiphon, Neomyzus, Macrosiphum*
Beetle	*Lilioceris, Diabrothica, Apocelus*
Bulb fly larvae	*Cumeues*
Stem borer	*Papaimema, Emboloecia, Neolasioptera*
Caterpillars	Various
Fungus gnat larvae	*Bradysia*
Scales	*Chrysomphalus*
Symphylids	*Scutigerella*
Thrips	*Ctenothrips, Hercenothrips, Frankliniella, Liothrips*
Lily weevil grubs	*Agasphaerops*
Mites, bulb	*Rhizoglyphus*
Slugs	*Lamix*

[a]From Gould (1957), Kiplinger and Langhans (1967), Westcott (1973).

Table VII

Diseases of Easter Lily[a]

Disease	Cause
Nonsystemic	
Bulb mold	*Cladosporium*
Bulb storage rot	*Penicillium, Rhizopus*
Damping off	*Pythium, Rhizoctonia*
Gall	*Corynebacterim*
Leaf blight ("fire")	*Botrytis elliptica*
Leaf spot	*Cercospora, Cercosporella, Heterosporium, Ramularia*
*Tepal spot	*Botrytis cineria, B. liliorum*
*Root rots	*Cylindrocarpon, Fusarium, Phytophthora, Pythium, Rhizoctonia*
Soft rot	*Erwinia*
Rust	*Puccinia, Uromyces*
Brown/black scale	*Colletotrichum*
Scale pitting/yellow bulb	*Rhizoctonia*
Scale rot	*Pseudomonas, Sclerotium*
*Stem and basal rot	*Fusarium, Macrophomina, Phytophthora, Rhizoctonia*
*Stem canker	*Rhizoctonia*
Stem lesion	*Pseudomonas*
Systemic	
Curl-stripe, fleck, mottle, rosette	Virus
Nematodes	
Leaf and bud	*Aphelenchoides*
Root knot	*Meloidogyne*
Root lesion	*Pratylenchus*

[a]From Gould (1957); Kiplinger and Langhans (1967); Raabe (1975); Kruyer and Boontjkes (1982); Horst (1979); Bald *et al.* (1979, 1983); Linderman (1985). Diseases commonly found during greenhouse forcing are preceded by an asterisk (*).

a. Sprouting. Shoots sometimes prematurely elongate out of bulbs. "Summer sprouting" occurs during field production during the summer. It varies in severity with cultivar and field growing conditions, and is not fully understood or easily managed. Because the summer sprout is the premature emergence of next year's flowering shoot, sprouted bulbs must be discarded. "Autumn sprouting" occurs near the time of bulb harvest, during bulb handling, or in shipping cases. It is especially severe after bulbs are exposed to cold conditions, either naturally or during storage, that breaks dormancy followed by warm conditions (Roberts *et al.*, 1985). Many autumn sprouts are short, and when carefully handled by the producer of greenhouse crops, will develop into a high quality potted plant.

b. "No Shows." When placed in a greenhouse for forcing some bulbs may die, some may not grow, while others may emerge, grow very slowly, and not flower in time. There are several known causes, including those related to storage conditions. Lack of oxygen (Stuart, 1968), high levels of carbon dioxide (Green, 1934; Thornton, 1941), low bulb moisture content (Stuart, 1954), and severe freezing are among the known causes of such difficulties.

c. Leaf Scorch. Leaf scorch to a large extent is circumvented by typical production procedures. The cause is high levels of fluoride and perhaps other elements in plants. Leaf tips or crescent-shaped segments of leaf margins turn yellow and die. Major sources of fluoride are fluorophosphate (e.g., superphosphate), irrigation water, or perlite in the potting mix. Contemporary growing mix preparation and irrigation and fertilization practices minimize fluoride-containing materials. Leaf scorch is especially prevalent within some Easter lily breeding lines. Some selections of 'Croft' and 'Ace' cultivars are quite susceptible, while selections of 'Nellie White' are more resistant (Roberts and Moeller, 1979).

d. Slow Growth and Forcing. Occasionally crops do not respond normally to forcing conditions. Possible causes are numerous, especially because of the elaborate handling the bulbs receive over several years. The forcing environment may not be optimum. Thermometers and thermostats should be calibrated and properly located, so the desired air and root zone temperatures are achieved. The flower inductive stimulus from cold treatment might have been erased by exposure of the bulbs to temperatures above 70°F.

e. Abnormal Leaves and Flowers. Abnormal growth patterns, leaf distortion, or both may occur because of virus infection, nutritional deficiencies or toxicities, high soluble salts, exposure to herbicides (Weller and Hammer, 1984), and abnormal temperatures. Aphid infestations can cause malformed flowers.

f. Bud Abortion and "Blasting." Death of young or well-developed flower buds has been attributed to the following causes: root injury causing water stress, high soluble salts, or root rots when flower buds are visible; low carbohydrate supply; inadequate light intensity (Mastalerz, 1965; Wilkins *et al.*, 1986) caused by shading or too close plant spacing, especially when plants are being forced at high temperatures; and application of growth retardant (Prince and Cunningham, 1989). The problem also seems to be related to ethylene as low as 1 ppm for 2 days in the greenhouse (Hitchcock *et al.*, 1932; Rhoads *et al.*, 1973). Ethylene can be synthesized by the crop during stress (Prince *et al.*, 1987).

g. Yellow Lower Leaves. Root system damage from disease or high soluble salts will result in yellow, perhaps wilted and dying lower leaves. Similar symptoms are caused by excessive rates of chemical growth retardants (Jiao *et al.*, 1986), especially when coupled with low phosphorus levels in plant leaves (i.e., less than approximately 1.6 mg phosphorus per gram dry weight) (Tsujita *et al.*, 1978). Low mineral nutrient levels in plants caused by termination of fertilizer application 2 to

4 weeks before marketing also promotes yellowing of lower leaves (Prince and Cunningham, 1989).

III. POSTPRODUCTION HANDLING

A. Grades

While no market grades are consistently used for Easter lilies, perceived value relates to pot size (4 to 10 inches in diameter), plant height (16 to 22 inches bottom of pot to top of flowers), plant form (healthy leaves, strong stems centered in the pot), number of stems in a pot (multiple stems are often found in larger pots), and number of flowers and buds on the plant (Fig. 6) . For 6- to $6\frac{1}{2}$-inch pot diameter plant grades of small (three buds or flowers), medium (four), large (five), and extra large (six) have been proposed.

B. Storage

To hold advanced plants and for easy crop distribution, lilies are sometimes held in 40°F cold storage for up to 2 weeks. Plants may be stored when the first bud is

Fig. 6. High quality lily, with adequate flower count and no loss of foliage from base of plant.

"puffy" white. Botrytis tepal spot is avoided with a registered fungicidal spray before placement in storage. However, even 3 to 7 days cold storage may cause leaf yellowing during postharvest use (Prince et al., 1987; Sassano and Weiler, 1988). Preliminary studies suggest that leaf yellowing can be minimized by avoiding high application rates of growth retardant during production, maintaining adequate mineral nutrient levels in the plants, and acclimating the plants in a low light, warm environment before returning stored plants to the greenhouse at higher light levels for at least 3 days before packing and shipping (Bailey, 1989; Prince and Cunningham, 1989). The problem has been minimized by treatments that inhibit effects of ethylene (e.g., STS application) (Prince et al., 1987). Leaf yellowing is also caused by 2 to 3 days of boxed shipment at 73°F (Prince et al., 1987).

C. Anther Removal

The yellow pollen stains the white tepals as plants are handled. For some markets, this is acceptable. For other markets, anthers are removed as the flower buds open but before pollen begins to shed to prevent yellow staining. It once was believed that anther removal increased flower longevity, but this belief has not been substantiated.

D. Bud "Blasting"

Some buds may die ("blast") before opening in consumer interior environments. Causes include application of growth retardant during production (Prince and Cunningham, 1989), postproduction cold storage (see Section B), and low interior light, water stress, or both.

E. Conserving Life

In low light postharvest conditions, few buds open, buds abort, and leaves die back. A brightly lit environment, adequate watering, and cool to moderate temperatures maximize product life. Flower life is extended by a spray application of 500 to 1000 ppm gibberellic acid to 2- to 4-inch flower buds (Kelley and Schlamp, 1964). The treatment delayed onset of petal necrosis by 1 to 2 days but is not applied commercially. Plants should be kept out of drafts to avoid extreme temperatures and drying. Watering needs should be ascertained daily. Removal of old flowers extends the period of acceptability of the plant.

REFERENCES

Ames, R. N., and Linderman, R. G. (1978). "The growth of Easter lily (*Lilium longiflorum*) as influenced by vesicular-arbuscular mycorrhizal fungi, *Fusarium oxysporum,* and fertility level." *Canad. J. Bot.* **56**(21): 2773–2780.

Anonymous. (1946). "Lessons of war and future possibilities stressed at Ohio Short Course." *Florists' Review.* **97**(2512): 31–36.

Anonymous. (1981). "New Easter lily released." *Ornamentals Northwest.* **5**(1): 17–18.

Anonymous. (1986). *Marketing California Ornamental Crops.* Federal-State Market News Service. USDA Agricultural Marketing Service.

Anonymous. (1988). *Floriculture Crops, 1897 Summary.* USDA Agric. Stat. Board., Stat. Rept. Ser. Publ. Sp Cr **6–1,** 44–45.

Bahr, F. (1922). *Commercial Floriculture.* New York: De La Mare, pp. 461–468.

Bailey, D. A. (1989). Personal communication. University of Arizona, Tucson.

Bailey, D. A., and Miller, W. B. (1989). "Whole-plant response of Easter lilies to ancymidol and uniconazole." *J. Am. Soc. Hort. Sci.* **114**(3): 393–396.

Bald, J. G., Lenz, J. V., and Paulus, A. O. (1979). "Stem lesion of Easter lilies—A complex disease. *California Agric.* **33**(3): 12–13.

Bald, J. G., Paulus, A. O., Lenz, J. V. (1983). "Control of field root and bulb diseases of Easter lily." *Plant Disease.* **67**(10): 1167–1172.

Blaney, L. T., and Roberts, A. N. (1966a). "Growth and development of the Easter lily bulb, Lilium longiflorum Thunb. 'Croft.'" *Proc. Amer. Soc. Hort. Sci.* **89:** 643–650.

Blaney, L. T., and Roberts, A. N. (1966b). "Influence of harvest date and precooling on leaf and stem elongation in the 'Croft' Easter lily (Lilium longiflorum Thunb.)." *Proc. Amer. Soc. Hort. Sci.* **89:** 651–656.

Blaney, L. T., and Roberts, A. N. (1967). "Bulb production." In *Easter Lilies,* edited by D. C. Kiplinger and R. W. Langhans. Ohio State University, Columbus OH and Cornell University, Ithaca NY, pp. 23–36.

Blaney, L. T., Roberts, A. N., and Lin, P. (1967). "Timing Easter lilies." *Florists' Review.* **140**(3624): 19.

Box, C. O., and Payne, R. (1967). "Natural cooling methods of handling bulbs." In *Easter Lilies,* edited by D. C. Kiplinger and R. W. Langhans. Ohio State University, Columbus OH and Cornell University, Ithaca NY, pp. 59–71.

Brierley, P. (1941). "Effect of cold storage of Easter lily bulbs on subsequent forcing performance." *Agric. Research.* **62**(6): 317–335.

Brierley, P., and Curtis, A. H. (1942). "Further studies of cool storage and other factors affecting the forcing performance of Easter lily bulbs." *Agric. Research.* **64**(4): 221–235.

Brumfield, R. G., Nelson, P. V., Coutu, A. J., Willits, D. H., and Sowell, R. S. (1981). "Cost of producing Easter lilies in North Carolina." *J. Am. Soc. Hort. Sci.* **106**(5): 561–564.

Bureau of the Census. (1970). *Horticultural Specialties. 1969 Census of Agriculture, Part 10.* U.S. Department of Commerce, Washington D.C.

Butterfield, H. M. (1947). "Production of Easter lily bulbs." *Calif. Agric. Ext. Ser. Circ.* **132:** 1–34.

Carlson, W. H., and DeHertogh, A. A. (1967). "A preliminary evaluation of various cooling techniques for forcing 'Ace' lily bulbs." *Michigan Florist.* **441:** 21, 24.

Chaplin, M. H., and Roberts, A. N. (1981). "Seasonal nutrient element distribution in leaves of 'Ace' and 'Nellie White' cultivars of the Easter lily, *Lilium longiflorum* L." *Comm. in Soil Science and Plant Analysis.* **12**(3): 227–237.

DeGraff, J. (1969). A new era in hybrid lilies. *Am. Hort. Soc. Bul.* **48**(3): 110–118.

DeHertogh, A. A. (1988). *Holland Bulb Forcer's Guide.* Hillegom, The Netherlands: International Flower-Bulb Centre, pp. C-49–61, 73–80.

DeHertogh, A. A., and Blakely, N. (1972). "Influence of gibberellins A_3 and A_{4+7} on development of forced *Lilium longiflorum* Thunb. cv. 'Ace.'" *J. Am. Soc. Hort. Sci.* **97**(3): 320–323.

DeHertogh, A. A., and Einert, A. E. (1969). "The controlled temperature forcing (CTF) method for potted Easter lilies, its concept, results and commercial adaptation." *Florists' Review.* **145**(3745): 25–27, 70–73.

DeHertogh, A. A., Rasmussen, H. P., and Blakely, N. (1976). "Morphological changes and factors influencing shoot apex development of *Lilium longiflorum* Thunb. during forcing." *J. Am. Soc. Hort. Sci.* **104**(4): 463–471.

DeHertogh, A. A., Roberts, A. N., Stuart, N. W., Langhans, R. W., Linderman, R. G., Lawson, R. H., Wilkins, H. F., and Kiplinger, D. C. (1971). "A guide to terminology for the Easter lily (*Lilium longiflorum* Thunb.)." *HortScience.* **6**(2): 121–123.

Emsweller, S. L. (1950). "Recent developments in lily breeding techniques." *Proc. Amer. Soc. Hort. Sci.* **56,** 498–508.

Erickson, R. O. (1948). "Cytological and growth correlations in the flower bud and anther of *Lilium longiflorum.*" *Amer. J. Bot.* **35:** 729–739.

Erwin, J. E., Heins, R. D., Berghage, R., Kovanda, B. J., Carlson, W. H., and Biernbaum, J. (1989). "Cool mornings can control plant height." *GrowerTalks.* **52**(9): 73–74.

Gent, M. P. N., Thorne, J. H., and Aylor, D. E. (1979). "Split-night temperatures in the greenhouse: The effects on the physiology and growth of plants." *Conn. Agric. Exp. Sta. Bull.* **781.**

Gianfagna, T. J., and Wulster, G. J. (1986). "Comparative effects of ancymidol and paclobutrazol on Easter lily." *HortScience.* **21**(3): 463–464.

Gianfagna, T. J., Wulster, G. J., and Teiger, T. S. (1986). "Effect of flowers on stem elongation in Easter lily." *HortScience.* **21**(3): 461–462.

Gill, D. L., and Stuart, N. W. (1967). "Effect of photoperiod in forcing Georgia Easter lilies." *Florists' Review.* **141**(3644): 19–20, 81–82.

Goldsberry, K. L., and Maffei, H. M. (1986). "Response of Easter lilies grown in rockwool, peat-lite and soil media to warm water irrigation." *Colorado Grhse. Growers' Assn. Research Bull.* **436:** 3–5.

Gould, C. J., ed. (1957). *Handbook on Bulb Growing and Forcing.* Mt Vernon, Washington: Northwest Bulb Growers Assn.

Green, D. E. (1934). Decay of lily bulbs during storage. *Royal Hort. Soc. Lily Yrbk.* **3,** 79–81.

Griffiths, D. (1930). *The Production of Lily Bulbs.* USDA Circ. **120,** 1–56.

Grueber, K., Healy, W., Pemberton, B., and Wilkins, H. (1987). "Metering lily bud development." *GrowerTalks.* **50**(11): 116–117.

Hammer, P. A. (1987). "Growing ideas." *GrowerTalks.* **51**(7): 98.

Hammer, P. A., and Hopper, D. A. (1989). "Modeling stem elongation of Easter lilies grown under various production schemes." *HortScience.* **24**(5): 785–788.

Hartley, D. E., Blaney, L. T., and Roberts, A. N. (1964). "Preparing 'Ace' lily bulbs for an early Easter." *Florists' Review.* **134**(3473): 17–18, 80–81.

Hasek, R. F., and Sciaroni, R. H. (1970). "The use of Osmocote formulations in commercial potted lily production." *Florists' Review.* **146**(3781): 45, 84–88.

Hasek, R. F., Sciaroni, R. H., and Farnham, D. S. (1971). "1970–1971 Japanese Georgia lily height trials." *Florists Review.* **149**(3849): 22–24, 62–64.

Healy, W. E., and Wilkins, H. F. (1984). "Temperature effects on 'Nellie White' flower bud development." *HortScience.* **19**(6): 843–844.

Heins, R., Erwin, J., Karlsson, M., Berghage, R., Carlson, W., and Biernbaum, J. (1987). "Tracking Easter lily height with graphs." *GrowerTalks.* **51**(8): 64, 66, 68.

Heins, R. D., Pemberton, H. B., and Wilkins, H. F. (1982). "The influence of light on lily (*Lilium longiflorum* Thunb.) I. Influence of light intensity on plant development." *J. Am. Soc. Hort. Sci.* **107**(2): 330–335

Hiraki, Y., and Ota, Y. (1975). "The relationship between growth inhibition and ethylene production by mechanical stimulation in *Lilium longiflorum.*" *Plant and Cell Physiol.* **16**(1): 185–189.

Hitchcock, A. E., Crocker, W., and Zimmerman, P. W. (1932). "Effect of illuminating gas on the lily, narcissus, tulip, and hyacinth." *Contrib. Boyce Thompson Inst.* **4:** 155–176.

Holmes, M. C. (1956). "The origin and history of the Easter lily." *Baileya.* **4**(1): 40–45.

Holmes, R. C., Aagensen, G. J., Elkins, D. M., and Coorts, G. D. (1975). "Foam application of A-Rest and Phosfon-L to 'Ace' Easter lily." *Florists Review.* **156**(4033): 22–23, 69.

Horst, R. K. (1979). *Westcott's Plant Disease Handbook.* New York: Van Nostrand and Reinhold.

Jiao, J., Tsujita, M. J., and Murr, D. P. (1986). "Effect of paclobutrazol and A-Rest on growth, flowering, carbohydrate and leaf senescence in 'Nellie White' Easter lily (*Lilium longiflorum* Thunb.)" *Scientia Horticulturae.* **30**(1/2): 135–141.

Kamel, H. A., Ibrahim, A. A., and Thabet, I. A. (1978). "Effect of planting depth and spacing of *Lilium longiflorum* bulbs." *Agric. Research Review.* **56**(3): 121–126.

Karlsson, M. G., Heins, R. D., and Erwin, J. E. (1988). "Quantifying temperature-controlled leaf unfolding rates in 'Nellie White' Easter lily." *J. Amer. Soc. Hort. Sci.* **113**(1): 70–74.

Kelley, J. D., and Schlamp, A. C. (1965). "Keeping quality, flower size and flowering response of three varieties of Easter lilies to gibberellic acid." *Proc. Amer. Soc. Hort. Sci.* **85**: 631–634.

Kiplinger, D. C., and Langhans, R. W., eds. (1967). *Easter Lilies.* Ohio State Univ., Columbus OH and Cornell Univ., Ithaca NY.

Kiplinger, D. C., Tayama, H. K., and Staby, G. (1971). "EL-531 on height of 'Ace' and No. 44 lilies." *Ohio Florists Assn. Bull.* **505**: 2.

Kohl. H. C., Jr. (1958). "Shorter-stemmed Easter lilies." *Calif. Agric.* **12**: 5.

Kosugi, K. (1952). "On the flower bud differentiation in Easter lily." *Hort. Assn. Japan.* **21**(1): 59–62.

Kruyer, C. J., and Boontjkes, J. (1982). "De warmeaterbehandeling van *Lilium longiflorum.*" *Bloembolencultuur.* **93**(25): 622–623.

Laiche, A. J., Jr. (1982). "Dual minimum temperature forcing of 'Ace,' 'Croft,' 'Harson,' and 'Nellie White' Easter lilies (*Lilium longiflorum,* Thunb.)." *HortScience.* **17**(6): 898–899.

Lange, N., and Heins, R. (1990). "The lowdown on how bulb size influences lily development." *GrowerTalks.* **53**(10): 52, 54.

Larson, R. A., and Kimmins, R. K. (1971). "Results with a new growth regulator." *Florists' Review.* **148**(3841): 22–23, 54.

Larson, R. A., Thorne, C. B., Milks, R. R., Isenberg, Y. M., and Brisson, L. D. (1987). "Use of ancymidol bulb dips to control stem elongation of Easter lilies grown in a pine bark medium." *J. Am. Soc. Hort. Sci.* **112**(5): 769–773.

Lattimer, M. (1954). "1954 lily studies." *Ohio Florists' Assn. Bull.* **303**: 4–5.

Lewis, A. J., and Lewis, J. S. (1980). "Response of *Lilium longiflorum* to ancymidol bulb-dips." *Scientia Horticulturae.* **13**(1): 93–97.

Lewis, A. J., and Lewis, J. S. (1981). "Improving ancymidol efficiency for height control of Easter lily." *HortScience.* **16**(1): 89–90.

Lewis, A. J., and Lewis, J. S. (1982). "Height control of *Lilium longiflorum* Thunb. 'Ace' using ancymidol bulb-dips." *HortScience.* **17**(3): 336–337.

Lin, W. C., and Wilkins, H. F. (1975). "Influence of bulb harvest date and temperature on growth and flowering of *Lilium longiflorum.*" *J. Am. Soc. Hort. Sci.* **100**(1): 6–9.

Lin, W. C., Wilkins, H. F., and Angell, M. (1975a). "Exogenous gibberellins and abscisic acid effects of growth and development of *Lilium longiflorum.*" *J. Am. Soc. Hort. Sci.* **100**(1): 9–16.

Lin, W. C., Wilkins, H. F., and Brenner, M. L. (1975b). "Endogenous promoter and inhibitor levels in *Lilium longiflorum* bulbs." *J. Am. Soc. Hort. Sci.* **100**(2): 106–109.

Linderman, R. G. (1985). "Easter lilies." In *Diseases of Floral Crops,* 2nd ed. New York: Praeger, pp. 9–40.

Liu, L., and Burger, D. W. (1986). "*In vitro* propagation of Easter lily from pedicels." *HortScience.* **21**(6): 1437–1438.

Lumsden, D. (1929). "The American florists' lily trade." In *Standard Cyclopedia of Horticulture,* edited by L. H. Bailey. New York: Macmillan, pp. 1865–1866.

McMurray, A. (1978). "Effects of water temperature on Easter lilies (*Lilium longiflorum*)." *N.C. Flower Growers' Bull.* **22**(2): 1–5.

Marousky, F. J. (1981). "Symptomology of fluoride and boron injury in *Lilium longiflorum* Thunb." *J. Am. Soc. Hort. Sci.* **106**(3): 341–344.

Marousky, F. J., and Woltz, S. S. (1977). "Influence of lime, nitrogen, and phosphorus sources on the availability and relationship of soil fluoride to leaf scorch in *Lilium longiflorum* Thunb." *J. Am. Soc. Hort. Sci.* **102**(6): 799–804.

Mastalerz, J. W. (1965). "Bud blasting in *Lilium longiflorum.*" *Proc. Am. Soc. Hort. Sci.* **87**: 502–509.

Miller, R. O. (1985). "Lilies." In *Ball Red Book,* edited by V. Ball. West Chicago IL: Ball Seed Co., pp. 575–596.

Miller, R. O., and Kiplinger, D. C. (1966a). "Interaction of temperature and time of vernalization on Northwest Easter lilies." *Proc. Am. Soc. Hort. Sci.* **88:** 635–645.

Miller, R. O., and Kiplinger, D. C. (1966b). "Reversal of vernalization in Northwest Easter lilies." *Proc. Am. Soc. Hort. Sci.* **88:** 646–650.

Miller, W. B. (1989). Personal communication. University of Arizona, Tuscon, AZ.

Miller, W. B., and Langhans, R. W. (1989a). "Carbohydrate changes of Easter lilies during growth in normal and reduced irradiance environments." *J. Am. Soc. Hort. Sci.* **114**(2): 310–315.

Miller, W. B., and Langhans, R. W. (1989b). "Reduced irradiance affects dry weight partitioning in Easter lily." *J. Am. Soc. Hort. Sci.* **114**(2): 306–309.

Nightingale, A. E. (1979). "Bulbil formation on *Lilium longiflorum* Thunb. cv. 'Nellie White' by foliar applications of PBA." *HortScience.* **14**(1): 67–68.

Nuttonson, M. Y. (1952). "Ecological crop geography and field practices of the Ryukyu Islands, natural vegetation of the Ryukyus, and agro-climatic analogues in the northern hemisphere." Washington, D.C.: *Am. Inst. Crop Ecol.*

Okada, M. (1952). "On propagation of *Lilium longiflorum*. II. Bulb division." *Engei Gakkai Zasshi.* **20:** 209–214.

Oliver, G. W. (1908). "The production of Easter lily bulbs in the United States." *USDA Bureau of Plant Industry Bull.* **120.**

O'Rourke, E. N., Jr., and Branch, P. C. (1987). "Observations on the relationship between degree-day summations and timing of Easter lilies." *HortScience.* **22**(5): 709–711.

Pertuit, A. J., Jr., and Kelly, J. W. (1987). "Timing of a lighting period for Easter lily bulbs prior to forcing." *HortScience.* **22**(2): 316.

Pfeiffer, N. E. (1935). "Development of the floral axis and new bud in imported Easter lilies." *Contrib. Boyce Thompson Inst.* **7:** 311–321.

Prince, T. A., and Cunningham, M. S. (1989). Production and storage factors influencing quality of potted Easter lilies. *HortScience.* **24**(6): 992–994.

Prince, T. A., and Cunningham, M. S. (1990). "Response of Easter lily bulbs to peat moisture content and the use of peat or polyethylene-lined cases during handling and vernalization." *J. Am. Soc. Hort. Sci.* **115**(1): 68–72.

Prince, T. A., Cunningham, M. S., and Peary, J. S. (1987). "Floral and foliar quality of potted Easter lilies under STS or phenidone application, refrigerated storage, and simulated shipment." *J. Am. Soc. Hort. Sci.* **112**(3): 469–473.

Raabe, R. D. (1975). "Increased susceptibilbity of Easter lilies to Pythium root rot as a result of infection by necrotic fleck virus complex." *Acta Horticulturae.* **47:** 91–96.

Rhoads, A., Trocano, J., and Brennan, E. (1973). "Ethylene gas as a cause of injury to Easter lilies." *Plant Disease Reporter.* **57**(12): 1023–1024.

Roberts, A. N. (1980). "1978–1979 growing seasons provide new insights on bulb maturity and forcing program for Easter lilies." *Ornamentals Northwest.* **4**(4): 13–14.

Roberts, A. N., and Blaney, L. (1967a). "History." In *Easter Lilies,* edited by D. C. Kiplinger and R. W. Langhans. Ohio State University, Columbus OH and Cornell University, Ithaca NY, pp. 7–10.

Roberts, A. N., and Blaney, L. (1967b). "Varieties and breeding." In *Easter Lilies,* edited by D. C. Kiplinger and R. W. Langhans. Ohio State University, Columbus OH and Cornell University, Ithaca NY, pp. 11–22.

Roberts, A. N., and Fuchigami, L. H. (1972). "New lily deflowering treatment shows promise." *Florists' Review.* **151**(3908): 25–26.

Roberts, A. N., and Moeller, F. W. (1979). "Evidence of genetic tendency to leaf scorch in *Lilium longiflorum* Thunb. 'Ace' inbred lines." *N. Am. Lily Soc. Yearbook.* **32:** 50–54.

Roberts, A. N., and Tomasovic, B. J. (1975). "Predicting bulb size and maturity in *Lilium longiflorum* Thunb. cultivars." *Acta Horticulturae.* **47:** 339–345.

Roberts, A. N., Stang, J. R., Wang, Y. T., McCorkle, W. R., Riddle, L. J., Moeller, F. W. (1985). "Easter lily growth and development." *Oregon State University, Agric. Exp. Sta. Tech. Bull.* **148.**

Roberts, A. N., Wang, Y., and Moeller, F. W. (1983). "Effects of pre- and post-bloom temperature regimes on development of *Lilium longiflorum* Thunb." *Scientia Horticulturae.* **18**(4): 363–379.

Roh, S. M. (1982). "Propagation of *Lilium longiflorum* Thunb. by leaf cuttings." *HortScience.* **17**(4): 607–609.

Roh, S. M., and Wilkins, H. F. (1973). "Influence of temperature on the development of flower buds from the visible bud stage to anthesis *Lilium longiflorum* Thunb. cv. 'Ace.'" *HortScience.* **8**(2): 129–130.

Roh, S. M., and Wilkins, H. F. (1974a). "Decay and dark reversion of phytochrome in *Lilium longiflorum* Thunb. cv. 'Nellie White.'" *HortScience.* **9**(1): 37–38.

Roh, S. M., and Wilkins, H. F. (1974b). "Red and far-red treatments accelerate shoot emergence form bulbs of *Lilium longiflorum* Thunb. cv. 'Nellie White.'" *HortScience.* **9**(1): 38–39.

Roh, S. M., and Wilkins, H. F. (1977a). "Comparison of continuous and alternating bulb treatments on growth and flowering in *Lilium longiflorum* Thunb." *J. Am. Soc. Hort. Sci.* **102**(3): 242–247.

Roh, S. M., and Wilkins, H. F. (1977b). "Influence of interrupting the long day inductive treatments on growth and flower numbers of *Lilium longiflorum* Thunb." *J. Am. Soc. Hort. Sci.* **102**(3): 253–255.

Roh, S. M., and Wilkins, H. F. (1977c). "Temperature and photoperiod effect on flower numbers in *Lilium longiflorum* Thunb." *J. Am. Soc. Hort. Sci.* **102**(3): 235–242.

Roh, S. M., and Wilkins, H. F. (1977d). "The control of flowering in *Lilium longiflorum* Thunb. cv. 'Nellie White' by cyclic or continuous light treatments." *J. Am. Soc. Hort. Sci.* **102**(3): 247–253.

Roh, S. M., and Wilkins, H. F. (1977e). "The effect of bulb vernalization and shoot photoperiod treatments on growth and flowering of *Lilium longiflorum* Thunb. cv. 'Nellie White.'" *J. Am. Soc. Hort. Sci.* **102**(3): 229–235.

Roh, S. M., and Wilkins, H. F. (1977f). "The physiology of dormancy and maturity in bulbs of *Lilium longiflorum* Thunb. cv. 'Nellie White.' II. Dormancy—Scale removal and light treatment of the bulbs." *J. Korean Soc. for Hort. Sci.* **18**(1): 101–108.

Sassano, N. P., and Weiler, T. C. (1988). Unpublished data. Cornell University, Ithaca NY.

Seeley, J. G. (1950). "Mineral nutrient deficiencies and leaf burn of Croft Easter lilies." *Proc. Am. Soc. Hort. Sci.* **56**: 439–445.

Seeley, J. G., and Monroe, J. (1976). "Incidence of leaf scorch of 'Ace' lilies as influenced by fluoride applications." *Florists' Review.* **159**(4112): 37, 50–52.

Shoushan, A. M. (1950). "Lily investigations." *Ohio Florist Assn. Bull.* **249**: 2–4.

Smith, D. R. (1963). "The influence of the environment upon initiation and development of *Lilium longiflorum* (Thunb.)." Ph.D. thesis, Cornell University, Ithaca NY.

Smith, D. R., and Langhans, R. W. (1962a). "The influence of day and night temperature on the growth and flowering of Easter lily (*Lilium longiflorum* Thunb. var Croft). *Proc. Am. Soc. Hort. Sci.* **80**: 593–598.

Smith, D. R., and Langhans, R. W. (1962b). "The influence of photoperiod on the growth and flowering of the Easter lily (*Lilium longiflorum* Thunb. var. Croft)." *Proc. Am. Soc. Hort. Sci.* **80**, 599–604.

Stewart, W. S., and Stuart, N. W. (1942). "The distribution of auxins in bulbs of *Lilium longiflorum.*" *Am. J. Bot.* **29**: 529–532.

Stuart, N. W. (1944). "USDA experiments with Creole lilies to determine blooming dates." *Florists' Review.* **94**(2439): 23–24.

Stuart, N. W. (1945). "Overcoming dormancy of lily bulbs by hot water treatment." *Florists' Review.* **96**(2492): 29–31.

Stuart, N. W. (1954). "Moisture control of packing medium, temperature, and duration of storage as factors in forcing lily bulbs." *Proc. Am. Soc. Hort. Sci.* **63**: 488–494.

Stuart, N. W. (1957). "Easter lily forcing." In *Bulb Growing and Forcing,* edited by C. J. Gould. Mt. Vernon, Washington: Northwest Bulb Growers Assn., pp. 75–78.

Stuart, N. W. (1967). "Present methods of handling bulbs." In *Easter Lilies,* edited by D. C. Kiplinger and R. W. Langhans. Ohio State University, Columbus OH and Cornell University, Ithaca NY, pp. 47–58.

Stuart, N. W. (1968). Personal communication. Beltsville MD: USDA, Ornamentals Section.

Stuart, N. W., Gill, D. L., and Hickman, M. G. (1961). "Height control of forced Georgia Easter lilies." *Florists' Review.* **129**(3344): 13–14, 40–41.

Tammen, J. F., Oglevee, J. R., Oglevee, E. J., and Duffy, L. (1986). "A new process for producing pathogen-free Easter lilies." *GrowerTalks*. **50**(3): 62, 65–66.

Thornton, N. C. (1941). "Effects of mixtures of oxygen and carbon dioxide on the development of dormancy in Easter lilies." *Proc. Am. Soc. Hort. Sci.* **38:** 708.

Thornton, N. C., and Imle, E. P. (1940). "Why a dwarf Easter lily?" *Florists' Exchange.* **94**(15): 9.

Tizio, M., and Seeley, J. G. (1976). "Leaf scorch of 'Ace' lilies as affected by nitrogen fertilizer source and fluoride application." *Florists' Review.* **159**(4115): 43, 85–87.

Trade Statistics and Analysis Branch. (1986). *US Foreign Agricultural Trade Statistical Report, Calendar Year.* USDA, Economic Research Service.

van Bonthuis, A. D., and van Tuyl, J. (1984). "Experimenten bij *Lilium longiflorum* 'White American.' Invoeled temperatuurbehandelingen van bol op bloementrek." *Vakblad voor de Bloemisterij.* **39**(41): 48–49.

van Meeteren, U., and de Proft, M. (1982). "Inhibition of flower bud abscission and ethylene evolution by light and silver thiosulfate in *Lilium.*" *Physiologia Plantarum.* **56**(3): 236–240.

van Nes, C. R. (1981). "Mogelijkheden bij invriezen van *Lilium longiflorum.*" *Bloembollencultuur.* **92**(6): 142.

van Tuyl, J. M. (1985). "Recenti sviluppi del miglioramento genetico del *Lilium.*" *Colture Protette.* **14**(1): 31–38.

van Tuyl, J. M., Franken, J., Jongerius, R. C., Lock, C. A. M., and Kwakkenbos, T. A. M. (1986). "Interspecific hybridization in *Lilium.*" *Acta Horticulturae.* **177**(2): 591–595.

Wang, Y. T. (1988). "Growth potential of the Easter lily bulb." *HortScience.* **23**(2): 360–362.

Wang, Y. T., and Breen, P. J. (1984). "Respiration and weight changes of Easter lily flowers during development." *HortScience.* **19**(5): 702–703.

Wang, Y. T., and Breen, P. J. (1986). "Growth and photosynthesis of Easter lily in response to flower bud removal." *J. Am. Soc. Hort. Sci.* **111**(3): 442–446.

Wang, Y. T., and Breen, P. J. (1987). "Distribution, storage, and remobilization of ^{14}C-labeled assimilate in Easter lily." *J. Am. Soc. Hort. Sci.* **112**(3): 569–573.

Wang, Y. T., and Roberts, A. N. (1983). "Influence of air and soil temperatures on the growth and development of *Lilium longiflorum* Thunb. during different phases." *J. Am. Soc. Hort. Sci.* **108**(5): 810–815.

Waters, W. E. (1966). "Photo-induction of flowering responses of the Easter lily." *Florida Flower Grower.* **3**(10): 1–2.

Waters, W. E., and Wilkins, H. F. (1967). "Influence of intensity, duration, and date of light on growth and flowering of uncooled Easter lily (*Lilium longiflorum,* Thunb. 'Georgia')." *Proc. Am. Soc. Hort. Sci.* **90:** 433–439.

Weiler, T. C. (1978). "Shade and ancymidol altered shape of potted *Lilium longiflorum* 'Ace.'" *HortScience.* **13**(4): 462–463.

Weiler, T. C., and Langhans, R. W. (1968). "Determination of vernalizing temperatures in the vernalization requirement of *Lilium longiflorum* (Thunb.) cv. 'Ace.'" *Proc. Am. Soc. Hort. Sci.* **93:** 623–629.

Weiler, T. C., and Langhans, R. W. (1972a). "Effect of storage temperatures on the flowering and growth of *Lilium longiflorum* (Thunb.) 'Ace.'" *J. Am. Soc. Hort. Sci.* **97**(2): 173–175.

Weiler, T. C., and Langhans, R. W. (1972b). "Growth and flowering responses of *Lilium longiflorum* Thunb 'Ace' to different daylengths." *J. Am. Soc. Hort. Sci.* **97**(2): 176–177.

Weiler, T. C., Boodley, J. W., Gortzig, C. F., Langhans, R. W., Seeley, J. G., Sanderson, J. P., Semel, M., Horst, R. K., and Daughtrey, M. (1988). *1989 Cornell Easter Lily Guidelines for New York State.* Ithaca NY: Cornell University.

Weller, S. C., and Hammer, P. A. (1984). "Susceptibility of Easter lily to glyphosate injury." *HortScience.* **19**(5): 698–699.

Westcott, C. (1973). *The Gardener's Bug Book.* Garden City, NY: Doubleday.

White, E. A. (1933). *The Florist Business.* New York: Macmillan, pp. 372–374.

White, H. E. (1940). The culture and forcing of Easter lilies. *Mass. Agric. Exp. Sta. Bull.* **376.**

Wiggans, S. C., Payne, R. N., and Ealy, R. P. (1960). "The effect of Phosfon, a growth retardant, on lilies, caladiums and abelias." *Oklahoma Expt. Sta. Proc.* Series **P-365.**

Wilkins, H. F. (1980). "Easter lilies." In *Introduction to Floriculture,* edited by R. A. Larson. New York: Academic Press, pp. 327–352.

Wilkins, H. F. (1985). "Lily leaf counting technique and time schedule." *Greenhouse Manager.* **4**(7): 104, 106–108, 111.

Wilkins, H. F. (1986). "The influence of 32°, 21°, 4°, or 27°, 18°, 10°, or 21°, 17°, 13°C light/dark temperature combinations during various growth phases of the Easter lily (*L. longiflorum* Thunb.)." *Acta Horticulturae.* **177**(1): 181–188.

Wilkins, H. F., and Roberts, A. N. (1969). "Leaf counting—A new concept in timing Easter lilies." *Minnesota State Florists' Bull.* **12**: 10–13.

Wilkins, H. F., and Roh, S. M. (1972). "The effect of bulb pre-emergence temperature treatment and subsequent shoot post-emergence light treatment on the growth and flowering of *Lilium longiflorum,* Thunb. 'Nellie White.'" *Plant Physiol.* **49**(Suppl): 54.

Wilkins, H. F., Grueber, K., Healy, W., and Pemberton, H. B. (1986). "Minimum fluorescent light requirements and ancymidol interactions on the growth of Easter lily." *J. Am. Soc. Hort. Sci.* **111**(3): 384–387.

Wilkins, H. F., Waters, W. E., and Widmer, R. E. (1968a). "Influence of temperature and photoperiod on growth and flowering of Easter lilies (*Lilium longiflorum* Thunb. 'Georgia,' 'Ace,' and 'Nellie White')." *Proc. Am. Soc. Hort. Sci.* **93**: 640–649.

Willkins, H. F., Widmer, R. E., and Waters, W. E. (1968b). "The influence of carbon dioxide, photoperiod, and temperature on growth and flowering of Easter lilies (*Lilium longiflorum* Thunb. 'Ace' and 'Nellie White')." *Proc. Am. Soc. Hort. Sci.* **93**: 650–654.

Wilkins, H. F., Widmer, R. E., and Waters, W. E. (1970). "An insurance policy: Lighting lilies at shoot emergence will overcome inadequate bulb procooling." *Florists' Review.* **147**(3806): 60–61, 109.

Wilson, E. H. (1925). *The Lilies of Eastern Asia.* London: Dalau and Co.

Wolf, S. T., and van Tuyl, J. M. (1984). "Hybridization of the liliaceae: Overcoming self-incompatibility and incongruity." *HortScience.* **19**(5): 696–697.

Woodcock, H. B. D., and Stearn, W. T. (1950). *Lilies of the World.* New York: Scribner.

Wulster, G. J., Gianfagna, T. J., and Clarke, B. B. (1987). "Comparative effects of ancymidol, propiconazol, triadimefon, and Mobay RSW0411 on lily height." *HortScience.* **22**(4): 601–602.

Yasui, K. (1973). "Histochemical changes in the apical meristem of *Lilium longiflorum* bulbs during storage." *J. Japan. Soc. Hort. Sci.* **42**(3): 271–279.

Zieslin, N., and Tsujita, M. J. (1988). "Regulation of stem elongation of lilies by temperature and the effect of gibberellin." *Scientia Horticulturae.* **37**(1–2): 165–169.

14

Hydrangeas

Douglas A. Bailey

I. INTRODUCTION

The florists' hydrangea, *Hydrangea macrophylla* subspecies *macrophylla* var. *macrophylla* (Thunb.) Ser. [Saxifragaceae], has been an important greenhouse crop for many years. Hydrangeas, or as they are better known internationally, hortensias, bear large spheres of white, pink, or blue flowers (Fig. 1). The cymose inflorescence is composed of sterile flowers with enlarged sepals as well as inconspicuous fertile flowers, which are buried beneath the sterile flowers. The species is endemic to the central Pacific coast on the Japanese island of Honshu where the plant is known as Temari-bana (Wilson, 1923). Another variety, var. *normalis,* coexists in the wild with var. *macrophylla.* This plant is known as Gakubana, and is the origin for the many "lacecap" cultivars grown as garden shrubs in Great Britain (Wilson, 1923). The lacecap inflorescence differs from florists' hydrangea, having a central area of small, fertile flowers surrounded by a circle of large, showy sterile flowers (Fig. 2). The lacecaps offer much potential as breeding material for new cultivars of florists' hydrangea.

Hybridization among florists' hydrangeas is easily accomplished, and there are over 500 cultivars available today. However, only a few cultivars are commercially important in North America (Table I). Of the cultivars listed, the top five probably account for over 80% of all hydrangeas produced in North America. Cultivar popularity differs with climatic region. For example, 'Todi' is most popular in the extreme northern United States and Canada because of its tolerance of cool, low light conditions during forcing. However, 'Rose Supreme' is preferred by many southern United States growers because it withstands high temperatures during forcing, though it is a vigorous growing cultivar.

Fig. 1. 'Merritt's Supreme,' a current cultivar.

Fig. 2. 'Tricolor,' a lacecap cultivar with potential as a potted florists' crop.

Table I

Hydrangea Cultivars Listed by Popularity in North America[a]

Popularity in North America	Cultivar	Sepal color description[b] Grown as a pink	Grown as a blue	Relative days to flower at 60°F night temperature[c]	Comments
1	'Merritt's Supreme'	Deep pink	Medium blue	88	Heat tolerant
2	'Todi' (Toddy)	Dark pink	Not recommended	88	Not heat tolerant
3	'Rose Supreme'	Rose pink	Light blue	103	Heat tolerant
4	'Kasteln'	Deep pink	Medium blue	95	
5	'Mathilde Gütges'	Pink	Light blue	95	
6	'Böttstein'	Light red	Not recommended	92	
7	'Kuhnert'	Not recommended	Medium blue	95	
8	'Sister Therese'	White	White	95	Best white color
9	'Red Star'	Brillant red	Medium blue	95	
10	'Schenkenburg'	Dark red	Not recommended	95	Best red color
11	'Wildenstein'	Pink	Not recommended	95	
12	'Brestenburg'	Not recommended	Dark blue	95	Best blue color
13	'Strafford'	Dark pink	Not recommended	99	Not heat tolerant
14	'Enziandom'	Dark pink	Deep blue	92	
15	'Merveille'	Light red	Not recommended	95	
—	'Blau Donau'	Pink	Clear blue	95	Heat tolerant
—	'Rosa Rita'	Light pink	Light blue	88	Not heat tolerant
—	'Dr. Steiniger'	Rose red	Not recommended	83	Earliest bloomer

[a]Modified from Bailey (1989a).

[b]The sepal color of hydrangeas (except for whites) can be modified through soil pH and nutrition manipulations. Color descriptions are given for plants grown for both pink or blue sepals.

[c]Based on using a 60°F night temperature up to 18 days prior to sale, then lowering night temperature to 4°F for color intensification. The amount of time required during forcing depends on temperature and light conditions.

Hydrangeas are forced into bloom primarily from Valentine's Day through Memorial Day; the main market holidays are Easter and Mother's Day. The most recent United States Census figures show a 42% decline in the number of hydrangeas sold wholesale from 1976 (2.6 million) to 1985 (1.5 million), but indicate a 75% increase in the average wholesale price, up from $2.07 to $3.65 (U.S. Dept. of Agr., 1977, 1986). The decline in production is perhaps due more to the longer production time and cost as compared to other floriculture crops than to a decline in popularity among consumers.

II. VEGETATIVE STAGE

A typical hydrangea production schedule mimics the natural development of the plant: vegetative growth and development occurs during the spring and summer months, inflorescence buds develop in autumn, plants are defoliated and placed into cold storage during late autumn, and flower-budded plants are forced into bloom in the spring.

A. Propagation

Hydrangeas are propagated from vegetative cuttings. Although some producers maintain their own stock plants, the majority of hydrangeas produced in North America originate from unrooted softwood cuttings obtained on the West Coast. Terminal cuttings (5 to 7 inches long) with three to four nodes each are received by growers from late April through early July. These cuttings are recut into a 2 to 4 inch long terminal cutting and a single node ("butterfly") cutting. The butterfly cutting can be cut in half longitudinally to form two leaf bud (single-eye) cuttings, but single-eye cuttings are slow to root and establish. Prepared cuttings are rooted in 72°F sand or other medium under intermittent mist and light shading to avoid water stress during propagation. Rooting is enhanced with the application of 10,000 ppm indolebutyric acid (IBA) in an IBA and talc formulation (Bailey and Weiler, 1984b). Under these conditions, terminal cuttings root in 3 to 4 weeks while butterfly cuttings require about 1 week longer (Koths *et al.,* 1973). The propagation date depends on the type of cutting (terminal or butterfly) and on the number of flowering shoots per plant desired on forcing (Table II).

Micropropagation techniques using shoot tip and apex cultures have been established for hydrangeas (Bailey *et al.,* 1986b; Stoltz, 1984), but few sources of tissue-cultured hydrangeas are available. The main use of micropropagation for hydrangeas has been in the development of virus-free material for use as stock plants (Allen and Anderson, 1980).

B. Summer Growth Phase

After cuttings have adequately rooted, they are planted into 4-inch pots for the summer growth phase. By maintaining the plants in small containers, growth is

Table II

Summer Culture Schedules for Dormant Plant Production of Various Grades[a]

Desired number of inflorescences per plant	Propagation date	Date of first pinch	Date of second pinch	Pot size during forcing[b]	Final spacing during forcing (in²/plant)
		Tip cuttings			
1	June 7–June 14	—	—	5 1/2–6 inch	64
3	May 7–May 14	June 21–June 28	—	7–8 inch	182
5	April 17–April 23	May 29–June 4	June 26–July 2	8 inch	210
		Butterfly cuttings			
2	May 7–May 14	—	—	6–6 1/2inch	144
3	April 24–April 30	June 26–July 2	—	7–8 inch	182

[a]Adapted from Shanks (1985) and Weiler (1980).
[b]'Rose Supreme' plants require the larger sized pots. The smaller size given for each plant type is appropriate for standard cultivars.

restricted, and size and shipping weight of dormant plants is minimized. The use of such small containers, however, requires contiual monitoring of soil moisture status to prevent leaf burn. The summer growth phase is usually conducted outdoors (Fig. 3). At more northerly latitudes where cool night temperatures would induce too early flower initiation, plants remain in covered, heated greenhouses during the summer. Plants are usually pinched to promote multiple branches, each capable of producing an inflorescence (Fig. 4). As with date of propagation, timing and number of pinches is dependent on original cutting type and product desired (Table II). The latest acceptable date for pinching plants is July 5. Plants pinched later than early July are unable to develop shoots mature enough to initiate inflorescences in autumn, and the incidence of blind shoots increases (Shanks, 1985).

An examination of the native habitat of florists' hydrangeas indicates that they have evolved in a maritime climate: moderate temperatures, moderate-to-high humidity, and extensive rainfall. Hydrangeas are damaged by high temperature and water deficiency. A constant moisture supply is essential for optimum plant growth. The growth medium used should have a high capacity for water retention as well as ample aeration; peat moss mixes are commonly used.

Another characteristic of the growth medium that must be addressed is pH and how pH affects aluminum and iron availability. Aluminum availability is the determinant of sepal color for nonwhite cultivars (Allen, 1943). Both pink and blue sepals contain the same anthocyanin pigment, delphinidin 3-monoglucoside (Asen *et al.,* 1957). When aluminum is present in sepals, it binds both with the pigment and a copigment, 3-caffeoylquinic acid. This aluminum/pigment/copigment complex causes the sepals to change from pink to blue (Takeda *et al.,* 1985). Aluminum availability is related to the pH of the growth medium solution: a low pH (5.0 to 5.5) results in high availability of aluminum and sepals can be blue (Allen, 1931), while

Fig. 3. Plants during outdoor summer growth in northern California.

a growth medium with high pH (6.0 to 6.5) restricts aluminum availability and sepals will be pink (Ulery, 1978). However, pH is not the only factor affecting aluminum availability and sepal color. Nitrogen (both ammonium-N and nitrate-N), phosphorus, and potassium levels in the growth medium can affect sepal coloration.

Iron availability also is affected by growth medium pH. Hydrangeas are susceptible to iron deficiency chlorosis, especially if the growth medium solution pH rises above 6.0. Growth media should be adjusted to appropriate pH for either pink or blue sepal color as recommended earlier. Reliance on acidic peat moss, as well as incorporation of aluminum sulfate or sulfur into media, lowers pH. Producers in regions with alkaline, highly buffered irrigation water must neutralize the bicarbonate buffer and lower water pH prior to use to help maintain desired media pH. Sulfuric, nitric, or citric acid can be used for both pink- and blue-flowering plants. Phosphoric acid can be used for pink-flowering plants but should not be used on

Fig. 4. Multibranched plant at the end of the summer vegetative growth phase.

blue-flowering plants as phosphorus restricts aluminum availability thereby preventing blue sepals. Iron chelate drenches correct iron deficiency chlorosis in hydrangeas; 3 to 5 ounces of 10% chelated iron per 100 gallons of water applied as an overhead drench is effective (Koths et al., 1985). Multiple applications may be required for continual control.

Hydrangeas use nutrients at a rapid rate, and maintaining adequate nutrient levels in the growth media (Table III) and plant tissue (Table IV) is essential for proper plant development. Nutrient effects on sepal color must be considered when selecting fertilizers. High nitrogen (Link and Shanks, 1952), high ammonium-N relative to nitrate-N (Asen et al., 1963), high phosphorus (Asen et al., 1959), and low potassium (Asen et al., 1959) levels are associated with pink sepals, and a 25–10–10 ($N–P_2O_5–K_2O$;Nitrogen–phosphorus pentoxide–potassium oxide) fertilizer source is recommended to create these conditions for pink-flowering plants (Shanks, 1985). Moderate nitrogen (Link and Shanks, 1952), low phosphorus (Shanks et al., 1950), and high potassium (Link and Shanks, 1952) levels are associated with blue sepals, and a 25–5–30 fertilizer source is recommended for blue-flowering plants (Koths et al., 1973). Superphosphate should not be incorporated in media for blue-flowering plants as phosphorus antagonizes aluminum uptake. White-flowering plants can be grown under nutrient recommendations for either pink- or blue-flowering plants.

As mentioned previously, hydrangeas grow best at moderate temperatures. During the summer production phase (production of dormant plants), growth is best wtih little or no shading at northern latitudes (42° to 59°N) (Litlere and Strømme, 1975), while plants benefit from moderate shade (20 to 50%) at southern latitudes (32° to 42°N) (Shanks, 1985). Shading is beneficial in reducing plant temperature and reducing transpirational water loss (Shanks, 1985). If hydrangeas are subjected to temperatures greater than 86°F for extended periods of time

Table III

Suggested Nutrient Concentrations for Hydrangea Media Extract[a]

| | Concentration in media extract (ppm) | | | |
| | Paste method[b] | | Spurway method | |
Nutrient	Pink sepals	Blue sepals	Pink sepals	Blue sepals
Ammonium-N	<15	<15	<6	<6
Nitrate-N	80–110	60–90	30–50	20–30
Phosphorus	6–10	2–6	6–12	1–5
Potassium	60–90	100–150	10–20	25–50
Calcium	>100	>100	>100	>100

[a]Adapted from Bing et al. (1981).

[b]Little work has been reported using paste extract for hydrangea nutrition monitoring, and the parts per million (ppm) given are intended as rough guidelines only.

Table IV

Suggested Interpretation of Foliar Analysis Values for Hydrangeas[a]

Nutrient	Deficient[b]	Low	Sufficient	High	Toxic[b]
			Percent dry weight		
Nitrogen	<1.50	1.50–3.00	3.10–6.00	—	—
Phosphorus	<0.20	0.20–0.35	0.36–0.80	—	—
Potassium	<1.00	1.00–2.00	2.10–4.50	—	—
Calcium	<0.60	0.70–1.00	1.10–1.50	—	—
Magnesium	<0.08	0.08–0.14	0.15–0.40	—	—
			ppm dry weight		
Boron	<25	26–55	56–100	100–199	>199
Iron	<60	61–99	100–160	—	—
Manganese	—	8–40	41–100	101–2499	>2499
Copper	—	<6	6–20	21–60	—
Zinc	<20	20–35	36–70	71–250	—
Molybdenum	—	<1	1–20	21–240	—

[a]Adapted from Bailey and Hammer (1988).
[b]Levels considered deficient coincided with visible symptoms appearing on plants. Levels considered toxic coincided with visible symptoms appearing on plants.

(more than 3 days), they develop thickened leaves that are narrow and sometimes mottled (Bailey and Hammer, 1989). The apexes are shortened and broadened in diameter. Weiler and Lopes (1974) first documented this problem and named it "hydrangea distortion." Plants with distorted apexes are unable to develop inflorescences. However, if temperatures decrease, plants resume normal growth, and a normal apex develops, capable of developing an inflorescence (Bailey and Hammer, 1989).

Hydrangeas are shrubs and can grow to heights of 3 to 7 feet. Although dwarf cultivars have been selected for greenhouse production, chemical height control is usually required during both production of dormant plants and forcing of flower-budded plants. Use of daminozide sprays has been the most common method of height control. During the summer growth phase, 500 to 7500 ppm is applied; taller-growing cultivars such as 'Rose Supreme' receive the higher concentration. Applications are made when shoots are 1 to 1½ inches long, and repeated applications may be made at 2- to 3-week intervals (Jung, 1964). During forcing of flower-budded plants, applications of 2500 to 5000 ppm daminozide are sprayed at 10- to 14-day intervals (Shanks, 1985). First applications are made when the newly expanding leaves are 1 to 2 inches long. Growth retardant treatments are not applied once inflorescence buds attain a ¾ inch diameter; applications after this stage reduce final inflorescence size. Other effective chemical growth retardant treatments for hydrangeas include: ancymidol sprays (50 to 100 ppm) (Tjia *et al.*,

1976), ancymidol drenches (0.3 to 0.6 grains active ingredient (a.i.)/per pot) (Tjia *et al.,* 1976), chlormequat drenches (4.6 to 24.7 grains a.i./per pot) (Anonymous, 1973), paclobutrazol sprays (50 to 100 ppm) (Scott, 1982), and uniconazole sprays (5 to 20 ppm) (Bailey, 1989b) (Fig. 5). As with daminozide, concentrations and number of applications for these growth retardants are dependent on cultivar and cultural conditions such as irradiance levels and temperature. Chlormequat sprays are ineffective on hydrangeas (Kohl and Nelson, 1966).

III. REPRODUCTIVE STAGE

A. Initiation and Development

Hydrangea inflorescences are terminal cymes composed of primary, second-ary, and tertiary axes bearing individual flowers (florets) (Litlere and Strømme, 1975). The fertile flowers develop during the autumn months and by early Nov-ember, dissected inflorescence buds contain fertile flowers differentiated to the G stage [gynoecia (pistils) are fully formed]; differentiation of sterile flowers does not occur until the forcing phase the following spring (Wiśniewska and Zawadzka, 1962). Both temperature and photoperiod are involved in stimulating inflorescence formation, and complete inflorescence formation (up to fertile flower G stage) requires 6 to 9 weeks of cool, short-day conditions of autumn (Shanks, 1985).

With night temperatures of 55° to 65°F, inflorescences readily initiate, regardless of photoperiod (Litlere and Strømme, 1975; Peters, 1975). Therefore, summer vegetative growth may require heated greenhouses in more northerly latitudes to prevent premature floral initiation (Post, 1959). With night temperatures of 66° to

Fig. 5. Chemical hieght control of 'Rose Supreme'; *left to right:* (1) control; (2–3) one or two sprays of 5 ppm uniconazole; (4–5) one or two sprays of 10 ppm uniconazole; and (6) one spray of 15 ppm uniconazole.

70°F, long photoperiods (greater than 14 hours) can delay (not prevent) floral initiation; short (8 hour) photoperiods allow rapid flower initiaiton (Litlere and Strømme, 1975; Peters, 1975). With night temperatures between 70° and 80°F, long photoperiods severely delay floral initiation, and a continuous (24 hour) photoperiod can prevent floral initiation for more than 16 weeks; 8 hour photoperiods allow rapid floral initiation (Bailey and Weiler, 1984a). Night temperatures above 80°F prevent floral initiation, regardless of photoperiod (Peters, 1975), so at southern latitudes plants are grown outdoors to (1) decrease growing temperatures to avoid hydrangea distortion and (2) maximize exposure to cool autumn temperatures to help promote floral initiation at that time.

Hydrangea floral initiation can be promoted by factors other than photoperiod and night temperature. Reducing nitrogen levels assists in flower bud formation (Dunham, 1948), so nitrogen fertilization is decreased concomitant with the onset of autumn to help assure rapid floral initiation (Ulery, 1978). High light conditions [greater than 2000 foot-candles (fc)] hasten floral initiation (Litlere and Strømme, 1975) and reduce the incidence of "blind shoots" (weak stems, incapable of developing an inflorescence) (Ray, 1946). Therefore, to reduce competition for light, plants should not be crowded during the floral initiation phase and little or no shading should be used over plants.

Some chemical growth retardants stimulate floral initiation in hydrangeas, even if plants are growing in an environment not conducive to initiation (i.e., 75°F night temperatures and 24-hour photoperiods) (Bailey et al., 1986a). Ancymidol (thirteen 100 ppm sprays applied weekly), paclobutrazol (five 100 ppm sprays applied every 2 weeks), and uniconazole (four 15 ppm sprays applied every 3 weeks) are all effective in stimulating hydrangea inflorescence initiation (Bailey, 1988; Bailey et al., 1986a). Undesirable side effects of these experimental treatments, however, include an increase in the forcing time and a reduction in inflorescence diameter on forcing (Bailey, unpublished data, 1989), and further research is necessary prior to attempting commercial use of growth retardants for floral initiation.

B. Dormancy

After stimulating floral initiation in autumn, short photoperiods during the winter and spring months inhibit inflorescence expansion and development of sterile flowers; inflorescences are considered "dormant." Commercially, dormancy is broken by subjecting plants to a 6-week period of cold (40° to 45°F) storage (Koths et al., 1973). Defoliation before treatment is essential to reduce the incidence of Botrytis bud rot during cold storage. If a longer storage period is required to delay flowering for a later market date, plants should be stored at 33° to 35°F (Shanks, 1985). An 8-week storage treatment at 52°F is also effective in breaking hydrangea bud dormancy (Shanks, 1985); storage at temperatures above 52°F is ineffective in allowing for rapid expansion of inflorescences (Shanks and Link, 1951). Experimentally, cold storage for breaking of dormancy has been replaced by subjecting budded plants to long (16 or more hour) photoperiods (Bailey and Weiler,

1984a; LeMattre, 1975). An alternative method for breaking bud dormancy that has been investigated is application of gibberellic acid (GA_3). Commercially acceptable plants are attainable with multiple GA_3 sprays to flower-budded plants at low (1 to 50 ppm) concentrations, but optimum number of applications and spray concentration vary with cultivar and greenhouse temperature (Shanks, 1985; Stuart and Cathey, 1962). Defoliation of the plants prior to GA_3 sprays increases the effectiveness of the treatments. Best results with substituting GA_3 for cold storage are achieved by using a combination of 4-weeks cold storage plus two sprays of 10 ppm GA_3 during forcing (Stuart and Cathey, 1962).

C. Defoliation and Cold Storage

Fertile flower development to the G stage is essential prior to placing plants into cold storage. Hand lens examination of dissected buds should be used to confirm complete development. Leaf removal is also essential prior to cold storage (Fig. 6). As plants enter into dormancy the leaves will fall off naturally, but this method requires multiple passes through the storage facility to remove fallen leaves. The most common method of defoliating hydrangeas is spraying with 7500 to 12,500 ppm 2-butyne-1,4-diol (Shanks, 1985). Complete leaf abscission occurs 7 to 10 days after application. Other methods that are described are gassing with canister-held ethylene (1 cubic foot per 1000 cubic feet of storage area) (Shanks, 1985), ethylene released from ripe apples (1 bushel per 400 cubic feet) (Post, 1959), or gassing with sodium methyl dithiocarbamate (10 teaspoons per 1000 cubic feet, mixed with enough water to cover the floor evenly (Koths *et al.,* 1973). Gassing treatments require the use of an air-tight storage facility for effective defoliation. Sprays of 10,000 to 15,000 ppm tributyl phosphorotrithioite also effectively defoliate hydrangeas (Kofranek and Leiser, 1958). Ethephon sprays at 1000 to 3000 ppm will stimulate defoliation, but treatments severely reduce plant height and

Fig. 6. Defoliated plants, ready for placement into a cooler.

inflorescence size on forcing (Shanks, 1969; Tjia and Buxton, 1976). Regardless of how defoliation is achieved, fallen leaves are removed to reduce the incidence of *Botrytis* during storage. Northern growers can use natural cooling in plastic-covered houses, but plants require protection from hard freezes, and storage houses require ventilation to prevent temperatures form rising above 50°F. Sunny days in conjunction with interior temperatures above 50°F will promote premature growth during storage (Snodgrass, 1988). Southern growers rely on refrigerated cooling during storage.

D. Greenhouse Forcing

Flowering plants are produced from dormant material that has received cold storage. The forcer may grow the plants up to this point, or precooled dormant plants may be purchased. Because the production of dormant plants takes place in 4-inch pots, plants must be repotted during the forcing phase. Transplanting should be delayed until new root growth is evident, about 2 to 3 weeks after placement into the forcing greenhouse, as transplanting prior to active root growth has been associated with increased root rot and leaf chlorosis (Kiplinger, 1945).

Forcing time is related to cultivar (see Table I) and forcing temperature (Table VI). Faster blooming cultivars such as 'Merritt's Supreme' and 'Todi' are grown for an early Easter, while slower cultivars such as 'Rose Supreme' are better suited for a late Easter or Mother's Day market. Hydrangeas are forced at 60°F night temperature (venting temperature set 10°F higher), but temperatures can be lowered or raised to "fine tune" flowering dates (Tables V and VI). Regardless of temperatures used during the first portion of forcing, the crop is finished at 54°F during the last 2 weeks to intensify sepal coloration and acclimate the plants to postproduction conditions (Fig. 7). Fertilization and irradiance also are reduced during the last portion of forcing to assist in acclimation and to prevent sunburning of flowers.

Even after growth retardation, staking of plants may be required. Hydrangea inflorescences are large and flowering shoots tend to be top-heavy. In many cases, each shoot requires a stake. Plants are staked prior to full expansion of inflorescences to prevent bending and breaking of shoots.

IV. PEST MANAGEMENT

Hydrangeas are attacked by a number of pests, and the most serious ones are listed in Table VII. Pests will vary with location, production phase (dormant plant production, cold storage, or greenhouse forcing), and control measures. The most effective pest management program prevents the occurrence of problems rather than remedying already-existing ones. Crop monitoring in conjunction with good cultural sanitation is essential for effective pest control; chemicals cannot and should not be relied on as the only control measure.

Table V

Hydrangea Production Schedules[a]

Days prior to sale	Date	Procedure
A. Sales date of April 8 for an April 15 Easter holiday[b]		
335	May 8	Stick shoot tip cuttings under mist
286	June 26	Pinch plants
272	July 10	Spray plants with 5000 ppm daminozide; repeat at 14 day intervals as needed for height control
212	Sept. 18	Reduce nitrogen fertilization by half to help stimulate inflorescence initiation
160	Oct. 30	Fertile flowers should be at G stage; apply defoliation treatment
153	Nov. 6	Place defoliated plants into 40° to 45°F dark storage
111	Dec. 18	Place plants in 54°F night temperature greenhouse to begin forcing
96	Jan. 2	Root growth should be evident; transplant into 7- to 8-inch pots
93	Jan. 5	Newly expanding leaves should be 1 to 2 inches long; spray with 2500 ppm daminozide; repeat at 10 to 14 days intervals as needed for height control
69	Jan. 29	Inflorescences should be visible; if not, increase night temperature to 58° to 60°F
33	March 6	Inflorescences should be 1½ inches in diameter; if not, adjust night temperature accordingly
18	March 21	Sepals should begin to show color; adjust night temperature to 54°F for color intensification
0	April 8	Sales date
B. Sales date of May 6 for a May 13 Mother's Day holiday[c]		
363	May 8	Stick shoot tip cuttings under mist
314	June 26	Pinch plants
272	July 10	Spray plants with 7500 ppm daminozide; repeat at 14 day intervals as needed for height control
212	Sept. 18	Reduce nitrogen fertilization by half to help stimulate inflorescence initiation
151	Nov. 8	Fertile flowers should be at G stage; apply defoliation treatment
144	Nov. 15	Place defoliated plants into 33° to 35°F dark storage
102	Jan. 24	Place plants in 60°F night temperature greenhouse to begin forcing
88	Feb. 7	Root growth should be evident; transplant into 8-inch pots
86	Feb. 9	Newly expanding leaves should be 1 to 2 inches long; spray with 5000 ppm daminozide; repeat at 10 to 14 day intervals as needed for height control
62	March 5	Inflorescences should be visible, but no more than ½ inch in diameter; adjust night temperature accordingly
31	April 5	Inflorescences should be 1½ inches in diameter; if not, adjust night temperature accordingly
18	May 6	Sepals should begin to show color; adjust night temperature to 54°F for color intensification
0	May 6	Sales date

(continued)

Table V *(continued)*

Days prior to sale	Date	Procedure
C. Shanks' (1985) minipot hydrangeas with sales date of June 25 for July 1 Canada Day		
154	Jan. 22	Stick shoot tip cuttings under 24-hour photoperiods; maintain 72°F night temperature
126	Feb. 19	Pot plants into 4½- to 5½-inch pots; maintain 24-hour photoperiods and 62°F night temperature
112	March 5	Reduce temperature to 52°F night temperature; place plants in 8-hour photoperiods for floral initiation
70	April 16	Increase night temperature to at least 62°F; use a 10 fc night break from 10:00 P.M. to 2:00 A.M.; spray with 25 to 50 ppm GA_3
0	June 25	Sales date

[a]Adapted from Bailey (1989a).
[b]Schedule based on producing 'Merritt's Supreme' plants with three inflorescences per plant.
[c]Schedule based on producing 'Rose Supreme' plants with three inflorescences per plant.

Probably the most severe pest is the two-spotted spider mite. Mite infestations spread rapidly once established and control depends on constant crop monitoring. Infested plants develop a stippled light-green or yellow pattern on the foliage; close examination of the underside of leaves reveals webbing, adult mites, nymphs, and eggs.

Botrytis is by far the most damaging and difficult disease to control in hydrangea production. Unfortunately, both hydrangeas and *Botrytis* do best under moist, humid conditions. Overhead watering should be avoided, especially during forcing when inflorescence buds are visible. *Botrytis* is a problem during every phase of

Table VI

Temperature Effects on Forcing of Hydrangeas[a]

Time interval	Night temperature (°F)		
	54	60	66
Days from start of forcing to bloom	112	88	80
Days from start of forcing to pea-sized inflorescence ($^3/_{16}$ inch diameter)	42	32	28
Days from pea-sized ($^3/_{16}$ inch) to bloom	70	56	52
Days from United States nickel-sized inflorescence ($^{13}/_{16}$ inch) to bloom	53	42	39
Days from United States silver dollar-sized inflorescence (1½ inch) to bloom	35	28	26
Days from first sepal color (drop night temperature to 54°F) to bloom	18	18	18

[a]The table is based on 'Merritt's Supreme' and other cultivars that bloom in 88 days using 60°F night temperature. The timings given assume a 54°F night temperature the last 18 days of forcing. Adapted from Koths *et al.* (1973) and Weiler (1980).

Fig. 7. Greenhouse forcing of 'Merritt's Supreme.'

production, especially during cold storage. Therefore, prompt removal of fallen leaves and other debris is crucial for control. Cultivars differ in susceptibility to *Botrytis,* and Powell (1973) gives the following rankings: 'Merveille' (most susceptible), 'Rose Supreme,' 'Todi,' 'Improved Merveille,' 'Regula,' 'Merritt's Supreme,' 'Sister Therese,' 'Strafford,' and 'Kuhnert' (least susceptible). However, even with the most intensive cultural control program and the least susceptible cultivar, fungicide applications are usually required for *Botrytis* control.

Hydrangeas are hosts for many viruses, but infection is less widespread today than in the past thanks to heat therapy and tissue culture technology for establishment of "clean" stock (Lawson and Horst, 1983). Virus and mycoplasma-like organism (MLO) infections are best avoided through purchasing assayed plant material from an established propagator. Scheduled indexing of stock plants is necessary for continued assurance of virus-free material.

V. POSTPRODUCTION CARE

Postproduction life span depends on finishing conditions in the greenhouse as well as shipping conditions, retailer care, and consumer care. "Hardening" of plants through lowering night temperatures, reducing irradiance, and reducing fertilization helps increase postproduction life span. A shipping temperature of 35° to 40°F is recommended for hydrangeas (Nell, 1990). Proper watering is the most essential element of hydrangea care. Plants will not recover from wilt, so a constant moisture supply in conjunction with a cool, well-lighted location is recommended to both the retailer and the consumer. Retailers should advise their customers that most hydrangea cultivars are winter-hardy to United States Department of Agriculture zone 7a (0°F) (Liberty Hyde Bailey Hortorium, 1976), so consumers in zones 7a to 10a should be successful in transplanting hydrangeas to their yard, given adequate soil and cultural conditions.

Table VII

Pests of Hydrangeas[a]

Pests	Genera or cause
Insects	
Aphids	*Aphis gossypii, Myzus circumflexus, M. persicae*
Four-lined plant bug	*Poecilocapsis lineatus*
Leaf-tiers	*Exartema ferriferanum, Udea rubigalis*
Rose-chafer	*Macrodactylus subspinosus*
Scale	*Lepidosaphes ulmi, Pulvinaria* species
Tarnished plant bug	*Lygus lineolaris*
Thrips	*Hercinothrips femoralis*
Whiteflies	*Bemisia tabaci, Trialeurodes vaporariorum*
Mites	
Two-spotted mite or red spider mite	*Tetranychus urticae*
Slugs	*Deroceras reficulatum, Limax* species
Snails	*Helix* species
Bacteria	
Bacterial wilt	*Pseudomonas solanacearum*
Fungi	
Blister rust	*Pucciniastrum hyangeae*
Bud rot	*Botrytis cinerea*
Gray mold	*Botrytis cinerea*
Inflorescence blight	*Botrytis cinerea*
Leaf spots	*Ascochyta hydrangeae, Cercospora arborescentis, Corynespora cassicola, Phyllosticta hydrangeae, Septoria hydrangeae*
Powdery mildew	*Erysiphe polygoni*
Root rot	*Armillaria* species, *Polyporus* species, *Rhizoctonia* species, *Sclerotium* species
Stem rot	*Polyporus* species, *Rhizoctonia* species, *Sclerotium* species
Mycoplasma-like Organisms (MLO)	
Hydrangea virescence	
Nematodes	
Leaf	*Aphelenchoides* species
Lesion	*Pratylenchus* species
Root-knot	*Meloidogyne incognita, M. hapla*
Stem	*Ditylenchus dipsaci*
Viruses	
Alfalfa mosaic	Cucumber mosaic
Hydrangea mosaic	Hydrangea ring spot
Tobacco rattle	Tobacco ring spot
Tobacco necrosis	Tomato ring spot
Tomato spotted wilt	

[a]Adapted from Lawson and Horst (1983), Pirone (1978), Thomas *et al.* (1983), and Weiler (1980).

VI. THE FUTURE

A most exciting aspect of the future for hydrangeas is the refinement of year-round production procedures, such as Shanks' (1985) outline for minipots (see Table V). This has become feasible with new understanding of the effects of temperature, photoperiod, and chemical growth retardants on hydrangea vegetative growth, floral initiation, and inflorescence expansion. The remaining task is to "fine tune" production recommendations and select the most appropriate cultivars for year-round production.

Hydrangea breeding also is an area of opportunity. Because hydrangea distortion susceptibility does differ with cultivar (Bailey and Hammer, 1989), breeding for resistance should be possible. Future breeding programs should also involve lacecap hydrangeas. The lacecap hydrangeas offer a different inflorescence style and force more rapidly than current florists' hydrangeas (Bailey, unpublished data, 1988). Incorporation of these traits into current cultivars could greatly increase the market opportunities for hydrangea producers and product diversity for consumers.

REFERENCES

Allen, R. C. (1931). "Factors influencing the flower color of hydrangeas." *Proc. Am. Soc. Hort. Sci.* **28:** 410.

Allen, R. C. (1943). "Influence of aluminum on the flower color of *Hydrangea macrophylla* DC." *Contr. Boyce Thompson Inst.* **13:** 221–242.

Allen, T. C., and Anderson, W. C. (1980). "Production of virus-free ornamental plants in tissue culture." *Acta Hortic.* **110:** 245–251.

Anonymous. (1973). "Good results achieved with growth regulators on hydrangeas." *Grower (London).* **80:** 647.

Asen, S., Siegelman, H. W., and Stuart, N. W. (1957). "Anthocyanin and other phenolic compounds in red and blue sepals of *Hydrangea macrophylla* var. Merveille." *Proc. Am. Soc. Hort. Sci.* **69:** 561–569.

Asen, S., Stuart, N. W., and Siegelman, H. W. (1959). "Effect of various concentrations of nitrogen, phosphorus, and potassium on sepal color of *Hydrangea macrophylla.*" *Proc. Am. Soc. Hort. Sci.* **73:** 495–502.

Asen, S., Stuart, N. W., and Cox, E. L. (1963). "Sepal color of *Hydrangea macrophylla* as influenced by the source of nitrogen available to plants." *Proc. Am. Soc. Hort. Sci.* **82:** 504–507.

Bailey, D. A. (1988). "Use of growth retardants for floral initiation of hydrangeas. 1988 Turfgrass and Ornamentals Res. Summary." *Univ. of Ariz. Coop. Ext. Sta. Bull. Ser.* **P–75:** 67–69.

Bailey, D. A. (1989a). *Hydrangea Production.* Portland, OR: Timber Press.

Bailey, D.A. (1989b). "Uniconazole effects on forcing of florists' hydrangeas." *HortScience* **24:** 518.

Bailey, D. A., and Hammer, P. A. (1988). "Evaluation of nutrient deficiency and micronutrient toxicity symptoms in florists' hydrangea." *J. Am. Soc. Hort. Sci.* **113:** 363–367.

Bailey, D. A., and Hammer, P. A. (1989). "Stimulation of 'hydrangea distortion' through environemntal manipulations." *J. Am. Soc. Hort. Sci.* **114:** 411–416.

Bailey, D. A., and Weiler, T. C. (1984a). "Control of floral initiation in florists' hydrangea." *J. Am. Soc. Hort. Sci.* **109:** 785–791.

Bailey, D. A., and Weiler, T. C. (1984b). "Rapid propagation and establishment of florists' hydrangea." *HortScience.* **19:** 850–852.

Bailey, D. A., Weiler, T. C., and Kirk, T. I. (1986a). "Chemical stimulation of floral initiation in florists' hydrangea." *HortScience.* **21:** 256–257.

Bailey, D. A., Seckinger, G. R., and Hammer, P. A. (1986b). "*In vitro* propagation of florists' hydrangea." *HortScience.* **21:** 525–526.

Bing, A. Boodley, J. W., Gortzig, C. F., Helgesen, R. G., Horst, R. K., Johnson, G., Langhans, R. W., Price, D. R., Seeley, J. G., and Williamson, C. E. (1981). "Cornell recommendations for commercial floriculture crops." *Cornell Univ. Agr. Bull.* 45–46.

Dunham, C. W. (1948). "The Culture and Flowering of Hydrangeas and Azaleas as Affected by Growth Habit." Master's thesis. Univ. Wis., Madison.

Jung, R. (1964). "The status of hydrangea growing today." *Florists' Rev.* **135**(3486): 13–14, 35–37, 40.

Kiplinger, D. C. (1945). "Well-grown hydrangeas are valuable for the spring holidays." *Florists' Rev.* **95:** 23–25, 29–30.

Kofranek, A. M., and Leiser, A. T. (1958). "Chemical defoliation of *Hydrangea macrophylla* Ser." *Proc. Am. Soc. Hort. Sci.* **71:** 555–562.

Kohl, H. C., Jr., and Nelson, R. L. (1966). "Controlling height of hydrangeas with growth retardants." *Calif. Agr.* **20:** 5.

Koths, J. S., Judd, R. W., Jr., and Maisano, J. J., Jr. (1973). "Commercial hydrangea culture." *Univ. of Conn. Agr. Ext. Bull.* **73–63:** 1–8.

Koths, J. S., Judd, R. W., Jr., Maisano, J. J., Jr., Griffin, G. F., Bartok, J. W., and Ashley, R. A. (1985). "Nutrition of greenhouse crops." *Univ. of Conn. Agr. Ext. Bull.* **85–2:** 1–16.

Lawson, R. H., and Horst, R. K. (1983). "Hydrangea diseases can be controlled." *Greenhouse Manager.* **1:** 66–71, 75, 77–80.

LeMattre, P. (1975). "Influence du facteur température sur la mise á fleur de l'Hortensia (*Hydrangea macrophylla*)." In *Phytotronics in Agricultural and Horticultural Research III,* edited by P. Chouard and N. de Bilderling. Paris: Gauthier-Villars, pp. 338–344.

Liberty Hyde Bailey Hortorium. (1976). *Hortus Third: A Concise Dictionary of Plants Cultivated in the United States and Canada,* 3rd ed. New York: Macmillan.

Link, C. B., and Shanks, J. B. (1952). "Experiments on fertilizer levels for greenhouse hydrangeas." *Proc. Am. Soc. Hort. Sci.* **60:** 449–458.

Litlere, B., and Strømme, E. (1975). "The influence of temperature, daylength, and light intensity on flowering in *Hydrangea macrophylla* (Thunb.) Ser." *Acta Hortic.* **51:** 285–298.

Nell, T. A. (1990). "Commercial transport of flowering potted plants: Keeping quality beyond the bench." *GrowerTalks.* **53**(9): 24–25, 27, 29, 31–32, 34, 37, 39.

Peters, J. (1975). "Über die Blütenbildung einiger Sorten von *Hydrangea macrophylla.*" *Gartenbau-wissenschaft.* **40**(2): 63–66.

Pirone, P. P. (1978). *Diseases and Pests of Ornamental Plants.* New York: Wiley.

Post, K. (1959). *Florist Crop Production and Marketing.* New York: Orange Judd.

Powell, C. C. (1973). "Botrytis blight of hydrangea." *Ohio Florists' Assn. Bull.* **528:** 3.

Ray, S. (1946). "Reduction of blindness in hydrangeas." *Proc. Am. Soc. Hort. Sci.* **47:** 501–502.

Scott, B. (1982). "Hydrangeas respond to new growth regulator." *N.C. Flower Growers' Bull.* **26:** 10–12.

Shanks, J. B. (1969). "Some effects and potential uses of ethrel on ornamental crops." *HortScience.* **4:** 56–58.

Shanks, J. B. (1985). "Hydrangeas." In *The Ball Red Book,* edited by V. Ball. West Chicago, IL: Geo. J. Ball, Inc., pp. 535–558.

Shanks, J. B., and Link, C. B. (1951). "Effects of temperature and photoperiod on growth and flower formation in hydrangeas." *Proc. Am. Soc. Hort. Sci.* **58:** 357–366.

Shanks, J. B., Haun, J. R., and Link, C. B. (1950). "A preliminary study on the mineral nutrition of hydrangeas." *Proc. Am. Soc. Hort. Sci.* **56:** 457–565.

Snodgrass, G. (1988). Personal communication. A. D. Mohr Farms Inc., Pleasant Hill, Mo.

Stoltz, L. P. (1984). "*In vitro* propagation and growth of hydrangea." *HortScience.* **19:** 717–719.

Stuart, N. W., and Cathey, H. M. (1962). "Control of growth and flowering of *Chrysanthemum morifolium* and *Hydrangea macrophylla* by gibberellin." *Proc. Intl. Hort. Congr.* **15:** 391–399.

Takeda, K., Kariuda, M., and Itoi, H. (1985). "Blueing of sepal colour of *Hydrangea macrophylla."* *Phytochemistry.* **24:** 2251–2254.

Thomas, B. J., Barton, R. J., and Tuszynski, A. (1983). "Hydrangea mosaic virus, a new ilavirus from *Hydrangea macrophylla* (Saxifragaceae)." *Ann. Appl. Biol.* **103:** 261–270.

Tjia, B., and Buxton, J. (1976). "Influence of ethephon spray on defoliation and subsequent growth on *Hydrangea macrophylla* Thunb." *HortScience.* **11:** 487–488.

Tjia, B., Stoltz, L., Sandhu, M. S., and Buxton, J. (1976). "Surface active agent to increase effectiveness of surface penetration of ancymidol on hydrangea and Easter lily." *HortScience.* **11:** 371–372.

Ulery, C. J. (1978). "Quality hydrangea production." *Ohio Florists' Assn. Bull.* **582:** 3–4, 9.

United States Department of Agriculture. (1977). "Flowers and foliage plants." *Crop Rpt. Board Stat. Rpt. Serv. SpCr* **6–1**(77).

United States Department of Agriculture. (1986). "Floriculture crops." *Crop Rpt. Board Stat. Rpt. Serv. SpCr* **6–1**(86).

Weiler, T. C. (1980). "Hydrangeas." In *Introduction to Floriculture.* edited by Roy A. Larson. New York: Academic Press, pp. 353–372.

Weiler, T. C., and Lopes, L. C. (1974). "Hydrangea distortion." *Focus on floriculture* (Purdue Univ.) **2**(2): 9.

Wilson, E. H. (1923). "The hortensias." *J. Arnold Arb.* **4:** 233–246.

Wiśniewska, E., and Zawadzka, Z. (1962). "The formation of inflorescence in *Hydrangea macrophylla* Ser. cv. 'Altona.'" *Acta Agrobotanica.* **11:** 157–165.

15

Cyclamen

Richard E. Widmer

Introduction to Floriculture, Second Edition

385

I. INTRODUCTION

Kyklos is Greek for circle, apparently referring to the leaf shape. The original species, *Cyclamen persicum,* is native to Palestine, Asia Minor, and islands of the Aegean and eastern Mediterranean seas, but it has not been found as far east as Iran (Blasdale, 1949). In their native habitat, cyclamen are dormant during the hot dry summers and new foliage develops in response to fall rains and cooler temperatures. Flowering follows and continues until terminated by dry summer heat. Seedlings may require 2 to 3 years to bloom under such conditions.

II. HISTORY AND TAXONOMY

Earliest mention of the species was probably in fifteenth century transcripts (Blasdale, 1949, 1951; Doorenbos, 1950a,b). It was introduced into Western Europe as a collector's item in the early seventeenth century. Breeding was begun in the middle of the nineteenth century when the plant started to achieve economic significance. Taxonomy of the genus remains partially unresolved, largely because of genetic heterogeneity. As many as 24 species have been described, but deHaan and Doorenbos (1951) list 15 species. The herbaceous plants consist of a cluster of bluish-green, heart-shaped, ovate leaves with silvery markings and crenate dentate margins, on long petioles arising from flattened tubers (Fig. 1). The tubers are frequently referred to as "corms" in commerce. Attractive sympetalous five-parted, strongly reflexed flowers are borne above the leaf canopy on scapes also arising from the tuber. The cyclamen is a pseudomonocot in that only one cotyledon is found in the embryo (Hagemann, 1959). The first true leaf develops directly opposite the cotyledon. Cotyledons closely resemble true leaves.

The cyclamen has been a leading year-round pot plant crop in northern Europe for many years. It is less popular in America, where sales have been limited primarily to late fall and winter, but production and demand are now overlapping to other seasons. Production and sales have been centered in northern areas and areas with a cool coastal climate. American interest in cyclamen is increasing with the development of improved cultivars and improved, accelerated production techniques. Research studies have centered in Europe, but some Japanese (Niizu, 1967), Canadian (Molnar and Williams, 1977), and United States (Stephens and Widmer, 1976) studies have been noted.

III. CULTIVARS

Early forms had small flowers of relatively pale color. British breeders first developed more intense petal color and broader, flatter corolla lobes. Double-sized flowers were noted in England and Germany by 1870 (Doorenbos, 1950b). Both diploid and polyploid cultivars are currently on the market. Large, intermediate, and

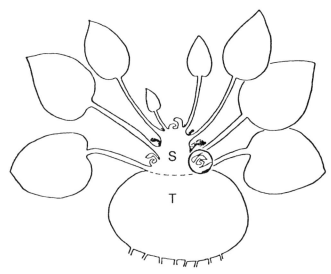

Fig. 1. Diagrammatic longitudinal section of cyclamen. The primary shoot(s) is shown atop the well-developed tuber (T). Axillary branch shoots are present at the lowest nodes (one such branch is circled on the right), and flower primordia are present in the axils of the upper leaves. The youngest leaf primordia exhibit hyponastic curvature and are folded over the shoot apex. (Courtesy of Marshall Sundberg, currently Department of Botany, Louisiana State University, Baton Rouge).

small, as well as common, fringed, creasted, frilled, and double-flowered forms, are now available in many colors (Wellensiek, 1952). Uniformity of cultivars is less than desired. Recently introduced F_1 hybrids are much more uniform, although not as much so as F_1 hybrids of many other plant species.

Primary objectives sought in today's cultivars include good germination; early, uniform flowering; floriferousness; attractive foliage; fast growth; compact, uniform plants; and above average life in the home and office. Selection within cultivars often results in greater differences in flowering time than found between cultivars. A large array of good cultivars, hybrid and nonhybrid, is now available.

New cultivars should be considered as they are released. Most seed is from European and, more recently Japanese, sources. Frequently, different cultivar names are used in the United States than overseas, although the cultivars are identical.

IV. PRODUCTION

Until the 1800s cyclamen was considered difficult to propagate and produce. Tubers were split to start new plants that were rested in the summer and production time was 2 years. By 1825, the Englishman John Wilmott started plants from seed and grew them in 15 months with no summer rest period (Doorenbos,

1950b). Some florists still grow cyclamen at temperatures around 50°F over a period of 15 months or more, but modern techniques make 8-month production schedules a reality (Widmer *et al.*, 1976). Figure 2 illustrates well-grown, fast-crop cyclamen.

V. PROPAGATION

Today propagation is primarily from seed that is sold by count rather than by weight. Propagation by tuber splitting is laborious and not practical, except to maintain specific clones for breeding purposes. Cotyledons cut from tubers will root and form new plants, but true leaves will not root.

Pollination of flowers is simple. Pollen is merely applied to the pistil with a label, pipe cleaner, or similar object. Once the ovule is fertilized, the peduncle continues to elongate and may bend or curve. Usually the seed pod is not lowered close to the ground, as is the case with many of the basic *Cyclamen* species. Seed ripens in 2 to 3 months. Menzel (1972b) found that small seed is not desirable, but this trait can vary with cultivar and numerous other factors. Many larger-flowered cultivars are tetraploids and have larger seeds. Diploids usually have smaller seed, may germinate more quickly, flower faster, and thrive better under home conditions (Wellensiek, 1961). Individual mother plants (seed-bearing parents significantly influence seed quality, germination, and subsequent growth (Noordegraaf, 1977). Cyclamen average approximately 3000 seeds per ounce.

How long is cyclamen seed viable? Massante (1964) stored cyclamen seed for 52 months at 36° and 50°F with no significant loss in germination. Cultivar and quality of the seed produced significantly influence the viable life of the seed. Fresh seed is preferable, but some sources claim that cyclamen seed is dormant for

Fig. 2. Nine-month-old plant 'Rosa von Zehlendorf TAS' grown in 4-inch pot illustrates the flowering potential of a well-grown specimen.

roughly 90 days after harvest. Industry sources (Widmer, 1983) report that the seed moisture level should be lowered to 5 to 7% to maintain viability. Holding the seed at 28% relative humidity for 90 days after harvest will lower the seed moisture level to the desired point. Then the seed should be placed in moisture-proof packages, usually foil and polyethylene lined. Recommended storage temperature is 40° to 45°F, although 60°F is almost as effective. Some distributors repackage in their own envelopes, a poor practice unless the proper moisture level is maintained. Growers are wise to purchase only a 1 year supply of cyclamen seed annually. No specific seed treatment consistently improves germination. Soaking the seed at room temperature for 12 hours is sometimes beneficial, but not essential (Anderson and Widmer, 1975; Lyons, 1980). Widmer *et al.* (1976) suggested that seed be sown at 3×3 inch spacing $\frac{1}{8}$ to $\frac{1}{4}$ inch deep in flats filled with moist, nutrient-enriched sphagnum moss peat (commonly referred to as peat moss in the industry) (Table I).

Sowing in plastic trays each having 50, $1\frac{3}{4}$-inch wide plugs is equally satisfactory, but the plugs should not be permitted to dry excessively. A similar mix recommended for seed germination and growth of some plants is commercially available. Fine or powdery peat should not be used. The peat should be compressed to two-thirds of its fluffed up volume prior to sowing. Some growers may prefer to broadcast the seed, plant in rows, or sow seed in peat disks or individual pots. Other lightweight, organic-type mixes can also be used.

Cyclamen seed germinates best at 66° to 68°F in the dark (Massante, 1964), whereas temperatures of 72°F and above may be inhibitory. Such conditions are best maintained in a temperature-controlled room with good air circulation, rather than in the greenhouse, where solar radiation elevates the germination medium temperature excessively. Flats should be watered after seed sowing, but allowed to dry for several hours prior to placement in the germination area. This practice

Table I

Nutrient Additions to the Sphagnum Peat Moss Medium Used in Cyclamen Production

Materials	Bushel		Cubic meter	Cubic yard
	Ounces	Grams		
Ground dolomitic limestone: 45% $CaCO_3$, 36% $MgCO_3$	7	200	7.5 pounds	9.7 pounds
Magnesium sulfate, 10% magnesium	0.7	20	12.2 ounces	1.0 pounds
Potassium nitrate, 13–0–44	0.25	7	4.2 ounces	5.5 ounces
Superphosphate, 0–20–2[a]	0.42	12	7.6 ounces	10.0 ounces
Slow-release fertilizer, 14–14–14	0.56	16	9.5 ounces	12.5 ounces
A slow-release microelements, mix as recommended on the package				

[a]Triple superphosphate (0–46–0) at an equivalent rate of phospate may be substituted for superphosphate.

lessens the probability of fungal growth on the medium or the container. Peat moss should be kept moist during germination. Cyclamen are sensitive to germination medium pH levels below 5.5; above 6.0 is preferable (Maatsch and Isensee, 1959). Sumitomo and Kosugi (1963) germinated cyclamen in petri dishes with controlled pH solutions. They reported 90% germination at pH 6.5, 79% at pH 5.5, and 74% at pH 8.0. Unpublished studies by Widmer also noted lower germination percentages in sphagnum peat moss at pH levels below 5.5 and especially below 5.0. Lime should be added before sphagnum peat moss is used for germination. Seeds will germinate poorly and fail to develop after germination because roots do not develop in the acidic, noncalcified medium (Fig. 3). Cyclamen seed will germinate if only lime is mixed in the moss, but the seedlings must then be fertilized within a few weeks after they are visible above the surface (Fig. 4) to encourage continued development.

Visible evidence of germination of inbibed seed starts below ground within 5 days (Anderson and Widmer, 1975). First the primary root penetrates the soil, then the hypocotyl begins to swell to form the tuber, and at 28 days, the thin hypocotyls are usually evident. The cotyledon blade is the last organ to emerge from the seed coat. Seedlings should be moved to a humid, shaded (February to November) greenhouse when the cotyledon petioles are stretching, to prevent excessive elongation. After the cotyledon blade has unfolded and a true leaf is evident, the plants can be transferred to a less humid area, but the 68°F night temperature should be continued for optimum growth (Menzel, 1972a).

Germination percentages up to 95% or more may be obtained, but 80 to 85% is more common. Seedlings that take more than 45 days to germinate are usually weak or crippled specimens and should be discarded. Up to 5% of the remaining plants may need discarding if growth is not normal. A maximum of 75 to 85% of the seeds can be expected to produce good plants. Usually hybrids produce fewer off-types for disposal. This factor is probably best attributed to the greater degree of genetical homogeneity of the hybrids.

Fig. 3. Cyclamen seedlings at 45 days. The two plants on the right were sown in peat moss to which lime and nutrients were added (note roots an inch long). The three plants on the left were sown in peat moss with no additions (note that only root stubs are present).

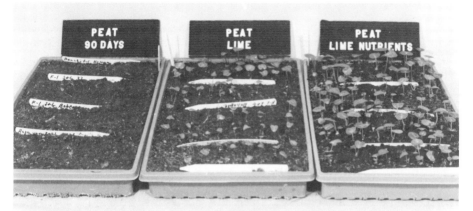

Fig. 4. Germination 90 days after seeding of four cyclamen cultivars sown in unaltered peat moss, peat moss to which pulverized limestone was added, and in peat moss enriched with pulverized limestone and nutrients as listed in Table I.

Storage of seedlings for up to 5 weeks is possible if desired. Maatsch (1958) placed seedlings with roots bedded in moist peat in polyethylene bags and stored them up to 5 weeks in a 38°F refrigerator with no negative effects. The seedlings required shielding from direct sunlight and drafts when first removed from storage. A 43°F storage temperature was not satisfactory.

Rather than struggling with germination, many smaller growers are now purchasing liners, 15 to 18-week old $2\frac{1}{4}$-inch plants grown in plastic trays, from specialists. They are grown to salable plants in 4 to 5 months. They may also buy 4-inch plants with flower buds rising, for sale 8 to 9 weeks later. This procedure helps southern growers avoid the undesirable effects of summer heat on the plants.

VI. VEGETATIVE GROWTH

Growth is slow in early stages, and an optimum environment should be provided to maximize growth. After seed has germinated, both the tuber and the cotyledon gradually increase in size. The first true leaves do not unfold simultaneously on all plants but are usually clearly evident within 80 to 90 days after sowing. After two true leaves have unfolded and five leaves have initiated, the rate of leaf initiation accelerates to about 1.3 per week and remains fairly constant through leaf number 17 (Sundberg, 1981a, 1981b, 1982).

The leaf count increases faster after leaf number 17 because of the production of the axillary branch shoots. An actual example of sequential leaf counts for two cultivars sown on December 7 is presented in Table II (Lyons, 1980).

A. Potting

At 17 weeks after sowing, plants of most cultivars should average approximately six to seven unfolded leaves or 13 to 15 leaf units in age (Stephens and Widmer, 1976). They are then ready for potting. If the seeds were spaced out at 3 × 3 inches in flats, plugs, pots, or benches when sown, no transplanting or spacing is needed prior to this stage. Plants grown in peat moss can be transplanted with a minimum of root disturbance. The tuber top is kept flush with the surface of the growth medium. The nutrient-enriched peat moss used for germination (Table I) is also recommended for potting. Some pea rock can be placed in the pot bottom for ballast, if desired. Plunging the tuber encourages development of wider-spreading plants. Tubers should not be plunged in heavier soils, which are not recommended. Light soil mixes consisting of a maximum of one-third loam plus peat moss, vermiculite, perlite, leaf compost, or similar materials may be used.

Quality specimens can be grown in 4-, 5-, 6-inch, and larger pots. Plastic pots with drainage openings are preferable to clay pots because they facilitate maintenance of uniform moisture levels in the soil. Smaller flowered cultivars are preferable for 4-inch pots and may also be grown in 5-inch pots. Intermediate and

Table II

Sequential Counts of Number of Unfolded Leaves per (Including 1 Cotyledon) Plant for Two Cultivars Sown on December 7

Days after seeding	'Swan Lake'	'Beautiful Helena'[a]
81	1.3	1.1
88	1.9	1.9
95	2.5	2.3
102	3.5	3.4
118	4.3	4.3
125	4.7	4.6
132	5.9	5.1
146	10.1	8.9
153	11.2	10.4
166	13.4	12.1
174	15.5	13.9
183	18.1	16.4
190	22.1	19.4
197	24.2	22.9
204	26.7	24.2
212	30.0	25.9
219	34.2	30.3
—	In bloom at 249 days	50.0
In bloom at 262 days	59.7	

[a]Minicultivars.

large-flowered cultivars are preferable for 5-inch and larger pots. Plants may be placed pot to pot until the foliage reaches the pot rim, but should be spaced promptly thereafter. Crowding of cyclamen plants results in elongated petioles and peduncles, leggy, weak plants and a significantly increased possibility of disease. A well-grown, 4-inch pot plant will ultimately require up to 100 square inches; a 5-inch pot plant, 144 square inches; and a 6-inch pot plant, 225 square inches of bench space. Production costs follow a similar ratio. Should a plant become too large for its pot, it can be repotted in a larger container, even when in full bloom, with no noticeable setback if the rootball is not broken.

B. Fertilization

Cyclamen are slow to exhibit conspicuous symptoms in response to low or high nutrient or salt levels in the growth medium. The first response is usually a decrease in the growth rate. Without control plants for comparison, the decrease may go unnoticed or be attributed to weather or other factors. Cyclamen require a constant, moderate supply of nutrients applied in proportion to plant size to maintain optimum growth. Usually, if the suggested nutrient-enriched peat moss mix is employed, no supplemental fertilizer is needed for 2 months after seed sowing. Analyzing the growth medium at intervals helps determine when fertilizer is required. Heavy watering accelerates and subirrigation (capillary mats) delays the need for fertilization. Capillary mats are not recommended in winter as plants may stretch during periods of dark (cloudy) weather, unless special care is used in watering.

Platteter and Widmer (1978) noted that vegetative characteristics of cyclamen are affected by the rate of application of nitrogen and potassium. Effects are similar at all growth stages beyond the seedling stage. The following suggestions are averages and may require modification for various cultural and environmental conditions. Before young plants are potted, they usually benefit when a 100 ppm nitrogen solution from a balanced fertilizer such as soluble 20–20–20 or 20–10–20 is applied every 3 weeks. Regular applications should begin 4 weeks after potting. For plants grown in peat moss and fertilized with every watering, the nitrogen source should be balanced for nitrate and ammonium forms. Apply at 100 ppm until the plants are potbound, then 150 and eventually 200 ppm nitrogen applications may be necessary as the plants develop (Fig. 5). Lower concentrations result in fewer and smaller leaves and smaller plants (Fig. 6). Slightly more (25 to 50 ppm) potassium than nitrogen may be required to obtain optimum leaf and plant size. European growers claim that failure to apply adequate potassium fertilizer increases plant susceptibility to plant pathogens. Chloride and sulfate sources of potassium are equally satisfactory.

Phosphorus is neither fixed nor held in a peat moss medium and is subject to leaching (Puustjarvi, 1976–1977). Thus, incorporation of phosphate fertilizer in the growth medium prior to planting is not adequate for the duration of the crop. Applications of 35 to 50 ppm phosphorus are recommended when plants

are fertilized with every watering, or higher concentrations if applied less frequently.

Foliage becomes pale green or chlorotic with excessive calcium levels, insufficient iron, or a high pH. An application of chelated iron is then recommended. Cyclamen grow best at a minimum soil pH of 6.0 for loam mixtures, but established plants grow well in peat moss with a pH as low as 5.0. Cyclamen respond favorably to levels of 1000 ppm carbon dioxide during the season when ventilators are kept closed. Plant growth and flowering are accelerated with the carbon dioxide injection.

C. Watering

Once the cyclamen wilts, especially in hot weather, some of the leaves will turn yellow within 24 to 36 hours. Peat moss should not be allowed to appear dry or to get as dry as a loam soil. Watering should be done in the morning whenever possible, to minimize *Botrytis* development. Overhead or subirrigation may be used, but the latter may cause excessive plant stretch in short days. An occasional leaching may be beneficial if salts accumulate in the medium and if subirrigation is used. A daytime relative humidity of 50 to 70% is considered desirable.

D. Temperature

A night temperature of 68°F is preferable until six to seven unfolded leaves per plant are present. A drop to 63° to 65°F is then recommended. Day temperatures should be 68°F (cloudy) 73° to 75°F (sunny). Pad and fan cooling are definitely recommended for most areas of the United States. Lack of cooling during a hot summer may delay the plants 1 to 2 months and decrease plant quality significantly. Some growers may be successful growing plants outdoors under lath shade during the summer, provided they can maintain adequate humidity levels and pest control.

E. Light

Plants should be shaded from April to October to provide a maximum of 4.3 klx. Full radiant energy the remainder of the year should not be excessive in most areas of the country. Symptoms of excessive radiant energy include hardened plants, pale, chlorotic foliage, and necrotic areas on the leaves. Inadequate radiant energy slows growth and causes weak, spindly petioles and weak floppy

---

Fig. 5. Effect of the quantities of nitrogen and potassium applied with every watering at three stages of plant development. Plants of cultivar 'Rosa von Zehlendrof TAS' were grown in a peat moss medium in 3-inch plastic pots for stage 1 and in 5-inch plastic pots thereafter. (A) End of stage 1. (B) End of stage 2. (C) End of stage 3.

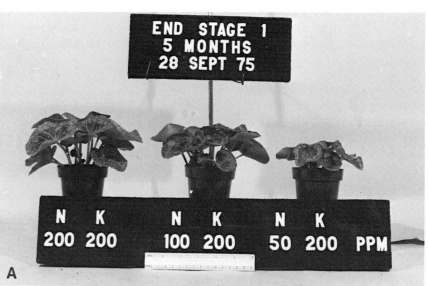

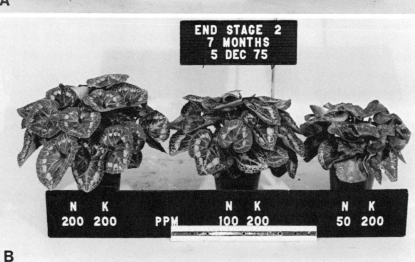

plants. Supplementing radiation energy with high intensity electric lighting of young plants in winter has accelerated plant development, but results have not been consistent or significant enough to warrant the cost when considering energy costs.

Good air circulation is desired to minimize diseases. Open mesh bench bottoms, low bench sides, proper spacing, and tube or pad and fan ventilation in season are all helpful. When old, bruised, yellow, and decomposing leaves need to be removed from the plant, they should be twisted or snapped off to avoid leaving undesirable stubs.

VII. FLOWERING

Cyclamen will not flower until a certain vegetative condition is attained (Hagemann, 1959, Sundberg, 1981a,b). Anything that encourages accelerated early plant development should hasten flowering. Flower buds initiate in the axils of the sixth and successive true leaves (Stephens and Widmer, 1976). Thus, floral initiation in the axil of the sixth leaf occurs between the initiation of the tenth and thirteenth leaf units. Initial growth of flower buds is extremely slow, and actual flowering often does not occur until the plants have 35 or more leaves unfolded, although this figure may vary with cultivar, pot size, and plant treatment (see Table II). Axillary branch shoots or growing points usually develop in the axils of the first five leaves (see Fig. 1).

A. Fertilization

Time of cyclamen flowering is affected by rates of application of nitrogen and potassium (Platteter and Widmer, 1978). High (200 ppm) potassium accompanied by lower (50 to 100 ppm) nitrogen applied with every watering to plants in a peat moss medium prior to the development of 15 unfolded leaves per plant can delay flowering at least a week. Application of 200 ppm nitrogen versus 50 ppm, and/or 200 ppm potassium versus 50 or 100 ppm with every watering when the plants had 15 to 35 unfolded leaves delayed flowering up to 2 weeks. Nutrient applications at any of these concentrations after the plants had 40 unfolded leaves did not influence flowering time. Nutrition within normal ranges does not alter total flower production. Excessively high nutrient levels can inhibit plant development at any stage.

Fig. 6. Effect of three different concentrations of nitrogen accompanied by a constant potassium concentration applied with every watering. Plants of cultivar 'Rosa von Zehlendorf TAS' were grown in a peat moss medium in 3-inch plastic pots for stage 1 and in 5-inch plastic pots thereafter. (A) End of stage 1. (B) End of state 2. (C) End of stage 3.

Neuray and Henrard (1966) and Niizu (1967) noted that a competition existed between leaves and buds. Thus, factors such as above-average but not excessive applications of nitrogen fertilizer and high growing temperatures that encouraged vegetative growth inhibit or delay flowering.

B. Temperature

Root temperatures of 55° to 65°F for 6 weeks at sometime during the plant stage from 6 to 40 unfolded leaves accelerated flowering up to 2 weeks or more over root temperatures of 75° and 85°F (Gembis, 1978; Stephens and Widmer, 1976). Final plant size was smallest at the lowest temperature and largest at the two highest temperatures. In a European cyclamen study, Menzel (1972a) reported that plants were smaller and flowered earlier at 60°F than at 68°F ambient temperature. Vegetative and reproductive plant responses form the basis for temperature recommendations presented in the vegetative growth section. Fick (1988) compared growth at 50°, 60°, and 70°F in various root and air temperature combinations during the two to twelve open leaf plant stage (Fig. 7). Plants grown at 50°F ambient temperature grew slower and were the last to flower. Plants in all other treatments were similar in plant growth and earlier flowering. Thus a 68° to 70°F root temperature combined with a 50°F ambient temperature is suggested for up to the first 6 months of crop growing time, provided this time span falls in the cool weather portion of the year. Moderately higher ambient day temperatures, because of solar radiation, do not delay the plants. A 65° to 68°F night temperature is suggested during warmer portions of the year. This program provides good quality plants, early flowering, and fuel economy. About 45 days before the scheduled flowering date, when plants of most cultivars grown in 5- or 6-inch pots have about 35 leaves, night temperature can be lowered to 62°F with a maximum day temperature of 70° to 74°F, especially during short days. A night temperature of 68°F at this time will result in bud abortion, small flowers, and poor plant quality.

C. Light

Neuray (1973) reported that cyclamen do not exhibit a clear photoperiodic response, but daylength and light intensity play a part in flower bud formation and flowering. Apparently, the greater the total light unit accumulation, if other environmental factors are proper, the more leaves and flowers develop and the earlier the flowering.

D. Gibberellic Acid

Application of gibberellic acid (GA_3) to young plants (Fig. 8) will hasten the start of flowering of many cultivars by as much as a month (Widmer et al., 1974, Lyons and Widmer, 1983a). In addition, flowering is more uniform. Treatment is most effective when applied 140 to 150 days after seed sowing when the plants have 10 to 12 unfolded leaves. At this time the first flower buds are barely visible (Lyons and Widmer, 1983b). A spray of 10 ppm GA_3 plus a nonphytotoxic wetting agent is directed on the crown of the plant. The F_1 hybrids are especially responsive. Quantity of spray applied per plant is a key point. Excesses should be avoided. One-quarter fluid ounce per plant (60 plants per pint of spray) is adequate. Growers who have not treated plants with GA_3 previously should treat a limited portion of the crop the

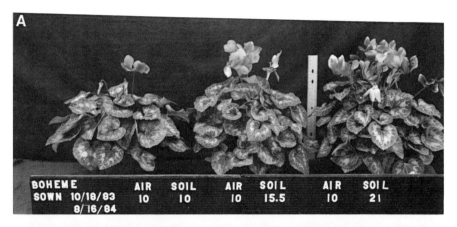

Fig. 7. Effect of varying air and root (soil) temperature during 2 to 12 open leaf plant stage on growth and flowering of plants of cultivar F_1 hybrid 'Boheme.' (A) Constant air temperature. (B) Constant root (soil) temperature. Seed sown on October 18, 1983, and plants photographed on August 16, 1984. Legend temperatures of 10°, 15.5°, and 21°C equivalent to 50°, 60°, and 70°F. Note 12-inch ruler in pictures.

first time, to adjust to the system and to determine cultivar response. Excessive applications can cause weakening of the peduncles. Application of GA_3 to the leaves instead of the crown provides a less uniform and less effective response.

VIII. KEEPING QUALITY

A. Plants

Well-grown specimens of modern-day cultivars have excellent keeping quality if properly watered and kept in a bright, cool (65° to 68°F nights) location. Individual flowers last about 4 weeks on plants in the greenhouse, office, or home. Plants should remain attractive for a minimum of 6 weeks in the home. Plants in plastic pots are preferable as the medium in them remains moist longer. Small specimens (4-inch pots) may not last quite as long unless given close attention.

Lack of fertilizer in the medium can shorten home life of the cyclamen. A slow-release fertilizer should be applied to the plants before they leave the greenhouse. Use of care cards is especially helpful as the care details for this plant are less well known than they are for chrysanthemums, poinsettias, and similar plants. The well-aerated peat moss medium decreases the probability of overwatering.

B. Cut Flowers

Cut cyclamen flowers are commonly used in some parts of Europe (Maatach and Isensee, 1960). Cultivars with long penducles are grown for this purpose. Flowers are removed from the corm with a snap or twist so that no stub remains to decay and encourage disease. Old, undesired leaves are removed in the same manner. If the peduncle is cut near the base, flowers will last up to 14 days (depending on the cultivar) in plain water. Without the cut, flowers may last only 2 days. Kohl (1975) reported that flower life varied by cultivar, and that a solution of 5% sucrose and 25 ppm silver ions (from silver nitrate) in deionized water substantially increased flower life.

IX. SCHEDULES

Temperatures and methods suggested herein are for fast culture production. Lower temperatures may be employed, as they have been for many years. Low temperatures slow plant growth to such an extent, however, that more total fuel may be required to produce a plant, and the total production cost is excessive, especially in northern climates (Widmer et al., 1983).

Average fast culture production time for cyclamen is shown in the following tabulation.

Fig. 8. A close-up of the crown of F₁ 'Swan Lake' cyclamen plant at 150 days after seed was sown. Recently initiated first flower buds are evident. This is the optimum time for application of GA₃ to accelerate flowering.

Final pot size (inch)	Time required (months)
3	6½
4	7½
5	8–8½
6	9–9½

Plants in small pots usually flower earlier than do plants in larger pots (Fig. 9). Flowering can be scheduled for any month of the year. Quickest production is from seed sown in February through April. Appropriate flower colors should be selected for individual seasons. In most areas, sales are most limited in June through August.

Small growers may find it more practical to purchase young plants from a specialist.

X. PROBLEMS

A. Diseases

Diseases can be troublesome. Growers should disinfest everything from potting bench to growing bench and pots to minimize difficulty. Some of the more common and important disease problems follow.

1. Fungi

a. Mold on Seed Flats. Excess moisture should be avoided and an appropriate fungicide such as Benlate drench applied at 1 tablespoon per 2 gallons of water. Captan inhibits cyclamen seed germination.

b. Crown Rot or Botrytis Blight (Botrytis cinerea). Crown rot is a soft decay of flowers and leaves, often in the crown. Affected parts are often covered with a downy gray mold. *Botrytis* development is encouraged by high humidity, poor air circulation, crowded plants, underfertilization, and a night temperature below 62°F. For disease control, environmental conditions should be improved, and an appropriate chemical spray or drench should be applied in the crown of the plant.

c. Fusarium Wilt (Fusarium oxysporum, F. cyclaminis). This vascular disease usually begins with yellowing at the base of the leaf blade or elsewhere on the leaf. The spots enlarge and discoloration may occur in roots and tubers, but outward symptoms are limited until the flowering stage when plants wilt quickly. The tubers remain firm. Everything, including the soil, should be pasteurized to control this organism. Fungicide sprays may protect healthy plants.

d. Stunt (Ramularia cyclaminicola). Stunt is less common than the aforementioned diseases. Symptoms include flowering below the leaves and reddish-brown discolored areas in the tuber. The growing medium should be pasteurized, and infected plants should be discarded. Young plants can be sprayed with an appropriate fungicide if brown areas are evident on the foliage.

e. Roots Rots (Pythium, Rhizoctonia, etc.). Appropriate drench material should be applied for control of root rots.

Fig. 9. Plants of F_1 'Swan Lake' cultivars in 4-inch (two on the *left*), 5-inch (*middle*), and 6-inch (*right*) pots. All plants are 32 weeks old and were grown in nutrient-enriched peat moss. Note that the larger the pot, the more time required for flowering.

2. Bacteria

Soft Rot (Erwinia carotovora). A sudden wilting and plant collapse characterizes this disease, and part of the tuber may become soft and slimy while the roots are intact. Petioles and flower scapes may also become soft and slimy. Hot weather encourages rapid progress of the rot. Proper spacing, avoiding splashing water, discarding diseased plants immediately, and applying an appropriate control chemical will reduce the severity of this disease problem.

3. Virus

Tomato Spotted Wilt Virus (TSWV). Although not new, TSWV has now become a major virus problem in greenhouse crop production. There is a wide range of symptoms and a wide range of hosts. On cyclamen, yellowed ring spots may develop into dry rings or spots of dead tissue (Nameth, *et al.,* 1988). Other possible symptoms that may vary by host include stunting, dark purple-brown sunken spots, stem browning, brown necrotic leaf areas and death. The virus is spread by thrips, including the western flower thrips, *Frankliniella occidentalis*. Control measures include eliminating thrips and weeds and discarding infected plants. There is no known chemical prevention or cure.

B. Insects and Mites

Approved insecticides only should be applied to control insect and mite infestations. Emulsion formulations frequently injure cyclamen leaves and flowers while wettable powder formulations of many insecticides are not phytotoxic.

1. Cyclamen Mites (Steneotarsonemus pallidus *Banks*)

These mites cause curling and distortion, some discoloration, and stiffening of young leaves and flowers. Mites are tiny, not visible to the naked eye, and semi-transparent with a brownish twinge.

2. Spider Mites, Red Spiders (Tetranychus urticae)

After infestation, a strippled yellow or brownish pattern develops on the foliage and, in advanced cases, forms webs of silky strands. When present, spider mites are usually found on leaf undersides and on the flowers.

3. Aphids

The most common is the green peach aphid (Myzus persicae Sulzer). Symptoms include crinkling and distortion of foliage, usually accompanied by shiny specks of honeydew, a sugary excretion. Aphids are visible to the naked eye.

4. Thrips

White, silvery, or brownish streaks develop in a somewhat stippled pattern. Thrips are small, narrow, elongated insects of various colors and are visible but often overlooked. They are often hidden in buds and flowers.

5. Fungus Gnats (Bradysia species and Sciara species)

Fungus gnats resemble fruit flies and reside primarily in and on the soil. Heavy infestations attack roots and decrease plant vigor. Chemical soil drenches are helpful.

C. Physiological Disorders

There are a variety of physiological disorders known to reduce the quality of cyclamen plants.

1. *Blasting of flower buds* may be caused by high temperatures, insufficient light, insufficient water, or excessive soil fertilizer levels.
2. *Delayed flowering* may be caused by growing the wrong cultivars, high or low temperatures, faulty nutrition, oversized pots, or insufficient light.
3. *Small flowers* may be caused by high temperatures, excessive soil fertilizer levels, or growing the wrong cultivar.
4. *Stretched plants* (too tall) may be caused by insufficient space per plant, excessive soil moisture, insufficient light, or high temperatures.
5. *Stunted plants* may be caused by stunt disease or excessive soluble salt levels in the soil.
6. *Weak growth* may be caused by high temperatures, disease, genetic variability, faulty nutrition, crowding, insufficient light, or application of excessive concentrations or quantities of GA_3.
7. *Hardened, dull blue-green, undersized foliage* may be caused by applying only the ammonium form of nitrogen.
8. *Wilting and soft plants* may be caused by dry soil, excess soluble salt levels in soil, extreme temperatures, poor light, or disease.
9. *Yellow or chlorotic foliage* may be caused by lack of nutrients, high pH, excessive light intensity, dry soil, or disease.

10. *Excessive leaf size* may result from applying above optimum levels of nitrogen potassium, or both, or excess soil moisture.

Symptoms of nutrient deficiency (Bussler, 1969) are varied.

1. Nitrogen: small, pale (chlorotic) foliage and, with extreme deficiency, weak growth.
2. Phosphorus: dark green, stiff leaves rich in anthocyanins, especially in petioles and on lower leaf surfaces.
3. Potassium: smaller leaves with necrotic dotted margins on older foliage; necrotic areas increase with deficiency severity, flower scapes are shorter than normal.
4. Calcium: flexing down of leaf and flower stalks and development of macroscopic brown streaks on edges of young leaves; roots remain short; tubers internally glassy with vessels partially brown.
5. Magnesium: as little as half of the normal flower production with no visible foliar symptoms.
6. Boron: young leaves thickened and irregularly curled, flower buds remain small and dry on short peduncles thickened near the base.
7. Iron: new foliage a pale green to greenish-yellow (Widmer).

A summary of suggestions for successful fast crop production is presented in Table III.

Table III

Key Suggestions for Successful Fast Crop Production

Select proper cultivars
Germinate seed in proper medium and environment
Maintain appropriate growing temperatures
Maintain appropriate humidity levels
Transplant on schedule and only once
Minimize root disturbance during transplanting
Use a light, well-aerated growth medium with good water retention
    capacity
Maintain uniform, adequate moisture levels in the growth medium
Fertilize regularly in proportion to plant size
Avoid excessive growth, medium soluble salt levels
Keep plants actively growing at all times
Maintain appropriate solar radiation energy levels
Use evaporative pad cooling in summer (in most areas)
Space plants properly
Provide for good air circulation
Apply GA_3 when appropriate
Use good sanitation practices at all times
Control insects and diseases promptly

REFERENCES

Anderson, R. G., and Widmer, R. E. (1975). "Improving vigor expression of cyclamen seed germination with surface disinfestation and gibberellin treatments." *J. Am. Soc. Hort. Sci.* **100**(6): 597–601.

Blasdale, W. C. (1949). "Early history of the Persian cyclamen." *Nat. Hort. Mag.* **28:** 156–161.

Blasdale, W. C. (1951). "Additional notes on the history of Persian cyclamen." *Nat. Hort. Mag.* **30:** 192–197.

Bussler, W. (1969). "Dungungsversuche zu Cyclamen." *Gartenbauwissenschft.* **34:** 495–510.

deHaan, I., and Doorenbos, J. (1951). "The cytology of the genus *Cyclamen.*" *Meded. Landbouwhogesch. Wageningen.* **51:** 151–166.

Doorenbos, J. (1950a). "Taxonomy and nomenclature of cyclamen." *Meded. Landbouwhogesch. Wageningen.* **50**(3): 17–29.

Doorenbos, J. (1950b). "The history of the Persian cyclamen." *Meded. Landouwhogesch. Wageningen.* **50**(3): 31–59.

Fick, B. J. (1988). Effect of air and root zone temperatures on growth and development of *Cyclamen persicum.* Master's thesis. Dept. Hort. Sci. Land. Archit., Univ. of Minnesota, St. Paul.

Gembis, J. H. (1978). Soil temperature effects on Cyclamen growth and flowering. Master's thesis. Dept. Hort. Sci. Land. Archit., Univ. of Minnesota, St. Paul.

Hagemann, W. (1959). "Vergleichende Marphologische, Anatomische und Entwecklungsgeschichtliche Studien on *Cyclamen persicum* Mill. sourie einigen cyclamen-arten. *Bot. Stud.* **9:** 1–88.

Kohl, H. C. (1975). "Cyclamen as cut flowers." *Calif. Agr. Ext. Flower Nursery Rep.* (March): 6.

Lyons, R. (1980). Role of gibberellin and auxin in the growth and development of *Cyclamen persicum.* Ph.D. diss. Dept. Hort. Sci. Land. Archit., Univ. of Minesota, St. Paul.

Lyons, R. E., and Widmer, R. E. (1983a). "Effects of gibberellic acid and naphthaleneacetic acid on petiole senescence and subtended peduncle growth of *Cyclamen persicum* Mill." *Ann. of Bot.* **52:** 885–890.

Lyons, R. E., and Widmer, R. E. (1983b). "Effects of GA_3 and NAA of leaf lamina unfolding and flowering of *Cyclamen persicum.*" *J. Am. Soc. Hort. Sci.* **108**(5): 759–763.

Maatsch, R. (1958). "Kuhllagerung von Cyclamen-Jungpflanzen." *Gartenwelt.* **58:** 51–52.

Maatsch, R., and Isensee, H. (1959). "Keimung von Cyclamen in Substraten mit verschiedengen p H-Werten." *Gartenwelt.* **64:** 59–363.

Maatsch, R., and Isensee, H. (1960). "Schnittblumenertrage bei Cyclamen V." *Gartenwelt.* **60:** 46–47.

Massante, H. (1964). "Unterschungen uber den Einfluss der Temperatur auf Lagerung und Kermung von Aierpflanzensamen." *Gartenwelt.* **64:** 291–293.

Menzel, K. (1972a). "Vorteilhafte Cyclamen-Entwicklunstemperaturen." *Dtsch. Gaertnerboerse.* **72**(19): 424–425.

Menzel, K. (1972b). "Vererbund der Fruh-und Spatbluhigkeit bei Cyclamen." *Dtsch. Gaertnerboerse.* **72**(30): 653–654.

Molnar, J. M., and Williams, C. J. (1977). "Response of *Cyclamen persicum* cultivars to different growing and holding temperatures." *Can. J. Plant Sci.* **57:** 93–100.

Nameth, S. T., Powell, C. D., and Lindquist, R. K. (1988). "Tomato spotted wilt virus: A serious threat to greenhouse crops." *Ohio Florists Assoc. Bull.* **704**(June): 2–3.

Neuray, G. (1973). "Bud formation in *Cyclamen persicum.*" *Acta Hortic.* **31:** 77–79.

Neuray, G., and Henrard, G. (1966). "L'influence de la lumiere sur la corissance du cyclamen." *Bull. Rech. Agron. Gembloux.* **2**(3): 536–550.

Noordegraaf, C. V. (1977). Personal communication. Res. Stn. Floric., Aalsmeer.

Niizu, Y. (1967). "Bud differentiation in cyclamen." *Nogyo Oyobi Engei Agr. and Hort.* **42**(8): 105–106.

Platteter, R. J., and Widmer, R. E. (1978). "Cyclamen responses to nitrogen and potassium regimes." *Annu. Meet. Am. Soc. Hort. Sci.* (Abstract 75) **235:** 368.

Puustjarvi, V. (1976–1977). "Phosphorus fertilization of sphagnum moss peat." *Peat Plant Yearb.,* 24–30.

Stephens, L. C., and Widmer, R. E. (1976). "Soil temperature effects on cyclamen flowering." *J. Am. Soc. Hort. Sci.* **101**(2): 107–111.

Sumitomo, A., and Kosugi, K. (1963). "Studies of Cyclamen I on the germination of seed." *Tech. Bull. of Faculty of Agr. Kagawa Univ.* **14**(2): 137–140.

Sundberg, M. D. (1978). Shoot and flower development in *Cyclamen persicum* Mill. Ph.D. Ph.D. diss. Dep. Bot., Univ. of Minnesota, St. Paul.

Sundberg, M. D. (1981a). "Apical events prior to floral evocation in *Cyclamen persicum* 'F-1 Rosamunde' *(Primulaceae)"* *Bot. Gaz.* **141**(1): 27–35.

Sundberg, M. D. (1981b). "The development of leaves and axillary flowers along the primary shoot axis of *Cyclamen persicum* 'F-1 Rosamunde' *(Primulaceae).*" *Bot Gaz.* **142**(2): 214–221.

Sundberg, M. D. (1982). "Leaf initiation in *Cyclamen persicum* (Primulaceae)." *Can. J. Bot.* **60**(11): 2231–2234.

Wellensiek, S. J. (1952). "The breeding of cyclamen." *Int. Hort. Congr.* **130:** 771–777.

Wellensiek, S. J. (1961). "The breeding of diploid cultivars of *Cyclamen persicum.*" *Euphytica.* **10**(3): 259–268.

Wellensiek, S. H., Doorenbos, J., van Bragt, J., and Legro, R. A. H. (1961). "Cyclamen, a description of cultivars." *Lab. Tuinbouwplantenteelt Landbouwhogesch. Wageningen.* **200.**

Widmer, R. E. (1983). "Cyclamen seed handling and germination." *Minn. State Florists' Bull.* **32**(5): 13–14.

Widmer, R. E., Stephens, L. C., and Angell, M. V. (1974). "Gibberellin accelerates flowering of *Cyclamen persicum* Mill." *HortScience.* **9**(5): 476–477.

Widmer, R. E., Platteter, R. J., and Gembis, J. (1976). "Minnesota fast crop cyclamen—1976). *Minn. State Florists' Bull.* April 1: 3–9.

Widmer, R. E., Stuart, M. C., and Lyons, R. E. (1983). "Seven month Cyclamen for Christmas." *Minn. State Florists' Bull.* **32**(2): 1–4.

16

Begonias

Meriam G. Karlsson and Royal D. Heins

I. INTRODUCTION

The genus *Begonia* L. has more than 1000 species that are native to a wide range of countries and climates (Heide and Rünger, 1985). Only a few of the many begonia species are commercially produced as flowering potted plants, foliage plants, or bedding plants.

Identification of *Begonia* L. species is difficult as many hybrids and cultivars have been given Latin names, causing confusion in the nomenclature. Hybrid group names and classifications such as rhizomatous, tuberous, or fibrous rooted begonias are often used instead of species names. The root-system based classification should be avoided because all begonias have a fibrous root system and rhizomes and tubers are in fact stem structures.

The *Begonia* hybrid groups most often used for commercial flowering potted plant production include *B.* × *cheimanta* 'Everett' (Christmas begonia or Lorraine begonia), *B.* × *hiemalis* 'Fotsch' (hiemalis begonia, elatior begonia or Rieger begonia) and *B.* × *tuberhybrida* Voss. (tuberous begonia). The wide range of flower types and colors available in commercial cultivars has greatly increased the importance of *B.* × *hiemalis* compared to *B.* × *cheimanta*. *B.* × *tuberhybrida* is propagated by tubers or by seed and sold as a bedding plant, hanging basket, or potted plant. *B. semperflorens* 'Link & Otto' (wax begonia) is propagated from seed and marketed as a bedding plant. Several begonia species have attractively colored and patterned leaves and are produced as foliage plants. Because of the commercial importance of *B.* × *hiemalis* and *B.* × *tuberhybrida*, this chapter will outline their environmental and cultural production requirements.

II. *Begonia* × *hiemalis*

A. Cultivar Development

The plant known as *B.* × *hiemalis* in the United States was given that name by K. A. Fotsch in 1933 (Heide and Rünger, 1985). The Western European countries of the International Union for the Protection of New Varieties of Plants (UPOV) agreed in 1979 that *B.* × *elatior* was the correct name of this plant. Despite the European decision, *B.* × *hiemalis* (Bailey and Bailey, 1976) continues to be the name used in the United States.

B. × *hiemalis* is a group of hybrid cultivars (Fig. 1) derived by crossing *B. socotrana* Hook. from southern Arabia and several tuberous species (*B.* × *tuberhybrida* complex) from Peru and Bolivia. These *B.* × *hiemalis* hybrids combine the short-day flowering characteristics of *B. socotrana* with the large and colorful flowers of *B.* × *tuberhybrida*.

The company Veitch & Sons of the United Kingdom produced the first *B.* × *Hiemalis* hybrid ('John Heal') in 1883. This company, together with another English

Fig. 1. Hiemalis begonia.

company, Clibrun & Sons, dominated the market until 1930. Dutch breeders introduced cultivars with double flowers at that time, and the Netherlands became the leading source of cultivars. Since 1955, Otto Rieger from Germany has released some of the most important cultivars for commercial production including 'Leuchtfeuer' (1956), 'Lachsorange' (1961), 'Goldlachs' (1964), 'Schwabenland' (1965), 'Aphrodite' (1967), 'Ballerina' (1974), and 'Nixe' (1975). The T-cultivars ('Tiara I,' 'Tiara II,' 'Tacora,' 'Toran,' and 'Turina') were developed by Dr. Doorenbos in the Netherlands. The miniature begonias ('Anita,' 'Valentina,' 'Melanie,' 'Lucy,' and 'Rosita') were also developed in the Netherlands. Mikkelsens Inc. of Ashtabula, Ohio has promoted the commercial production of hiemalis begonia in the United States. The breeding program at Mikkelsens, Inc. continues to release new cultivars. Among the cultivars developed by Mikkelsens, Inc. are 'Appleblossom,' 'Chantilly,' 'Cheers,' 'Enchantment,' 'Guinevere,' 'Lancelot,' and 'St. Helena.' During recent years, additional cultivars have been introduced by Daeehnfeldt of Denmark ('Barbara,' 'Connie,' 'Dorthe,' 'Grete,' and 'Ninon'). Ljones of Norway has introduced the selections of 'Aida,' 'Rosli,' 'Nelson,' and 'Charm.'

Current breeding is aimed at developing cultivars that can be produced during a larger proportion of the year with improved transportability, longer postharvest life, and increased disease resistance, especially toward powdery mildew.

B. Flower Induction Requirements

The short-day photoperiodic flowering response in *B.* × *hiemalis* is influenced by temperature. Heide and Rünger (1985) reported *B.* × *hiemalis* as an obligatory short-day plant at temperatures above 75°F, while at lower temperatures flower initiation occurred either under short or long photoperiods. Welander (1979) observed similar flower initiation responses to daylength, although the shift from quantitative to obligate short-day response occurred at 68°F instead of 75°F.

The critical daylength for flower initiation has been reported as 12.5 to 13 hours (Heide and Rünger, 1985), and 10 hours (Molnar, 1974). There appear to be large differences in photoperiodic response among cultivars, and some cultivars may even be day neutral. Short days (10 hours of light) for an extended period will induce dormancy and tuber formation in some cultivars (Hilding, 1982). Tuber formation will also occur at low temperatures (54° to 59°F) independent of photoperiod (Hilding, 1982).

Flower number is influenced by the duration of the short-day treatment. The number of flowers increased as the number of short days was extended from 7 to 14 days (Molnar, 1974). Two weeks of short days at 9 to 10 hours of light are recommended for flower initiation in commercial production during winter conditions at northern latitudes, and 3 weeks of short days are recommended during the summer (Heide and Rünger, 1985; Hilding, 1982; Lemper, 1980). The longer duration of short days during the summer is required to reduce plant height and leaf size. An alternative to 3 weeks of short days in the summer is to apply 2 weeks of short days followed by growth regulator applications (chlormequat) to control growth.

Long-day conditions (16 hours of light) are necessary to stimulate vegetative growth prior to flower initiation. The leaf unfolding rate under long days was one leaf every 2 weeks and was independent of season (Powell and Bunt, 1978). Under short days, the leaf unfolding rate decreased to one leaf every 4 weeks under high light and one leaf every 14 weeks under low light conditions. In addition to the slower leaf unfolding rate, the fully expanded leaves under short days were smaller than leaves formed under long days. Leaves initiated under short days and transferred during cell elongation to long days were also small. The daylength affected the number of initiated cells in each leaf but did not influence cell extension. The lower cell number decreased the potential for large leaf development under short days.

The desired final plant size determines timing for initiation of the short-day treatment. Plants having at least three shoots with either three well developed leaves or with 2 inch long shoots were traditionally considered suitable for short-day treatment in Europe (Lemper, 1980). Vegetative growth for 3 to 6 weeks under long days after transplanting is normally required to reach the desired plant size for short-day treatment. The period of vegetative growth before the start of short-day treatment is related to final pot and plant size.

Continued plant development after short-day treatment requires about 3 weeks of long days to reach marketing size. Extended short-day treatment results in slow leaf and shoot growth, and the plants will eventually go dormant. Many new cultivars develop terminal inflorescences and stop growing under prolonged short photoperiods.

C. Environmental Requirements

1. Light

High light conditions cause sun scald (marginal leaf desiccation and burning), vegetative hardening, and growth reduction. Temperature affects the level of high light tolerance in begonia. The plants can safely be exposed to short periods of 3000 foot-candles (fc) at temperatures below 64°F, to 2000 fc at 70°F, and to 1500 fc at 81°F (Mikkelsen, 1973).

Long days (16 hours of light) are required before and after the flower initiation period. Plants will perceive photoperiodic light when exposed to light levels of 15 to 20 fc or more. Light interruptions soon after initiation or toward the end of the dark period have best long-day effects.

Supplemental lighting is beneficial during production periods of low light conditions. Plants receiving 200 fc of supplemental lighting prior to short-day treatment developed faster, and the short-day treatment could be initiated earlier than plants grown under natural lighting (Lemper, 1980). Supplemental lighting from high pressure sodium lamps is recommended during the winter in Scandinavia (Hilding, 1982). Under Scandinavian conditions, plants are irradiated with 500 to 600 fc for 20 hours per day prior to flower initiation, for 10 to 12 hours per day during the short-day treatment, and for 16 to 24 hours per day after flower initiation. Necessary supplemental lighting levels decrease as natural irradiance levels increase. For example, recommended lighting levels in the Netherlands are 250 to 300 fc.

2. Temperature

The three developmental phases of begonia have different temperature requirements (Hilding, 1982). Recommended temperature during the vegetative phase from planting to the start of short days is 66°F. During short days, a higher temperature (70°F) is required for complete and fast flower initiation. The fastest development of flowers after initiation occurs at 75°F. Such high temperatures, however, cause undesirable internode and inflorescence elongation, small and poorly colored flowers, and large leaves. For commercial production, 64°F is recommended after completing short-day treatment to obtain good plant and flower quality. Further temperature reductions to the 59° to 63°F range may be desired to intensify flower color as plants approach marketing.

3. Water

The fibrous root system of hiemalis begonia can easily be damaged by uneven moisture content in the media or high soluble salt concentrations. Frequent watering may be necessary to keep the plants evenly moist. Wetting of the foliage should be avoided, and a tube watering system or subirrigation (capillary mat or ebb and flow irrigation) are suitable irrigation methods. Capillary mat or ebb and flow irrigation combined with occasional watering from the top have given good results (Hilding, 1982). Top watering is used to ensure a more uniform salt distribution in the pot.

4. Nutrition

Hiemalis begonia does not require high nutrient levels. Good production results have been achieved using 100 to 150 ppm nitrogen and 50 to 125 ppm potassium as constant liquid feed (Nelson et al., 1978). For an artificial medium, 200 ppm nitrogen, 190 ppm phosphorus (P_2O_5) and 200 ppm potassium (K_2O) have been recommended until the middle of the short-day treatment. Fertilizing is then discontinued for 1 week and resumed with 100 ppm nitrogen, 190 ppm P_2O_5, and 150 ppm K_2O every other watering until flowering (Mikkelsen, 1973). A phosphorus deficiency during early development causes stunted growth. Adequate phosphorus levels are critical during early development because the plants do not recover after a phosphorus deficiency (Nelson et al., 1977). Micronutrients should be added to plants grown in soilless media.

Nelson et al. (1977) developed the following summary of seven nutrient deficiency symptoms in B. × hiemalis 'Schwabenland Red,'

a. Chlorosis is a dominant symptom.
    b. Chlorosis interveinal
        c. Interveinal chlorosis on older leaves followed by light tan necrotic spots within chlorotic areas that expand until leaf dies . . . . *Magnesium*
        cc. Interveinal chlorosis on younger leaves . . . . *Iron*

    bb. Chlorosis not interveinal
        c. Lower leaves uniformly yellow, then purplish-yellow, and finally necrotic . . . . . *Nitrogen*
        cc. Margins of canopy leaves yellow, then murky green-brown, and finally necrotic; all symptoms spread towards the leaf center . . . . *Calcium*

aa. Chlorosis not a dominant symptom.
    b. Necrosis begins along the margin of lower leaves and progresses inward . . . . . *Potassium*
    bb. Plants stunted but normal green . . . . *Phosphorus*
    bbb. Rust color, striations and cracks develop on young leaf petioles and peduncles perpendicular to their axes; internodes shortened and lateral shoots prolific; young leaves brittle, crinkled around rust color spots that turn necrotic; chlorosis and necrosis spreading inward from the margin of young leaves . . . . *Boron*

5. Gases

Carbon dioxide is beneficial to B. × hiemalis. Plants grew larger, and the short-day treatment could be applied earlier with carbon dioxide enrichment during early development (Lemper, 1980). Increasing carbon dioxide concentration from 300 to 3000 ppm resulted in better branching, more leaves, more buds and flowers, shorter production time, and better quality (Lindemann, 1973). Concentrations above 2000 ppm are reported to cause plant damage (personal communication, Ruud Veenenbos, 1989). Recommended concentration for carbon dioxide enrichment is 600 to 900 ppm (Mortensen and Ulsaker, 1985).

High humidity should be avoided in begonia production. The potential for infection and development of powdery mildew (*Xanthomonas*) and foliar nematodes increases with high or fluctuating relative humidity. Cultural practices that minimize high and fluctuating relative humidity are good air circulation, no watering in the afternoon, and no wetting of the foliage.

Begonia plants are sensitive to air pollutants. Severe injuries will develop in susceptible cultivars of hiemalis begonia at low ozone concentrations (25 to 50 ppm ozone for 4 hours a week; Reinert and Nelson, 1979). Characteristic symptoms of ozone damage are red-brown to brown pigmented spots and a bronze appearance of the upper leaf surface. Flowers can also develop necrotic lesions on the petals.

D. Cultivation

1. Propagation

Hiemalis begonia is vegetatively propagated by leaf or terminal cuttings. Generally, cultivars with poor development from leaf cuttings are propagated by terminal cuttings. The production time can be shortened by direct planting of a terminal cutting in the final pot. The 'Aphrodite' series of cultivars is primarily propagated by terminal cuttings while the 'Schwabenland' cultivars are propagated by leaf cuttings. Rooted cuttings of many cultivars are available from commercial propagators. Most cultivars in Europe are now propagated from terminal cuttings (personal communication, Ruud Veenenbos, 1989).

a. Terminal Cutting Propagation. The stock plants should be vegetative. A terminal cutting harvested on a reproductive stock plant will not develop into a high quality potted plant. Long days (more than 16 hours of light) and high temperatures (72° to 75°F) are required to keep the stock plants vegetative. Cuttings are harvested often to avoid aging of shoots. Terminal cuttings with 1 and ½ expanded leaves are planted in small pots (1 and ½ inch) or in final pots filled with a slightly fertilized medium. During the rooting process, long days and high temperatures (72°F) are necessary to keep the cuttings vegetative. The cuttings are ready for transplanting or delivery after 4 to 5 weeks (Fig. 2).

b. Leaf Cutting Propagation. A leaf cutting has to form both roots and shoots to develop into a growing plant (Fig. 3). Shoot formation is promoted on cuttings taken from reproductive stock plants. Stock plants maintained at 4 weeks of short days (12 to 13 hours of light) will give cuttings with satisfactory shoot formation (Hentig, 1978). Days with the dark period longer than 14 hours will induce stock plant dormancy.

Stock plant growing temperatures affect the rate of cutting production and the ability of the harvested leaf cutting to form shoots. Both cutting production and leaf unfolding rate increase at higher temperatures while temperatures below 59°F

Fig. 2. Propagation of hiemalis begonia from a terminal shoot cutting.

Fig. 3. Propagation of hiemalis begonia from a leaf cutting.

encourage shoot formation of the cutting (Gidsleröd, 1974; Heide, 1964). The recommended temperatures for stock plant production vary with cultivars, and temperatures of 60° to 68°F have been suggested (Hentig, 1976; Hilding, 1982).

The smaller leaves at the top of the stock plant are suitable for leaf cuttings. Older leaves produce cuttings with poor shoot formation. The leaf cuttings are harvested by breaking the leaf cutting from the stem at the base of the petiole.

Proper temperature and adequate light conditions are important during rooting of leaf cuttings. The ideal rooting temperature is 64°F and supplemental lighting of 200 fc should be provided if the natural average light levels are below 200 fc (Hilding, 1974). Short-day treatment may improve shoot formation during the rooting of leaf cuttings. The short-day treatment should be no longer than 2 weeks to avoid the development of flowers. Some cultivars, however, may initiate flowers with less than 2 weeks of short days. Knowledge of cultivar response is important. The original leaf on the cutting should be maintained with the cutting as long as possible during the rooting process. The cuttings should develop three to five strong shoots with good roots over a period of 10 to 13 weeks and can be transplanted at that time.

c. Propagation by Tissue Culture. Tissue culture is utilized to propagate large quantities of plants rapidly. High mutation rates and less desirable plant habits are difficulties encountered using tissue culture compared to propagation by terminal and leaf cuttings.

Plants propagated by tissue culture can originate from several different plant parts. Techniques have been developed to propagate hiemalis begonia from vegetative buds (Reuther, 1980), petiole segments (Hilding and Welander, 1976), and flower stem segments (Appelgren, 1978). Propagation from vegetative buds is labor intensive and is primarily used to develop stock plants free of *Xanthomonas begoniae*.

Leaf petioles from stock plants grown at 64° to 68°F under long days are used for tissue culture propagation. The petioles are sterilized, cut in ⅛ inch long pieces, and placed in test tubes on agar containing macro- and micronutrients. After the development of shoots, the petioles are subdivided and placed on agar in a second set of test tubes to form roots. The test tubes are placed in an environment of 300 fc, 70°F, and a daylength of 18 hours. The petiole pieces with roots and several shoots are removed from the test tubes after 14 to 16 weeks and planted in a medium commonly used for begonia production. Time required to produce a flowering plant after this transplanting is similar to the production time of plants propagated by leaf cuttings.

Flower stems for tissue culture are harvested prior to branching from stock plants grown at 59°F (winter conditions) or 64° to 70°F (summer conditions). After sterilization, the flower stems are divided into ¹⁄₁₂ inch long pieces and placed on agar in test tubes. An environment of 70°F and 200 fc for 18 hours per day is required for explant development in the test tubes. The developing shoots can be transplanted into a peat medium after 10 weeks. One stock plant can give rise to a large number of plants because each flower stem may produce up to 250 plants.

2. Medium and Planting

Hiemalis begonia requires a fast draining, well-structured medium. Most media used for begonia production have a large proportion of peat. A mixture consisting of 70% peat and 30% vermiculite is an example of a suitable medium. The medium should be sterile, lightly fertilized, and have a pH of 5.5 to 6.0.

Plants received from the propagator should be planted so the top of the root ball is ¼ inch above the surface of the potting medium. This planting procedure will decrease the risk for stem rots.

3. Spacing

Plant spacing influences quality and profitability of begonia production. Among the factors determining suitable spacing are method of propagation, desired final plant size, production under natural or artificial daylengths, and season. Crowded plant spacing will result in undesirable stem elongation and poor quality. The plants can initially be grown pot to pot but should be spaced before the leaves touch, about 3 to 4 weeks after planting. A final spacing of two plants per square foot (4-inch pots) has been recommended for pinched or strong growing cultivars and four plants per square foot (4-inch pots) for unpinched or weak growing cultivars (Hilding, 1982). Suitable final spacing for plants in 5- and 6-inch pots is 9×9 inches and 11×11 inches, respectively.

4. Support

No support is normally required for plants grown in 4-inch pots. Larger plants in 6-inch pots will need staking.

5. Pinching

Pinching results in larger plants and more uniform shoot development. No pinching is required for small plant production from either leaf or terminal cuttings. A soft pinch almost immediately after potting will result in large plants with more uniform and concentrated flowering. Dominating shoots on plants propagated from leaf cuttings should be pinched no later than the first week of short days. Plants propagated from terminal cuttings are usually only pinched if the cuttings were inadvertently induced to flower or plants are grown in 5- or 6-inch pots.

Terminal cuttings can be harvested when plants are pinched 3 to 5 weeks after potting. The pinched shoots should have at least two remaining leaves to ensure continued plant development. This type of pinch will delay flowering 4 to 5 weeks.

6. Growth Regulators

Chlormequat (Cycocel) and ancymidol (A-Rest) are effective in controlling height of $B. \times hiemalis$. The most commonly used growth regulator for commercial production is Cycocel. Cycocel concentrations commonly used for spray applications are in the range of 500 to 1000 ppm (Hilding, 1975; Sandved, 1972). Under high light conditions and vigorous growth, a higher concentration (1500 ppm) may be required for adequate height control. Leaf chlorosis will occur after high concentration spray applications. If the chlorosis is not too severe, the plants normally outgrow this Cycocel-induced chlorosis by time of marketing.

Several Cycocel applications during development may be required for desired plant growth. Goldschmidt (1974) recommends one 500 ppm spray application at start of the short-day treatment and a second 1000 ppm application 3 to 5 weeks later. Plants propagated by terminal cuttings and pinched plants are treated by Cycocel when the developing shoots are about 2 inches in length (Hilding, 1975). Plants propagated from leaf cuttings have a tendency to elongate relatively late, and a Cycocel application may be necessary immediately after completion of the short-day treatment. Cycocel applications late in development inhibit flower elongation, and plants develop undesirably with flowers below the leaf canopy.

Height control in hiemalis begonia can also be achieved by Bayleton. Besides acting as a growth regulator, this fungicide provides good powdery mildew control. Bayleton can be applied late in plant development without causing damage to flowers or leaves.

Cytokinins have been used to improve shoot formation in leaf cuttings. Shoot formation after a cytokinin treatment is highly dependent on cultivar, and both positive and negative responses have been observed (Davies and Moser, 1980). Applications of too high cytokinin concentrations result in delayed shoot and root formation. In commercial production, cytokinin applications should be used with caution and only after considerable testing.

A growth regulator that combines chemical pinching and height control is dikegulac (Atrinal). Spray applications of Atrinal on hiemalis begonia reduce apical growth, increase branching, and give more compact growth (Agnew and Campbell, 1983). Plants are treated 2 weeks after potting or at pinching. The increased branching caused by the Atrinal treatment delays flowering, and only cultivars with poor shoot formation should be considered for treatment. Desired and undesired effects of new and unknown chemicals should be examined in small plant trials prior to implementation in the production schedule, and only growth regulators that have label clearance should be used.

E. Problems

1. Diseases

Bacterial leaf spot is caused by *Xanthomonas begoniae*. The symptoms are small, circular, greasy spots on the underside of the leaf. As the disease progresses, the spots become translucent in light.

The best control of bacterial leaf spot is preventive measures. Clean stock plants and cuttings, avoidance of high temperatures and relative humidity, dry foliage, and immediate removal of infected plants are methods used to prevent disease. Infected plants are impossible to cure.

Botrytis blight and stem rot is caused by *Botrytis cinerea*. Infected leaves and stems show brown spots and as the disease progresses, gray fungus growth becomes visible. Flowers may also be infected. High humidity and water on the foliage favor spreading and development of botrytis blight.

Powdery mildew is caused by *Oidium begoniae* and occurs frequently in hiemalis begonia production. Leaves and shoots are covered with a white fungus growth, and severe infections can result in death of flowers and marring of leaf surfaces. Uneven media moisture content, high relative humidity, and temperatures below 61°F favor the development of this disease. Good control can be received by evaporating sulfur from sulfur pots.

Root and stem rots are caused by *Rhizoctonia solani*, *Pythium debaryanum*, and *Thielaviopsis basicola*. The best control measures of these diseases are good sanitation and sterilization of media, benches, flats, and pots.

2. Insects

Poor root initiation and development may be caused by fungus gnats. Small, brownish leaves, short internodes, and dried flower buds are characteristics of cyclamen mite infested plants. Crinkled and light green leaves are sometimes caused by aphids. Thrips generate long, white streaks on leaves, and whitefly infested plants are sticky with an occasional black covering with sooty mold.

3. Physiological Disorders

Oedema can be a problem under conditions with low light and variable humidity. Small, water-soaked blisters develop on the leaf surface that subsequently turn brown and corky. Small and pale flowers develop under high temperatures and hard leaves with red edges under low temperatures and high light intensities.

4. Other

Foliar nematodes (*Aphelenchoides species*) are serious pests causing yellow-brown spots between leaf veins and defoliation. A film of water is required for nematode movement, and dry foliage lessens the opportunities for nematode infestation.

F. Harvesting, Handling, and Marketing

Hiemalis begonias in the marketing stage are sensitive to ethylene. Ethylene exposure for 24 to 72 hours in combination with 3 days darkness result in flower and bud drop and significantly reduces marketing value (Höyer, 1985). Plant quality is not affected after 3 days of dark transportation without ethylene exposure. Flowers are most sensitive, and significant flower drop occurs even after 24 hours of 0.1 ppm ethylene. A silver thiosulfate (STS) spray is effective in preventing flower drop during marketing (Moe and Fjeld, 1985). Recommended application rate is 0.5 to 1.0 mM STS.

G. Scheduling

Table I describes the scheduling parameters for various types of propagation: terminal cutting, leaf cutting, and forcing.

Table I

Scheduling for Propagation and Production of Hiemalis Begonia

Cultural phase	Growing time for cultural phase (weeks)	Cultural procedure	Temperature (°F)	Photoperiod
Propagation		Plant terminal cutting	72	Long days
	4–5	↓		
		Transplant	66	Long days
Propagation		Plant leaf cutting	64	Short day is possible, but long day preferred
	1–2	↓		
		Start long days	64	Long days
	10–12	↓		
		Transplant	66	Long days
Forcing		Transplant[a]	66	Long days
	3–6	↓		
		Start short day	70	Short days
	2–3	↓		
		Transfer to long day	64, finish at 59–63	Long days
	3–5	↓		
		Flower		

[a]Short days can be started immediately after transplanting for production of small hiemalis begonias in 4-inch pots.

III. *Begonia × tuberhybrida* COMPLEX

A. Cultivar Development

The *B. × tuberhybrida* cultivars are derived through hybridizations and selections of begonia species from the Andes of South America including *B. boliviensis*, *B. Clarkei*, *B. Davisii*, *B. Pearcei*, *B. rosiflora*, and *B. Veitchii* (Bailey and Bailey, 1976; Ewart, 1985). *B. × tuberhybrida* cultivars are placed in groups based on plant habit, flower shape, or flower color. Among these groups are the single group with large, single flowers; the Crispa group with large, single flowers and frilled tepal margins; the Camellia group with large, double flowers of various solid colors, the pendula or hanging-basket group with trailing growth habit and many single or double flowers; and the multiflora group with bushy, compact plants and many, relatively small, single or double flowers (Bailey and Bailey, 1976).

The 'Non-Stop' series of cultivars was introduced during the 1970s by Benary Seeds of Germany (Ewart, 1985). These 'Non-Stop' cultivars are propagated by seed, have uniform growth over a wide range of environmental conditions, and have semidouble flowers in many colors. During the 1980s, the 'Non-Stops' had become popular for production of bedding plants, potted plants, and hanging baskets.

The hybrid cultivar series 'Patio,' 'Pavilion,' 'Memory,' 'Clips,' 'Spirit,' and 'Musical' are commonly used in commercial production in addition to the 'Non-Stop' series (Ewart, 1985). Plants of the 'Patio,' 'Pavilion,' and 'Memory' series have large, camellia-type flowers. The 'Clips' series has medium small flowers. 'Spirit' is a series with similar characteristics to the 'Non-Stop' series. The 'Musical' series is used for hanging basket production, and plants have medium large, double flowers.

B. Flower Induction Requirements

Flower initiation in tuberous begonia is dependent on the formation of new leaves (Heide and Rünger, 1985). Long days (more than 12 hours of light) are required for leaf and flower initiation. Under photoperiods of less than 12 hours light, growth ceases and tuber formation starts. Although the photoperiod is the "triggering" process for tuber formation, low temperatures enhance the short day induced dormancy (Oloomi and Payne, 1982). Once tuber formation has started, tuberous begonia cannot be induced to resume active growth and flowering again. The tops of the plants die down, and the tubers stay dormant for several weeks before new sprouting can take place.

C. Environmental Requirements

1. Light

Maximum light levels should not exceed 2500 to 3500 fc. During the summer, shading of the greenhouse is required to avoid burning of the plants and to improve temperature control.

Long days (more than 12 hours of light) are required for flower initiation. Daylength extension or night interruptions under natural short days result in similar plant growth and number of flowers (Fonteno and Larson, 1982; Oloomi and Payne, 1982). Cyclic lighting with 6 minutes of light per 30 minutes of dark for 5 hours has been used successfully. Light levels of 10 fc are sufficient for light perception by the plant.

2. Temperature

Recommended temperatures after transplanting are 72°F days and 64°F nights (Ewart, 1985). Larger plants and more flowers will develop at 72°F day and 64°F night temperatures compared to 79°F day and 64°F night temperatures (Fonteno and Larson, 1982). To increase flower size and intensify flower color after the buds are formed, the plants can be finished at 57° to 61°F.

Low or high temperatures result in undesirable plant development. Temperatures below 63°F encourage tuber formation and limit plant growth and flowering (Fonteno and Larson, 1982). Temperatures above 79°F result in small plants and a reduced number of flowers.

Expected time to flower after transplanting (½ inch plug seedling) is 10 to 12 weeks for potted plants and 12 to 14 weeks for hanging baskets at 72°F day and 64°F night temperatures (Ewart, 1985).

3. Water

Tuberous begonias have a sensitive, fibrous root system. Uneven, medium moisture content or high salt levels can easily damage the roots. A good irrigation system is capillary mats where the pots and mats are thoroughly moistened at each watering.

4. Nutrition

Tuberous begonias require moderate levels of fertilizer. The nitrogen level should not exceed 150 ppm when plants are fertilized at every irrigation.

5. Gases

There is no information available on the effects of carbon dioxide enrichment in tuberous begonia production. It is likely, however, that growth increases at elevated carbon dioxide levels.

D. Cultivation

1. Propagation

Seed propagation is the main propagation method of tuberous begonia, although they can be propagated by tubers. Seeds are very small with approximately 1,000,000 seeds per ounce (35,000 seeds per gram). A peat-lite medium at pH 6.0 is suitable as a germination medium. The small seed size makes hand sowing difficult, and an automatic seeder is the preferable way of sowing. Seeds are sown directly on the surface of a thoroughly moistened medium without covering the seeds.

Tuberous begonia seedlings require higher humidity and a longer growth period than most other seedlings. Covering seed flats with clear plastic will ensure high humidity levels. The plastic should not be removed until the first true leaf has a width of at least ¼ inch unless the humidity of the greenhouse can be maintained at a high level.

The media temperature during germination should be maintained at 70° to 75°F (Ewart, 1985). Germination takes about 2 weeks at this temperature. A few days after germination is completed, the temperature can be lowered to 64°F.

Light is required for germination. Seedlings respond to photoperiod immediately after germination. The daylength must be longer than 12 hours at emergence to ensure proper development and flowering. Supplemental lighting from high pressure sodium or fluorescent lamps at 500 fc for 24 hours a day increases germination uniformity and accelerates early growth.

Suitable seedling size for transplanting is attained when the third true leaf is at least 1 inch wide. Depending on temperature and light conditions during propagation, 7 to 10 weeks are required to reach proper seedling size for transplanting.

Plug seedlings of tuberous begonia are available from commercial propagators and are usually 6 to 9 weeks old when purchased. Purchase of seedlings from a reliable source is a good alternative for growers without suitable facilities for germination and seedling development.

2. Medium

A medium with a high content of peat and perlite that has good aeration is suitable for tuberous begonias.

3. Pinching

Pinching is not required to produce quality plants of most tuberous begonias in 4 to 5 inch pots. High quality hanging baskets can also be produced without any pinching by planting four or five plug seedlings in a 10-inch basket. The first flower of the plants normally develops in the direction the leaves point. Allowing the leaves to point outward from the basket at planting will give more symmetrical potted plant or hanging basket. Large-flowered cultivars in larger pots may require pinching 2 to 3 weeks after transplanting to develop into quality plant products. At least four leaves should be left on the plants at pinching, and flowering is delayed about 2 weeks (Ewart, 1985).

4. Growth Regulators

Growth retardation is normally attained by a chlormequat (Cycocel) spray treatment at 500 ppm. Cycocel may cause leaf yellowing. Bayleton is also reported to give height control in addition to controlling powdery mildew.

E. Problems

1. Diseases

The major disease problem is powdery mildew caused by *Oidium begoniae*. High relative humidity and irregular water regimes favor the occurrence and development of this disease. Good air circulation, good humidity control, and keeping the plants evenly moist decrease the powdery mildew occurrence.

Botrytis blight (*Botrytis cinerea*) is common under cool and humid conditions. Good sanitation and avoidance of high humidity and water on the foliage will decrease the spread and development of botrytis blight.

Prevention is the best control of bacterial leaf spot (*Xanthomonas begoniae*). Spreading and development of bacterial leaf spot is reduced by avoiding high temperature and humidities and by removing infected plants immediately.

Root and stem rots (*Pythium debaryanum* and *Rhizoctonia solani*) may cause damping off during early seedling growth. The use of sterile materials and recommended fungicides during seed germination and development will reduce the occurrence of root and stem rots.

2. Insects

Cyclamen mites, spider mites, and thrips attack growing tips and flower buds. Their small size and ability to hide int he plant cause problems for early detection and control. They cause reddish-brown corklike growth and flowers that fail to open. Whiteflies, mealybugs, and aphids may also be problems in the production of tuberous begonia.

3. Physiological Disorders

Poor seedling development in the flat or plug tray may be caused by short-day conditions, low humidity, or nutrient levels. Tuberous begonia seedlings require high humidity for a longer period than most other seedlings. If the humidity is too

Table II

Scheduling for Propagation and Production of Tuberous Begonias

Cultural phase	Growing time for cultural phase (weeks)	Cultural procedure	Temperature (°F)	Photoperiod
Propagation		Sow seed (½ inch plug)	70	24 hour light
	2	↓		
		Germiniation	70	Long days
	5–8	↓		
		Transplant		
Forcing in 4-inch pots	8–10	Transplant ↓	72 day, 64 night Finish at 57–61 night temperature	Long days
		Flowering		
Forcing in hanging baskets		Transplant to 1 ¼-inch plug ↓	72 day, 64 night	Long days
	4			
		Transplant to 10-inch basket ↓	72 day, 64 night. Finish at 57–61 night temperature	Long days
	8–10	Flowering		

low, the seedlings will develop slowly or die shortly after germination. The seedlings are kept in the flat for a long time period, and depletion of nutrient levels may result in slow growth. Soluble sale levels need to be closely monitored and two to three fertilizer applications at the rate of 100 ppm nitrogen may be required to avoid slowed growth.

Tuberous begonia will produce tubers under short days and low temperatures. Once tuber formation has started, the plants will not revert to shoot growth and flower formation. In contrast to hiemalis begonia, long days (more than 12 hours of light) and temperatures above 63°F are required continuously from germination to flowering of tuberous begonia to prevent tuber formation.

F. Harvesting, Handling, and Marketing

Tuberous begonias are shipped when the first flowers appear above the foliage. A STS treatment at the concentration of 0.3 mM shortly before shipment will prevent flower and bud drop during marketing (Aimone, 1983; Ball, 1985).

G. Scheduling

Table II describes the scheduling parameters for propagating tuberous begonias.

REFERENCES

Agnew, N. H., and Campbell, R. W. (1983). "Growth of *Begonia* × *hiemalis* as influenced by hand-pinching, dikegulac, and chlormequat." *HortScience.* **18**: 201–202.

Aimone, T. (1983). Nonstop production. *GrowerTalks.* **46**(11): 38, 40, 42.

Applegren, M. (1978). "Formering av hiemalisbegonis *in vitro.*" *Gartneryrket.* **32**: 911–912.

Bailey, L. H., and Bailey, E. Z. (1976). *Hortus Third, a Concise Dictionary of Plants Cultivated in the United States and Canada.* New York: Macmillian.

Ball, V. (1985). "Tuberous begonias." In *Ball Red Book,* (edited by V. Ball) 14th ed. Reston, VA: Reston, pp. 359–361.

Davies F. T., Jr., and Moser, B. C. (1980). "Stimulation of bud and shoot development of Rieger begonia leaf cuttings with cytokinins." *J. Am. Soc. Hort. Sci.* **105**: 27–30.

Ewart, L. C. (1985). "Tuberous rooted begonia." In *Bedding plants III, A Manual on the Culture of Bedding Plants as a Greenhouse Crop,* (edited by J. W. Mastalerz and E. J. Holcomb) 3rd ed. Pennsylvania Flower Growers, University Park, PA: pp. 420–422.

Fonteno, W. C., and, Larson, R. A. (1982). "Photoperiod and temperature effects on NonStop tuberous begonias." *HortScience.* **17**: 899-901.

Gisleröd, H. (1974). "Forsøk med bladstiklinger av hiemalisbegonia." *Gartneryrket.* **64**: 499–502.

Goldschmidt, H. (1974). *Markwichtige Blütenbegonia. Reihe: Gartnerische Praxis 41.* Berlin: Paul Parey.

Heide, O. M. (1964). "Effects of light and temperature on the regeneration ability of begonia leaf cuttings." *Physiol. Plant.* **17**: 789–804.

Heide, O.M., and Rünger, W. (1985). "Begonia." In *Handbook of Flowering,* volume II, edited by A. H. Halevy, Boca Raton, FL: CRC Press, pp. 4–23.

von Hentig, W. -U. (1976). "Zur vermehrung von elatiorbegonien 'Riegers Schwabenland' und 'Riegers Aphrodite.'" *Gartenwelt.* **5:** 95–100.

Von Hentig, W. -U. (1978). "Zur vermehrung von elatiorbegonien. Weitere ergebnisse mit riegersorten." *Gb+Gw.* **78**(9): 193–195.

Hilding, A. (1974). "Inverkan av dagslängd och temperatur vid förökning av höstbegonia (*Begonia* × *hiemalis*) med bladsticklingar." *Lantbr. högsk. meddn. A.* **209:** Uppsala.

Hilding, A. (1975). "Inverkan av temperatur, toppning och retarderande medel på tillväxt och utveckling av höstbegonia, *Begonia* × *hiemalis.*" *Lantbr. högsk. meddn. A.* **252:** Uppsala.

Hilding, A. (1982). *Produktion av Begonia* × *elatior. Trädgård 220.* Swedish University of Agricultural Sciences Research Information Centre, Alnarp.

Hilding, A., and Welander, T. (1976). "Effects of some factors on propagation of *Begonia* × *hiemalis in vitro.*" *Swedish J. Agric. Res.* **6:** 191–199.

Høyer, L. (1985). "Bud and flower drop in *Begonia elatior* 'Sirene' caused by ethylene and darkness." *Acta Hort.* **167:** 387–394.

Lemper, J. (1980). "*Begonia elatior.* Wirkung von licht, temperatur und kohlendioxid auf wachstum und blütenbildung." *Gb+Gw.* **80**(11): 252–255.

Lindemann, A. (1973). "Wachstumsbeeinflussung durch anreichung der gewachshausluft mit kohlendioxyd (CO_2) unterschiedlicher dosierung." *Zierpflanzenbau.* **19:** 778–779.

Mikkelsen, J. C. (1973). "Production requirements for quality Rieger begonia." *GrowerTalks.* **37**(6): 3–9.

Moe, R., and, Fjeld, T. (1985). "Keeping quality of pot plants as influenced by ethylene." *Gartner Tidende.* **48:** 1580–1583.

Molnar, J. M. (1974). "Photoperiodic responses of *Begonia* × *hiemalis* cv. 'Rieger.'" *Can. J. Plant Sci.* **54:** 277–280.

Mortensen, L. M., and Ulsaker, R. (1985). "Effects of CO_2 concentrations and light levels on growth, flowering and photosynthesis of *Begonia* × *hiemalis* Fotsch." *Scientia Hortic.* **27:** 133–141.

Nelson, P. V., Krauskopf, D. M., and Mingis, N. C. (1977). "Visual symptoms of nutrient deficiencies in Rieger Elatior begonia." *J. Am. Soc. Hort. Sci.* **101:** 65–68.

Nelson, P. V., Krauskopf, D. M., and Mingis, N. C. (1978). "Nitrogen and potassium requirements of Rieger begonia (*Begonia* × *hiemalis* Fotsch)." *J. Am. Soc. Hort. Sci.* **103:** 603–605.

Oloomi, H., and Payne, R. N. (1982). "Effects of photoperiod and pinching on development of *Begonia* × *tuberhybrida. HortScience.* **17:** 337–338.

Powell, M. C., and Bunt, A. C. (1978). "Leaf production and growth in *Begonia* × *hiemalis* under long and short days." *Scientia Hortic.* **8:** 289–296.

Reinert, R. A., and Nelson, P. V. (1979). "Sensitivity and growth of twelve elatior begonia cultivars to ozone." *HortScience.* **14:** 747–748.

Reuther, G. (1980). "Elatiorbegonia I. Weitere untersuchungen zur gewinnung von befallsfreien elitepflanzen durch gewebekultur." *Gb+Gw.* **80**(39): 876, 880–881.

Sandved, G. (1972). "Effekt av CCC til hiemalisbegonia." *Gartneryrket.* **62:** 508.

Welander, T. (1979). "Begonia elatior hybrid. *Begonia* × *hiemalis* Fotsch. Begonia Lorraine hybrid. *Begonia* × *cheimantha* Everett." Avd. för Prydnadsväxtodling, Alnarp, Sweden.

17
Kalanchoe

Alton J. Pertuit, Jr.

I. HISTORY

Kalanchoes make attractive, long-lasting potted plants for the home, requiring little care. In the United States, under naturally occurring daylengths (really night lengths), kalanchoes flower in greenhouses in January (i.e., they are short-day plants or long-night plants). The genus *Kalanchoe* (Kal-an-*ko*-e) is derived from the native name for a Chinese species. The species name (*blossfeldiana*) is for Robert Blossfeld, a German hybridizer who in 1932 introduced it in Potsdam from its native Madagascar. It presently is the number one potted plant in Denmark, and, during the last 20 years, it has rapidly increased in importance in commercial pot plant production in the United States (Bailey, 1928; Bailey and Bailey, 1976; Hillman, 1962; Johnson and Smith, 1986, Pemberton, 1980; Schwabe, 1985).

The increase in demand for kalanchoes in the United States is mainly due to the introduction of new hybrids (Love, 1976) with more handsome foliage and a broader flower color range, the result of breeding programs (e.g., 'Mace,' 'Telstar,' and other asexually propagated cultivars) and the importation of Swiss seed cultivars (e.g., Melody and Exotica). Previously there had been very few choices: 'Tom Thumb' and its improved types (e.g., 'Vulcan' and 'Yellow Darling'). Although the Swiss seed types were excellent in their color range, they were not uniform in growth, form, or color. Most were too tall and too heat sensitive (i.e., even though they are short-day plants, they will not initiate flowers under inductive long nights if the temperatures are too hot), precluding their year-round production in some areas. The asexually propagated cultivars Linda, Jean, and Exotic Gold were among the first cultivars that resulted from crosses of these types (Irving, 1972). Today, Mikkelsen, Inc. of Astabula, OH is a primary source of kalanchoes (all asexually propagated) in the United States (Fig. 1). These include their own cultivars (e.g., Bonanza, Mikkel, Attraction, Bingo) and the Wyss (formerly Wyss/Grob) imports, among others (Anonymous, 1990a,b).

II. BOTANICAL INFORMATION

Kalanchoe Blossfeldiana Poelln. is a glaborous, succulent (its leaves and stems have a low surface to volume ratio) herb to subshrub, and it may develop woody tissue with age. It is native to Madagascar, a large island (approximately $1/16$ the size of the United States) in the Indian Ocean, off the southeast coast of Africa. It is a dicot and a member of the Crassulaceae (Orpine) family, which includes about 30 genera and 1500 species. *K. Blossfeldiana* has an upright growth habit with opposite leaves arranged in a ranked manner (i.e., when viewed from above, there are four ranks of leaves at right angles to one another like a cross). Leaves are obtuse to acute, sinuate to crenate (upper half), 1 to 3 inches long, and taper to petioles

Fig. 1. *Kalanchoe* 'Attraction' is a typical example of one of the many high quality, vegetatively propagated cultivars available today. It has a 9-week (summer) or 12-week (winter) response. (Courtesy of Mikkelsen, Inc.; Ashtabula, OH.)

of about 1 inch long (Bailey, 1928; Bailey and Bailey, 1976; Hillman, 1962; Longman, 1984).

The inflorescence is a cyme (a determinate, branched inflorescence usually broad or flat topped—a "branched umbel") of the dichasial type [a falsely dichotomous cyme—the axis bears a terminal flower between two "equal" branches, each branch repeating the process one or more (in a compound dichasial cyme) times or each bearing only a terminal flower (a simple dichasial cyme)]. When the cyme is large, it becomes a cincinnus (a coiled cyme with its branches developing left and right rather than in one direction). Always, the lowermost floret that terminates the primary axis matures first. Floret colors (spectrum) range from red through yellow (Anonymous, 1990a,b; Bailey and Bailey, 1976; Irwin, 1972).

III. PROPAGATION

Kalanchoes may be propagated from seed or asexually via terminal or leaf petiole cuttings. Today, they are propagated mostly from terminal cuttings of named cultivars (Love, 1980). This ensures a uniform product and more rapid production (Fig. 2).

A. Seed

Kalanchoe seeds are very small (1 to 2½ million per 1 ounce). It is recommended that they be pressed (lightly) into the germinating medium (Post, 1959). The medium in the germinating flat should be watered before sowing the seed in rows, as opposed to broadcasting. Row sowing is recommended because if disease problems occur (damping off), the problem can generally be confined to one row. Any well-drained, well-aerated, highly organic growing medium [e.g., 1:1 (v/v) peat:sandy compost (Post, 1959) or number four grade vermiculite (Love, 1980)] is suitable for germination. It should be sterilized before use, preferably by steam [i.e., medium heated to 180°F at its coldest location for 30 minutes (Post, 1959)], and its pH should be in the 6.0 to 6.5 range (Hesse, 1985). Seeds germinate in about 10 days in light at 70°F (Hesse, 1985). After seeds are sown, the

Fig. 2. *Kalanchoe* production in Denmark, where it is the number one flowering pot plant. (Courtesy of Roy A. Larson.)

flat should be placed in a clear, plastic bag in indirect yet adequate sunlight, and the bag removed as soon as the seedlings begin growth (Love, 1980).

Seeds sown in March will make excellent 4-inch plants by December. It is recommended that seeds not be sown before March because of their photo-periodic response (Post, 1959); however, a greater benefit from this recommendation may be because of the increased photosynthesis that would occur with the later sowing because of increased light exposures (daylengths and light intensities). Seedlings with only a few pairs of leaves and grown under short days cannot be photoperiodically induced to initiate flowers. It has been reported that seedlings need about eight pairs of leaves (the apical meristem contains three pairs) before they become photoperiodically sensitive. In general, as they age fewer short days are required for floral initiation (Schwabe, 1985). Seedlings grow slowly, even under optimum conditions (long days LD); therefore, sowing earlier than March when the days are short is not advisable. To ensure that they remain vegetative, the seedlings should receive a night light break [0.110 klx (10 foot-candles—fc) at plant level (Hartman et al., 1981)] of 4 hours in the middle of the night (Schwabe, 1985). Fluorescent lights inhibit internode elongation, but incandescent lights are more efficient in preventing floral initiation; however, if 4 hours of night lighting are imposed, either type will be satisfactory.

Seedlings should be ready for transplanting in about 7 weeks (Love, 1980), at which time the night temperature should be lowered to 60°F (Hesse, 1985) and the plants grown in full sunlight to prevent stretching (Anonymous, 1972). Plant spacing within the flats should be 2-inch centers (Post, 1959). Flat planting makes watering easier, and when the plants begin to touch one another, each should be transplanted into a 2 ½-inch pot and later into its 6- or 6 ½-inch finishing pot. When transplanting into flats and later into pots, a growing medium (adjusted to pH 6.0 with dolomitic limestone) of peat moss, pine bark (soil conditioning grade), and builder's sand [1:1:1 (v/v/v)] is recommended (Pertuit, 1973a), but any highly organic, well-drained growing medium will do. About 2 weeks after transplanting, the fertilization program [400 ppm nitrogen (N), 200 ppm phosphorus pentoxide (P_2O_5), and 300 ppm potassium oxide (K_2O) (Anonymous, 1990b), should begin, repeated at each watering, and continued until about a month before flowering. Irwin (1972) suggests that fertilization cease when the first floret opens. For a nitrogen source, it is recommended that calcium nitrate instead of ammonium nitrate be used, particularly in winter (Anonymous, 1990b). High phosphorus can induce zinc deficiency in kalanchoes (Love, 1980), resulting in chlorosis and fasciation (i.e., leaves, stems, or both fuse) (Nelson, 1981). Usually kalanchoes grown from seed will flower within 3 ½ months from the start of short days (Anonymous, (1972).

B. Vegetative

Most Kalanchoes produced in the United States are from asexually propagated terminal cuttings of named cultivars. A grower may obtain these from his or her stock plants or from specialists. A license from the licensor is required to propagate patented cultivars, and a royalty fee must be paid for each cutting propagated. Specialists offer unrooted cuttings and 72 cell pack "starter plantlets."

Stock plants must, of course, be kept under noninductive (i.e., short night) conditions. This may be accomplished by applying a night break as described for seedling kalanchoes (see Section III A). If a light break is properly imposed, temperature will not affect its inhibition of floral initiation, and relatively low intensities work satisfactorily (Schwabe, 1985).

Stock plants should be kept in vigorous condition (i.e., pest and disease free). They should be grown at near optimum temperatures and light intensities: 60°F nights with day temperatures about 72°F and about 49.5 klx [4500 fc (Anonymous, 1990b)]. The goal is to grow these stock plants under conditions conducive to the maximum growth in the minimum time.

Stock plants should never be allowed to experience water stress, and whenever watering (or fertilizing) enough liquid must be applied so that some runs out the bottom of the pot. Watering should be done early enough in the day so that any water that gets on the foliage has time to evaporate before dark, when the temperature falls. This reduces the chance of diseases occurring. Some kalanchoe cultivars have cupped or concave-like foliage that retains water, not allowing it to run off. Such cultivars should probably be grown using some type of tubing system or on water mats.

Healthy stock plants can easily produce a new crop of shoots for terminal cuttings on a monthly basis. They should be fertilized regularly with the fertilization program previously described for kalanchoes grown from seed (Anonymous, 1990b). Also, insect and disease prevention programs should be maintained (see Section VIII).

Because cultivars vary in internode length (i.e., some naturally produce longer internodes than others), it is difficult to say exactly how long a terminal cutting should be. Terminal cuttings with only a couple of pairs of leaves (excluding those pairs near the apical meristem with no visible internode between them) are ideal. One can strip off a pair of leaves below these to produce a stem for sticking in the rooting medium. It takes a few weeks for terminal shoot cuttings to root and, once rooted, a few weeks for them to start growing after transplanting. Cuttings root extremely easily. There is really no need for a stem dip in an auxin before sticking, but, if an auxin is used, it should be at a low concentration. Excellent results have been reported with a propagation medium (v/v) of peat moss and perlite (Love, 1976); however, the same growing medium previously described for seed-grown kalanchoes (Pertuit, 1973a) also produces excellent rooted cuttings.

Intermittent mist is ideal for rooting kalanchoes. It permits the cutting to be maintained at high light intensities without wilting, allowing more photosynthesis to occur than would be possible at lower light intensities; hence, the cuttings produce more food at a faster rate and, therefore, root faster. Still, light shading may be necessary during the seasons of greatest light intensities (i.e., late spring, summer, and early fall). Terminal cuttings should be spaced in the flat (or bed) in rows, with cuttings just far enough apart so their leaves barely touch. Until the cuttings have calloused, the mist cycle should be set for 6 seconds per 3 to 5 minutes during summer (i.e., when temperatures are hot and the light intensities high) and about half of this (6 seconds per 6 to 10 minutes) in winter (Love, 1980). Once cuttings have calloused (about 7 to 10 days), the misting frequency should be reduced. The rooting medium should ideally be 70°F, which may be accomplished via heating cables (thermostatically controlled) during cooler weather.

During rooting, it is recommended that the cuttings receive a light break as has been recommended for seedlings (see Section III, A) and stock plants. It is not known when terminal cuttings become photoperiodically sensitive, as far as floral induction is concerned. They may not be sensitive until they develop roots, as is the case with chrysanthemum cuttings. A night break, therefore, may be unnecessary at this early stage. Until this has been determined, a night break during propagation must be recommended.

Unrooted cuttings may be stuck directly in their finishing pots for rooting. This procedure reduces labor but requires a lot of space for a long time, especially if they are marketed in 6-inch pots. The procedure has been done without intermittent mist, simply by misting the plants with a hose as necessary during the day until they no longer require this attention. The finished produce is almost the same as rooting the cuttings and then transplanting each into its finishing pot. When transplanting, a rooting cutting can be adjusted deeper in the finishing pot, making the final product not only more attractive but also shorter than when the cutting is rooted directly in its finishing pot.

IV. PLANT CULTURE

A. Growing Media

With kalanchoes, as with most plants, it is very important that the growing medium be well drained and well aerated (Laurie et al., 1958). Several growers recommend a medium containing half peat moss (Anonymous, 1990; Post, 1959), while others have used less peat than this (Love, 1980; Pertuit, 1973a). As previously stated, a highly organic, well-drained medium is recommended.

The pH should be in the 6.0 to 7.0 range (Love, 1980). Because zinc deficiency is aggravated by a high phosphorus content, a pH of 6.0 to 6.5 would be ideal. In this range, zinc is highly available, and phosphorus is also available but not excessively so (Nelson, 1981). The pH should be adjusted with dolomitic limestone (Love, 1980; Pertuit, 1973a,b), which contains calcium as well as magnesium.

Soil is becoming more difficult to obtain. Today, many commercial soilless mixes are available for ready use. Some of these commercially available mixes are quite suitable, and there is no need for "soil sterilization," as described in Section III, A. One should only purchase proven media from a reputable, experienced company that has tested the product and is noted for its quality control.

B. Potting

Most kalanchoes in the United States are marketed in 4- to 6 ½-inch pots (Hesse, 1985). Plastic pots work well with the loose mixes in which kalanchoes are grown. They are often cleaner than clay pots, and plants in them require less frequent watering.

Kalanchoes vary in growth rate and foliage size. It is recommended that those cultivars with relatively small foliage be finished in smaller containers, and those with larger foliage be finished in larger containers (Love, 1980). Mikkelsen (Anonymous, 1990a) has developed schedules for producing 6-inch pots of kalanchoes with three cuttings per pot. These are not pinched and will produce a finished product in 2 weeks less time than is required for a single, pinched cutting grown in a 6-inch pot. Because the grower must pay a royalty fee for every cutting, the cost of growing three cuttings in a single pot (instead of one) increases production costs. The question is whether 2 more weeks of production time is more expensive than the cost of two additional cuttings. This, of course, could vary with the season. Quality also is a consideration. The retail price might be greater when three cuttings (i.e., compared to one) are in the pot. If so, this could overcome the handicap of increased cost of cuttings when three cuttings are used. The market will determine which method is more profitable.

When transplanting, leaves should not be covered with the growing medium. Often at transplanting, however, the bottom pair of leaves is removed and the cutting planted deeper than if these leaves remained. These leaves are generally smaller, and the additional stem allows for more rooting along the stem. The end result can be a more attractive plant.

C. Spacing

The key to spacing is experience. There are no general recommendations that can be made to cover the many variables that dictate the spacing of kalanchoes. Among the factors that must be considered are the cultivar; whether plants are pinched or not (and when); whether one, two, or three cuttings will be planted per pot; the season; the production area location; the projected size of the finished product; how long plants remain vegetative; and, whether or not growth retardants will be applied. Specialists (e.g., Mikkelsen) have schedules (Anonymous, 1990a) that are helpful, yet even they do not adjust to the variation in production time due to location. In the summer, days are much longer in the northern United States and in winter much shorter than in the South. Kalanchoes produced in these two areas in these seasons cannot be expected to require the same amount of production time. Some obvious generalizations can be made: the larger the finishing pot size, the more production time and space will be required.

D. Pinching

Soft pinching or the removal of only a couple of pairs of leaves at the shoot apex is often part of a production schedule. One should not assume that pinching will always increase the number of flowering shoots or that it will always reduce plant height. In some schedules with some cultivars, pinching neither decreases plant height (Pertuit, 1973a) nor increases the number of flowering shoots (Lyons, 1987; Pertuit, 1973a); in others it increases the number of flowering shoots but has no effect on plant height; and in other cases, it may reduce plant height and have no effect on the number of flowering shoots (Pertuit, 1973a). Also, it does not always reduce cyme height (Pertuit, 1973a). Pinching, therefore, should not be universally recommended for kalanchoe production because it is not always necessary for axillary growth and development. Some kalanchoe cultivars grow very much like the China aster (Post, 1959) where lateral development occurs without pinching.

Some growers employ pinching as a means of obtaining cuttings for propagation and future production, which is one way of eliminating the need for maintaining stock plants (Love, 1980). When this is done, a license must still be obtained and royalty fees paid (see Section III, B).

E. Light Intensity

Light intensity recommendations for kalanchoe growth and development have been mentioned. High light intensities play an important part in floral initiation in *Kalanchoe Blossfeldiana*, the effect not basically photosynthetic nor clearly understood. For example, kalanchoes, (short-day plants) will *not* initiate flowers if grown at 68° or 77°F in continuous darkness. If, however, in darkness, they are subjected to a one second flash of high intensity incandescent light (in the presence of carbon dioxide) every 24 hours, they will initiate flowers. Obviously, not much photosynthesis occurs during one second; therefore, it has been suggested that the high light intensity requirement of kalanchoe for floral initiation is related to additional factors other than photosynthesis (Schwabe, 1972).

Kalanchoes grown under identical conditions, except for light intensity, will initiate more flowers at the higher intensity, and it has been suggested that this is caused by a higher carbohydrate level produced at the higher intensity. Also, if the length of the photoperiod is shortened (light intensity remaining the same) fewer flowers will be initiated. These effects are not, however, strictly the effect of light intensity on carbohydrate level, although at first glance they may seem to be. Kalanchoes exposed to lower light intensities just before the dark period will initiate fewer flowers than those exposed to higher intensities. This would be expected, considering the light intensity–carbohydrate level effect just mentioned. If, however, the length of the dark period is increased (i.e., reducing the length of the photosynthesis even further), the number of flowers can be increased to its previous number. This cannot be explained on the basis of the light intensity–carbohydrate level relationship alone. Also, it is doubtful that this is the result of a change in the balance of the two phytochrome forms. It has been suggested that inhibitory auxins are involved and that their levels are increased under low light intensities (Schwabe and Papafotiou, 1987; Schwabe et al., 1985).

With regard to light intensities, it can be stated that in kalanchoes high light intensities promote floral initiation. Greater intensities are conducive to more floral initiation. In some cases, reduced flowering due to low light intensity exposures can be restored by increasing the length of the dark period (Schwabe, 1954). The reasons for all of this are not exactly clear.

F. Watering

Watering procedures, practices, and requirements have previously been described (see Section III, B). Although kalanchoes are succulents, they should not be subjected to water stress during production. They do, however, withstand periods of water stress extremely well, and this ability makes them one of the best house plants.

G. Temperature

Optimum temperatures for germination, rooting, and production of kalanchoes have previously been discussed (see Section III, A,B).

During inductive conditions, high night temperatures must be avoided to prevent "heat delay" (i.e., failure to initiate flowers when grown under inductive nights because of high night temperatures). Research has shown that kalanchoes grown at 80°F days under 12 hour day and 12 hour night cycles (i.e., clearly inductive photoperiodic conditions) responded several ways, depending on the temperature during the dark exposures: (1) if the first 3 hours of the dark exposure were 86°F, the plants would not flower; (2) if the last 3 hours of the dark exposure were 86°F, the plants flowered; and (3) if the entire dark exposure was 86°F, the plants did not flower (Hillman, 1962). This clearly illustrates how heat delay can be a problem in kalanchoe production. Covering with black cloth, usually employed for long night implementation, can help promote heat delay because of the heat buildup that occurs under it. Even though the night temperatures may subsequently cool off later in the dark period, the hottest part of the dark period under black cloth will be in the beginning (i.e., during late afternoon), and this is when high temperatures promote heat delay.

For earliest floral initiation, it is important to maintain the dark period at its optimum recommended temperature (60°F, see Section III, B). When grown at 70°F days, cool (50°F) as well as warm nights (70°F) can prolong flower initiation, when compared to 60°F nights, the extent of this effect being cultivar-dependent (Pertuit, 1977).

Under noninductive photoperiodic conditions (e.g., when vegetative growth is desired), temperature does not affect floral inhibition of kalanchoes (see Section III, B), and the presence of carbon dioxide is not necessary for it to be effective (Spear, 1959).

H. Fertilization

Some fertilization recommendations have already been discussed (see Section III, A). Alternatives to constant feeding with a completely soluble fertilizer include weekly applications (540 to 720 ppm nitrogen–phosphorus–potassium) and the use of a slow-release fertilizer (e.g., Osmocote) as a supplemental source (Love, 1976). Lower nitrogen levels have been reported to stimulate crassulacean acid metabolism activity in kalanchoes (Ota, 1988) (see Section V).

In addition to its unusual zinc requirement (see Section III, A), iron, manganese, copper, boron, and molybdenum are also important micronutrients required by kalanchoes. These may be mixed into the growing medium as fritted trace elements, or they may be applied as a soluble formulation (Love, 1980).

I. Photoperiod

1. Floral Induction and Initiation

Kalanchoes are, of course, considered a classic example of the short-day plant. It is, however, really the night (dark period) that controls flower initiation; therefore, kalanchoes are short-day plants because they initiate flowers when exposed to nights (dark periods) longer than some critical length.

Floral initiation in kalanchoe is not always as definitive as with most other plants. Kalanchoes, if not clearly induced to remain vegetative and, at the same time, not clearly induced to initiate flowers, produce a shoot intermediate between vegetative and reproductive growth. The branching habit is a reproductive one with a sparse number of florets intermingled with miniature leaves. When even less induced to flowering than this, it will initiate no florets but produce a reproductive type branching habit with leaves of reduced size (Hillman, 1962).

The critical night length for *Kalanchoe Blossfeldiana* is about 12 hours. On a 24 hour cycle, the same number of flowers will be produced whether the plant is grown under a 20 minute day or an 11 hour day. Only 2 long nights are required for floral initiation (minimum induction); however, as the number of long nights increase, the number of flowers initiated increases exponentially until 14 long nights are imposed, at which time the rate of increase in the number of flowers that are initiated declines (Schwabe, 1985). For commercial production, a minimum of 40 long nights is recommended (Anonymous, 1990b; Love, 1976); however, Seeley (1952) produced excellent plants with only 3 weeks of short days, followed by 7 weeks of long days. This regime also restricted the unsightly elongation of the terminal cyme and produced a more compact plant.

In introductory horticultural, botany, and even plant physiology courses, we learn that the night, not the day, is really what controls flower induction of short-day plants, as has been previously stated. We learn that one can prove this by changing the cycle from a 24 hour one. In the case of a plant with a 12 hour critical length, the photoperiod can be increased well beyond the 12 hour "critical length," and the plant will still flower as long as the night exceeds 12 hours, thus proving that it is the night that controls floral initiation. Kalanchoe is an exception to this. It has an upper limit of daylength at which floral initiation will occur. Floral initiation will not occur if the light period exceeds 17 hours, even though the dark period far exceeds the critical length (Schwabe, 1985)!

2. Floral Development

Once flowers are initiated, they will reach anthesis regardless of the length of the night (Love, 1980). They may not, however, develop at the same rate under long nights as under short nights.

Kalanchoes are manipulated to flower for a projected date in much the same way that we do for mums, also short-day plants. Kalanchoes are not, however, as easily programmed as chrysanthemums. For example, a 10-week response greenhouse mum [i.e., one that flowers 10 weeks after short days (long nights) are first imposed], will flower in about 10 weeks, regardless of the season. This 10-week period includes floral induction, initiation, and development. At some locations, growers have to switch chrysanthemum response groups with the season to maintain high quality. A 10-week response kalanchoe cultivar in summer may become a 13-week response cultivar in winter (e.g., Fortyniner, Garner, Sensation); and, even then, there may be some variation in flowering (Anonymous, (1990b). This is probably caused by light intensity differences between the seasons (see Section IV, E). It is imperative, therefore, that the grower be familiar with the response times of the cultivars and their seasonal variations if one intends to produce crops for particular dates. Plant age at the time of floral induction will also influence total response time. When grown side by side under naturally occurring photoperiods, results show that older kalanchoe plants flower earlier than younger plants (Pertuit, 1973a; Younis, 1955).

V. CRASSULACEAN ACID METABOLISM

Kalanchoes have an extremely "flexible photosynthetic system" (Osmond, 1978b). This system is regulated in part by factors such as light intensity, water, temperature, and photoperiod. They are CAM plants because of their cells, which contain chloroplasts, have the ability to absorb carbon dioxide *in the dark*, synthesize it into malic acid, and *store* this malate in their vacuoles. In light, the malic acid is decarboxylated and this released carbon dioxide is used in photosynthesis. Some of this internally released carbon dioxide is, however, lost because their stomates are not fully closed during the day. Temperature greatly influences stomatal aperture control (Osmond, 1978a). Higher day temperatures (78°F) result in inadequate stomatal closure, allowing the escape of carbon dioxide that could be used in carbohydrate production. Kalanchoe's acidification–deacidification photosynthetic capacity is only part of its carbon dioxide fixing ability. They also fix carbon dioxide in light, like most C_3, C_4 green plants. Environmental conditions, such as photoperiod and stress, can activate their CAM activity, as can any environmental condition that reduces translocation activity and results in photosynthetate accumulation, particularly in leaves (Chaturvedi and Zabka, 1972; Osmond, 1978a,b; Ting and Hanscom, 1977).

It should be pointed out that although CAM activity is generally associated with succulent stems and leaves, not all succulent plants are capable of CAM. Also, CAM activity is sometimes observed in nonsucculent organs. The ability of succulents to maintain high water potentials in extremely low water potential environments is due to their morphological makeup, not their physiological CAM abilities. It is true, however, that CAM plants are more plentiful in more arid regions (Schwabe, 1985).

With CAM plants, plant water potential controls stomatal opening and closing. When it is high, the stomata respond to the plant air water vapor pressure differences and internal carbon dioxide concentrations, *not light*. Following water stress, mature leaves show a significant increase in dark carbon dioxide fixation and malate synthesis, accompanied by deacidification and stomatal closure (relatively speaking) in the light. Compared to water requirements for carbon dioxide fixation in the C_3/C_4 schemes, dark carbon dioxide fixation is extremely efficient in terms of water cost. Once the stress is removed, CAM activity is reduced and, in some cases, ceases (Schwabe, 1985).

Once kalanchoes are induced to complete flowering, the plants will not revert to their previous short-day carbon dioxide pattern (i.e., net carbon dioxide uptake in light and production in dark) if given short nights. They will continue a pattern of net carbon dioxide uptake in dark and production in light (Spear, 1959).

There are several extremely interesting similarities between strong CAM activity and floral initiation in kalanchoes. Spear (1959) summed these up as follows: both are promoted by short days; both are inhibited by interrupting the dark period; the peak of carbon dioxide fixation occurs 11 to 12 hours after the beginning of the dark period, and this is the critical time for flower initiation; carbon dioxide fixation ceases after 16 hours of darkness, and this is the optimum dark period for flower initiation; and, using 96 hour cycles with a long dark period interrupted at various intervals, he found that treatments that inhibit flower initiation also inhibit CAM activity. With all these similarities, it is still impossible to say whether there are physiological relationships between CAM activity and floral initiation in kalanchoe or whether these similarities are merely casual.

Intense CAM activity in kalanchoes occurs from floral initiation through anthesis. Also, high day temperatures during this period enhance carbon dioxide loss by restricting stomatal closure. It follows, then, that day temperatures from the point of floral induction may be significantly more important in kalanchoe production than has been realized.

VI. HEIGHT CONTROL

Cultivars vary in vigor and some stretch too much, especially in summer. Height of the plant affects its quality; therefore, height must be controlled. Height control may be accomplished in several ways: cultivar selection, physical control (e.g., choice of propagation date, pinching), environmental control [e.g., manipulation of the vegetative and reproductive cycle, choice of supplemental light (incandescent versus fluorescent)], and/or chemical plant growth regulator application(s). All these factors must be kept in mind, for, only when they are correctly balanced among themselves will quality plant result.

Growth retardant application to potted kalanchoes has been show to enhance their quality by reducing plant height and compressing the flowering head; however, none are registered for use on kalanchoes. Nevertheless, several growth retardants have successfully been incorporated into kalanchoe production, while others require additional testing before definite conclusions can be reached. Retardants that have been successfully applied to kalanchoes include B-Nine (daminozide), A-Rest (ancymidol), and Florel (ethephon). They have been applied as foliar sprays, "soil" drenches or both. It takes about three foliar sprays to produce an amount of retardation similar to that produced by one soil drench. Atrinal, although tested, cannot confidently be recommended for kalanchoes because of reports of no effect (Lyons and Hale, 1987) or erratic results among cultivars (Nightingale, 1985).

B-Nine (Daminozide) works well as 2500 to 5000 ppm sprays applied 2 weeks after pinching. B-Nine sprays also are recommended when flower buds begin to show (Hesse, 1985). When applied at this later stage, the main effect would be a reduction in peduncle (flower stem) length, not plant foliage height. The number of spray applications employed depends on the B-Nine spray concentration applied, the cultivar grown (e.g., more for Cinnabar, Fortyniner, and other vigorous cultivars), and the environmental conditions (i.e., more would be needed under conditions that promote axis elongation such as high temperatures and low light intensities) (Anonymous, 1990a,b). A-Rest (ancymidol) has been used effectively as a foliar spray or as a soil drench (Carlson *et al.*, 1977; Pertuit, 1973a). Two 50 ppm foliar sprays (applied a week apart to allow runoff) will reduce *total* plant height (i.e., foliage height above soil and cyme length above foliage added together) when applied as much as 8 weeks before or 5 weeks after short day commencement; however, concentrations less than 50 ppm are not always effective. When foliar sprays are applied after short days begin, the reduction in plant height is mainly a result of a reduction in cyme length rather than a reduction in foliage height (Pertuit, 1973a). This suggests that one can "tailor" the plant (i.e., control the proportion of foliage height to cyme length) by timing the foliar spray applications. As a soil drench, the best time to apply A-Rest is 2 weeks after short days commence. A rate of 50 ml or 0.50 mg A-Rest per 4-inch pot applied at this time will reduce total plant height, mainly by a reduction of plant foliage height. When a soil drench of 0.75 mg (50 ml per 4-inch pot) is applied at this time, A-Rest will reduce cyme length (Carlson *et al*, 1977).

Florel (ethephon) sprays of 250 ppm applied twice (a week apart) about a month after short days begin will reduce total plant height by reducing cyme length, not plant foliage height (Fig. 3). If this spray concentration is increased to 500 ppm, foliage plant height and cyme length will be reduced. If the cultivar grown is an

Fig. 3. *Kalanchoe* 'Thor.' Control plant (on *left*)compared to plant (on *right*) that received two, 500 ppm Florel foliar sprays about 4 weeks after floral inductive photoperiods had begun. Florel slightly reduced the height, its main effects being a reduction in cyme height and an evening up the inflorescence, producing a more attractive plant.

extremely tall, vigorous type, a concentration greater than 500 ppm applied about 2 weeks after short days begin might be considered (Pertuit, 1973a). At lower pH values the prepared Florel and water mixture is more effective because ethylene gas is freely released. Distilled water should be used to prepare the Florel and water mixture if the pH of the water is near neutral, pH of 7 or above.

Sumagic (uniconazole) and Bonzi (paclobutrazol), two more recently released growth retardants, may be effectively useful on kalanchoes; however, research data are currently lacking on these. No suggestions, therefore, can be made for their use on kalanchoes.

It is difficult to predict exactly how much a chemical growth retardant will control kalanchoe height because numerous production factors also affect height.

VII. POSTGREENHOUSE HANDLING

A. Marketing Stage

Recommendations on the stage of inflorescence development for sale vary somewhat. Some recommend that florets should be fully opened before sale (Hesse, 1985; Love, 1980) because florets do not readily open under low light conditions. Others (Anonymous, 1990b) say that the key is that the central floret of most flowering branches must be open (i.e., once this stage is reached, the entire inflorescence will open). Regardless, even if they are marketed at the early full flowering stage, they will easily last 4 to 6 weeks in the home environment with minimum care.

B. Seasonal Demand

Like many other flowering plants, the greatest demand for kalanchoes is in late winter and spring. With the advent of the new cultivars of different colors, however, demand in general is increasing on a year-round basis. As with other flowering potted plants, the smaller sizes (e.g., 4-inch pot) are sold in the mass markets and the larger size (6-inch pot or larger) are sold by garden centers and florists.

C. Care during Marketing

As previously mentioned, kalanchoes should receive high light intensities (see Section III, B) whenever possible. Temperature (see Section III, B) is also important. In addition to proper light and temperature, the plants should not be crowded, and they should be in a well-ventilated location. This will forestall disease development and the yellowing of the lower leaves due to lack of light. Of course, all this means that the plants should be unsleeved for display. Kalanchoes should be displayed in areas free from ethylene. It is reported that this gas can prevent the florets from opening and cause premature flower fading and desiccation (Love, 1980). Mastalerz (1977) reports that fruits such as apple and pear can produce particularly harmful amounts of ethylene, oranges and lemons moderately harmful amounts, and grapes and grapefruit nonharmful amounts.

It is recommended that kalanchoes be kept well watered; however, these tough plants are able to withstand relatively dry conditions rather well.

D. Consumer Care

Few plants are as long lasting in the home environment as kalanchoes. They always should be kept in as bright a location as possible. A reading lamp, for supplemental lighting, will help them retain their original color (Irwin, 1972).

VIII. PRODUCTION PROBLEMS

A. Insects

The primary insect that attacks kalanchoes is the common (citrus) mealybug. Although a problem, the green peach aphid is easily controlled. Brown soft scale can be found on stock plants. The cabbage looper occasionally is a problem, and the cutworm is an evasive pest. CAUTION: Kalanchoes are extremely sensitive to many of the insecticidal carriers (e.g., xylene), damage often being so severe that the plants cannot be marketed. To be safe, a wettable powder is suggested.

1. Citrus Mealybug [Plancoccus citri (Risso)]

The mealybug, although a polyphagous sucking insect, seems to have a propensity for kalanchoes. Once they infest, they are particularly difficult to control because they congregate in locations that are hard to spray (e.g., in the a bases of petioles). The adult female (0.12 inch long, 0.06 inch wide) is oval, yellow when newly molted, but later pink to orange-brown and covered with a white, fluffy wax. Parthenogenetic populations have been reported, although males are commonly found. Each female is capable of producing 300 to 600 eggs in a waxy sac. This species has been reported to produce six generations per year in greenhouses, although generations usually overlap. Eggs hatch in 6 to 10 days, and they mature in 6 to 8 weeks (Baker, 1978; Nakahara and Williams, 1980; Pirone, 1978).

2. Green Peach Aphid [Myzus persicae (Sulzer)]

The aphid, like the mealybug, is a sucking insect. Aphids are light to dark green, sometimes pink, with red eyes, and they may or may not have wings (Baker, 1978). Not that common on kalanchoes, they generally appear in the spring and summer. Their ability to reproduce in the greenhouse is amazing. A plant may be covered with them in a couple of days. They are parthenogenic: females give birth to females. They secrete a sugary, sticky honeydew that, when it falls on lower foliage, can become a growing medium for the sooty mold fungus, turning the foliage black. Aphids are also vectors of numerous plant viruses. In the greenhouse, they are capable of producing a new generation about every week.

3. Brown Soft Scale [Coccus hesperidum (Linnaeus)]

The brown soft scale is the most polyphagous Coccidea species. The ovoviviparous females (0.06 to 1.8 inches long, 0.04 to 0.16 inch wide) are yellowish-brown, ovate, convexed, and hard. Males are rarely found. In warm greenhouses, they will produce about six generations per year. They seem to congregate along leaf margins but really are not that selective. They also produce honeydew, leading to the development of the sooty fungus (Gill, 1977). Once established, they are difficult to control.

4. Cabbage Looper [Trichoplusia ni (Hubner)]

The cabbage looper is the most prominent of the Lepidoptera larvae that chew on kalanchoes. The larva usually appears from late spring to early fall. The semi-nocturnal, grayish to dark brown adult moth has a wingspan of about 1.3 inches and is capable of producing over 200 hemispherical eggs. The full grown larva (1.4 to 1.6 inches long) is light green with two dorsal white stripes and two (larger) lateral white stripes (Baker, 1978; Pirone, 1978).

5. Cutworm (Various Genera)

Like the cabbage looper, these Lepidoptera mostly appear from late spring until early fall, but, if unchecked, may be found throughout the year in the greenhouse. Night lighting attracts them into the greenhouse. Unlike the cabbage looper, the caterpillar (1.8 inches long) hides in the soil during the day and at night climbs the plant, chewing the leaves and stem, sometimes cutting the plant off at the soil line. Each adult female is capable of producing 500 eggs, and the larvae develop in 3 to 4 weeks. Baits are quite effective in controlling them (Baker, 1978; Pirone, 1978).

B. Diseases

1. Powdery Mildew (Spaerotheca himuli var. fulginea)

Kalanchoe cultivars vary in their susceptibility to powdery mildew; those with thicker, darker leaves are most resistant. This disease is more prevalent at cool temperatures and high humidity. It may be controlled by proper heating and ventilating.

The mycelium appear as grayish-white, powdery masses and can be spread to other plants by wind and careless watering. Young leaves become curled and stunted, then dry out and die (Pirone, 1978).

2. Crown Rot, Wilt (Phytophthora cactorum)

Black lesions appear at the soil line. The black rot then extends up the stem through the plant, which wilts (Pirone, 1978). This disease seems to be more of a problem in kalanchoes started from seed (Love, 1980), possibly because they are more likely to be overwatered or transplanted too shallowly when small.

Proper growing medium selection (see Section IV, A) and sterilization (see Section III, A) should help prevent this.

C. Physiological Disorders

1. Oedema (Wart Disease)

In locations of high humidity, oedema may occasionally occur on kalanchoes. The problem may first be noticed several days following overcast conditions when the humidity was exceptionally high, the light intensity relatively low, and the growing medium fairly wet (Pirone, 1978). Cells in the stem and leaves are not able to release their water fast enough to the atmosphere under these conditions. Pressure builds within the cells, and they burst, after which they appear water soaked and darker green. Subsequently, callous develops at the injured location, producing an unsightly corky layer of cells.

2. Heat Delay

Warm temperatures during the first part of the dark period can inhibit floral initiation even though the plants are exposed to clearly inductive photoperiodic conditions (see Section IV, G). It is recommended that the dark period be given at 60°F to preclude the heat delay effect.

3. Failure to Flower (and Fewer Flowers)

Failure to flower may result from heat delay (see Section VIII, C, 2), an inadequate dark exposure, or both of these. The length of the dark period should be at least 12 hours, and it is recommended that plants be given at least 40 consecutive days with long nights (see Section IV, I, 1).

Kalanchoes initiate fewer flowers if grown under low light intensities and if the length of the dark period is too long under high light intensity conditions (see Section IV, E). Under low light intensity conditions, long, long nights (i.e., longer than that which is necessary for floral initiation) can increase the number of flowers to a level equal to the number initiated under night lengths considered optimum for that light intensity (see Section IV, E).

In summary, for high quality plants, grow kalanchoes under 4500 fc with 60°F nights, 72°F days (see Section III, B). For induction, initiation, and development, increase the dark period to 12 hours (see Section IV, I, 1).

REFERENCES

Anonymous. (1972). "Kalanchoe In *The Ball Red Book*, edited by V. Ball, Reston, VA: Reston, p. 372.
Anonymous (1990a). *Flowering Schedule for Mikkelsens Pot Kalanchoes*. Ashtabula, Ohio: Mikkelsen, Inc.

Anonymous. (1990b). *Mikkelsens 'Bonanza' Hybrid Kalanchoes.* Astabula, Ohio: Mikkelsen, Inc.

Bailey, L. H. (1928). "Kalanchoe," In *The Standard Cyclopedia Of Horticulture,* vol. II, New York: Macmillian pp. 1731–1732.

Bailey, L. H., and Bailey, E. Z. (1976). *Hortus Third.* New York: Macmillian.

Bailey, R., and McDonald, E., ed. (1974). *The Good Housekeeping Illustrated Encyclopedia of Gardening.* New York: Hearst.

Baker, R. (1978). "Insect and related pests of flower and foliage plants." *N.C. Agric. Ext. Svc.* N.C. State University, Raleigh, N.C.

Brandon, P. C. (1967). "Temperature features of enzymes affecting crassulacean acid metabolism." *Plant Physiol.* **42:** 977–984.

Broertjes, C., and Leffring, L. (1972). "Mutation breeding of kalanchoes." *Euphytica.* **21:** 415–423.

Brulfert, J., Gurrier, D., and Querioz, O. (1973). "Photoperiodism and enzyme activity: Balance between inhibition and induction of the crassulacean acid metabollism." *Plant Physiol.* **51:** 220–222.

Carlson, W. H., Schnabel, S., Schnabel, J., and Turner, C. (1977). "Concentration and application time of ancymidol for growth regulation of *Kalanchoe blossfeldiana* Poellniz. cv. 'Mace.'" *HortScience.* **12**(6): 568.

Chaturvedi, S. N., and Zabka, G. (1972). "Studies in dark fixation of carbon dioxide in Kalanchoe. *Ann. Bot.* **41:** 493–500.

Cockburn, W. and McAulay, A. (1974). "The pathway of carbon dioxide in crassulacean plants." *Plant Physiol.* **55:** 87–89.

Crews, C. E., Vines, H. M., and Black, C. C., Jr. (1975). "Postillumination burst of carbon dioxide in crassulacean acid metabolism plants." *Plant Physiol.* **55:** 652–657.

Gill, R. J., Nakahara, S. and Willliams, M. L. (1977). "A review of the genus *Coccus* Linnaeus in America north of Panama (Homoptera: Coccoidea: Coccidae)." In *Occasional Papers In Entomology,* vol. 24. State of California, Department of Food and Agriculture, Sacramento, pp. 18–24.

Gregory, F. G. (1954). "The interrelation between CO_2 metabolism and photoperiodism in Kalanchoe." *Plant Physiol.* **29:** 220–229.

Hartman, H. T., Flocker, W. J., and Kofranek, A. M. (1981). "Floriculture: Greenhouse flowering plants." In *Plant Science. Growth, Development, and Utilization Of Cultivated Plants.* Englewood Cliffs, Prentice-Hall, pp. 369–484.

Hesse, P. (1985). "Kalanchoe." In *The Ball Red Book* edited by V. Ball, Reston, VA: Reston, pp. 565–567.

Hillman, W. S. (1962). *The Physiology Of Flowering.* New York: Holt, Rinehart, and Winston, pp. 4-5, 22, 26-27, 92, 94, 112.

Irwin, J. (1972). "Try a 'new' crop—Kalanchoes." *Florists' Rev.* **151**(3917): 23, 56–58.

Johnson, A. T., and Smith, H. A. (1986). "Kalanchoe." In *Plant Names Simplified.* Buckenhill, England: Landsmans Bookshop, Ltd., p. 58.

Kenyon, W. H. A., Holaday, S., and Black, C. C. (1981). "Diurnal changes in metaboite levels and crassulacean acid metabolism in *Kalanchoe daigremontiana* leaves." *Plant Physiol.* **68:** 1002–1007.

Laurie, A., Kiplinger, D. C., and Nelson, K. S. (1958). "Kalanchoe." In *Commercial Flower Forcing,* New York: McGraw-Hill, pp. 390–391.

Levi, C. and Gibbs, M. (1975). " Carbon dioxide fixation in isolated Kalanchoe chloroplasts." *Plant Physiol.* **56:** 164–165.

Link, C. B. (1978). "Kalanchoe—an outstanding flowering pot plant." *The Md. Flor.* **214:** 3–5.

Longman Dictionary Of The English Language. (1984) London: Longman Group Ltd.

Love, J. W. (1976). "Kalanchoe production." *N.C. Flower Grower's Bull.* **20**(2): 1–3.

Love, J. W. (1980). "Kalanchoe." In *Introduction To Floriculture,* edited by Roy A. Larson, New York: Academic Press, pp. 409–434.

Lyons, R. E., and Hale, C. L. (1987). "Comparison of pinching methods on selected species of *Columnea, Kalanchoe,* and *Crassula.*" *HortScience.* **22**(1): 72–74.

Manzitti, C. (1978a). "New kalanchoe hybrids. Pt. I: A breed that's easier to produce than pronounce." *Florist.* **11**(8): 70–74.

Manzitti, C. (1978b). "New kalanchoe hybrids: Pt. II: A production cookbook for consistent crop results." *Florist.* **11**(9): 61–63.

Mastalerz, J. W. (1977). "Kalanchoe," In *The Greenhouse Environment.* New York: Wiley, pp. 243–245.

Mikkelsen, J. (1975). "A-B-C of kalanchoe culture." *Ohio Flor. Assoc. Bull.* **550:** 7.

Nakahara, S., and Williams, M. L. (1980). "The soft scale insects found in the continential United States (Homoptera: Coccidae). A syllabus prepared for the 1974 Coddidology Training Session." Beltsville, MD: USDA.

Nelson, P. V. (1981). *Greenhouse Operation And Management,* 2d ed. Reston, VA: Reston, pp. 260, 287, 312.

Nightingale, A. E. (1970). "The influence of succinamic acid 2,2-dimethylhydrazide on the growth and flowering of pinched vs. unpinched plants of the kalanchoe hybrid 'Mace.'" *J. Am. Soc. Hort. Sci.* **95**(3): 273–276.

Nightingale, A. E., Cross, S., and Longnecker, M. T. (1985). "Dikegulac alters growth and flowering of kalanchoe." *HortScience.* **20**(4): 722–724.

Osmond, C. B. (1978) "Crassulacean acid metabolism: A curiosity in context." *Plant Physiol.* **29:** 379–414.

Osmond, C. B., and Bjorkman, O. (1975). "Pathways in CO_2 fixation in CAM plant *Kalanchoe daigremontiana.* II. Effects of O_2 and CO_2 concentrations on light and dark CO_2 fixation. *Austral. J. Plant Physiol.* **2:** 155–162.

Ota, K. (1988). "Stimulation of CAM photosynthesis in *Kalanchoe blossfeldiana* by transferring to nitrogen-defficent conditions." *Plant Physiol.* **87:** 454–457.

Pemberton, Brent. (1980) "Height control of Kalanchoe blossfeldiana." *Florists Rev.* **165**(4284): 46, 48, 51.

Pertuit, A. J., Jr. (1973a). "The effects of terminal pinching and chemical growth regulation of *Kalanchoe Blossfeldiana,* v. Poellnitz." *Univ. Ga. Res. Bull.* **132.**

Pertuit, A. J., Jr. (1973b). "The effects of temperature during dark exposure and date of exposure to naturally-occurring daylengths on growth and flowering of Kalanchoe Blossfeldiana, v. Poellintz." *Univ. Ga. Res. Rep.* **170.**

Pertuit, A. J., Jr. (1977). "Influence of temperature during long-night exposures on growth and flowering of 'Mace,' 'Thor,' and 'Telstar' kalanchoes." *HortScience.* **12**(1); 48–49.

Pirone, P. P. (1978). "Mealybugs." In *Diseases and Pests of Ornamental Plants,* 5th ed., New York: Wiley, p. 62.

Post, K. (1959). "Kalanchoe." In *Florist Crop Production And Marketing.* New York: Orange-Judd, pp. 590–592.

Ranson, S. L., and Thomas, M. (1960). "Crassulacean acid metabolism." *Ann. Rev. Plant Physiol.* **11:** 81–110.

Schwabe, W. W. (1954). "The effects of light intensity on the flowering of *Kalanchoe blossfeldiana* in relation to the critical daylength." *Physiol. Plant.* **7:** 745–752.

Schwabe, W. W. (1972). "Flower inhibition in *Kalanchoe blossfeldiana.* Bioassay of an endogenous long-day inhibitor and inhibition by ($\pm$) abscissic acid and xanthorin." *Planta.* **103:** 13–23.

Schwabe, W. W., and Papafotiou, M. (1987). "Inhibition of flowering—effects in the leaf and on translocation of the stimulus. In *Manipulation Of Flowering,* edited by J. G. Atherton. Boston: Butterworths, pp. 351–359.

Schwabe, W. W. (1985). "Kalanchoe blossfeldiana." In *Handbook Of Flowering,* vol. III, edited by Abraham H. Halevy. Boca Raton, FL: CRC Press, pp. 217–235.

Seeley, J. G. (1952). "Long days after budding improve kalanchoes." *Pa. Flower Growers Bull.* **15:** 1–3.

Spear, I. (1959). "Metabolic aspects of photoperiodism." In *Photoperiodism,* edited by R. B. Withrow. Washington, D.C.: *Amer. Assoc. Adv. Sci.,* pp. 289–300.

Sutton, B. G. (1975). "The path of carbon in CAM plants at night." *Austral. J. Plant Physiol.* **2:** 377–387.

Ting, I. P., and Hanscom, Z. (1977). "Introduction of acid metabolism in *Portulacaria afra.*" *Plant Physiol.* **59:** 511–514.

Winter, K. (1980). "Carbon dioxide and water vapor exchange in crassulacean acid metabolism plant *Kalanchoe pinnata* during a prolonged light period." *Plant Physiol.* **66:** 917–921.

Wynn, T. (1981). "Effects of O_2 tension and temperature during light uptake of CO_2 on the dark release of CO_2 by excised C_3 and C_4 leaves. *Plant Physiol.* **68:** 1253–1256.

Younis, A. F. (1955). "Studies on the photoperiodism of *Kalanchoe blossfeldiana*. 1. Effect of age on response to short-day treatment." *Physiol. Plant.* **8:** 223–229.

Zabka, G. G., and McMahon, E. (1965). "Relationships among CO_2 dark fixation, succulence, flowering, and organic formation in *Kalanchoe blossfeldiana* variety 'Tom Thumb.'" *Can. J. Bot.* **43:** 447–452.

Zawarvi, M. A., and Irving, R. M. (1968). "Interaction of triiadobenzoic acid and indole-acetic acid in the developemnt and flowering of kalanchoe." *Proc. Am. Soc. Hort. Sci.* **93:** 610–617.

18

Geraniums

William C. Fonteno

I. HISTORY

Geraniums are important spring greenhouse crops for both pot culture and landscape use. There are approximately 280 species, native mostly to South Africa. The important cultivated geraniums fall into four major groups. The most widely cultivated is the "zonal" or "fish" geranium, *Pelargonium* × *hortorum* (Bailey). These cultivars have a characteristic dark band or "zone" within each leaf. The "regal," or "show" geranium (*P.* × *domesticum* Bailey) is grown to a much lesser extent but is gaining in popularity because of its beautiful and unusual flowers. The "ivy" geranium (*P. peltatum* L. and its derived forms) has a trailing habit, making it favorable for hanging baskets. The fourth group (scented) is a series of species (or their derivatives) having pungent or fragrant leaves.

Geranium production has steadily increased over the past several years. According to United States Department of Agriculture's figures, in 28 states reporting, there were over 46 million potted geraniums sold in 1977, the vast majority of them in 4-inch pots. Most were zonals (*P.* × *hortorum*), grown as spring bedding plants. In 1985, more than 79 million pots were sold, with 27 million propagated from seed. Also in 1985, about 3.5 million flats of seedling geraniums were sold. In 1988, over 95 million pots (55 million from cuttings, 40 million from seed) were reported sold. The future for geraniums looks bright, with new cultivars, colors, and better performance in the landscape.

All four major groups can be propagated from cuttings. Zonals can also be propagated from seed. In fact, there are large differences in production techniques for the seed propagated geranium. They will be treated separately when appropriate.

II. CLASSIFICATION

A. Vegetatively Propagated

1. Zonal Geraniums

The most common geranium is *P.* × *hortorum* (Fig. 1). The exact origin is unknown. This is a hybrid that has resulted from intercrossing of several species within the subgenus Ciconium. The principle contributors have been identified as *P. zonale, P. inquinans, P. scandens,* and *P. frutetorum.* The matter of cultivars differs considerably among growers, markets, and sections of the country. Traditionally, a color mix of 60 to 70% red, 20% pink or salmon, and the balance in white has satisfied most market demands. In 1990, this color mix has shifted to 45% red, 30% salmon, 15% pink, and 10% white.

Most cultivars are obtained from specialty propagators or their licensees. Geraniums are very susceptible to bacterial stem rot and leaf spot (caused by *Xanthamonas pelargonii*), diseases that almost ended the production of geraniums in the 1950s. Specialty propagators use culture-indexing techniques and very strict sanitation to reduce disease occurrence.

Fig. 1. Zonal geraniums, propagated from cuttings, produce large showy pot plants for patio or garden plantings.

There is a series of variegated cultivars with green and white foliage known as "Brocades." These have red, pink, or salmon flower colors and provide variety to traditional zonals.

2. Ivy-Leaved Geraniums

The ivy-leaved geranium, *P. peltatum*, is a member of the subgenus Dibrachya. The trailing habit of these cultivars provides an effective flowering hanging basket for spring sales (Fig. 2). They can also be used as cascading plants in window boxes and as wall coverings. They are used successfully as ground covers or accent plantings in sheltered locations, and they provide an opportunity to use geraniums in full or partial shade plantings.

Fig. 2. The cascading habit of ivy-leaved geraniums makes them ideal for hanging baskets, window boxes, and wall coverings.

3. Regal Pelargoniums

P. × domesticum is a member of the subgenus Pelargonium. This is a hybrid species, with parentage much more complex than *P. × hortorum*. Regals have been considered the most beautiful of the cultivated geraniums for over 100 years (Fig. 3). Also known as 'Martha Washington' or 'Lady Washington' geraniums, they have not been as popular as the zonals because of problems in cultivation and difficulties in floral initiation and flowering.

B. Seed Propagated

Seed propagated geraniums became a reality with the introduction of the first commercial cultivar, 'Nittany Lion Red,' in 1963 and the series, 'New Era' and 'Carefree' in the late 1960s. These first F_1 hybrids of *P. × hortorum* had several problems. Compared to their counterparts propagated from cuttings, they were less floriferous, shorter lasting, and took longer to flower. The flowers also shattered easily in shipment and germination often was poor. Since then, seed firms have produced a large number of new cultivars with improved germination, compact growth, better flowering, that are more shatter resistant, and have decreased flowering time. They come in a wide array of colors from deep red to pure white.

Seed propagated geraniums have certain advantages over vegetatively propagated plants. Seed geraniums are very predictable and easy to schedule. Their

Fig. 3. The strikingly beautiful flowers of *Pelargonium* × *domesticum* make these geraniums truly the most "regal."

growth habit is very compact, which allows close spacing. They can be grown in cell packs as well as 4-inch pots. The plants are vigorous, freely branching, and bloom until frost. These varieties do not require stock plant production, are usually initially disease free, and are readily available from seed firms. They also do well in the home landscape because they are able to endure more environmental stress than vegetatively propagated geraniums (Fig. 4).

III. PROPAGATION

A. Vegetatively Propagated

1. Stock Plants

One option for growers is to propagate their own cuttings from stock plants. These are usually started from culture-indexed cuttings obtained from specialty propagators. Stock plant cuttings can be started as early as May to August for cuttings the next spring, though usually stock plants are received after the Christmas crops are finished. Rooted cuttings are placed in 1 to 1.25 gallon containers or directly in benches. Spacing should be one plant per square foot. Bench planting provides the easiest management of nutrition and moisture, but the use of containers allows better air movement among plants and fewer disease problems. Plants must be pinched regularly to develop as many shoots as possible.

Fig. 4.   Seed propagated geraniums generally perform well in the landscape because they are more stress tolerant than their vegetatively propagated cousins.

For fewer cuttings, stock plants are started in November and December in 6-inch pots. Plants are not pinched; rather, cuttings are removed, rooted, and potted as soon as they are available. This continues throughout the winter months, with early cuttings becoming stock plants. Up to 40 cuttings can be produced for each cutting bought as a stock plant.

Proper sanitation is extremely important in geranium propagation. Geraniums are very susceptible to bacterial diseases, which can be quickly spread during propagation. A common practice in the past was to hold a few plants over from last year's crop for stock plants the next year. Stock plants now are usually started from cuttings obtained from specialty propagators who employ culture-indexing to produce pathogen-free plants.

High light and warm temperatures produce more cuttings per stock plant. Light levels between 32 and 54 klx should be used. Low light can lead to soft, succulent cuttings that often do not root well and are more susceptible to disease. During naturally occurring low light intensity times of the year, supplemental high intensity discharge (HID) lighting at 3 to 6 klx for 18 hours per day can increase cutting production by 30 to 100%.

Ivy geraniums do not tolerate high light. Maximum intensity levels range from 21 too 38 klx depending on the cultivar. High light intensity results in bleached foliage, reddened petioles, stunted growth, and increased chances for edema.

Cutting production can be improved 20 to 30% with foliar application of ethephon (Florel). Rates of 350 to 500 ppm will shorten internodes, reduce leaf size, and cause normally dormant lateral shoots to arise from the main stem. Uniform applications with minimum run-off can start 3 to 4 weeks after planting and continue monthly through mid-December. Applications should be made at the same temperature each time, as high and low temperature variations can cause different effects. Also, cuttings should not be taken for at least 10 days after application to avoid delays in rooting.

2. Cuttings

Cuttings are removed with a knife or by snapping the cuttings off with the fingers (preferred). Knives should be sterilized with disinfectant between stock plants. Cuttings snap off easier by hand when plants are turgid, such as in the morning, with the stock plants watered the day before. Terminal cuttings 2 to 3 inches long should be taken. Each cutting should have at least three full and perfect leaves when placed in the propagation medium. Smaller cuttings generally take longer to root, while larger cuttings fall over easily, can reduce air movement, and wilt quickly.

Propagation media should be pathogen free. Pasteurization of bulk media with steam at 160°F for at least 30 minutes will remove harmful organisms. Other propagation media such as expandable peat pellets, Oasis foam, rockwool cubes, and "rubber dirt" can be used.

The rate of rooting is influenced by temperature in the root zone. Optimum temperature for rooting is 75°F. Root zone temperatures should be between 70° and 80°F. Air temperatures should fall in the range of 62° to 70°F at night with a 5 to 18 degree rise during the day, depending on light intensity. Bottom heating is critical when using intermittent misting, as propagation units can be 5 degrees lower because of evaporative cooling. Misting cuttings can be done in the early stages of rooting to maintain turgidity. However, geraniums should not be over misted. Foliar disease problems are minimized when foliage is kept dry.

Cuttings should be rooted under as high light as possible to ensure the fastest rooting time. Zonal geraniums should be rooted under 37 to 48 klx, while ivy-leaved geraniums need 20 to 40 klx. Cuttings should begin to callus in 7 to 8 days and be sufficiently rooted in 12 to 21 days or longer, depending on light intensity and time of year. If bottom leaves begin to yellow, application of 20–10–20 fertilizer at 200 ppm nitrogen (13.5 ounces per 100 gallons) every few days will correct the problem, once the cuttings are rooted.

B. Seedling Geraniums

1. Breeding

Breeding of hybrid geraniums has resulted in a wide array of cultivars with differing characteristics: tall or dwarf, single or double flowers, and different zonal patterns. Breeders have developed all colors except blue and yellow. Some of the more popular series of cultivars have been 'Carefree,' 'Orbit,' 'Sprinter,' 'Flash,' and 'Ringo,' but new series are introduced frequently. The cultivars within each series are of similar growth habit and usually have flower colors of red, salmon, and white.

2. Sowing and Transplanting

Geranium seed have very hard and impenetrable seed coats, which must be cut or softened before germination can occur. Seed from reputable companies have been cleaned and scarified (usually with acid). Fresh seed should germinate greater than 80%, with some cultivars germinating up to 90% under optimal conditions. To assist in mechanical seeding operations, seed can be made more uniform in size and shape by encasing (pelletizing) them in either clay or vermiculite. Pelletized seed improve uniformity of distribution in the seed trays, but are more costly, reduce germination percentage, and require 2 to 4 more days to germinate.

Light intensity and duration have little effect on germination, but bottom heat is essential for rapid, uniform germination. Optimal medium temperatures are between 72° and 77°F, with temperatures above 86°F and below 68°F decreasing uniformity and germination. Percent germination and time for germination will also depend on night temperature. For example, at 50°F it can take 16 days to obtain 80% germination, while at 70°F it might take only 9 days to get 91%.

Germination media should be well aerated, somewhat fine structured, and free of large particles. Commercial germinating mixes are available that consist of combinations of peat, vermiculite, perlite, or all of these. A combination of 50% peat and 50% vermiculite will also work well. These media should be sterile and have either low or no fertilizer added. Containers for germination should be seed flats or plug trays.

Once sown, the medium must be kept constantly moist so the seed can imbibe enough moisture. Medium water content must be high but not high enough to cause dripping of free water from the bottom. Placing flats or trays under intermittent mist will provide the necessary moisture and humidity. Also, high humidity tents can be used with good success.

Seedlings can be transplanted 10 days after sowing. Plants must be carefully handled by the cotyledons to avoid damage to the delicate stems or small root systems. At 14 to 21 days, the first one to two true leaves will appear. Seedlings should be transplanted by 21 days after sowing to prevent possible delays in flowering time.

IV. CULTURE

A. Vegetative Stage

After rooting, cuttings are placed in 3-, 4-, 5-, or 6-inch pots (one cutting each) and finished. Seedlings are either sold in the germinating plug flats, or transplanted into bedding plant flats (generally 48 plants per flat) or 4-inch pots.

1. Temperature

Temperature has a dramatic effect on the growth rate of geraniums. Geraniums can be produced under temperatures of 50° to 85°F. Commercial production of geraniums is best at warmer temperatures. Temperatures of 50°F or lower for 12 hours cause a buildup of anthocyanin (reddish pigment) in the leaves. Temperatures above 85°F for 12 hours or more can cause chlorosis (yellowing) of the youngest leaves. As night temperature drops below 60°F, growth and development slows down. At 55°F night, growth is even slower and flowering is delayed, and at 50°F growth virtually stops. At day temperatures above 85°F for 12 hours or more, heat stress develops and growh is markedly reduced. Cultivars differ in the level of response to temperature. It is possible to have some cultivars showing symptoms of high or low temperature stress while others seem unaffected.

Optimal day temperature depends on the plant's ability to maximize photosynthesis and is, therefore, dependent on light intensity. On bright days, day temperatures of 70° to 75°F are optimal, with night temperatures ranging from 60° to 65°F. This combination should result in a crop of geraniums grown in 4-inch pots in 6 to 8 weeks. On dull, cloudy days, the ability of the plant to photosynthesize is reduced and, therefore, temperatures can be lowered to 65°F day and 55° to 60°F night.

In European production, new transplants (up to 8 to 10 days old) are kept at temperatures higher than 59°F, preferably 65°F. Fast cropping can be done with 64° to 68°F day and 58° to 61°F night temperatures. Day temperatures can run as high as 77°F, depending on light intensity.

If the greenhouse is equipped with environmental monitors, such as computers, the cropping program can be modified to a 24-hour average program. Research has shown that mean daily temperature is a better measure of growth and development of geraniums than night temperature alone (Tayama, 1988). As mean daily temperature increases, time to anthesis decreases. A mean daily temperature can be achieved by adjusting the night temperature according to the greenhouse temperature of the previous day. This allows the day temperature to fluctuate, with growers taking advantage of high temperatures and high light intensity and compensating at night so the mean daily temperature stays constant. This can result in energy savings up to 50% with no loss of cropping time.

While a mean daily temperature of 61°F is best, it is possible to work within a range up to 80°F. Above 80°F heat stress develops and should be avoided. A high mean daily temperature could mean a low (50°F) night temperature. Care should be taken to control humidity, especially at night. Dew forming on the leaves can enhance *Botrytis* formation. Plants grown under low temperatures should be kept at low relative humidities, if possible, to decrease chances for disease problems.

Cropping time can be reduced with root zone heating. After potting, root zone temperatures of 70° to 75°F promote root growth. Once the roots have reached the outer edges of the medium, a root zone temperature of 65° to 70°F can be maintained for the duration of the crop.

2. Light

a. Intensity. Geraniums are high light requiring plants. Light levels of 32 to 54 klx are practical for production. Full sunlight can be used as long as leaf temperatures do not exceed 85°F for long periods, as this leads to heat stress and a breakdown of chlorophyll in the young leaves (solarization). Reddening of the petioles and stem can also be an indication of too much light.

Low light levels cause growth to slow or even stop. Geraniums show no assimilation of carbon dioxide when light levels are below 1.5 klx. Fair to good growth occurs at 5 klx. Supplemental lighting can be used. High-pressure sodium lamps (400 watt) can be placed 6 feet above plant level to provide 6 klx (illuminate for 16 hours). Many growers will not choose to use supplemental lighting but will reduce temperatures to compensate for lower light levels. Lower light intensity and cooler temperatures will cause a significant delay in flowering, but may be a more economical approach in certain markets. As light levels change during a season, temperature and watering should be adapted to compensate for these changes.

Spacing also affects light intensity at plant level in the greenhouse. Close spacing reduces light to the lower portions of the plants. This can lead to yellowing and the eventual loss of the lower leaves. Also, close spacing reduces air movement and can increase chances of foliar diseases such as *Botrytis*. For top quality plant production, 4-inch pots should be spaced two to three pots per square foot. Plants in 6-inch pots do best at 1.5 pots per square foot (9-inch centers). Slightly tighter spacing can be achieved at all pot sizes, but decreased plant quality will result.

Irregular spacing can lead to asymmetrical growth. Geranium shoots tend to grow into open spaces on the bench. If spacing is tight or irregular, plants will still appear to be evenly spaced on the bench but may not be growing in the centers of the pots.

Ivy-leaved geraniums generally require lower light intensities for production. Generally, they do best under light levels between 20 and 40 klx. Because light intensity affects leaf temperature, both light intensity and temperature must be considered to produce top quality plants. At high air temperatures (>85°F) plants should be grown at the lower limits of light intensity. (However, high temperatures coupled with low light intensity can result in weak, spindly growth.) Under more moderate temperatures (65° to 80°F), plants can be grown near the maximum light level. If temperatures can be maintained lower than optimum, light levels can be increased. This light–temperature relationship differs among cultivars. Ivy-leaved geraniums can be placed into three groups, based on the production light levels (Table I).

Seedling geraniums should be grown under maximum light intensity. High light reduces the time plants are in the vegetative stage and hastens floral development. Low light reduces plant size, causes thin leaves, and reduces sugar and starch accumulation in the leaves, regardless of production temperatures.

b. Duration. Photoperiod does not have any direct effect on vegetatively propagated geraniums. Accumulated irradiance (intensity over time) affects several morphological characteristics, such as plant height, number of branches, and plant quality. Plants grown under high accumulated irradiance are generally short-

Table I

Light Levels for Ivy-Leaved Geraniums[a]

Light level	Cultivars
High (32–37 klx)	'Balcon Imperial'
	'Balcon Princess'
	'King of Balcon'
	'Gallilee'
	'Grenchen'
	'Salmon Queen'
	'Peppermint Candy'
Medium (27–32 klx)	'Balcon Royale'
	'Decora Pink'
	'Decora Red'
	'Cornell'
	'Decora Lavender'
	'Beauty of Eastbourne'
	'Double Lilac'
	'White'
Low (21–27 klx)	'Amethyst'
	'Harvard'
	'Sugar Baby'
	'Sybil Holmes'
	'Yale'

[a]Modified from Masterlerz and Holcomb (1982).

er, more highly branched, and have a greater fresh weight than plants grown under lower irradiance conditions. Accumulated irradiance directly influences floral initiation of seedling geraniums and will be discussed later in this chapter.

3. Media

To obtain rapid, consistent growth, the medium must provide good drainage and aeration while supplying adequate moisture and nutrients, and be pathogen free. Growers can choose from many commercially available mixes on the market today or mix their own. The exact components of the mix are not as important as the physical and chemical properties they produce. There is no magical mixture that is best for growing geraniums. The choice of media depends on many factors, such as the availability of materials, size and type of container for growing, how the materials are handled by the grower, and the method of watering. For the consumer, media should be low in soluble salts, high in aeration and water retention, slow to dry but also hard to overwater.

In the 1970s, the mark of a good grower was the mix produced. From the standpoint of cost, ease of use, labor, and potential problems, most growers will benefit from using a commercially available mix. Geraniums, however, can do well in properly made "home" mixes.

Components of mixes are sphagnum peat moss, vermiculite, perlite, aged pine bark, calcined clay, sand, and mineral soil. Commercial mixes usually contain two to four of these components, excluding soil. Most commercial mixes also contain some fertilizers such as calcium nitrate, potassium nitrate, superphosphate, trace elements, and a wetting agent.

Many growers find it useful to put a portion of pasteurized field soil in the mix. Mineral soil can add nutrient buffering capacity, reduce watering frequency, and add weight to the mix, all of which can be beneficial when growing in containers 4 inches or larger. Mineral soil, however, can be extremely variable, must be pasteurized before use, and performs poorly in small containers and flats. In light soils, a 1:1:1 soil:peat:sand or perlite works well. With heavier clay soils, a 2:3:3 or 3:5:5 ratio (on a volume basis) can be used.

Soilless mixes should have a total porosity of at least 80%, while mixes with soil should be at least 70%. The pH should be 5.5 to 6.5 for zonal and brocade geraniums, 5.3 to 5.8 for ivy-leaved geraniums, and approximately 6.0 for regal pelargoniums. When using at least 25% mineral soil (by volume), pH should be on the high side of the range. Plants can be grown in pure peat, but peat must never be allowed to dry out. Consequently, geraniums grown in peat frequently receive excess water, which causes elongated internodes, large leaves, and height control problems.

4. *Watering*

Watering is one of the most influential cultural practices on all greenhouse crops. There is an old greenhouse axiom, "The person on the end of the hose controls the profits." Watering is still more of an art than a science. This can be especially true for geraniums.

Geraniums should always be watered thoroughly enough so that 10 to 20% of the volume applied to the surface will drip out of the bottom of the container (Nelson, 1985). This translates to approximately 4 to 8 ounces per 4-inch pot and 6 to 12 ounces per 6-inch pot. This will provide thorough wetting of the medium as well as keep soluble salts concentration within acceptable levels.

Generally, the more water applied to geraniums the greater the incidence of disease, such as *Botrytis*. Also, more geraniums probably are lost by overwatering when young than for any other reason. Geraniums should be watered early in the day, and care should be taken to keep water off of the foliage. Watering may not be necessary on cloudy days. Geraniums should be kept slightly on the "dry side" but never allowed to dry out. Automated watering systems provide several advantages over hand watering, such as lower cost, reduced labor, more uniform and accurate watering, and less incidence of disease. Watering should be done early in the day, preferably in the morning. Under conditions of high humidity, watering late in the day can give rise to a foliar abnormality called edema, which will be discussed in more detail later.

Container size affects the frequency of watering. Flats and containers less than 4 inches in diameter have small volumes and are short, which means they are more susceptible to improper watering than larger containers. Depending on environmental conditions, plants should be inspected daily when first planted and twice daily when over 2 weeks old. They must be inspected frequently, but only watered as needed.

Geraniums are quite drought tolerant. They have the ability to reduce their stomatal openings and metabolic functions, effectively shutting down under such conditions. Research has shown that seedling geraniums will exhibit this trait when water is withheld for as little as 24 hours (Milks, 1986). This author has even withheld water for as long as 6 weeks, and the plants revived and began to grow again once water was reapplied (personal observations). This same trait has allowed geraniums to be dug in the fall and hung in basements all winter and planted again the following spring, with the resumption of growth.

This drought tolerance can work against a grower trying to grow "fast crop" geraniums. Once a plant has experienced drought stress, photosynthesis is effectively shut down and never returns to its previous level (Armitage et al., 1983). Plants grown in 4-inch pots took 3 days to recover to maximum photosynthesis, but the level was much lower than before stress occurred. If plants are stressed too frequently, significant delays can occur in finishing (Fig. 5). Under normal conditions, proper watering is the most influential cultural practice for successfully growing geraniums.

5. Nutrition

Zonal and ivy-leaved geraniums respond favorably to high fertility. A balanced fertilizer such as 15–15–15 works well for constant liquid feeding or when applied weekly. For constant liquid feeding, 200 ppm nitrogen and potassium has been successful for mixes containing at least 25% mineral soil. For soilless mixes 250 to

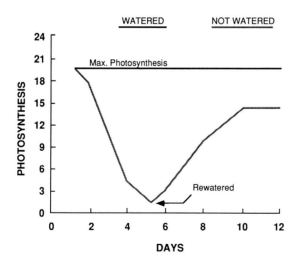

Fig. 5. The effect of water stress on net photosynthesis of seed geraniums. (From Armitage, 1985.)

300 ppm nitrogen and potassium can be used. Weekly fertilizations should be made with 500 ppm nitrogen and potassium. Plants that are fertilized weekly respond well to an additional top dressing of a slow release fertilizer such as Osmocote 14–14–14 when appllied at the rate of 1 teaspoon per 6-inch pot. Application of magnesium sulfate (epsom salts) at 2 pounds per 100 gallons every 4 weeks can improve foliage quality. Regal geraniums can be fertilized at 200 ppm nitrogen at every watering under bright conditions but should be fertilized at only 100 to 500 ppm nitrogen during low light situations.

Seedling geraniums can be fertilized at 150 ppm nitrogen from emergence with no damage. High fertility, however, will produce larger plants which may not be desirable. Plug trays can be fertilized starting 1 week after sowing at 150 to 200 ppm nitrogen. After transplanting, plants should be fertilized at 150 to 200 ppm nitrogen until well established (indicated by the roots reaching the sides and bottom of the container) when the rate can be increased to 300 ppm nitrogen. To avoid soluble salts buildup, clear water (no fertilizer) should be used at least once every four waterings. By weeks 8 to 10, seedling geraniums grown in soilless mixes may show chlorosis of the upper foliage, indicating an iron deficiency. Application of iron chelate at 1 teaspoon per gallon will overcome the chlorosis.

All geraniums need fertilizers high in nitrate nitrogen with a lesser percentage in ammonium. High ammonium can cause yellowing of the foliage and reduce plant growth. Ammonium should be less than 30% of the total nitrogen with the rest as nitrate nitrogen. Problems arising from ammonium toxicity generally occur under low light conditions or cool temperatures. Nitrogen deficiency in geraniums results in small, light colored leaves, small plants, and woody stems.

Phosphorus and potassium deficiencies generally do not occur in geraniums, but magnesium deficiency can be a problem. Deficiency symptoms of magnesium are interveinal yellowing of lower foliage, leading to marginal burn. If using a home made mix, corrections can include using dolomitic limestone in the mix and incorporating magnesium sulfate (epsom salts) at 2 to 4 pounds per cubic yard. Epsom salts can also be applied as a drench as stated above for all mixes. Care must be taken not to mix the epsom salts with other fertilizers, to avoid insoluble precipitates in the tank.

Iron deficiencies are rare in vegetatively propagated geraniums. However, symptoms can occur from anything that interferes with iron uptake by the roots such as overwatering, poor drainage, cold soils, high soluble salts, high pH, soil-borne diseases, insects, and nematodes.

Fertilization of all geraniums should be terminated before sale. Fertilization up to time of sale results in high soluble salts in the container, and reduces shelf-life of flats and pots. Seedling geraniums should receive clear water, beginning when flower buds first show color. Vegetatively propagated geraniums should receive clear water 2 to 4 weeks before sale.

6. Carbon Dioxide

Elevated levels (1000 to 1500 ppm) of carbon dioxide have been shown to influence geranium plant vigor, earliness, and overall quality (Mastelerz and Holcomb, 1982). As a result, there also seems to be a reduction in the incidence of *Botrytis*. Injection is most effective in closed greenhouses with no outside ventilation. The usual practice is to begin injection 1 hour before sunrise and stop in midafternoon or when a temperature of 80°F is reached, whichever occurs first. Young geranium plants are most sensitive to enriched carbon dioxide levels. Although injecting carbon dioxide for the entire crop will produce the largest plants, the most economical time for increased carbon dioxide levels is during the first 4 to 6 weeks of production.

7. Growth Regulation

Chemical growth regulators are not commonly used for stock plant production as they increase the time between cuttings. However, gibberellic acid (GA) can be used to promote stem elongation on geranium trees. Five weekly applications of 250 ppm beginning 2 weeks after planting has been successful.

B. Reproductive Stage

1. Temperature

For zonal and ivy-leaved geraniums, temperature does not directly affect floral initiation. Once flower buds are initiated, however, temperature has a profound effect on the rate of bud development. Flower bud development has been found to be directly related to the average daily temperature. As mean daily temperature increased, time to flowering decreased (Armitage et al., 1981).

High quality seedling geraniums can be produced under 60° to 62°F night temperatures and 70°F day temperatures. Time from visible bud to full flowering for seedling geraniums is also affected by temperature, like the vegetatively propagated geraniums. This is approximately 25 to 30 days for most cultivars and can be hastened or delayed by carefully adjusting night temperatures. In early spring, a 1 degree drop in night temperature can result in a 1 day delay in flowering. During late spring, altering night temperatures has less of an effect because of significantly higher day temperatures (Armitage, 1985).

Floral initiation in regal geraniums is controlled by exposure to cool temperature. After transplanting and subsequent plant establishment, plants are given a cool treatment of 35° to 50°F for at least 4 weeks, but preferably for 6. After the cooling period, plants should be kept at night temperatures of at most 58°F until buds are visible (about pea size). Then plants can be finished in approximately 8 to 10 weeks at 60°F night temperatures. Specialty propagators also provide "conditioned" young plants that have had their cold treatment. These plants can be transplanted and finished without any precooling treatment by the grower.

2. Light

Although seedling geraniums are not photoperiodic, they are "light accumulators" (Craig and Walker, 1963). The transition from juvenile to reproductive growth is genetically controlled in seedling geraniums, but accumulated light hastens the transition. Sensitivity to accumulated light is strongest during the first six-to-eight leaf stage. During this time high light intensities and long durations reduce time to flowering. After visible bud, light influences bud size and quality but no longer influences timing (Armitage and Wetzstein, 1984).

In areas with low winter light levels, supplemental lighting reduces time to flower, increases branching, and produces compact growth. High pressure sodium lamps (HPS) and metal halide (MH) lamps have proven to be best as supplemental light sources. Low pressure sodium is not recommended. Plants can be lit for up to 6 weeks from time of sowing in the seed flat or plug tray, for 18 hours per day. After 6 weeks, the benefits of supplemental lighting become less economical. Natural light intensities of 27 to 54 klx are commonly used commercially, but higher light levels will induce faster flowering.

Because vegetatively propagated zonal and ivy-leaved geraniums are propagated from mature stem tissue that has the ability to flower, the transition from the juvenile to reproductive phase has already occurred. These "reproductive" cuttings will develop flowers regularly while continuing to develop vegetative shoots. Although accumulated light has an indirect effect on flowering as it influences plant size, temperature is the controlling factor from visible bud to anthesis. Conversely, low light levels can delay flowering by slowing overall plant development that supports flowering.

Although cool temperatures promote flower initiation in regal geraniums, long photoperiods and high light intensities cause elongation of the peduncle and shorten the time for floral development. Because geraniums are light accumulators, regal geraniums benefit more from extended daylengths of 16 to 18 hours (>43 klx) than from traditional interrupted night lighting (10 P.M. to 2 A.M.) After dark, plants do best with extended light levels of at least 11 klx. Under natural high light levels (40 klx), regal geraniums have been successfully finished with extended day lighting (J. Robert Oglevee, unpublished data).

3. Growth Regulators

Most cultivars of vegetatively propagated geraniums can be successfully grown without the use of chemical growth regulation. However, growth regulators can prove useful in certain situations. Just as ethephon can be used to increase the number of cuttings in stock plant production, it can be used at the same rate (350 to 500 ppm Florel) for cultivars that are not as freely branching as others. Application must stop 4 to 6 weeks prior to sale to prevent further flower bud abortion.

Height on zonal geraniums can be controlled with foliar sprays of chlormequat (Cycocel). Rates vary from 750 to 1500 ppm, depending on application method. At 1500 ppm, plants should be sprayed 14 days after planting, with a second application for more vigorous cultivars 14 days later. Many growers prefer to use lower concentrations applied more often. At 750 ppm Cycocel should still be applied 14 days after planting but repeated three to four times at weekly intervals. Cycocel not only controls height but promotes earliness and produces darker leaf color.

Cycocel is a powerful growth regulator and can cause unwanted side effects. Under the best of conditions, Cycocel may cause yellowing of the foliage, particularly at higher rates, from which the plants eventually recover. Under poorer conditions, Cycocel can cause plant stunting, leaf distortion, and delayed flowering. Plants should not be sprayed in bright sunlight. Spraying on a cloudy day or early in the morning is best. The medium should be moist and the plants turgid. Plants should be well fertilized and not show signs of stress or nutrient deficiency. Temperatures should be at least 60°F for 24 hours when spraying. Plants should be sprayed with a light mist or to glisten. Runoff should be avoided.

Cycocel can also be used as a drench at 1500 ppm, applying 4 ounces of dilute solution per 4-inch pot. Medium should be moist, and pots not watered for 24 hours afterward. Drenching eliminates the chances for yellowing foliage and has a greater residual effect, but is more costly for both materials and labor. Foliar applications of ancymidol (A-Rest) at 200 ppm have also been shown to be effective (Tayama, 1988).

Regal geraniums can get quite large when cropping under high temperatures and supplemental lighting. Cycocel (used as above) is useful to keep plants compact and prevent excessive leaf expansion. However, some of the new compact cultivars require no growth regulator treatment.

Seedling geraniums are commonly treated with growth regulators to keep plants compact (Armitage *et al.*, 1978). Cycocel can be applied as a foliar spray at 1500 ppm 5 weeks from sowing and again 1 week later. The first application should occur when three to four true leaves have expanded. For plug production, 750 ppm Cycocel can be applied when two to three leaves have expanded (approximately 14 days from sowing) and again at 1500 ppm 28 days after sowing. Under high temperatures, another application at day 35 may be necessary. Ancymidol, at 200 ppm, is also effective as a drench or spray. As with Cycocel, plugs should be treated with half dosages (100 ppm) for the first two applications.

The need for growth control on seedling geraniums varies with cultivar. Some of the more vigorous cultivars may need more than the stated number of applications, while some of the newer dwarf cultivars do not need chemical growth regulation.

New chemical growth regulators have potential for geraniums. Paclobutrazol (Bonzi) has been applied to seedling geraniums at 2.5 ounces per gallon (78 ppm), first when the plants are approximately the size of a half-dollar and again 10 days later, with excellent results. Vegetatively propagated geraniums can also be treated with paclobutrazol, but with a reduced rate of 6 to 16 ppm (0.2 to 0.5 ounces per gallon). Sumagic (uniconazole) has been shown to give effective height control on zonal geraniums using 20 ppm, first when new growth is 1.5 to 2 inches long and again 14 days later (Roy A. Larson, unpublished data).

Gibberellic acid can be used for finishing plants, with a very weak spray solution (2 to 3 ppm) applied directly to the first two to three florets when they show color to increase floret size and longevity (Tayama, 1988).

The practice of using the difference between day and night temperatures to control height (DIF) has proven effective on many crops. Geraniums also respond to this treatment but not as strongly as most other species. To prevent extending cropping time, the mean 24 hour temperature should be the same as for normal production. Lowering the temperature 10° to 15 °F (low to mid-50°Fs) for 2 hours at dawn will provide the greatest effect (Fig. 6) (Royal D. Heins, personal communication, 1991). However, DIF should be viewed as another tool and not used as the sole method of plant growth regulation.

V. PESTS, DISEASES, AND PHYSIOLOGICAL DISORDERS

A. Insects and Mites

Fortunately, geraniums are not overly burdened with insect problems in production. Geraniums do not "attract" specific pests like some other crops, and a good,

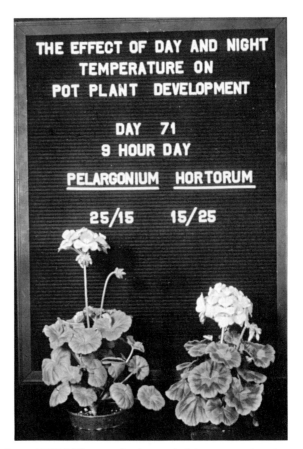

Fig. 6. The effect of DIF (difference in day and night temperatures) on growth of zonal geraniums. Plant on the *left* was grown under 77°F day and 59°F night temperatures, while the plant on the *right* was grown under 59°F days and 77°F nights.

clean greenhouse operation will generally not have many insect, mite, or related pest problems. However, geraniums still are susceptible to several pests.

The green peach aphid and the melon-cotton aphid can be troublesome to geraniums. These soft-bodied insects have sucking mouthparts that remove fluids and reduce plant vigor. Most are wingless, but some winged adults can form. These insects are sluggish and easily seen, especially on new growth. Aphids also secrete a sticky, sweet substance (honeydew) that attracts a sooty mold fungus under humid conditions.

There are two kinds of whiteflies (greenhouse whitefly and sweet potato whitefly) that attack geraniums. Whiteflies also have piercing-sucking mouthparts, remove fluids, and secrete honeydew. Whiteflies are winged and are usually found on the undersides of upper leaves. Gently tapping a plant infested with whiteflies results in several taking flight and eventually settling back onto the plant in a manner similar to falling snow. Whiteflies are attracted to yellow sticky traps, and such traps should be used to monitor whitefly populations. Populations can build up resistance to pesticides, so insecticides should be alternated for maximum effectiveness.

Thrips are very small insects with rasping mouthparts that feed on flowers and foliage of geraniums. Open flowers and expanding buds seem to be particularly favorite targets. Symptoms can be silver streaking on leaves and flowers, distorted leaves or flowers, and spotty pigment loss in flowers. The western flower thrip is by far the most noxious species of thrip in greenhouses. A more serious problem than actual thrip damage is the vectoring and transmission of the virus that causes tomato spotted wilt. While geraniums are not particularly susceptible to tomato spotted wilt virus, many other greenhouse crops are. Control of thrips can be very difficult. Large populations are virtually impossible to eliminate during production; control should start before geraniums are planted.

Spider mites are very small and difficult to see. They are generally found on the underside of leaves and can produce a "sand-blasted" appearance on the leaves with their rasping mouthparts. As populations build, characteristic webbing can be seen on the underside of leaves and between leaves or stems. Spider mites are most troublesome during hot and dry conditions. Keeping geraniums on the "dry side" can increase spider mite damage if plants become too stressed.

Caterpillars, such as cabbage loopers, leaf rollers, and cut worms, are larval stages of moths. They have chewing mouthparts and damage is very characteristic and can be easily seen. Cabbage loopers ingest leaf sections; leaf rollers "roll" or "tie" leaves together with webbing and feed from inside; cut worms usually cut plants off just above the medium surface. All of these caterpillars can be easily controlled with chemicals or biological controls.

Fungus gnats and shore flies are often confused with each other, have similar feeding habits, and can be more of a nuisance than a problem. Fungal gnats are dark gray or black. Shore flies are black, with reddish eyes, white spots on the wings, and short antennae. Adult fungal gnats do no damage to plants and are found on the medium surface or around the pots. Larvae, however, feed on fungi, decaying organic matter, and on some living tissue, although damage is usually very minor. Shore fly adults and larvae feed on algae. Although these pests are more nuisances than harmful, they may transmit disease.

All of the above pests can be controlled with pesticides. Growers should check with their state agricultural extension service for specific recommendations on chemicals, rates, and application methods. Pesticides cannot provide total control of insects, mites, and related pests. They should be used as part of an integrated program of pest control. Weeds are harbingers of many pests and should be eliminated both inside and immediately outside of the greenhouse.

Many insects are attracted to yellow. Yellow sticky traps, placed at plant height, can be used to monitor pest populations, so control can begin before large populations build up. Also, yellow clothing should be avoided, as this also attracts insects that can be transported by unsuspecting greenhouse workers and customers. "Pet plants" should not be allowed in production areas. Protective fine mesh screening can be used over vents to prevent insect entry into the greenhouse. Pesticides should be rotated to avoid resistance. A regular preventative program targeted at specific pests is essential for control of greenhouse pests.

B. Diseases

Although geraniums can be grown disease-free, there are several pathogens that can infect geraniums. As mentioned in the introduction, diseases almost eliminated geraniums as a greenhouse crop in the 1950s. Until the introduction of culture-indexed cuttings, epidemics of bacterial blight, *Xanthomonas pelargonii*, were common. Even today bacterial blight has the potential for causing severe losses. This bacterium attacks stems, leaves, and cuttings, with stem rot and wilt being the most common. The pathogen will be very active at temperatures between 70°F and 80°F, while higher or lower temperatures suppress bacterial growth. Plants receiving high nitrogen levels display symptoms faster than plants receiving moderate amounts of balanced nutrients.

Bacteria are commonly spread from an infected stock plant through cuttings. Sometimes (particularly in cool weather) plants are symptomless and yet can spread the disease through propagation. Knives used for propagation are ideal carriers of the pathogen. For this reason, knives should be disinfested in 70% methanol. Turgid shoots can be snapped off with fingers if disease transmission is feared.

Other greenhouse practices can quickly spread the disease. Overhead watering can easily spread the pathogen, especially among closely spaced pots. Water on the foliage should be avoided. Picking dead leaves and flowers from infected plants can inoculate the fingers for further transmission, so hands should be washed. Insects, such as whiteflies, have been shown to carry bacteria (Tayama, 1988).

Control of bacterial blight starts with culture-indexed stock plants. Plants held over from the previous season and used as stock plants have high potential for being sources of infection. Media should be pasteurized, and containers should be new or sterilized before use. Growing areas, whether benches or floors, should be sterilized. Hand tools and equipment should be disinfested.

Good sanitation is very important during production. Chemicals cannot substitute for good growing practices and are ineffective. Plants should be searched continually, and infected plants disposed of immediately.

Other bacterial diseases occur on geranium. Southern bacterial wilt (*Pseudomonas solanacearum*) causes initial wilting of lower foliage, followed by yellowing, blackening of the stem at the soil line, and plant collapse, similar to *Xanthomonas*. Bacterial leaf spot (*Pseudomonas cichori*) produces large (5 to 20 mm) black lesions on the leaves, surrounded by yellowing tissue. Bacterial fasciation, caused by *Corynebacterium facians*, is the development of many short, fleshy, thick, aborted stems with misshapen leaves. This fasciated growth can be 1 to 3 inches in diameter. The pathogen can produce other nonfasciated symptoms, such as a gall-like growth on the stem at or just below the medium surface.

All bacterial diseases are best controlled by starting with disease-free plants or cuttings, using good sanitation practices, and removing infested plants.

While most vegetatively propagated geraniums are susceptible to these bacterial diseases, seedling geraniums enjoy a greater degree of tolerance and are generally not infected with them. This does not mean that they are immune, but the chances of these diseases occurring in hybrid geraniums are greatly reduced.

Botrytis blight (*Botrytis cinerea*) or gray mold is the most common disease problem encountered in the commercial production of all geraniums. This fungus attacks leaves, stems, and flowers, and it can live on plant parts that have fallen onto the ground, or were even deposited in trash cans. Losses can occur in all phases of production, from stock plants through flowering.

Control involves altering the environmental conditions optimum for disease development, proper plant maintenance, and a preventative spray program. Free water should be eliminated from leaf and flower surfaces. Watering should be done early in the day. Under bench or bottom heating will dry the foliage. Greenhouse ventilation and heating should be synchronized to reduce the relative humidity to less than 75%. Plants should be properly spaced to allow adequate air movement, and inspected daily. Plant parts, such as old flowers, damaged leaves, and stems, should be removed. A weekly prophylactic fungicide program also will provide a good disease program.

Geranium rust (*Puccinia pelargonii-zonalis*), as the name of the pathogen implies, is most infectious to zonal geraniums. Ivy-leaved and regal geraniums are less affected, with many cultivars being immune. Because of the rapidity with which rusts can spread, this disease can be almost "explosive." The disease organism produces very small, pale yellow spots, mainly on the undersurface of leaves. Heavily infected leaves contain many pustules and fall off prematurely. The disease is spread by spores that occur in these pustules. Once the pustules erupt, the spores are carried aloft by the wind and can be spread quickly over large distances.

The best control is to start with pathogen-free cuttings. Plants and cuttings should be inspected on arrival. if a single pustule is found, the shipment should be rejected. Other cultural practices, as mentioned for controlling *Botrytis*, should also be followed for rust prevention.

All geraniums are susceptible to root and crown rots. *Pythium ultimum, Rhizoctonia solani,* and *Thielaviopsis basicola* are soil-borne pathogens that can devastate seedlings, cuttings, and finishing plants. These diseases attack either the basal stem or root system, causing wilting and death of the leaves. Using good sanitation procedures and pasteurizing the media, reduces the chance of these diseases occurring. Chemical drenches applied soon after potting or benching have been effective. Healthy, vigorously growing plants are less likely to express the disease, compared to stressed, weak, and poorly grown plants.

There are 12 viruses known to infect geraniums. None of these viruses have been shown to cause death of geraniums. They can be very destructive, but their economic impacts on geranium production have been difficult to assess. Disease symptoms can be acute or chronic. Acute symptoms are plant stunting, leaf vein yellowing, streaking, clearing, or banding; yellow to black spots and ring spots; distorted and/or crinkled leaves; and leaf mottling or mosaics. Less detectable, chronic symptoms include reduction in flower number; slower rooting, and reduced cutting production; and irregularity in production. Symptom expression can be severe or relatively mild under certain environmental conditions. Many commercial cultivars contain viruses, and symptoms may "come and go" in a crop or during a season.

Viruses cannot be directly controlled. Culture-indexed plants are screened for viruses. Insects are vectors for viruses; reducing these populations reduces the chance of infection.

C. Physiological Disorders

Edema is a leaf disorder brought on by an imbalance in water relations during production. This can occur on all geraniums, but ivy-leaved geraniums are most susceptible. Symptoms appear at first as small, water-soaked blisters, generally on the underside of the leaves, but they can also appear on stems or petioles. These blisters enlarge and burst, leaving small wounds. The wounds heal over with a permanent brown, corky layer.

Edema is most likely to occur during periods of cool, cloudy weather and when there is high moisture content in the medium. With warm, moist media and cool moist air, water moves into the plant quickly but does not move out through leaf evapotranspiration at a similar rate. As a result, the leaves become overly turgid, swell, and burst. This condition is prevented by providing proper spacing for adequate air movement, watering early in the morning, and venting the greenhouse late in the day to exhaust the highly moist air to prevent high humidity and condensation at night.

Solarization (heat stress), as mentioned earlier, is caused by high leaf temperatures resulting in chlorosis. This yellowing usually occurs in developing foliage during periods of high leaf temperature stress. High leaf temperature is most common under high light and high temperatures. However, high light and cool air temperatures in the early spring as well as low light (from shading) and high air temperatures in early summer can induce symptoms. Solarization is more likely to occur when air temperatures are greater than 85°F. Seedling geraniums are more heat tolerant but can show symptoms when temperatures exceed 90°F.

Reducing leaf temperature through shading and with temperature control can help prevent the problem. Cultivars differ greatly in their heat tolerance. In the same greenhouse, some cultivars may show severe chlorosis, while others show no symptoms. In southern locations or during extremely warm periods, heat-tolerant cultivars are advised.

In seedling geraniums the biggest physiological problem is petal shatter (Armitage et al., 1980). This can occur anytime after the flowers are open. This was more prevalent in older cultivars, but it is still a major problem. Although not as sensitive, geraniums from cuttings also can experience petal shatter. Petals fall off when exposed to ethylene, even in very small amounts. Exposure can be reduced by good plant grooming. Fallen flowers, leaves, and any other decomposing plant tissue give off high levels of ethylene. Greenhouse heaters that are either improperly vented or are burning inefficiently give off ethylene. Petal shattering can also produce side-effects on other crops. Petals that fall from ivy-leaved geraniums in hanging baskets can land on plants grown below, and disease organisms, such as *Botrytis*, can infect a second crop.

The plants' sensitivity to ethylene can be reduced by lowering temperatures as plants begin to flower. Lowering temperatures to 40° to 55°F will improve flower color, extend shelf life, and make maintenance easier. Plants shipped in full flower have a greater tendency to shatter than those shipped with less than half of the florets open.

Applications of silver thiosulfate (STS) when flowers show color can reduce the incidence of shattering. However, STS also significantly increases the susceptibility of some cultivars to Pythium root rot, and also does not have label clearance.

VI. MARKETING

A. Scheduling

Fast-cropping 4-inch geraniums (nonpinched) can occur in as little as 4 to 6 weeks from time of planting rooted cuttings. However, schedules of 7 to 10 weeks are more common. Six and 6.5-inch plants generally take 3 weeks longer to produce than 4-inch plants. Growing a pinched crop takes 2 to 3 weeks longer than growing nonpinched crops. Unrooted cuttings, directly stuck in 3-inch pots, take 7 to 10 weeks to finish.

Regal geraniums can be forced in 10 to 12 weeks when plants have been precooled. Unconditioned plants need a cold treatment of 4 to 6 weeks. After visible bud, plants can be finished in 8 to 10 weeks (Anonymous, 1990).

The ability to schedule accurately seedling geraniums varies from year to year. Because floral initiation is dependent on accumulated light levels, weather can increase or decrease the timing. Once visible bud is reached (about the size of a pea), flower timing is directly controlled by temperature. Because it takes 25 to 30 days from visible bud to flowering at 62°F night and 70°F days for most cultivars, growers can speed up or delay flowering with temperature. Seed companies can provide specific schedules for most cultivars.

Seedlings grown as plugs need approximately 2 to 3 weeks in the propagation area and another 3 to 4 weeks in production to be ready for transplanting. An additional 7 to 10 weeks may be needed after transplanting plugs into finishing flats.

B. Shipping and Storage

Geranium cuttings should be rooted soon after they are taken. If necessary, they can be stored in the dark at 41°F and 95% relative humidity (Tayama, 1988). Rooted cuttings store for 3 weeks, while unrooted cuttings can be stored for 2 weeks. Low pressure or hypobaric storage (1/10 atmosphere) can extend storage life for rooted and unrooted cuttings by 2, and 3 to 4 weeks, respectively. Disease control can be improved by dipping cuttings in a fungicide. Cuttings should be dry prior to storage.

Geraniums should be shipped in cool (40° to 65°F) temperatures. High temperatures increase ethylene production, which can lead to petal shatter and leaf yellowing. Plants held in dark containers for longer than 48 hours become chlorotic.

VII. POSTGREENHOUSE CARE

Geranium losses are often highest after production because of improper placement and handling. In outdoor retail areas, plants should be shaded with at least 50% shade cloth or wooden lath. Full sunlight can dry out plants and cause heat stress.

Ventilation is very important to prevent heat and ethylene buildup around the plants. Leaf temperatures can rise very high in confined spaces. Plants should be moved away from buildings for better air circulation. Plants deteriorate rapidly if put too close together. Plants placed directly on asphalt or concrete in the full sun with tight spacing will fare very poorly.

Ventilation is also improved by raising pots and flats off of the ground or floor. Placing plants on benches not only improves air circulation but enables customers to see the plants.

Sales areas should be cleaned constantly, with plant grooming occurring throughout the day. Dead leaves and old flowers not only are eye sores, but can also harbor pests and give off ethylene. Plants should be watered early in the morning to allow adequate time for the foliage to dry and to prevent heat stress during the day.

Responsible, knowledgeable people must be used in the retail area. Work assignments must be reasonable. A person assigned to the care and grooming of plants cannot also be expected to run the cash register. According to research, most of the problems related to post-greenhouse care are people-related, not plant-related (Armitage, 1982). Care tags, informing customers of the proper maintenance procedures for maximum longevity and beauty of the product, should accompany each purchase.

REFERENCES

Anonymous. (1986). *Geraniums*. Connellsville, PA: Oglevee Associates, Inc.

Anonymous, (1988a). *Bonzai*. Supplemental Labeling. ICI America.

Anonymous. (1988b). *Oglevee Manual*. Connellsville, PA: Oglevee Associates, Inc.

Anonymous. (1990). *Fischer Geraniums Manual*. Publication No. 50 P 928. Homestead, FL: Fischer Geraniums USA, Inc.

Armitage, A. M. (1982). "Keeping quality of bedding plants: Whose responsibility is it?" *Florists' Review*. December: 24–35, 39.

Armitage, A. M. (1985). *Seed Propagated Geraniums*. Portland, OR: Timber Press.

Armitage, M., Carlson, W. H., and Flore, J. A. (1981). "The effect of temperature and quantum flux density on the morphology, physiology and flowering of hybrid geraniums." *J. Am. Soc. Hortic. Sci.* **106**: 643–647.

Armitage, A. M., Heins, R. H., Dean, S., and Carlson, W. H. (1980). "Factors affecting petal abscission in the seed-propagated geranium." *J. Am. Hortic. Sci.* **105**: 562–564.

Armitage, A. M., Tsujita, M. J., and Harney, P. M. (1978). "Effects of Cycocel and high intensity lighting on flowering of seed propagated geraniums." *J. Hortic. Sci.* **53**: 147–149.

Armitage, A. M., Vines, H. M., Tu, Z. P., and Black, C. C. (1983). "Watering relations and net photosynthesis in hybrid geranium." *J. Am. Soc. Hortic. Sci.* **108**: 310–314.

Armitage, A. M., and Wetzstein, H. Y. (1984). "Influence of light intensity on flower initiation and differentiation in hybrid geranium." *J. Am. Soc. Hortic. Sci.* **109**: 114–116.

Craig, R., and Walker, D. F. (1963). " The flowering of *Pelargonium hortorum* Bailey seedlings as affected by cumulative solar radiation." *Proc. Am. Soc. Hortic. Sci.* **83**: 772–776.

Mastalerz, J. W., and Holcomb, E. J., eds (1982). *Geraniums III. A Manual on the Culture of Geraniums as a Greenhouse Crop*. Penn. Flower Growers.

Milks, R. R. (1986) "Culture and water relations of *Pelargonium* × *hortorum* Bailey 'Ringo Scarlet' seedlings established with limited root volumes." Ph.D. Diss. North Carolina State Univeristy, Raleigh.

Nelson, P. V. (1985). *Greenhouse Operations and Management*, 3rd ed. Reston: Reston, VA.

Tayama, H. K., ed. (1988). "Tips on growing zonal (vegetatively propagated) geraniums." *Ohio Coop. Extension Ser. Bull.* **FP–765**.

19

Other Flowering Pot Plants

P. Allen Hammer

Introduction to Floriculture, Second Edition
Copyright © 1992 by Academic Press, Inc.
All rights of reproduction in any form reserved.

I. INTRODUCTION

Many flowering potted crops grown in the greenhouse have been termed "minor" because they are not grown in large numbers across a wide geographic area. In a particular greenhouse or region, however, a "minor" crop becomes a "major" crop when it is grown in large numbers; therefore, the term "other" pot crops seems more appropriate. Another very important aspect of these other pot crops is the frequent lack of detailed and published research results. In many cases, it appears that these crops are not grown in large numbers because of the lack of cultural information. This deficiency, however, does not correct the need for a diverse product line for the consumer. Market need should be of prime importance when these other flowering potted crops are considered.

Sachs *et al.* (1976) presented criteria for evaluating new pot plant species (Table I). They proposed an interesting scheme to evaluate the suitability of species for pot plant culture, using a rating system. The rating system may not have universal appeal, but it would be well to evaluate each greenhouse crop according to the proposed criteria. It can provide insight into the strong and weak points of each crop, particularly when comparisons are made among present and potential pot crops. Research and development expenditures could be aimed at improving these weak areas. A simple example might be deciding between two very similar cultivars except that cultivar A requires growth retardant application whereas cultivar B is naturally of proper dimensions for pot plant. From a production point of view, cultivar B would be easier and less costly to produce and from a research point of view, why develop a lot of growth retardant response data for cultivar A when cultivar B is available? Very few decisions on crop selection are this simple. One should evaluate each potential pot plant crop carefully, however, before selecting which pot crops to grow.

Table I

Selected Criteria for Evaluating New Pot Plant Species[a]

Propagation time and special environment requirements during propagation
Greenhouse production (forcing) time
Seasonal or year-round production possible
Special environmental requirements for production, flowering, or both
Special medium and fertility requirements
Natural growth habits relative to pot size
Requirement and response to growth regulating chemicals
Freedom from insects and diseases
Floral qualities
Foliar qualities
Market potential
Postharvest life

[a]Modified from Sachs *et al.* (1976).

II. FLOWERING POTTED PLANTS

A. *Calceolaria herbeohybrida*

Calceolaria herbeohybrida Voss (Scrophulariaceae) is a group of cultivars generally grown as flowering pot plants in the greenhouse and was probably derived from *Calceolaria crenatiflora* Cav. It is sometimes called the "pocket-book plant" because its flowers resemble inflated pouches (Bailey Hortorium Staff, 1976).

Moe (1977b) has divided *Calceolaria* cultivars into four groups according to flower and plant size: *Grandiflora*, with 1.5- to 2-inch wide flowers on 12- to 16-inch plants; *Grandiflora primula compacta*, with 1.8- to 2-inch wide flowers on 8-inch plants; *Multiflora*, with 1.2- to 1.6-inch wide flowers on 10- to 12-inch plants; and *Multiflora nana*, with 0.8- to 1.2-inch wide flowers on 12-inch plants. Many cultivars are available, with continued improvements being made. The F_1 hybrida are more uniform in size and color, bloom 4 to 5 weeks earlier, and are more widely grown than open pollinated selections.

Calceolaria are seed propagated. The seed are small, with 17,000 to 40,000 seeds per ounce. They germinate in 8 to 10 days at 64° to 68°F. Seed are sown on the surface of the germination medium and are not covered. Damping-off can be a serious problem, so sanitation should be practiced and watering carefully controlled.

It has generally been accepted that *Calceolaria* require a period of temperatures below 60°F for flower initiation (Poesch, 1931; Post, 1937). Studies with the newer cultivars have shown *Calceolaria* to be a long-day plant, particularly at high irradiance (Johansson, 1976; Runger, 1975; White, 1975). Runger (1975) found *Calceolaria* 'Zwerg Meisterstuck' to be a long-day plant with a critical daylength of 14 to 15 hours when grown at high irradiance. At low irradiance (winter), flowering in long days must be preceded by treatment in cool temperatures of 50°F or short days at 60° to 65°F. Daylength had little effect during the cool temperature treatment. Johansson (1976) observed flowering during long days, but not during short days, at 60°F. Temperature and daylength affected plant quality as characterized by number of flowers and plant height (Table II).

Root zone heating has been shown to increase *Calceolaria* plant shoot fresh and dry weights and flower number with no effect on days to anthesis (White and Biernbaum, 1984b). Plants were grown at ambient 50° to 54°F air temperature with root zone heating of 68° to 72°F. Root zone heating also increased the foliar concentration of most macro- and micronutrients (White and Biernbaum, 1984a). *Calceolaria* grown in greenhouses with infrared or forced hot-air heating systems showed no differences in plant growth. Water relations were also not affected by the heating system as measured by stomatal resistance or water potential (Panter and Hanan, 1987).

Several schedules can be proposed for the production of blooming *Calceolaria* plants without the traditional lengthy exposure to cool temperature (Tables III to V). Of particular importance in the schedules are the following: (1) incandescent

Table II

Effect of Temperature and Daylength on Flowering of *Calceolaria* 'Harting's Red'[a]

Temperature (°F)	Daylength (hours)	Days[b] to flower	Number of flowers per plant	Height at anthesis (inches)
48	9	111	75	3.2
	12	73	104	3.9
	16	67	177	4.3
	24	46	113	4.7
54	9	150	55	3.5
	12	74	81	3.9
	16	46	93	4.3
	24	35	95	3.9
59	9	—	—	—
	12	94	64	4.7
	16	34	68	5.1
	24	31	63	5.5

[a]Cultivar was grown in a growth chamber with 6 klx Cool White fluorescent light for 9 hours per day. Daylength extension was with 0.060 klx of incandescent light. 'Portia' responded similarly except it is earlier flowering. Modified from Johansson (1976).
[b]From start of treatment to visible flower buds.

lighting (12 W/m^2) should be used to extend the daylength (Runger, 1975); (2) plants must have developed four to five pairs of leaves before long-day exposure; and (3) cultivar selection is important. Whatever schedule one uses, long days should be provided during flower initiation and early development to increase floriferousness.

Chlormequat has been used as an effective height control agent for *Calceolaria*. Two foliar sprays of 400 ppm, with the first applied when the flower buds were about

Table III

Traditional Schedule for the Production of Flowering *Calceolaria* in 15-inch Pots[a]

Operation[b]	Time of year	Night temperature[c] (°F)	Daylength
Sow seed	Early Sept.	64	Natural
Transplant to flats	Late Sept.	59–64	Natural
Transplant to 4-inch pot	Early Nov.	55	Natural
Transplant to 5-inch pot	Early Dec.	55	Natural
Begin cool temperature treatment	Early Dec.	45–50	Natural
End cool temperature treatment	Mid-Jan.	55	Natural
Flowering	Late March		

[a]Modified from Reiss (1974) and White (1975).
[b]The time from seeding to flower is 25 to 29 weeks.
[c]Day temperature should be below 64°F.

Table IV

Schedule for 'Hartings' and 'Portia' *Calceolaria*[a]

Operation	Time of Year	Temperature (°F)
Sow seed	Late July	64
Transplant to flats	Mid-Aug.	59
Transplant to 5-inch pots	Late Sept.	59
Begin long days (18 hours)[b]	Late Oct.	55
Flowering	Mid to late Jan.	

[a]'Portia' is earlier; thus seeding can be delayed 2 weeks if the two cultivars are flowered together. Proposed by Wikesjo (1976).

[b]Plants must have developed four to five leaf pairs before long days are given.

in diameter and the second 2 weeks later, reduced plant height by 18%. Single sprays at 800 ppm gave the same height control but resulted in some foliar phytotoxicity (Johansson, 1976). White (1975) reported that a chlormequat drench at 3000 ppm at the time of visible flower buds reduced plant height by 50% without injury. Chlormequat did not affect flowering time.

Major insect pests of calceolarias are whiteflies, aphids, and mites. Chemical control is usually necessary. Good sanitation should be practiced to avoid stem rot

Table V

Proposed Schedule for Fast Crop Production of 4-inch Pot Flowering *Calceolaria*[a]

Operation[c]	Time of year	Night temperature[b] (°F)	Daylength (hours)
Winter flowering[b]			
Sow seed	Late Sept.	64	18
Transplant to flats	Mid-Oct.	64–70	18
Transplant to 4-inch pot and begin short days[c]	Mid-Nov.	59–64	8
Begin long days	Late Dec.	55	18
Flowering	Late Jan.		
Summer flowering[d]			
Sow seed	Early April	64	18
Transplant to flats	Mid-April	64–70	18
Transplant to 4-inch pot and begin short days	Mid-May	59–64	8
Begin long days	Early July	59–64	18
Flowering	Late July		

[a]Cultivars 'Portia OE,' 'Hartings,' 'Lenz,,' 'Zwerg Meisterstuck,' and 'Yellow with Red Spots' in 4-inch pots. From White (1975) and Runger (1975).

[b]Time from seeding to flowering 17 to 19 weeks.

[c]Plants should have at least four to five pairs of leaves before short days begin.

[d]Greenhouse should be shaded to provide 54 klx. Evaporative cooling is essential. Summer flowering should be on a trial basis and will probably work only in cooler regions.

and *Botrytis* infections. The plants should not be planted too deeply, and a well-drained growing medium is essential.

B. *Campanula isophylla*

Campanula isophylla Moretti (Campanulaceae), sometimes knows as "Italian bell," "Star-of-Bethlehem," or "falling stars," is grown as a flowering potted plant or hanging basket. Its many flowers are borne erect in short, corymbose panicles. Cultivars are 'Alba' with white flowers, 'Caerulea' with blue flowers, and 'Mayi' with large grayish-pubescent flowers (Bailey Hortorium Staff, 1976). A cultivar, 'Blå,' with blue flowers, is popular in Norway.

Campanula isophylla is a long-day plant with a critical daylength of 16, 15, and 14 hours at night temperatures of 54° to 59°F, 64°F, and 70°F, respectively. Plants, once flowering, must be maintained under long days for continued development of flowers (Heide, 1965). Propagation is by cutting from stock plants maintained vegetatively under short days (12 hours or less). Root formation and growth are greatly inhibited on cuttings in flower, even when the cuttings are treated with the rooting hormone indolebutyric acid (IBA) (Moe, 1977b).

In a growth chamber study, Moe (1977a) showed that stock plants produced more cuttings with increased fresh and dry weight when grown at an irradiance of 10 klx and 900 ppm carbon dioxide than when grown at lower irradiance and carbon dioxide levels. Cuttings rooted better and produced more vigorous plants when taken from stock plants grown at increased irradiance and carbon dioxide. He recommended that stock plants be grown at 59° to 64°F with a 12-hour daylength using cool white fluorescent lamps (1 to 5 klx), as supplemental light, and 900 ppm carbon dioxide. Supplemental fluorescent light would not be required in much of the United States because natural light levels reach 10 klx.

Cuttings are 1 to 1½ inches long with five to six expanded leaves after the lowest two to three leaves are removed. Cuttings are dipped for 5 seconds in 1000 to 1500 ppm IBA solution and rooted in a 64°F air and 70°F media temperatures. They are ready for potting in 2 to 3 weeks. Plants are generally produced in 4-inch pots with one cutting per pot. After potting they are grown for 6 to 9 weeks at 64°F under short days (12 hours). Plants are then exposed to long days (16 to 18 hours) to initiate flowering. Hildrum (1968) reported that the addition of fluorescent lighting produced more vigorous plants than incandescent lighting and that shoots were shorter on the blue-flowered cultivar but not on the white-flowered cultivar when grown under fluorescent lamps. Fluorescent lighting was recommended to extend the daylength.

Daminozide (B-Nine) is a more effective growth regulator for *Campanula isophylla* than is chlormequat (Cycocel) (Brundert and Stratmann, 1973; Lavsen, 1967). A spray application of 2500 to 5000 ppm daminozide is recommended, whereas 10,000 ppm is phytotoxic. Shoots on the cultivar Blå (blue), tended to grow upright rather than pendant when treated with daminozide (Hildrum, 1968). The cultivar 'Alba' (white) is more vigorous than 'Blå' (blue), so it has a greater requirement for height control. The application of daminozide is recommended 1 week after the start of long days (Moe, 1977b).

During long days, the best-quality plants are produced when grown at a constant 64°F, day and night temperature. Plants flower 10 to 12 weeks from the start of long days.

A leaf spot on *Campanula isophylla* caused by *Ascochyta bohemica* has been reported (Garibaldi and Gullino, 1973). *Fusarium culmorum* infected plants have also been reported (von Wachenfelt, 1968), and great care should be taken to avoid propagating plants suspected of infection.

C. *Capsicum* Species and *Solanum pseudocapsicum*

Capsicum Species L. (Solanaceae), commonly known as Christmas peppers, and *Solanum pseudocapsicum* L. (Solanaceae), commonly known as Jerusalem or Christmas cherry, are grown as Christmas potted plants for their attractive fruit. Christmas pepper fruit is yellow, purple, orange, or red and of various sizes and shapes. The same plant may have more than one color of fruit at one time. Jerusalem cherry has scarlet or yellow globose, persistent fruit. It is of Old World origin, but it has naturalized in the tropics, subtropics, and United States Department of Agriculture zone 9 in the United States (Bailey Hortorium Staff, 1976). Jerusalem cherry fruit has been reported to be toxic, and their sales have been declining because of this reputation.

Christmas peppers have 9000 seeds per ounce. They are generally sown from late April to early May and germinate in 12 to 21 days at 70° to 80°F. They are grown in 2½-inch pots until large enough to be transplanted into the final pot. One plant per 4-inch pot, three plants per 5-inch pot, and four plants per 6-inch pot make a nice display (Rigdon and Wolfram, 1976). Plants are pinched when there are two or three nodes of growth and again when the new growth is 2 to 3 inches long, but not after early July. They should be well fruited by early December. Because fruit set is important, the plants should be grown where they are exposed to wind or flower movement for pollination. Some growers lightly brush over the plants with a stick to assure self-pollination. A night temperature of 60° to 65°F night is commonly used.

Jerusalem cherries have 12,000 seeds per ounce. They germinate in 15 days at 70°F, but will germinate at temperatures of 55° to 85°F and do not require light for germination (Table VI). They are sown in mid-February for a well-fruited plant in early December. They are transplanted to 2½-inch pots and remain there until they are large enough to transplant to the final pot. They are seldom grown in pots smaller than 5½-inches because of plant size. Plants are pinched beginning at two to three nodes of growth and again when the new growth is 2 to 3 inches long. They will not bear fruit for Christmas sales if pinched after mid-July.

Davis (1978) suggested the following schedule to produce a Christmas cherry tree for Christmas. Seed are sown in mid-February and transplanted to 2½-inch cell packs when large enough to handle. When roots fill that container, they are

Table VI

Percentage Germination of Several Seed-Propagated Pot Plants at Eight Temperatures with and without Light[a]

Genus (cultivar)	Light treatment	Percentage germination at temperature (°F)							
		50	55	60	65	70	75	80	85
Solanum pseudocapsicum	Dark	0	92	84	84	88	96	88	84
('Masterpiece')	Light	0	96	80	92	96	76	92	96
Exacum	Dark	0	0	1	0	1	1	3	0
('Tiddly Winks')	Light	0	0	100	100	100	100	100	98
Primula malacoides	Dark	0	0	0	0	0	2	0	0
('White Giant')	Light	0	0	8	12	26	10	0	0
Primula obconia	Dark	0	0	0	0	8	2	0	0
('Fasbender's Red')	Light	0	0	42	48	48	42	2	0

[a]Modified from Cathey (1969a,b).

transplanted to 6-inch pots. When plants are 4 to 6 inches tall that are treated with two foliar applications of gibberellic acid (GA) at 250 ppm, applied 10 days apart. A wooden stake should be placed in the pot to support rapid growth. Generous applications of nitrogen are made during this rapid growth period. When the rapid growth and elongation stop, all the bottom foliage is removed so only six leaves remain at the top of the plant. It is pinched at this time, and again whenever the new growth is 2 to 3 inches long. Again, the last pinch should not be made after mid-July for Christmas sales. To split the timing on the finishing of a crop, half of the crop can be pinched on July 1 and half on July 17.

Christmas cherries are generally grown at a 50° to 55°F night temperature. They are commonly grown outdoors in cold frames in the summer to aid in pollination for fruit set. It is important not to allow the roots to grow into the soil on which plants are placed, as root injury can cause loss of leaves and fruit when they are moved back into the greenhouse in the fall. Growers should be careful not to over- or underwater the plants during this reduced growth period in the fall, to avoid loss of leaves and fruit.

Before the plant is shipped, tips of shoots without fruit are pruned off to improve the appearance of the plant. The fruit is long-lasting, but long or rough shipping will cause some fruit abscission.

As mentioned previously, fruit of Jerusalem cherry has been reported as toxic, but there are no proven cases of such toxicity (Kingsbury, 1967). Solanaceous alkaloids have been isolated from the fruit, and its sole use as an ornamental plant should be emphasized.

D. *Clerodendrum thomsoniae*

Clerodendrum thomsoniae Blaf. (Verbenaceae), known as bleeding-heart vine, has attractive red and white flowers (large, white persistent, calyx and crimson corolla) (Fig. 1). It is a woody, twining, evergreen shrub native to west tropical Africa (Bailey Hortorium Staff, 1976).

No known commercial cultivars of *Clerodendrum* are available. Hildrum (1973) showed that plants from commercial greenhouses varied considerably in growth and flowering characteristics. He selected a clone that flowered profusely on short shoots. Beck (1975) reported a Wisconsin clone that had very little flower abscission, a problem with the European clone. It would be advisable to carefully select a superior clone for pot plant production because much variability does exist.

Early research suggested initiation of flower buds in *Clerodendrum* seems not to be affected by daylength, but flower development is delayed under long days. When long days were established by low-intensity irradiance from incandescent lamps, few flowers developed, and the stems elongated considerably even at a 16-hour daylength. However, daylength extension with fluorescent (20 W/m^2) lamps resulted in plants with short, floriferous shoots, even at a 24-hour daylength. Gibberellic acid delayed flower development (Hildrum, 1973).

Studies with the Wisconsin clone suggests *Clerodendrum* is a facultative short-day plant. Flower production was reduced as daylength was increased and light intensity was decreased. The red to far-red ratio of the spectrum affected flowering. Plants had fewer flowers with a low-red to far-red ratio, suggesting light quality is an important factor to consider in production (Koranski *et al.*, 1987).

Fig. 1. *Clerodendrum thomsoniae.*

Clerodendrum thomsoniae is vegetatively propagated from vigorous stock plants. Stock plants should be grown in full light at 70°F night temperature under long days (16 to 20 hours). Iron chlorosis develops at a pH above 6.3; therefore, soil pH should be maintained at 5.0 to 5.5 for optimum growth. Use of an acid fertilizer is recommended. Iron sulfate additions also have been effective (Beck, 1975; Wendzonka, 1978).

Stock plants should be renewed every 6 to 8 months. Plants grown from cuttings taken from older stock are of poorer quality than those taken from younger plants (Beck, 1975).

Single-node cuttings will root in 10 to 14 days under mist at an air temperature of 70°F and medium temperature of 72°F (Beck, 1975; Hildrum, 1972). Long days during propagation are recommended. A rooting hormone will speed rooting, but it is not essential. Beck (1975) suggested defoliation of the cutting before sticking to enhance uniformity of axillary bud break. Wendzonka (1978) suggested a small, "hard wood," single-node cutting 1 inch long as best for the most rapid growth and flowering.

Once rooted, one to three cuttings are potted in 4-inch or larger pots and grown under long days at a night temperature of 70°F. If a single cutting is used per pot, pinching is required when the shoots are 1 to 2 inches long. Hildrum (1972) has recommended a soft pinch, while Vereecke (1974) has recommended a pinch just above the first pair of leaves.

Short days (10 hours) should start at the time pinching is begun. Temperatures under black cloth coverings in summer should not be higher than 70°F because higher temperatures enhance shoot growth and inhibit flowering. Night temperatures below 70°F result in very slow growth. Ancymidol (A-Rest) reduces internode length and promotes flowering. Artificial short days are not essential when the plants are treated with ancymidol (Hildrum, 1972). A drench application of ancymidol at 0.15 mg per 4-inch pot (Beck, 1975) or 0.3 mg per 5½-inch pot (Sanderson and Martin, 1975), when the new shoots are 2 to 3 inches long gave the best results. Vereecke (1974) and Noordegraff *et al.* (1975) showed that two sprays of anycmidol (100 to 200 ppm), applied 10 days apart, were satisfactory. Daminozide (B-Nine) and chlormequat (Cycocel) gave poor results.

Production time from propagation to flowering is 12 to 14 weeks during summer months and 16 to 18 weeks in the winter, even with daylength control.

Plants infected with tobacco ring spot virus have been reported (Khan and Maxwell, 1975). Because *C. thomsoniae* is asexually propagated, plants suspected of infection should be discarded. There appear to be few other disease problems with *C. thomsoniae*. *Botrytis* has been observed on older leaves and flowers, but it is easily controlled. Whitefly is the major insect pest on *C. thomsoniae*, and chemicals controls are usually necessary.

Exposure of flowering plants to high temperature and low light levels can cause flower-bud and flower abscission. Fumigation with insecticides also can cause abscission. Shipping of flowering plants in darkness at high temperatures for more than 1 day will also cause flower abscission. High light or low temperature (below 60°F) will reduce the amount of flower abscission (Hildrum, 1972).

E. *Exacum affine*

Exacum affine Balf. f. (Gentianaceae) is the annual species of *Exacum* grown in the greenhouse as a flowering pot plant. It is commonly known as the German or Persian violet. The most popular cultivar has bluish flowers but a white-flowered selection is available. "Hortus Third" lists *Exacum macranthum* as the best of the genus with large, rich blue flowers, but it is probably not grown in the greenhouse because it is a biennial (Bailey Hortorium Staff, 1976). *Exacum affine* flowers are about one-half inch in size and have a mild fragrance.

Jim Irwin, of Canyon, Texas, a leading propagator, grows one plant per 6-inch pot year-round (Ball, 1975b). Production in smaller pots (3- to 4-inch) for spring sales is also popular (Ball, 1978).

Exacum affine is generally propagated by seed although it can be propagated by terminal cuttings. There are approximately 1,000,000 seeds per ounce. Major cultivars are 'Tiddly Winks,' 'Elfin,' 'Starlight Frangranie,' 'Blithe Spirit,' 'Blue Champion,' 'Jill,' 'Lou,' 'Bloom Rite,' 'Blue Double,' 'White Double,' and 'Midget.' 'Midget,' 'Petite,' 'Little Champs,' and 'Improved Elfin' are well suited for small containers because they are genetically compact. *Exacum* can be germinated over a wide range of temperature (60° to 68°F) but requires light for germination (Table VI) (Cathey, 1969b). Young seedlings grow slowly but do not require much space (Ball, 1975a,b; Kamp and Nightingale, 1977). Night temperatures of 60° to 65°F and day temperature of about 75°F are desirable during the production period.

Exacum has been successfully propagated from terminal and lateral buds *in vitro* (Torres and Natarella, 1984). Shoots were not successfully initiated from callus cultures. This technique may be useful in a superior cultivar that lacks uniformity in seed propagation.

Kamp and Nightingale (1977) devised a production schedule. Seed are sown January 1 and transplanted to 2-inch pots. After approximately 5 weeks, they are shifted to 6-inch pots and placed pot-to-pot for 3 weeks. They are then spaced on 12-inch centers and are salable in 6 weeks. Other schedules suggest a late fall seeding for spring sales in small pots or a December–January seeding for summer sales. Three plants per 5-inch pot have also been suggested to reduce production time (Ball, 1978). Cost of plants and production time must be considered with such a practice. Schedules will be affected by regional climatic differences.

It has been shown that *Exacum* flowers in relation to the total radiant energy the plants receives and not photoperiod. Crop time is increased in winter and can be reduced with supplementary irridation. Plants were too compact and brittle when grown under 16 hours of fluorescent light at 345 μmol m^{-2}s^{-1} while 183 μmol m^{-2}s^{-1} of light produced high quality plants (Holcomb and Craig, 1983).

Some light reduction has been suggested in the summer. Salable plants in 6-inch pots require 7 to 8 weeks in summer while 12 to 14 weeks might be required in midwinter (Larson, 1981). B-nine at a spray application of 2500 ppm prior to visible buds has been used during the winter months to reduce plant elongation in low light conditions (Furry and Albrecht, 1986). *Exacum* plants do not require pinching.

Gibberellic acid has been applied to *Exacum* in an attempt to influence flowering. Plants treated with 125, 250, and 375 ppm did have more flowers than untreated plants, but plant quality was significantly reduced (Neumaier *et al.*, 1987). Additional research is suggested for GA concentrations lower than 125 ppm and used in combination with a growth retardant to overcome the undesirable plant stretching. A double-flowered *Exacum* selection with poor flower opening characteristics was treated with GA to promote flowering. When plants with buds 0.24 to 0.35 inches in diameter were treated with two spray applications of 100 ppm GA at 2 week intervals or one spray application of 250 or 500 ppm GA, plants were salable in 2 to 3 weeks after initial treatment (Kofranek *et al.*, 1984).

A problem of calcium deficiency has also been shown on *Exacum*. The meristem and a few upper leaves rapidly wilted and died (Holcomb and Craig, 1983). Botrytis stem canker is a serious disease problem with *Exacum*. *Exacum* is also susceptible to tomato spotted wilt virus (Jones and Moyer, 1987). A disease prevention program is essential to avoid and control these pests. *Exacum* is susceptible to whiteflies, mealybugs, broad mites, and a number of worms (Baker, 1981). Western flower thrips have also been a serious pest. It can be used as an outdoor bedding plant and will tolerate full Texas sun until mid-July, but it will flower until frost in partial shade (Kamp and Nightingale, 1977).

F. *Hibiscus*

Hibiscus rosa-sinensis L. (Malvaceae) is a very popular potted plant in certain markets. Its production is probably limited because of very high light and temperature requirements for growth and flowering. Another handicap is the 1 day longevity of open flowers.

There are a number of cultivars that are propagated from 4- to 5-inch terminal cuttings rooted under intermittent mist or white plastic tents. Cuttings will be rooted in 4 to 5 weeks under proper conditions.

Potted hibiscus are generally grown as single-branched plants in a 4-inch pot. Cuttings are given soft pinches 2 to 3 weeks after potting. When grown under full sunlight and 70° to 85°F plants flower in 4 to 6 weeks. Growth will be retarded and flowering delayed at temperatures less than 70°F.

Hibiscus plants require high fertility, in the 300 ppm nitrogen and potassium range, at each watering. They are also subject to iron chlorosis when the pH of the growing medium exceeds 6.5.

Bud and leaf abscission are serious problems during dark storage or shipping of hibiscus. This problem can be helped by minimizing shipping time (3 days or less) and using shipping temperatures of 50° to 60°F (Gibbs *et al.*, 1989). Temperatures below 50°F can be considered as chilling temperatures for hibiscus.

Hibiscus plants can also be as garden or patio plants, but cannot withstand cold temperatures. It is difficult to get hibiscus plants to flower again indoors, as high intensity light is necessary. Red spider mites, aphids, and whiteflies are major insect pests of hibiscus. Control measures are essential during production.

G. *Impatiens walleriana*

Impatiens walleriana (Balsaminaceae), commonly known as New Guinea impatiens, had become a very important potted flowering crop in the late 1980s (Fig. 2). There are many excellent cultivars available with large flowers in a large array of colors. Foliage color can also be selected to add interest to the crop. Many of the newer cultivars do not require pinching because they are free branching. The 'Kientzler' series from Paul Ecke Ranch and the 'Sunshine' series from Mikkelsens, Inc. are perhaps the most popular.

Propagation is by vegetative cuttings. New Guinea impatiens require a well-drained growing media. Mixes should have good water holding capacity but must have good aeration. New Guinea impatiens are slow to start when first transplanted and are very sensitive to overfertilization and overwatering for the first 2 weeks after transplanting. Fertilizer rates should be 150 ppm nitrogen and potassium or less during this stage and the media should be allowed to dry out before watering. After this period fertilizer rates should not exceed 250 ppm nitrogen and potassium and plants should not be allowed to dry out. Optimum night temperature is 66°F to 68°F or slightly higher (68° to 70°F) during the early establishment period. Day temperature should be maintained at 75°F and should not exceed 85°F. New Guinea

Fig. 2. New Guinea impatiens, with a good assortment of flower and foliage colors.

impatiens require high light for optimum growth and foliage coloration. Shading is required in late spring and summer in most areas, however. A light intensity of 4000–6000 foot-candles (fc) of light has been considered as an optimum. Most cultivars require 8 to 12 weeks from transplanting to a marketable state, but a few cultivars do require a longer production schedule.

Major insect pests are aphids, spider mites, thrips, and fungus gnats. Chemical pesticides are necessary to control these pests. The major disease problems are Botrytis, root and stem rots (Pythium, Rhizoctonia, and Phytophthora), and tomato spotted wilt virus. *Botrytis* can be avoided with fungicide application and environmental control. Cool (55° to 65°F), humid greenhouse conditions should be avoided. Tomato spotted wilt virus has become a major disease problem. The virus has a very wide host range and many other floriculture crops are very susceptible to it. All plant material coming into the greenhouse should be closely examined and New Guinea impatiens cuttings or plants should be virus free. The virus is spread by western flower thrips so it is essential that thrips be excluded or eradicated from the greenhouse.

It has already been mentioned that New Guinea impatiens are free-branching and do not require pinching. They also are quite compact if grown under appropriate light conditions, and growth regulators are not needed unless vigorous cultivars are grown in small containers.

New Guinea impatiens can be grown in 4-, 5-, or 6-inch pots, hanging baskets, or patio containers. One cutting is planted in 4 or 5 inch pots, one to three cuttings in 6-inch pots, three cuttings in an 8-inch basket and four to five cuttings in a 10-inch basket. Marketable size will be attained more quickly with the higher number of cuttings per container, but cost of plant material must be considered. Most of the cultivars are patented, so royalties must be paid on all cuttings.

Adequate spacing is essential, as leggy plants are produced under crowded, shaded conditions, and high quality plants are not attained.

New Guinea impatiens can be used for interiorscapes, if lighting is adequate, and also in exterior plantings. Some shade is required outdoors for optimum growth and flowering. Originally New Guinea impatiens were regarded as quite drought-tolerant, but plant beauty is eventually lost if dry conditions persist.

H. *Pachystachys lutea*

Pachystachys lutea Nees. (Acanthaceae), sometimes erroneously called the "golden shrimp plant," is more popular in Europe than it is in the United States. Flowers have a white corolla about 2 inches long, subtended by gold yellow bracts, cordate to 1 inch long, borne on large, terminal upright spikes about 4 inches long. *Pachystachys* has dark green, glossy, narrowly ovate foliage. In Peru, its native habitat, it is a small shrub (Bailey Hortorium Staff, 1976).

Pachystachys is easily rooted from single-node cuttings. Mist and bottom heat probably speed propagation but are not essential (Holland, 1975).

Earliest flowering and best growth have been obtained at a night temperature of 72°F minimum (Pedersen et al., 1973; Pedersen, 1975). With bench heating, Moes (1976) recommends a bench temperature of 68° to 72°F and ambient temperatures of 60° to 65°F during the day and 55°F at night. Joiner et al. (1977) showed that reduced light intensity delayed flowering and decreased the number of flowers per plant (Table VII). More inflorescences and darker leaves were obtained with 200 ppm nitrogen and potassium at each watering as compared to half that rate (Pedersen, 1975). Salable Pachystachys plants can be produced from cuttings in approximately 100 days if they are grown in 4-inch pots, are pinched, and are given high light intensity, 70° to 75°F night temperatures, and adequate fertilizer.

As most of the plants are grown in a small pot size (4-inch), chemical growth retardants are required for a proper plant and pot balance. Ethephon reduces the number of lateral branches, and plants remain vegetative (Adriansen, 1974; Holland, 1975). A single spray application of daminozide at 1000 ppm when shoot length is 3 to 4 inches long increases the number of inflorescences and reduces plant height (Adriansen, 1974), but the same treatment was not effective in a study conducted in Florida (Joiner et al., 1977). Ancymidol sprays were not effective (Holland, 1975; Joiner et al., 1977). Pedersen et al. (1973), however, reported a 25 to 50 ppm spray of ancymidol to be more effective than chlormequat. Chlormequat reportedly had been the most effective growth retardant (Adriansen, 1974; Hermann, 1975; Holland, 1975; Joiner et al., 1977). Generally, a 1500 to 3000 ppm drench was most effective, but several 800 to 1500 ppm spray applications were also effective but caused some chlorotic leaf blotches. Applications of growth retardants were made when new growth was 3 to 4 inches long.

Postharvest care of Pachystachys has not been studied in great detail. The high light requirement for growth suggests postharvest life would be extended under high light conditions. Pachystachys has been criticized as being brittle for packing and transport but Holland (1975) does not believe this is a problem. Pachystachys are subject to a number of insect pests with whiteflies of greatest concern.

Table VII

Response of Pachystachys lutea to Reduced Light Levels[a]

Light treatment	Shoot length (inches)	Number of nodes	Number of flower spikes per plant	Number of days to flower
Control (Florida greenhouse)	12	5.6	7.7	83
40% shade	13	6.3	2.7	100
60% shade	12	6.6	2.2	115

[a]Modified from Joiner et al. (1977).

I. *Primula*

Several species of *Primula* (Primulaceae) have been grown as flowering pot plants. The most widely grown species are *malacoides* Franch, the annual primrose; *P. veris* L., the hardy perennial primrose with several subspecies; and *P. vulgaris* Huds. and *Primula × polyantha* Hort., called Polyanthus, which is a hybrid group with parentage of *P. veris, P. elatior,* and *P. vulgaris. Primula sinensis* sab. ex. Lindl. and *Primula × kewensis* W. Wats have been grown in the past as pot plants. *Primula obconica* Hance. is grown, but some greenhouse employees or customers could be allergic to it, as it does have a subescence that causes a skin rash (Bailey Hortorium Staff, 1976).

Many *Primula* selections, cultivars, and hybrids are available (Fig. 3). Several seed sources should be consulted before making the decision on which species to grow. *Primula vulgaris* is the most popular in Sweden although some *P. × polyantha* is also grown (Wikesjo, 1975). General comments about the culture of *Primula*, will be discussed, though these are some unique traits of different species.

Primula are seed-propagated. They germinate best at temperatures of 60° to 70°F, and there seems to be a light requirement (Cathey, 1969b; Thompson, 1967, 1969). The light requirement is apparently not required for *P. × polyantha* (Anonymous, 1977a). Seed do germinate best when sown on the surface and left uncovered (Turner and Heydecker, 1974). Germination does not occur at temper-

Fig. 3. *Primula malacoides.*

atures exceeding 70°F, making sowing in summer a questionable practice (Table VI). For 3- to 4-inch pot production, seeds are generally sown in either June or August when temperature controlled germination chambers are available and transplanted to 2-inch pots after 6 to 8 weeks. Some growers will transplant directly to the finishing pot while others transplant to the finishing pot when forcing starts. *Primula* require low light (probably 32 klx maximum) and should be shaded most of the year. They are exposed to 40° to 45°F until three to four flower buds are clearly visible. During this period, it is best to keep plants on the dry side. When flower buds are visible, plants can be forced (January–March) at a temperature of 55°F. Forcing time is usually 2 to 4 weeks. Forcing before the flower buds are visible will result in poor flowering and excessive leafiness. Forcing at temperatures above 55°F will result in elongated, weak flower stalks. *Primula malacoides* should be given short days during the low-temperature, flower initiation period for best flowering (Smith, 1969; Zimmer, 1969).

Overwatering (poorly drained soil) can result in disease problems. Leaf spots from *Rumularia* fungus have been reported. Aphids, red spider mites, and caterpillars are major insect pests.

Primula species provide variety for the consumer. Plants should be maintained in a cool place out of direct sunlight for increased longevity. Some selections are fragrant, and the perennial types can be planted in the garden, adding consumer value to the plant.

J. *Rhipsalidopsis gaertneri*

Rhipsalidopsis gaertneri [Regel] Moran (Cactaceae), commonly known as Easter cactus, is an epiphytic cacti with leafless and jointed stems. Flowers are very showy and bright red, and are 2 to 2½ inches in diameter.

Easter cactus plants are not grown in large numbers because of irregular and often unreliable flowering. Flower initiation has been reported to require a temperature of 63°F or lower. The optimum for profuse flowering is 50° to 60°F for 60 to 80 days (Peters and Runger, 1971). Later research by Boyle *et al.*, (1988) showed that flowering was greatly improved when 2 or more weeks of short days (8 hours of light) were followed by long days (natural photoperiod followed by incandescent lighting from 1600 to 2200 hours). It is not known if flower induction or initiations occurs during the short-day period. These plants were grown at 64°F. This technique shows great promise in improving the reliability of *Rhipsalidopsis* as a flowering pot plant.

The application of 100 ppm benzylaminopurine (BA) 14 days after the start of long days stimulated flower bud formation, but reduced flower size, increased flower bud abortion, and reduced overall plant quality. Application of silver thiosulfate (STS) did not reduce BA-induced flower bud abortion (Boyle *et al.*, 1988). *Rhipsalidopsis* is affected by few insect pests. However, mealybugs and scale insects can be a problem.

K. *Rosa*

Several *Rosa* species L. (Rosaceae) are grown as flowering potted plants for the Easter and Mother's Day holiday market or as spring-flowering pot plants. Even with the very large number of different roses available, few are suitable for pot culture. The major groups used are polyantha, floribunda, and miniature roses (Table VIII).

The polyantha group is by far the best for pot culture relative to plant size, shape, and flower display. The miniature types have not received much attention, but are certainly well suited for the smaller pot sizes (4-inch) and should play an increasing role in the market (Dubois *et al.*, 1988).

Production of potted roses in the northern United States would begin with the arrival and potting of dormant 2-year-old budded plants (miniatures are not budded) in early January for the Easter market and late January for the Mother's Day market. Plants are available in three grades: "X," at least one strong cane; "XX," at least three strong canes; and "XXX," at least four strong canes. Most producers force "XXX" plants because they make a more uniform, attractive potted plant. Plants should not be delivered to the grower until the time of potting unless they can be stored at 31°F with very high relative humidity. Plants should be unpacked on arrival, immersed in water for several hours or wet thoroughly, and covered with moist burlap for a day (Ball, 1975a). This will aid in new root growth and lateral shoot initiation when potted. Plants are usually grown in 7-inch pots because of the finished plant size.

Before or right after potting, plants are pruned to remove dead branches and to reduce the length of good canes to 7 to 8 inches. Canes should be pruned to an "eye" (bud) that is toward the outside of the pot. This will direct the new top growth to the outside of the pot. Plants are pinched when new growth is 3 to 4 inches long, with the last pinch 6 to 7 weeks before Easter and 5 to 6 weeks before Mother's Day. Previous recommendations suggested an initial temperature of 45° to 50°F (Laurie *et al.*, 1969). Work by Moe (1970) and grower experience (Clark, 1978) recommend 63° to 65°F for more uniform and well-proportioned plants, even though there were fewer flowers per shoot at the higher forcing temperatures. Finished plants were also shorter at 63° to 65°F. Moe (1970) showed Phosfon was not effective for height control of 'Margo Koster' and 'Morsdag'; flowers appeared faded after plants were treated with daminozide. The best plants were produced with two spray applications of chlormequat. The first application was made when the shoots were 2 inches long (15 to 20 days after cutback) and the second 10 days later. The spray concentration should be less than 2000 ppm to avoid foliar injury.

Moe (1970) found that 'Morsdag' grew more vigorously when budded on *Rosa multiflora* Thunb. than on *Rosa multiflora* 'Japonica,' which suggests some interesting possibilities in using root stock to control growth.

Moe (1973) proposed a scheme of growing potted roses from cuttings. Single node cuttings with one attached five-leaflet leaf were taken from stems with flower buds in color or in bloom. The upper most cutting had a higher percentage of rooting and more shoots than basal cuttings. A reduction in leaf area decreased

Table VIII

Rose Cultivars Commonly Forced as Flowering Potted Plants

Cultivar	Commercial synonyms	Date of origination or introduction	Flower color	Remarks
Polyantha				
'Dick Koster'		1929	Deep pink	Sport or 'Anneke Koster'
'Margo Koster'	'Sunbeam'	1931	Salmon	Slightly fragrant, sport of 'Dick Koster'
'Mothersday'[a]	'Fetes des Meres' 'Morsdag' 'Muttertag'	1949	Deep red	Sport of 'Dick Koster'
'Triomphe Orleanais'		1912	Cherry-red	Slightly fragrant
'Tammy'[b]			Clear pink	Sport of 'Mothersday'
Floribunda				
'Carol Amling'		1953	Deep rose-pink, edged lighter	Sport of 'Garnette'
'Garnette'		1951	Garnet-red, base light lemon-yellow	
'Marimba'		1965	Pink	Sport of 'Garnette,' slightly fragrant
'Roswytha'		1968	Pink	Sport of 'Carol'
'Thunderbird'		1958	Rose red	Sport of 'Skyland,' vigorous
'Bright Pink Garnette'[b]			Bright deep pink	Sport of 'Garnette'
Miniature				
'Chipper'		1966	Salmon-pink	Slightly fragrant
'Cinderella'		1953	Satiny white, tinged pale flesh	Fragrant (spicy)
'Pixie'	'Little Princess' 'Princesita'	1940	White, center faint Hermosa pink	Slightly fragrant
'Red Imp'	'Maid Marion' 'Mon Tresor'	1951	Deep crimson	Slightly fragrant
'Scarlet Gem'		1961	Orange, scarlet	Slightly fragrant
'Starina'		1965	Orange, scarlet	
'Sweet Fairy'	'Scarlet Pimpernel'	1946	Apple-blossom pink	Fragrant

[a]Correct spelling according to Modern Roses (1969).
[b]Not listed in Modern Roses (1969).

root formation and growth. Dips in 500 to 2000 ppm IBA solution for 5 seconds increased rooting but decreased the percentage of bud break. Indoleacetic acid (IAA) and naphthaleneacetic acid (NAA) were not effective in promoting rooting. An attractive potted plant was produced with three to five cuttings per 4- to 5-inch pot, following the schedule presented in Fig. 4.

Powdery mildew is the most serious disease in the production of potted roses. Fungicide applications and humidity regulation by temperature are necessary to control this disease. Mites and aphids are major insect pests and must be controlled with appropriate pesticides.

Temperature during postharvest handling is important, as shown by Maxie *et al.* (1974). When potted 'Mothersday' plants in flower were boxed and placed in the sun for 4 hours, unopened buds failed to open and subsequently abscised from the plant. There were no adverse effects on open flowers. Air temperature inside the box was recorded within the range of 70° to 100°F. They observed the same response in the field with several days of high temperatures. 'Margo Koster' did not show this response of bud abscission from high temperature in the field.

L. *Schlumbergera*

Schlumbergera bridgesii Lofgr. (Cactaceae), commonly known as the Christmas cactus, and *Schlumbergera truncata* Moran, the Thanksgiving cactus, are epiphytic cacti with flat-jointed stem segments (phylloclades) and are native to Brazil. They are both similar except *S.truncata* has margins of joints sharply two to four serrate (Fig. 5). The natural flowering periods also differ, with *S. truncata*

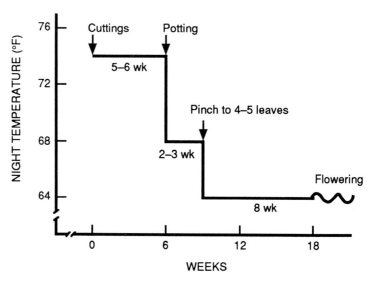

Fig. 4. Schedule for producing potted roses from cuttings. (From Moe, 1973.)

Fig. 5. Stem segments from *S. bridgesii* or Christmas cactus (*left*), and *S. truncata* or Thanksgiving cactus (*right*).

flowering in November and *S. bridgesii* flowering in December (Bailey Hortorium Staff, 1976).

Much confusion exists over the identity of the species used in research. *Schlumbergera truncata* was formerly names *Epiphyllum truncatum* and *Zygocactus truncatus*. These names are used in the literature with the common name Christmas cactus, but it would appear from photographs and descriptions that most of the work was done with Thanksgiving cactus. Most of the new hybrids that are called Christmas cactus in the trade also appear to be Thanksgiving cactus, *S. truncata*.

A large producer of Christmas cactus (Thanksgiving cactus) in the United States is B. L. Cobia Co. Inc., Winter Garden, Florida. They have an extensive research and breeding program and hold patents on selections from their breeding program. In their breeding program they are able to grow these cacti from seedling to flowering size in 17 months (Patch, 1977; Wade *et al.*, 1985).

Roberts and Struckmeyer (1939) reported that Christmas cactus (*Zygocactus truncatus*, probably *S. truncata*) flowered under short days (9.5 to 10 hours) but not long days (16 to 18 hours), at a night temperature of 63° to 65°F. At 55°F plants flowered under both short and long days, whereas at 70° to 75°F the plants did not flower under either short or long days. Runger (1961) reported similar findings with *Zygocactus* 'Weihnachtsfreude' ('Christmas Cheer') (probably *S. truncata*, based on photographs) except at the higher temperature. At 88°F, flowering required daylengths of less than 12 hours, whereas at 60°F flowering occurred regardless of daylength. He found the critical daylength to change gradually between 60° and 88°F, and 50°F prevented flower initiation. Vegetative growth took place only at temperatures above 70°F and under long days. Later, Runger (1968) found that under short days a 60°F night temperature was optimum for flowering, and, when combined with a higher day temperature, more flower buds were formed. Under long days, flower initiation was similar to that under short days when the day temperature was 50° or 55°F, but day temperatures of 75° and 88°F almost completely suppressed flower initiation even when combined with low night temperature. Long days inhibited bud development until the buds became visible. Runger (1970) also reported that the rate of development of the buds increased with increasing temperatures (treatments went to 75°F). Temperatures of 55°F inhibited early development after short days were first started. Buds aborted at 50°F. When the flower buds were ¼ inch long or longer, high temperature (75°F) was shown to cause the abortion of buds.

Poole (1973) also reported Christmas cactus (*Zygocactus truncatus* 'Christmas Cheer,' probably *S. truncata*) flowered at temperatures as high as 90°F when given short days for 6 or more weeks. Flower buds were visible 3 to 4 weeks after the start of short days and flowering time was not affected whether the plants received 6, 9, or 12 weeks of short-day treatment. Temperatures above 90°F probably caused flower abscission and increased the time to flower by 3 to 4 weeks.

From these studies, it would appear that optimum conditions for flowering would be 6 weeks of short days (8 hours) at a 55°F night temperature and 70° to 72°F day temperatures. The night temperature could be raised to 65°F after the initiation period to hasten flower development. High day or night temperatures (≥75°F) during the development period should be avoided to prevent bud abortion. Pruning the plants 6 to 8 weeks before short days start improves uniformity of flowering by increasing uniformity of maturity of shoots when flower initiation begins. Some growers suggest running plants dry at the beginning of short days to increase the number of flower buds. If this practice is followed, it should be discontinued during the development and flowering period as it will delay flowering and reduce flower size. A schedule for flowering Thanksgiving cactus, *S. truncata*, for Christmas in 4-inch pots is presented in Table IX.

Phylloclade numbers were increased by 150% when BA was applied during long days during vegetative growth (Heins *et al*., 1981; Ho *et al*., 1985). The application of a 100 ppm spray of BA increased flower bud number by 40% when sprayed 2 weeks after the start of short-day treatment.

Although these plants are cacti, they should be kept moist if vigorous growth is to be realized. Water stress during flower initiation also reduces the number of flower buds formed (Heins *et al*., 1981). A well-drained soil high in organic matter is recommended.

Knauss (1975) reported that several pathogens are capable of causing severe basal stem and root rot of Christmas cactus (*S. truncata*). He studied *Phytophthora*

Table IX

Proposed Schedule for Flowering *Schlumbergera truncata* for Late November to Mid-December[a]

Operation	Time of year
Stick cuttings[b]	Jan.–Feb. (64°–70°F night temperature, long day)
Prune terminals	Late June
Begin short days (8 hours)	Early to mid Sept. (55°F night, 70°–72°F day)
Begin natural day	Mid-Oct. to late Oct. (65°F night, 70°–72°F day)
Flower	Late Nov. to mid-Dec.

[a]*Schlumbergera bridgesii* would probably follow a similar schedule.
[b]Propagation can be delayed until May on some of the faster-growing free-branching hybrids, particularly for 4-inch pot production.

parasitica and *Pythium aphanidermatum* and reported that either can produce a root and basal stem rot. Disease development was more rapid with *Pythium*, and phylloclade abscission more prevalent, with *Phytophthora. Fusarium oxysporum* has also been isolated from *Schlumbergera* with basal stem and root rot (Mitchell, 1987). Soil pasteurization and fungicides help control the pathogens.

Schlumbergera bridgesii and *S. truncata* have few insect problems. Mealybugs and scales are the two most important pests. Both are difficult to control when populations are large, so early detection and elimination are essential.

Schlumbergera truncata plants remain in flower 4 to 6 weeks (Poole, 1973), with a flower longevity of 6 to 9 days (Patch, 1977). The main post-greenhouse problem perhaps is flower abscission from exposure to high temperatures (>75°F). Plants are usually not shipped great distances with many open flowers because the blooms are fragile. Foliar application of STS with silver concentration of 2 mM or less significantly reduced flower and bud abscission of *Schlumbergera* plants for as long as 4 weeks after application (Cameron and Reid, 1981). Plants were sprayed to runoff at the tight bud stage. Plants were experimentally stressed with exposure to ethylene and dark storage. Although STS is effective, it does not have label clearance for use on the crop.

M. *Senecio × hybridus*

Senecio × hybridus Regel (Compositae) is a member of one of the largest genera of flowering plants, estimated at 2000 to 3000 species. The florists' cineraria is a perennial but is grown as an annual. It apparently originated in England as hybrids between *S. cruentus* and *S. heritieri* and possibly other species in the Canary Islands. It is a very showy flowering pot plant and has many daisy-like flowers available in a wide range of colors and flower patterns (Bailey Hortorium Staff, 1976). Plants are sometimes divided into grandiflora and multiflora cultivars. The grandiflora group has fewer, larger flowers and blooms earlier than the multiflora group (Moe, 1977b). Finished plant size (width and height) also varies tremendously among cultivars (Fig. 6). Careful cultivar selection should be made, partially based on desired final plant size and pot size.

Cineraria plants are seed-propagated. Seeds are small, with 150,000 seed per ounce, and are easily germinated at 65° to 68°F. Seeds are usually sown in September, October, and November. Seedlings are transplanted to cell packs in flats or 2-inch pots as soon as they are large enough to handle. They should be grown at 64°F at this stage. When they become crowded, they should be transplanted to 4- to 6-inch pots for finishing. Most plants in the United States are finished in 6-inch pots at present, but smaller pots are now being used for mass market outlets. There has been some research on micropropagation of cineraria. There was a large clone-dependent variation in the growth and development of micropropagated plants (Gertsson, 1988 a,b).

Cool temperature (below 55°F) is required for flowering of cineraria (Hildrum, 1969; Post, 1942), thus most production is for the late winter to early spring market. January flowering can be achieved by selecting early-flowering cultivars and sowing seed in early August (Moe, 1977b). Although various cultivars have different critical temperatures and length of exposure time for flower initiation, 50° to 55°F for 6 weeks appears to be optimum (Hildrum, 1969). Plants should be grown at 64°F for 8 to 10 weeks before the cool treatment is given to have adequate leaf area for optimal response to low temperature (Moe, 1977b).

Daylength is reported to have no effect on flower initiation (Hildrum, 1969), but Hammer (1975) observed 90% flowering in 'Starlet' and 'Early Dwarf Erfurt' when grown under long days at 64°F minimum temperature. Plant quality is poor, how-

Fig. 6. *Senecio* (cineraria), a cool temperature crop.

ever. Additional photoperiod research with all cultivars is needed and may suggest that long-day plants can be selected.

Gibberellic acid can replace part of the cool temperature requirement; however, flower stems were elongated, and plant quality was very poor (Moe, 1977b).

Chemical height control is generally not recommended for cineraria. Chlormequat is not effective although daminozide spray applications of 5000 to 10,000 ppm when buds were first visible limited shoot elongation. It will delay flowering if applied during flower initiation (Moe, 1977b).

After the cold treatment, an additional 9 weeks are required until flowering. During this period, long days and a temperature of 52° to 55°F accelerate flower development. Temperatures above 55°F reduce plant quality. Flowering can be delayed by forcing at a cooler temperature. Some producers use this technique to expand the time of flowering. Some reduction in light intensity is often needed when plants are timed for late-spring flowering.

High fertility levels, particularly nitrogen, should be avoided as the leaves may become too large for the finished plant. A constant fertilizer program of 100 ppm nitrogen and potassium is probably adequate. A high nitrate nitrogen fertilizer is preferred over ammoniacal nitrogen because of the cool growing temperature. Cinerarias can be adversely affected by overwatering but frequent watering is necessary because of the large leaf area. Plants will wilt on bright, sunny days even if the medium is moist (Wilkins, 1974).

Mosaic and streak viruses can cause losses in cineraria production (Jones, 1944; Singh et al., 1975). Both can be seed transmitted and great care must be taken in seed production, so seed should be purchased from reliable sources. Mosaic and streak can also be transmitted by aphids and thrips, respectively. Diseased plants should be discarded as soon as they are noticed (Forsberg, 1975). Phytophthora erythroseptica has been identified as causing wilt of the laminae of larger basal leaves and epinasty of the petioles (Lucas, 1977). If these symptoms are observed, a pathologist should be consulted as little information is available on control.

Major insect problems with cineraria are mites, aphids, whiteflies, and caterpillars. An integrated pest management approach is necessary to control these pests.

Cineraria plants generally have a short post-greenhouse life because of the dry, warm conditions that prevail in most homes. They do better when held at temperatures below 55°F and at high humidity. They require frequent watering because of their large leaf area. Drying of the growing medium is probably the most severe post-greenhouse problems.

N. *Stephanotis*

Stephanotis floribunda (Asclepiadaceae) is grown as a twining pot plant in Europe (Wikesjo, 1982). Very fragrant, white flowers are borne in leaf axils in clusters of six.

Table X

Miscellaneous Flowering Potted Plant Species

Scientific name	Common name	Family	Description	Method of propagation	Container size (inches)	Growing Temperatures (°F)	Comments
Abutilon spp.	Flowering maple	Malvaceae	Soft-woody shrubs, maple-like leaves; flowers red, orange, and yellow hanging like bells	Seed, cutting	4, 8 to 10, hanging basket	60 min.	Mealybugs, thrips, aphids, and spider mites are serious pests
Aechmea spp.	Living-vase Urn plant	Bromeliaceae	Epiphytic herbs, leaves stiff and spiny; flowers on spike with colored bracts	Seed, division	4 to 6	60 min.	Keep cup filled with water. Ethylene treatment will promote flowering
Allamanda cathartica	Golden-trumpet	Apocynaceae	Climbing vine, milky sap, dark green shiny leaves; yellow, trumpet-shaped flowers	Cutting	6	65	Cool day/night temperatures of 63/52°F promotes flowering
Angiozanthos spp.	Kangaroo-paws	Haemodoraceae	Horizontally growing stems with vertically folded elongated leaves; flowers are wooly and range from yellow-green to bright red-green	Seed, division, and tissue culture	4 to 6	52 max.	Cool day/night temperatures of 63/52°F promotes flowering
Anthurum Scherzeranum	Pigtail plant	Araceae	Dark green leaves, blades leathery, oblong-obovate; spathe red heavily peppered white, spadix yellow	Seed, division, and stem cutting	4 to 6	65	Long lasting spathe
Aster novi-belgii	Aster	Compositae	Composite flowers are violet, blue, pink, red, and white	Seed, cutting	4 to 6	60	Seome asters require long days for flowering. Aster yellows a serious disease problem

Scientific name	Common name	Family	Description	Propagation		Temperature	Remarks
Bougainvillea spectabilis, Bougainvillea glabra	Paperflower	Nyctaginaceae	Trailing bushy, woody plant, B. glabra bright purple bracts, B. spectabilis—light brick—red bracts	Cutting	4, 8 to 10 hanging basket	65	12 hour or shorter day length, for most rapid and prolific flowering. Growth retardants also promote early flowering and compact plants
Calathea spp.	Calathea	Marantaceae	Short underground stems with attractively marked leathery leaves, inflorescence a dense, bracted spike or raceme borne on a scape	Division, sometimes cutting	4	55	Sensitive to fluoride
Catharanthus roseus	Madagascar periwinkle	Apocynaceae	Glossy green leaves, flowers rose-pink, varying to mauve and white	Seed	4	65	14–15 weeks from sowing to flowering. Can grow at 60°F 3 weeks after transplanting
Crossandra infundibuliformis	Firecracker flower	Acanthaceae	Dark green, shiny leaves; flowers salmon-orange on vertical spikes	Seed	4 to 5	65	Flower 7–8 months after seeding, flower over an extended time period
Dahlia spp.	Dahlia	Compositae	Tuberous-rooted plants, large, showy flowers available in a wide variety of flower types and colors	Seed, cutting	4	65	Flower in 4 months from seeding, dwarf cultivars best for pots. Subject to tomato spotted wilt virus
Dianthus caryophyllus, Dianthus chinensis	Carnations	Caryophyllaceae	Leaves small, bluish, with waxy surface; flowers in many colors, some fragrant	Cutting	4 to 6	52	'Colorado Majestic' mountain series bred and selected for pot production

(continued)

Table X *(continued)*

Scientific name	Common name	Family	Description	Method of propagation	Container size (inches)	Growing Temperatures (°F)	Comments
Euphorbia fulgens, Euphorbia Milii	Crown of thorns	Euphorbiaceae	Woody, spiny, climbing shrub, leaves mostly on young growth; small bracts pink or red	Cutting	4	60	An unusual plant that creates great interests when in flower
Fuchsia spp.	Lady's eardrops	Onagraceae	Growth upright or pendulous; flowers showy, mostly pendulous, rather fleshy various shades of red to purple to white	Mostly from cuttings	4	65	Long-day plant requiring daylength greater than 12 hours for flowering, allow 7–8 weeks from last pinch to flowering
Gardenia jasminoides	Cape jasmine	Rubiaceae	Evergreen shrub, dark green shiny leaves; flowers white, double, fragrant	Cutting	4 to 6	60–65	Soil pH should be below 6.0. Night temperature should be below 65°F for flowering. Foliar chlorosis is a major problem with gardenia
Hippeastrum hybrid	Amaryllis	Amaryllidaceae	Funicate truicate bulbs, leaves strap-shaped; large flowers of various colors on a hollow, leafless scape	Division, seed	6	70	Forcing time is 6–8 weeks. Temperature lowered to 60°F at color.

	Common name	Family	Description	Propagation			Comments
Justicia Brandegeana	Shrimp plant	Acanthaceae	Evergreen shrub; flowers in drooping spikes, bracts bronze to yellow-green	Cutting	4	60	Sometimes referred to as *Beloperone*. An interesting plant but not very showy
Justicia carnea	Plume flower	Acanthaceae	Upright plant with oval, grey green leaves; flowers purple-pink on short spikes at terminals	Cutting	4	65	Sometimes referred to as Jacobinias. Are now classified as *Justicia*
Lantana Camara	Shrub verbena	Verbenaceae	Hairy shrub, leaves rough above; flowers orange-yellow or orange, changing to red or white in flat topped heads. Various other colors are available	Mostly cutting, seed	4	65	Whitefly and spider mites should be monitored regularly
Mandevilla Sanderi	mandevilla	Apocynaceae	Woody vine, green, shiny leaves; large dark-pink, trumpet shaped flowers	Cutting	4 to 6	65	Beautiful plant, slow growing. Also called Dipladenia
Passiflora caerulea	Passion flower	Passifloraceae	Vines climbing by tendrils; flowers large white developing in leaf axils, corona filamentous purple at base, white in middle, blue at apex	Cutting, seed	4	60	Vigorous vine that will require support and training. An unusual flower form
Schizanthus × *wisetonensis*	Butterfly flower Poor man's orchid	Solanaceae	Annual, thin stems, leaves pinnately dissected; flowers terminal in many colors	Seed	4	50	Pinching improves branching. Some research suggests long days promote earlier flowering

Propagation is by single-node softwood cuttings. Stock plants should be selected for prolific flowering and short internodes. Cuttings can be rooted in peat or peat-lite media with low nutrients at 75°F. Rooting generally occurs at nodes so cuttings should be stuck with the nodes slightly below the rooting medium surface. Rooting occurs in 4 to 6 weeks with only one shoot developing per cutting.

Cuttings are potted in final containers with a support system for the vining growth. Plants are grown over the summer and fall of the next year. A growing temperature of 64°F is recommended and high summer temperatures should be avoided.

Plants are held at 55°F from November until early February as a rest period. Forcing begins in early February. Plants are grown at a constant 84°F with artificial long days of 14 hours for 8 to 10 days to initiate flowers. Long gays are continued after this treatment, but the temperature is reduced to 64°F until they are sold. Temperatures below 54°F cause flower bud abortion.

A salable plant should have 15 or more flower clusters with the oldest cluster buds 1 inch long. Although this is a 12 to 15 month crop, it should have some potential as a specialty crop, particularly for June weddings.

O. Miscellaneous Flowering Potted Plant Species

There are many other crops one could grow. These are listed in Table X, with some major requirements for culture.

REFERENCES

Adriansen, E. (1974)."Retardering of Pachystachys lutea med Ar.-85, Ethrel og CCC." Tisskr. Planteavl **78**(3): 331–341.

Anonymous. (1977a). "Strong future seen for primrose sales." Grower. **87**(14): 794–795.

Anonymous. (1977b). Flower and Foliage Plant. Crop Reporting Board, SRS, USDA, Washington, D.C.

Anonymous. (1989). New Guinea Impatiens—Cultural Information. Paul Ecke Ranch: Encinitas, CA.

Bailey Hortorium Staff. (1976). Hortus Third. New York: Macmillan.

Baker, J. R. (1981). "Insect pests of exacum." N.C. Flower Growers Bull. **25**(4): 9–10.

Ball, V., ed. (1975a). The Ball Red Book, 13th ed. West Chicago, IL. Geo. J. Ball, Inc.

Ball, V. (1975b). "Ohio—'75." Grower Talks. **38**(11): 17.

Ball, V. (1978). Six other pot plants-Exacum-interesting. Grower Talks. **41**(10): 17–18.

Beck, G. E. (1975). "Preliminary suggestions for the culture and production of Clerodendrum." Ohio Florists' Assoc. Bull. **547**: 6–7.

Boyle, T. H., Jacques, D. J., and Stimart, D. P. (1988). "Influence of photoperiod and growth regulators on flowering of Rhipsalidopsis gaertneri." J. Am. Soc. Hort. Sci. **113**(1): 75–78.

Brundert, W., and Stratmann, S. (1973). "Einsatz von zusatzlicht und wuchshemmitteln als kulturhilfe bei Campanula isophylla, Moretti." Dtsch. Gaertnerboerse. **73**(1): 4–6.

Cameron, A., and Reid, M. S. (1981). "The use of silver thiosulfate anionic complex as a foliar spray to prevent flower abscission of Zygocactus." HortScience. **16**(6): 761–762.

Cathey, H. M. (1969a). "Guidelines for the germination of annual, pot plants and ornamental herb seeds-1." Florists' Rev. **144**(3742): 21–23, 58–60.

Cathey, H. M. (1969b). "Guidelines for the germination of annual, pot plants and ornamental herb seeds-2." Florists' Rev. **144**(3743): 18–20, 52–53.

Clark, S. (1978). Personal communication. Andrews Greenhouses, Andrews, IN.

Davis, W. E. (1978). Personal communication. Sandpoint Greenhouse, Fort Wayne, IN.

Dubois, L. A. M., and DeVries, D. P. (1988). "Comparison of the plant habit of pot roses propagated *in vitro* and by cuttings." *Acta Horticulturae*. **226**: 611–613.

Forsberg, J. L. (1975). *Diseases of Ornamental Plants*, Spec. Publ. No. 3 (Rev.) Urbana, IL: Univ. of Illinois.

Furry, M. Z., and Albrecht, M. L. (1986). "Fragrant exacum is pot plant alternative." *Greenhouse Grower*. **4**(9): 54–55.

Garibaldi, A., and Gullino, G. (1973). "Malattie nuove o poco note delle piante da fiore e ornamentali in Italia." *Not. Mal. Piante*. **88/89**: 53–71.

Gertsson, U.E. (1988a). "Development of micropropagated plants from different clones of *Senecio ×	hybridus* in relation to BAP concentration and temperature *in vitro*." *J. of Hort. Sci*. **63**(3): 489–496.

Gertsson, U. E. (1988b). "Influence of macronutrient composition, TIBA and dark treatment on shoot formation and nitrogen content in petiole explants of *Senecio × hybridus*." *J. of Hort. Sci*. **63**(3): 497–502.

Gibbs, M. M., Blessington, T. M., Price, J. A., and Wang, Y.-T. (1989). "Dark-storage temperature and duration influence flowering and quality retention of hibiscus." *HortScience*. **24**(4): 646–647.

Hackett, W. P., Sachs, R. M., and DeBie, J. (1971). "Growing bougainvillea as a flowering pot plant." *Florists Rev*. **150**(3886): 21, 56–57.

Hammer, P. A. (1975). Unpublished research, Purdue University, W. Laf. IN.

Heide, O. M. (1965). "*Campanula isophylla* som langdagsplante." *Gartneryrket*. **55**: 210–212.

Heins, R. D., Armitage, A. M., and Carlson, W. H. (1981). "Influence of temperature, water stress and BA on vegetative and reproductive growth of *Schlumbergera truncata*." *HortScience*. **16**(5): 679–680.

Henrard, G. (1976). "Quelques facteurs influencant la croissance et le development du *Bougainvillea glabra*." *Ball. Rech. Agron. Gembloux*. **11**(1–2): 101–120.

Hermann, P. (1975). "Optimale nahrstoffversorgung und gartenbau-Cycocel erganzen sich. Ergebnisse aus versuchen zu *Pachystachys* und *Pelargonium* F_1—hybriden." *Gartenwelt*. **75**(24): 507–508.

Hildrum, H. (1968). "Virkning av iyskvalitet og B-nine pa vekst og glomstring hos *Campanula isophylla*, Moretti." *Gartner Tidende*. **84**: 491–493.

Hildrum, H. (1969). "Factors af fecting flowering in *Senecio cruentus* D.C." *Acta Hortic*. **14**: 117–123.

Hildrum, H. (1972). "New pot plant–*Clerodendrum thomsonae* Balf." *N.Y. State Flower Ind. Bull*. Nov./Dec.: 3.

Hildrum, H. (1973). "The effect of daylength, source of light and growth regulators on growth and flowering of *Clerodendrum thomsonae* Balf." *Sci. Hortic*. **1**(1): 1–11.

Holcomb, E. J., and Craig, R. (1983). "Producing exacum profitably." *Greenhouse Grower*. **1**(11): 18, 57.

Holland, R. (1975). "Lee Valley EHS trials reveal the right qualities for a good future in Pachystachys." *Grower*. **83**(14): 709.

Ho, Y.-S., Sanderson, K. C., and Williams, J. C. (1985). "Effect of chemicals and photoperiod on the growth and flowering of Thanksgiving cactus." *J. Am. Soc. Hort. Sci*. **110**(5): 658–662.

Johansson, J. (1976). "The regulation of growth and flowering in *Calceolaria × speciosa* Lilja." *Acta Hortic*. **64**: 239–244.

Joiner, J. N., Gruenbeck, E. R., and Conover, C. A. (1977). "Effects of shade and dwarfing compounds on growth and quality of *Pachystachys lutea*." *Fl. Flower Grower*. **14**(5): 1–4.

Jones, L. K. (1944). " Streak and mosaic of cineraria." *Phytopathology*. **34**: 941–953.

Jones. R. K., and Moyer, J. W. (1987). "Exacum, a host for tomato spotted wilt virus." *N.C. Flower Growers Bull*. **31**(5): 8–9.

Kamp, M., and Nightingale, A. E. (1977). "Exacum a durable, low-maintenance crop." *Florists' Rev*. **161**(4171): 98–99.

Khan, M. A., and Maxell, D. P. (1975) "Identification of tobacco ring spot virus in *Clerodendrum thomsoniae.*" *Phytopathology.* **65**: 1150–1153.

Kingsbury, J. M. (1967). *Poisonous Plants of the United States and Canada.* Englewood Cliffs, N. J.: Prentice-Hall.

Knauss, J. F. (1975). "Control of basal stem and root rot of Christmas cactus." *Fl. Flower Growers Bull.* **12**(12): 3.

Kofranek, A. M., Halevy, A. H., Moc, Y., and Kubota, J. (1984) "Opening double *Exacum affine* flowers with gibberellic acid." *Florist's Rev.* **174**(4506): 33–34.

Konjoian, P. S. (1988). "Scheduling New Guinea impatiens for profit." *Ohio Florists' Assoc. Bull.* **709**: 4–5.

Koranski, D. S., Struckmeyer, B. E., Beck, G. E., and McCown, B. H. (1987). "Interaction of photoperiod, light intensity, light quality and ancymidol on growth and flowering of *Clerodendrum thomsoniae* Balf. 'Wisconsin.'" *Sci. Hortic.* **33**: 147–154.

Larson, R. A. (1981). "Commercial production of exacum." *N.C. Flower Growers Bull.* **25**(4) 1–4.

Laurie, A., Kiplinger, D. C., and Nelson, K. S. (1969). *Commercial Flower Forcing,* 7th ed. New York: McGraw-Hill.

Lavsen, E. R. (1967). "Vaekstretarderende stoffer til *Campanula isophylla.*" *Gartner Tidende.* **83**: 71.

Lucas, R. (1977). "New plant disease record," *N.Z. J. Agric. Res.* **20**: 253–254.

Maxie, E. C., Hasek, R. F., and Sciaroni, R. H. (1974). "Keep potted roses cool." *Flower Nursery Rep., Univ. Calif.* March: 9–10.

Mitchell, J. K. (1987). "Control of basal stem and root rot of Christmas and Easter cacti caused by *Fusarium oxysporum.*" *Plant Disease.* **71**: 1018–1020.

Modern Roses. (1969). No. 7 Compiled by Int. Regist. Auth. Roses, Am. Rose Soc., McFarlane Co., Harrisburg, PA.

Moe, R. (1970). "Growth and flowering of potted roses as affected by temperature and growth retardants." *Meld. Nor. landbrukshoegsk.* **49**: 1–16.

Moe, R. (1973). "Propagation, growth and flowering of potted roses." *Acta Hortic.* **31:** 35–50.

Moe, R. (1977a). " Effect of light, temperature and CO_2 on the growth of *Campanula isophylla* stock plants and on the subsequent growth and development of their cuttings." *Sci. Hortic.* **6**: 129–141.

Moe, R. (1977b). "*Campanula isophylla* Moretti culture—Cineraria—*Calceolaria herbeohybrida* Voss." *Minn. State Florists' Bull.* April: 1–6.

Moes, E. (1976). "Temperature til *Pachystachys lutea.*" *Gartner Tidende.* **92**(18): 269–270.

Neumaier, E. E., Blessington, T. M., and Price, J. A. (1987). "Effect of gibberellic acid on flowering and quality of double persian violet." *HortScience.* **22**(5): 908–911.

Noordegraaf, C. V., Kuip, J., and Sytsema, W. (1975). "Gaat de clerodendron een interessante potplant worden?" *Vakbl. Bloemisterij.* **30**(50): 14–15.

Panter, K. K., and Hanan, J. J. (1987). "Stomatal resistance and water potential of Calceolaria grown under two different heating systems." *J. Am. Soc. Hort. Sci.* **112**(14), 637–641.

Patch, F. W. (1977). "It's Christmas every day for Cobia cacti." *Florists' Rev.* **161**(4177): 28.

Pedersen, A. M. (1975). "Standarddyrkning et *Pachystachys lutea* Nees." *Tidsskr. Planteavl.* **79**(4): 474–480.

Pedersen, A. M., Adriansen, E., and Moes, E. (1973). "Forsog pa sohus med *Pachystachys.*" *Gartner Tidende.* **89**(45): 638–639.

Peters, J., and Runger, W. (1971). "Blutenbildung von *Rhipsalidopsis gaertneri.*" *Gartenbawvissen.* **36**: 155–174.

Poesch, G. H. (1931). "Forcing plants with artificial light." *Proc. Am. Soc. Hortic. Sci.* **28**: 402–406.

Poole, R. T. (1973). "Flowering of Christmas cactus during the summer." *HortScience.* **8**(3): 186.

Post, K. (1937). "Further responses of miscellaneous plants to temperature." *Proc. Am. Soc. Hortic. Sci.* **34**: 627–629.

Post, K. (1942). "Effects of daylength and temperature on growth and flowering of some florist crops." *Cornell Agric. Exp. Stn., Bull.* **787**: 1–10.

Reiss, W. (1974). "Calceolarias and cinerarias." *Ohio Florists' Assoc. Bull.* **537**: 2.

Rigdon, K., and Wolfram, N. (1976). "May sowings." *GrowerTalks.* **39**(12): 27.

Roberts, R. H., and Struckmeyer, B. E. (1939). "Further studies of the effects of temperature and other environmental factors upon the photoperiodic responses of plants." *J. Agric. Res.* **59**(9): 699–709.

Runger, W. (1961). "Uber den Einflub der temperatur und der tageslange auf die blutenbildung von *Zygocactus* 'Weihnachtsfreude.'" *Gartenbauwissenschaft.* **26**: 529–536.

Runger, W. (1968). "Uber den Einflub diurnal und einmal weckselnder temperatur wahrend kurztag— und langtagperioden auf die blutenbildung von *Zygocactus* 'Weihnachtsfreude.'" *Gartenbau-wissenschaft.* **33**: 149–165.

Runger, W. (1970). "Einflub von temperatur und tageslange auf die blutenentwicklung von *Zygocactus* 'Weihnachtsfreude.'" *Gartenbauwissenschaft.* **35**(17): 379–386.

Runger, W. (1975). "Flower formation in *Calceolaria* × *herbeohybrida* Voss." *Sci. Hortic.* **3**: 45–64.

Sachs, R. M., Kofranek, A. M., and Hackett, W. P. (1976). "Evaluating new pot plant species." *Florists' Rev.* **159**(4116): 35–36, 80–84.

Sanderson, K. C., and Martin, W. C. (1975). "Cultural concepts for growing *Clerodendrum Thomoniae* Balf. as a pot plant." *Proc. Fla. State Hortic. Soc.* **88**: 439–441.

Singh, S., Verma, V. S., and Padma, R. (1975). "Studies on a mosaic disease of *Senecio cruentus* L." *Gartenbauwissenschaft.* **40**(2): 67–68.

Smith, D. R. (1969). "Controlled flowering of *Primula malacoides*." *Exp. Hortic.* **20**: 22–34.

Thompson, P. A. (1967). "Germination of the seeds of natural species." *J. R. Hortic. Soc.* **92**: 400–406.

Thompson, P. A. (1969). "Some effects of light and temperature on the germination of some *Primula* species." *J. Hortic. Sci.* **44**: 1–12.

Torres, K. C. and Natarella, N. J. (1984). "*In vitro* propagation of *Exacum*." *HortScience.* **19**(2): 224–225.

Turner, Y. J., and Heydecker, W. (1974). "The germination of polyanthus seeds." *Seed Sci. Technol.* **2**: 293–303.

Vereecke, M. (1974). "Chemical control of growth and flowering in *Clerodendrum thomsonae*, Balf." *Rijksuniversiteit, Gent.* **39**(4): 1597–1602.

von Wachenfelt, M. A. (1968). "Fusarium-rota pa *Campanula isophylla*." *Vaextskyddsnotiser.* **32**: 45–49.

Wade, J. M., White, R. S., Miller, H. A., and Vickers, D. H. (1985). "Phytochemical differences between a *Zygocactus* hybrid cultivar and its parental types." *Phytoiogia* **57**(2): 107–126.

Wendzonka, P. (1978). "Clerodendrum hanging baskets." *Focus Floric.* **6**(2): 6–7.

White, J. W. (1975). "Calceolarias—A year-round crop." *Pa. Flower Growers Bull.* **283**: 1, 6–9.

White, J. W., and Biernbaum, J. A. (1984a). "Effects of root-zone heating on elemental composition of Calceolaria." *J. Am. Soc. Hort. Sci.* **109**(3): 350–355.

White, J. W., and Biernbaum, J. A. (1984b). "Effects of root-zone heating on growth and flowering of Calceolaria." *HortScience.* **19**(2): 289–290.

Wikesjo, K. (1975). "Production programs for potplants and cutflowers in Sweden." *Hortic. Advis. Bul.* **89**.

Wikesjo, K. (1976). "Calceolarias for winter blooming." *Focus Floric.* **4**(3): 1–8.

Wikesjo, K. (1982). "Cultivation of *Stephanotis floribunda*." *Minn. State Florist Bull.* **31**(2): 5–7.

Wilkins, H. F. (1974). "Cineraria (*Senecio cruentus*)." *Minn. State Florists' Bull. Dec.* 13–14.

Zimmer, K. (1969). "Zur blutenbildung bei *Primula malacoides*." *Gartenwelt.* **69**: 137–138.

20

Bedding Plants

William H. Carlson, Mark P. Kaczperski, and Edward M. Rowley

Introduction to Floriculture, Second Edition
Copyright © 1992 by Academic Press, Inc.
All rights of reproduction in any form reserved.

I. INTRODUCTION

The bedding plant industry has grown steadily for the last 45 years. A major factor in this growth has been the increasing population and its expanding use of bedding plants. No other flower commodity has been as stable in its growth rate or in its demand by the consumer.

The term *bedding plant* no longer applies only to those plants grown for planting into outdoor flower beds. Over the years it has been broadened to include any herbaceous plant that is primarily used in the home landscape. Flowering plants, herbs, ground covers, perennials, small fruits—and even some woody ornamentals—can be found in such a listing.

Today, the home landscape includes, in addition to garden beds, planters for porches, patios, and window boxes—both indoors and out. It even includes special artificially lighted indoor planters. The bedding plant grower supplies plants for all of these situations.

Perhaps the best definition of a bedding plant would be "any plant, usually herbaceous, started under controlled conditions and then purchased and grown by a consumer." This definition includes an array of different plants such as tomatoes, strawberries, and garden chrysanthemums, in addition to traditionally grown flowering annuals and perennials. It also differentiates between started plants and finished ones, such as fully flowered pot mums, poinsettias or harvested cut flowers.

II. HISTORY

A. Bedding Plants

There is no recorded history of the bedding plant industry, and available information is scattered. Early bedding plant growers were probably vegetable growers who grew extra plants for consumer sale before starting crops for field planting.

We do know that bedding plant growers have been in the United States since colonial times. Dutch gardeners were reported in 1655 to be growing several crops, such as marigolds and violets, in addition to many other types of flowers and vegetables. This may have been the beginning of the bedding plant trade, though many colonists probably sowed their own seed for the many garden flowers they used.

According to Ball (1976), bedding plants were definitely part of early American horticulture. Many lovely gardens, such as the Williamsburg Palace Gardens in Virginia, attest to this. As early as 1789, bedding plants, such as geraniums and myrtle, were advertised for sale in Philadelphia. Richard Morris, a great American gardener of the early 1800s, suggested that tender annuals including ageratum, lobelia, and begonias, could have their season extended by starting the plants early in the greenhouse, and then setting them outside when the weather permitted.

The market for bedding plants further expanded around the time of the Civil War when mass plantings became popular, thus giving rise to the term *bedding plant*. As the nation entered the twentieth century the market for bedding plants grew rapidly. By 1923, bedding plants were the major source of income for florists.

Tremendous growth of the bedding plant industry occurred after World War II. With great strides made in transportation because of the war effort, the fresh vegetable market was able to locate in California where year-round production was possible. The produce could be shipped to the rest of the country via air or truck. As a result, many former vegetable producers in the east and the midwest turned to bedding plant production as a means of filling their vacant greenhouses. Since this time, the growth of the bedding plant industry has outpaced that of other floricultural crops.

The wholesale value of bedding plants increased 94%, from $16.9 million to $32.8 million, between 1949 and 1959. From 1959 to 1970, there was another 88% increase to $61 million, and from 1970 to 1977, an 85% increase to $113 million.

In 1976, the United States Department of Agriculture's Crop Reporting Board began to include bedding plants in its survey of 28 states. This report indicated that $681,599,000 worth of flower and vegetable plants were produced in 1987. The majority of bedding plant production is centered in only a few states (Table I).

Table II gives a typical breakdown of species raised by bedding plant producers as reported in a 1988 survey of members of the Professional Plant Growers

Table I

Wholesale Values of Flowering, Foliar, and Vegetable Bedding Plants for the Ten States Leading in Production in 1987[a]

State	Flower and foliage growers	Value ($)	Bedding geranium growers	Value ($)	Vegetable growers	Value ($)	Total value
California	121	96,969	19	2,230	56	7,062	106,261
Michigan	341	31,825	83	1,669	318	4,952	38,446
Ohio	384	25,313	203	3,335	350	4,062	32,710
Texas	109	19,418	39	2,878	87	4,745	27,041
Pennsylvania	425	19,192	82	2,459	365	3,414	25,065
Florida	51	14,707	NA[b]	NA	NA	NA	19,631
New York	257	11,775	75	1,631	233	2,698	16,104
North Carolina	123	7,829	40	3,425	111	1,419	12,673
Wisconsin	181	7,985	31	311	174	1,626	9,922
Colorado	69	7,585	14	562	53	1,111	9,258

[a]Date in thousands of dollars. From Floriculture Crop 1987 Summary. National Agricultural Statistics Service, Agricultural Statistics Board. USDA, Washington, DC.
[b]NA Not available.

Table II

Percentage of Crop Made Up by Various Species of Bedding Plants[a]

Ranking	Plant name	Percentge of crop	Ranking	Plant name	Percentage of crop
1	Impatiens	12.3	16	Cabbage	1.6
2	Cutting geranium	11.5	17	Pepper	1.6
3	Petunia	9.7	18	Celosia	1.2
4	Seed geranium	8.3	19	Coleus	1.2
5	Begonia	5.2	20	Dahlia	1.2
6	Marigold	5.0	21	Dianthus	1.2
7	Tomato	3.9	22	Garden mum	1.2
8	Pansy	3.4	23	Lobelia	1.2
9	Salvia	2.4	24	Verbena	1.2
10	Vinca	2.4	25	Zinnia	1.2
11	Ageratum	1.6	26	Aster	0.8
12	Alyssum	1.6	27	Browallia	0.8
13	Dusty miller	1.6	28	Phlox	0.8
14	Portulaca	1.6	29	Strawberry	0.8
15	Snapdragon	1.6	30	Others	11.9

[a]Adapted from Voigt (1989).

Association. Most species percentages have remained fairly constant over the past several years, though impatiens have made large gains while petunias have lost some popularity. Species mixes, however, vary greatly from one geographic area to another.

B. Seeds

Flowering plants as we know them did not exist in their present form four decades ago. The growth of the bedding plant industry closely paralleled that of the seed industry. Growers used new cultivars as they were developed. Traditionally, breeders would work extensively on developing a certain species. As new cultivars were introduced, the species would become very popular with the public and give the bedding plant industry a boost. The most important species developed, often said to be responsible for the modern bedding plant industry, was the petunia.

Petunias were known to be cultivated in America as early as 1880. About this time, Mrs. Theodosia Shepherd began breeding 'California Giant' or 'Superbissima' petunias. These were very popular because of their large flowers, which often exceeded 5 to 6 inches in diameter.

Several seed companies began developing new types of petunias. The early 1930s saw the introduction of the fringed types from Germany and Japan. These had the drawback of not producing true from seed, necessitating propagation by cuttings.

Many petunia cultivars were developed over the next 20 years, although plants were often of poor quality. A breakthrough occurred when a 1950 All-American Selection (AAS) named 'Fire Chief' was introduced. This light red, multiflora cultivar was the first open-pollinated petunia. The door was opened for the introduction of 'Ballerina,' the first hybrid grandiflora in 1952, and 'Comanche,' the first hybrid multiflora in 1953.

Originally, no real market was envisioned for petunias because of the high cost of the seed. But, through the persistence of several breeders, this problem was overcome and several hundred cultivars have been developed over the past 40 years. The petunia, since the beginning of the modern bedding plant industry, has always been numbered among the top five bedding plants.

In the late 1950s through the 1960s, many breeders began to work on the marigold. As the cultivars improved, marigolds became very popular with consumers furthering the growth of the industry. The cycle was repeated with seed propagated geraniums in the 1970s and with impatiens and begonias in the 1980s. The introduction of these new cultivars spurred growth of the industry.

All these improved varieties would not be available commercially, however, without the ability of seed producers to provide quality seed at a reasonable cost. Again, it is a situation where the seed and bedding plant industries have worked together.

Modern technology has also played an important role in the development of the bedding plant industry. Wooden flats have given way to plastic cell packs with individualized compartments. Growing media has changed from field soil and field soil mixes to scientifically developed artificial mixes. The grower now has many modern conveniences not available only a few short years ago. Among these are automatic seeders and transplanters, improved structures and benching systems, and computers for environmental control. Wise use of modern technology enables the grower to produce the crop more efficiently and economically and, therefore, for a greater profit.

III. BEDDING PLANT FAMILIES

For purposes of identification, different plants within a species are designated as cultivars. For example, a red petunia would have a cultivar name such as 'Red Joy,' whereas a white one may be called 'Snow Cloud.' All the different cultivars of petunia would compose a species. The common garden petunia comes from the species known as *hybrida*, which is a hybrid genus, obtained by crossing two original species, *P. axillaris* and *P. violaceae. Petunia* × *hybrida*, along with other species of petunia, comprise a genus. One who deals with bedding plants seldom encounters a need for classification more general than family. In fact, genus is usually sufficient.

The meaning of the word "cultivar" varies. With tomatoes, for example, the term is used to designate various types in the species *esculentum*. These include the varieties *cerasiforme* (cherry tomato), *commune* (common garden tomato), *grandifolium* (potato-leaved), and *pyriforme* (pear tomato). Then, with disregard to strict use of terms, the cultivar *commune* is divided into different garden cultivars, such as 'Better Boy,' 'Early Girl,' and 'Rutgers.' The foregoing discussion is simply intended to remind the reader of inconsistencies sometimes encountered in horticultural naming.

Named cultivars are sometimes further divided into strains. An improved strain of a cultivar may have larger or more intensely colored flowers, less conspicuous seed pods, more uniform fruit, or any number of other characteristics, which may be considered desirable, not found in the original strain.

Seventy species represent the bulk of bedding plants sold, but over 250 species contribute overall. The 70 species represent 25 families. Those 25 families with some representative species may be found in Table III.

Table III

Bedding Plant Families

Family name	Representative species
Amaranthaceae	Celosia, gomphrena, amaranthus
Apiaceae	Celery, dill, parsley
Apocynaceae	Vinca
Asteraceae	Aster, ageratum, african daisy, cosmos, chrysanthemum, dahlia, dusty miller, marigold, zinnia
Balsaminaceae	Impatiens, New Guinea impatiens, garden balsam
Begoniaceae	Fibrous begonia, tuberous begonia
Boraginaceae	Forget-me-not, various herbs
Brassicaceae	Cole crops, alyssum, candytuft, stocks
Capparidaceae	Cleome
Caryophyllaceae	Carnation, dianthus, baby's breath
Convolvulaceae	Morning glory, sweet potato
Cucurbitaceae	Cucumber, melons, squash, gourds
Fabaceae	Sweet peas, garden peas, beans, peanuts
Geranaceae	Pelargoniums
Lamiaceae	Coleus, salvia, various herbs
Liliaceae	Chives, garlic, onions, lilies, tulips
Lobeliaceae	Lobelia
Onagraceae	Fuchsia, clarkia, godetia
Polemoniaceae	Phlox
Portulacaceae	Portulaca
Scrophulariaceae	Snapdragon, foxglove
Solanaceae	Petunia, tomato, pepper, potato, tobacco, browallia, eggplant, nicotiana
Tropaeolaceae	Nasturtium
Verbenaceae	Verbena, lantana
Violaceae	Pansy, viola, sweet violet

IV. PLANNING FOR FUTURE CROPS

The plan for next year's crop should begin with the current year's production. Beginning growers will depend on the experience of others until sufficient personal experience is accumulated. Because personal experience is best, a good record should be kept of planting dates, quantities, germination times, special problems, and observations. Notebooks should also be kept in locations where important information can be recorded by employees. From all this information a complete, permanent record should be established to plan future crops. Future planning includes the ordering of seed and supplies, such as containers, labels, and growing mix or its ingredients.

Any of several methods can be developed and modified by the grower to tailor record keeping needs to a specific operation. Seed ordering charts can often be found in commercial seed catalogs, or the grower can develop a chart similar to the one shown in Figure 1. Special conditions will determine the number of columns and their headings for each grower. Information on number of flats grown should come from daily transplanting records.

The entire crop can be outlined in advance using action plans based on the 52-week crop calendar year (Fig. 2). Each step for producing a given crop is written down, along with when the step is to be performed and the person responsible for the step. Confusion is avoided, and each step can be performed at the proper time.

Wholesale catalogs usually list the average number of seeds per ounce for each species of plant. This number can be used to determine the number of plants expected from a given amount of seed. For example, at 80% yield of usable seedlings, 1/64 ounce of petunia seed at 256,000 seeds per ounce, would yield $256,000 \times 1/64 \times 0.80 = 3,200$ plants. Thirty two hundred plants at 72 plants per flat would yield $3,200/72 = 44$ flats plus 32 plants. Similar calculations can be made for other types. However, the seeds per ounce stated by the catalogs are not absolute. In several species such as petunia and salvia, the number of seeds per ounce may vary by cultivar. If in doubt as to the number of seeds in a given quantity, check with the supplier of the seeds before ordering.

Growing space will determine the number of seeds to order. Space allowed to various types of plants should be apportioned according to the market (see Table II). Some plants must be kept inside during cold weather, while others can tolerate some light frost. Therefore, petunias, pansies, onions, snapdragons, asters, the cole crops, and many perennials can be moved into cold frames after they are started, opening space inside the greenhouse for starting additional plants. Knowledge of the number of flats that can be moved out by certain dates will aid in planning both the seed order and planting schedule.

A large wall calendar with ample space for notes is helpful in planning the season. The entire year can be scheduled on a week by week or even day by day basis. Expected arrival dates of seeds and various supplies, sowing, transplanting, and expected shipping dates can all be recorded for easy viewing. This procedure will enable the grower to prepare work schedules, ensures that enough people are

VARIETY	AMT OF SEED SOWN	DATE SOWN	DATE TRANS.	PROPOSED TRANSPLANTING TYPES OF CONTAINERS USED			ACTUALLY TRANSPLANTED TYPES OF CONTAINERS USED			INVENTORY AS OF:			NUMBER SOLD			REMARKS

Fig. 1. Example of a seed ordering chart. From George J. Ball, Inc., Seed Department.

ACTION PLAN

Plan prepared on: Sept. 15, 1990
Prepared by: W. H. Carlson

Title: Alyssum
Action Plan No. Bedding Plant 1
Purpose: To produce 650 flats of alyssum

Projected sowing date: Week 2
Projected completion date: Week 9
Total crop time (# of weeks): 7

Procedure	Person responsible	Date due	Completed
1. Order seed. 650 flats (36 cell) at 80% germination = 41,120 seed. Need 0.47 oz. total. 35% blue, 30% pink, 35% white	Grower	Week 51	
2. Order tags, media for seed flats and prefilled flats.	Grower	Week 51	
3. Take pH reading for soil.	Grower	Week 2	
4. Fill and moisten seed flats.	Seeder	Week 2	
5. Sow seed.	Seeder	Week 2	
6. Apply Banrot to seed flat.	Seeder	Week 2	
7. Cover flat with polyethylene.	Seeder	Week 2	
8. Place in germination area.	Seeder	Week 2	
9. Germination complete (70°–75°F for 15 days).	Seeder	Week 4	
10. Place in holding area at 60°F day and night.	Seeder	Week 4	
11. Fertilize with 100 ppm 20–20–20.	Grower	Weekly	
12. Transplant. 60° day temperature, 60° night temperature.	Grower	Week 6	
13. Apply Banrot to flats.	Grower	As needed	
14. Watch for pest problems. Likely insects for this crop: spider mites, whitefly, thrips. Use sticker boards to identify. Diseases: Pythium, Rhizoctonia.	Grower	Daily	
15. Spray as necessary.	Grower	Weekly	
16. Take soluble salt and pH readings. Plot on chart.	Grower	Weekly	
17. Load and sell.	Shipper	Week 9	

Fig. 2. Sample of an action plan for planning a bedding plant crop.

available at peak times when needed, and necessary supplies are available when required.

The person responsible for ordering should be aware that most things must be ordered far in advance, even up to 1 year for some supplies.

V. PROPAGATION

With few exceptions, most bedding plants are now started from seed. The most notable exceptions are zonal geraniums, chrysanthemums, some perennials, ground covers, and small fruits. This discussion will be confined to seed propagation.

Fresh, high-quality seed purchased from a reputable supplier will provide the best opportunity for success in producing a high quality crop. Whenever possible,

F_1 hybrid seed should be selected over nonhybrid seed. Hybrid seed offers many advantages, including higher germination percentage, increased plant vigor, and more prolific flowering. On occasion, market conditions may demand that older, nonhybrid cultivars be grown. In most cases, consumers desire these older cultivars simply because they are unaware of the benefits of the newer cultivars. With a little effort, however, the consumer can be convinced of the advantages of hybrid plants.

The increased handling and production costs of hybrid seeds increase their price when compared to nonhybrid seed. This increase, though, is outweighed by the better performance of the hybrid seed. Economy-priced seed should be used with caution and only on a small scale.

Many cultivars of each species are usually available for purchase through commercial seed companies. Not all cultivars, however, perform the same for all geographical locations. Some cultivars of a species may be well suited for the cooler season found in the northern areas of the country, while others excel in the hot climate of the South. The grower can make an informed decision as to which cultivars to grow by making personal observations of their performances at field trials and in local gardens.

A. Seed Storage

Ideally, a grower will calculate the exact amount of seed needed for a season with none left to be stored for the following year. However, this is not always practical. Unexpected germination percentages, production problems, and market conditions will alter seed requirements.

Temperature and humidity are important concerns when seed must be stored from 1 year to the next. Harrington (1968) recommended that the temperature in degrees Fahrenheit plus the relative humidity in percent should not exceed 100. Thus, at 50°F storage, relative humidity should not exceed 50%; at 60°F, relative humidity should be no more than 40%, etc. Best results are obtained when seeds are stored between 35° and 40°F in a sealed container while in the original, unopened seed packets.

Some seed can be kept viable for many months or years. Other seed can lose viability in a year, and if they do germinate, the seedlings will be less vigorous. Among these are columbine, aster, cornflower, cleome, delphinium, kochia, larkspur, linum, lunaria, phlox, salvia, torenia, verbena, periwinkle, and viola. Seeds of these species should never be saved from one growing season to the next.

A germination test will indicate the viability of stored seed. One such method follows:

1. Count out an exact number of seed: 50, 100, etc.
2. Place seed between two pieces of blotting paper, paper towel, or soft muslin. A plastic container makes a good growth chamber.
3. Moisten the seed and maintain temperature at 70° to 75°F.

4. When the seeds germinate, count the seedlings to determine the percentage of germination. Do this by dividing the number of seeds sown into the number of seeds germinated.
5. If germination is poor, below 60%, it will probably not be worth the time and effort to germinate the seed.
6. If germination is fairly good, sow more seed than is needed to make up for the poorer percentage.

If stored seed is to be used for the upcoming season, determine its germination percentage well in advance so that replacement seed can be ordered if necessary. Discovering poor germination at a point too late for replacement will result in lost profits. Quality seed is the cornerstone of a quality crop.

B. Scheduling

Sowing dates depend on market dates, as well as geographical location. Because of great variations in plant growth and bloom dates due to natural environmental conditions, sowing schedules cannot be determined on a national scale. Conditions vary from year to year even in one location with early springs and rainy spells upsetting schedules. At best, a schedule can only estimate the time and events between sowing and marketing.

The sowing schedule also depends on whether plants are to be sold green or in bloom. If the market will tolerate green plants, considerable greenhouse time can be saved in producing them. Attempts should be made to educate consumers of the desirability of using green plants, the major advantage being less transplant shock due to overgrown or interlaced roots at transplanting. Although there are significant advantages in marketing a green crop, most bedding plants are marketed in flower, and it is the instant color that sells the crop.

Temperature is one factor that can be controlled that will influence scheduling. Each species has an ideal medium temperature for germination and early seedling growth (Table IV). Variations in temperature will delay germination, which in turn delays flowering. Note that by temperature, we mean temperature in the medium, not the air in the greenhouse. A thermostat placed at eye level will be easy to read but will not be an accurate indicator of temperature in the flat. Soil thermometers are needed to monitor temperatures and to maximize germination percentages.

Sowing dates are determined by first determining the expected marketing date for the plants. The days to flower or sale are then counted back from the marketing date to determine the sowing dates. Schedules should be flexible enough to allow for the time needed for sowing, transplanting, and shipping, because none of these tasks are performed in only 1 day. Flexibility is also necessary for delayed shipping of the plants because of inclement weather or unfavorable market conditions. Actual growing experience and good record keeping will provide the best guidelines for future crops for a given geographical location.

Table IV

Seed Information and Production Schedules for Selected Annuals

Common name	Scientific name	Approximate seeds per ounce	Germination light requirement[a]	Germination soil temperature (°F)	Germination time (days)
African Daisy	Dimorphotheca auratiaca	10,000	D	60–70	7–15
Ageratum	Ageratum houstonianum	200,000	D or L	70–80	7–10
Alyssum	Lobularia maritima	90,000	D or L	75–79	7–15
Amaranthus	Amaranthus tricolor	45,000	D or L	70–75	8–10
Aster (China)	Callistephus chinensis	12,000	L	70–80	8–10
Baby's breath	Gypsophila elegans	25,000	D or L	70–80	10–14
Begonia (fibrous)	Begonia × semperflorens	2,000,000	L	70–75	14–21
Begonia (tuberous)	Begonia × tuberhybrida	2,000,000	L	65	15–20
Browallia	Browallia speciosa	125,000	L	75	7–10
Candytuft	Iberis coronaria	9,500	D or L	70	7–14
Celosia (crested)	Celosia cristata	34,000	L	75	5–10
Celosia (feathered)	Celosia plumosa	39,000	L	75	5–10
Chinese forget-me-not	Cynoglossum amabile	5,000	D	60–70	5–10
Clarkia	Clarkia elegans	90,000	L	65–70	5–14
Cleome	Cleome spinosa	12,500	D or L	60° (night) 85° (day)	7–21
Coleus	Coleus × hybridus	110,000	L	65–75	10–15
Cornflower	Centaurea cyanus	7,000	D	65–70	10–15
Cosmos	Cosmos bipinnatus	5,000	L	70–75	5–14
Dahlia	Dahlia × hybrida	2,800	D or L	79–80	7–10
Dusty miller	Senecio cineraria	90,000	L	72–75	10–15
Forget-me-not	Myosotis aplestris	44,000	D	55	10–14
Gaillardia	Gaillardia pulchella	15,000	L	70–80	15–20
Garden pinks	Dianthus chinensis	25,000	D or L	70–75	5–7
Geranium	Pelargonium × hortorum	6,200	L	70–75	5–12
Gomphrena	Gomphrena globosa	5,000	D	70–80	14–20
Impatiens	Impatiens wallerana	52,000	L	70–75	15–18
Lobelia	Lobelia erinus	1,000,000	D or L	75–80	6–20
Marigold (African)	Tagetes erecta	9,000	D or L	75–80	5–8
Marigold (French)	Tagetes patula	9,000	D or L	75–80	5–8
Morning glory	Ipomoea purpurea	1,000	D or L	80	7–14
Moss rose	Portulaca grandiflora	280,000	D or L	75–80	7–10
Nasturtium	Tropaeolum major	175	D	65	10–15
Nicotiana	Nicotiana alata	200,000	L	70	7–14
Pansy	Viola × wittrockiana	20,000	D	63–68	7–14
Periwinkle	Catharanthus roseus	10,000	D	75–85	10–15

(continued)

Table IV *(continued)*

Common name	Scientific name	Approximate seeds per ounce	Germination light requirement[a]	Germination soil temperature (°F)	Germination time (days)
Petunia	*Petunia × hybrida*	210,000	L	75–79	3–10
Phlox	*Phlox drummondii*	14,000	D	65	10–15
Pot marigold	*Calendula officinalis*	3,000	D	70–80	5–10
Salpiglossis	*Salpiglossis sinuata*	125,000	D	70–75	14–21
Salvia	*Salvia splendens*	7,500	L	70–75	10–17
Satinflower	*Godetia grandiflora*	100,000	L	61	15
Snapdragon	*Antirrhinum majus*	180,000	L	70–75	10–15
Stock	*Mathiola incana*	18,500	L	70	7–10
Strawflower	*Helichrysum bracteatum*	45,000	D or L	70–80	5–14
Sweet pea	*Lathyrus odoratus*	400	D	55–61	14–35
Verbena	*Verbena hybrida*	10,000	D	75–79	14–18
Zinnia	*Zinnia elegans*	2,500	D	81	5–10
Vegetables					
Broccoli	*Brassica oleracea italica*	9,000	D or L	68–86	3–10
Brussels sprouts	*Brassica oleracea gemmifera*	9,000	D or L	68–86	3–10
Cabbage	*Brassica oleracea capitata*	9,000	D or L	68–86	3–10
Cantaloupe	*Cucumis melo reticulatus*	1,200	D or L	68–86	4–10
Cauliflower	*Brassica oleracea botrytis*	9,000	D or L	68–86	3–10
Celery	*Apium graveolens*	72,000	L	65–70	10–21
Collard	*Brassica oleracea acephala*	4,500	D or L	68–86	3–10
Cucumber	*Cucumis sativus*	1,000	D or L	68–86	3–7
Eggplant	*Solanum melogena esculentum*	6,000	D or L	68–86	10–14
Lettuce (head)	*Lactuca sativa*	25,000	L	68–70	5–7
Lettuce (leaf)	*Lactuca sativa*	25,000	L	68–70	5–7
Onion	*Allium cepa*	9,000	D or L	68	6–10
Pepper	*Capsicum frutescens*	4,500	D or L	68–86	6–14
Squash (summer)	*Cucurbita pepo melopepo*	400	D or L	68–86	4–7
Tomato	*Lycopersicum esculentum*	9,500	D or L	68–86	5–14
Watermelon	*Citrulius vulgaris*	300	D or L	68–86	4–14

[a]D, dark; L, light.

C. Germination Media

Just as high quality seeds are important to good germination, so is the medium in which the seeds are germinated. Best results are obtained when a light, sterile medium is used that provides the seeds with adequate amounts of moisture and oxygen. Peat-like mixes are now regularly used with excellent results.

Several good germinating mixes are available commercially, or the grower may elect to use a regular commercial growing mix or even a homemade mix. A good germinating mix will have a fine texture, allowing for even filling of the small cavities of plug trays.

Soil-based media are not recommended for germination because of the problems of sterilization and inconsistency among soil sources. If a soil-based mix is used, it should be sterilized with steam and not chemicals, because chemical sterilization will impede the germination of many bedding plant seeds.

A good germination medium should be free of weeds, insects, and disease organisms. Because commercial mixes may settle during shipment, they should be thoroughly mixed before use to provide a proper balance of air, moisture, and nutrients. If the medium holds too much water, the seeds will rot. If it does not hold enough water, the seeds may dry out and fail to germinate. The pH of the medium should be between 5.5 and 6.5. Many species will have reduced germination or fail to germinate all together if the pH is out of this range. In addition, the K value of soluble salts, when mixed 1:2 with distilled water, should be less than 80 to prevent burning of the emerging roots.

Selecting the medium that is best for an operation depends on economic factors, availability, specific practices in culture and experimentation. One should never forget the importance of trying new procedures on a small scale before investing time and money in their full scale use.

D. Sanitation

Good sanitation practices should be employed throughout the life of the crop. The easiest way to reduce or eliminate many production problems is to eliminate their opportunity to occur. Although good sanitation is not a guarantee against insect and disease problems, poor sanitation will almost always ensure disaster.

Before sowing the seed the grower should be sure that all flats, tools, and the growing medium are free of organisms that may be harmful to the seedlings. Commercial germinating and growing mixes are generally free of these organisms. Bagged mixes should be stored where the bags will not be damaged. Torn bags can be easily infested with weed seeds or disease organisms. Bulk mixes should be prepared on a clean surface and stored in a protected area until use.

Some growers may prefer to mix their own soil-based mix. Soil-based mixes are not recommended because of the inconsistency and lack of reliable soil sources. When these types of mixes are to be used they first must be sterilized. Several methods are available for soil sterilization, the most popular being steam. Large-scale steaming is accomplished with commercially available steaming carts or specially adapted dump trucks. Whatever the method used, the soil should be steamed to 180°F at the coldest spot for 30 minutes. Aerated steam can be used to heat the soil to 160°F at the coldest spot for 30 minutes.

Other sterilization methods include chemical or electrical sterilizers. Methyl bromide is the most common chemical method, but it may cause reduced germination in many bedding plants. Salvia and carnations will not grow in methyl bromide-treated soil, even if germinated in another medium. Electrical sterilizers use heating elements to raise the temperature of the soil to a desired level. This process, however, is usually not economical on a large scale.

Newly purchased flats are generally sterile and can be used as is. However, if flats are saved from year to year, they should first be cleaned to remove any adhering debris and then soaked in a disinfectant diluted with water at label rates for about 10 minutes. Chlorine bleach can be used for the same purpose, soaking the flats in a 1:10 bleach:water solution for 30 minutes. Tools and benches can also be disinfected with the same solutions.

Warm, humid conditions in the greenhouse are ideal for disease organisms to thrive. Problems can be prevented through sanitation practices. If diseases occur despite precautions, immediate measures should be taken, such as treating infested areas with fungicides so that the problems can be kept under control.

E. Media Testing

The growing medium should be tested by a soil testing laboratory before the crop is planted. The pH of the medium and soluble salts levels are the most important tests. Amendments can be used to adjust pH if necessary. The addition of limestone to the medium will raise its pH, while acids or acid-producing substances can be used to lower the pH. If the soluble salts level is too high, the medium should be leached before planting to avoid burning the seedlings. A good test will also analyze the nitrogen, potassium, phosphorus, calcium, and magnesium content of the medium.

Medium problems are more easily corrected if discovered before the flats are filled. Most problems discovered after the seed have germinated result in malformed plants that cannot be salvaged.

When soil-based media are used, the soil should be free of herbicides. The soil should be tested by germinating bean or tomato seeds in it. The plants should be grown until the second or third set of true leaves have expanded. If plants are malformed, another soil source should be selected.

Remember, media tests should be performed *before* the crop is affected.

F. Seed Sowing

Two primary options are available to the grower for sowing seed. These are using an automatic seeder or sowing the seed by hand. In any case, the grower should strive to use as sterile a technique as possible. All flats should either be new or properly cleaned and sterilized before use. Bags of premixed media should not be opened prior to use. Homemade media should be prepared on a clean surface, pasteurized if necessary, and used as soon as possible. Bench tops and all tools should be sterilized before use. Extra attention to sanitation from the beginning will lessen the likelihood of problems developing later.

The easiest and most accurate method of sowing seed is to sow into plug trays using an automatic seeder. A plug tray is a seed flat divided into individual cells. Plug trays may contain between 50 and 800 cells per tray. Sowing into plugs has many advantages over sowing in standard seed flats, but there are some disadvantages (Table V). With a little extra attention by the grower, however, plug production of high quality bedding plants can be very profitable.

Several types of automatic seeders are commercially available. Most rely on the same general principles of vacuum or compressed air. A manifold, a plate or drum with a series of nipples or small openings, picks up seed from a seed tray (Fig. 3). The seed are held in place by a vacuum, and then transferred to a plug tray, which is automatically moved through the seeder. The vacuum is released, and the seeds drop to the medium surface. Manifolds with different sized openings are available to accommodate different sized seeds. Extremely small seed, such as begonias or petunias, are available in pelleted form and odd-shaped seeds, such as marigold, may be de-tailed to better accommodate their use in a seeder. Many seed companies have introduced "hi-tech" or "refined" seed especially for use with automatic seeders. These seed are selected by size, shape, and weight to increase germination and seedling vigor.

Some growers may not wish to invest in an automatic seeder. Seeding can be performed successfully by hand using seedling flats. Seedling flats with preformed

Table V

Advantages and Disadvantages of Using Plugs for Starting Bedding Plants

Advantages	Disadvantges
Mechanical seeder sows faster and more acurately than manual sowing	Special equipment required; can be costly
Little or no transplanting shock to young seedlings	Small soil volume in which to grow seedlings; greater chance for error
Plants can be held for longer time before transplanting with little or no damage to seedlings	More space required in the germination area, which is heated to a warmer temperature than other growing areas
Less spread of disease through the seedling flat	Requires a competent operator
Individual plants are easier to handle, decreasing transplanting time	
Saves overall greenhouse space and allows for extra cropping of the same area	

Fig. 3. Vacuum seeder for sowing plug trays.

rows are preferred over open bottom flats because diseases will be less likely to spread through this type of flat. With either type flat, a shallow depression is made in the medium surface to guide placement of the seed. Several inexpensive tools, such as a handheld, vibrating seed trough, are available to aid in placement of the seed.

When planting small seed by hand, such as petunia or begonia, a small amount of sugar or silica sand can be mixed in with the seed. The sand or the sugar will help the sower monitor the rate of sowing. Some growers display a tendency to oversow flats. As a result, seedlings quickly crowd each other and are difficult to separate at transplanting. A general rule to follow is to sow enough seed to produce no more than 1000 seedlings per 11- × 21-inch flat. Seed geraniums can be sown at a lower rate to produce 200 seedlings per flat.

No matter which sowing method is used, several steps should always be followed. Plug trays or seed flats should be loosely filled with a good growing medium. The medium should not be packed down, as this reduces oxygen content. The flats should be well watered before sowing. Watering after sowing may wash away many seeds resulting in reduced germination. Most seeds, except those that are extremely small, should be covered with a thin layer of fine vermiculite to maintain moisture around the seed. Seed of some plant species require light for germination; a thin layer of fine vermiculite will allow enough light to penetrate to the seed while still maintaining the moisture around the seed. Planted seed flats should be moved from the sowing area to the germinating area as soon as possible. They should be moved carefully to prevent the seed from bouncing out of the seed flats. Once in the germination area, high humidity should be maintained to prevent desiccation of the germinating seeds. This can be accomplished by covering the seed flats with a suitable material, such as clear polyethylene, or placing the seed flats under mist.

Sowing must be done carefully. Haphazard methods will waste expensive seed and increase the chances of other problems developing in the crop. The grower must pay close attention to each step in producing the crop to ensure success.

G. Conditions for Germination

After the seeds are sown, ideal conditions should be maintained as nearly as possible. Seeds need heat, air, and moisture to germinate. Some species also require light.

Most species do not have the same ideal conditions for germination. Some species may require very warm temperatures for germination, others cooler temperatures. In most commercial operations, providing several germinating temperatures is impractical if not impossible. Most seeds will germinate satisfactorily if medium temperatures are maintained, between 70° to 75°F, in the germination area. Those species requiring warm temperatures can be placed closer to the heat source, while those requiring cooler temperature should be farther from the heat source.

Bottom heat is the method of choice when germinating seeds. Supplying heat from below can be accomplished in several ways. Steam or hot water can be piped under the benches. Hot air can be forced through large, plastic tubes placed under benches. Electric heating cables can be placed on the benches under the seed flats. An important factor to consider with any method is that proper medium temperatures must be maintained, with less attention to air temperatures. Thermometers should be placed in the medium to assure that optimum temperatures are achieved.

If seed flats are germinated in the greenhouse under polyethylene, caution should be used to ensure that temperatures do not exceed 85° to 90°F, especially on sunny days. Temperatures above 90°F will damage or kill the developing seedlings. If necessary, the plastic should be raised or removed to keep the temperature in an acceptable range.

Seed flats properly moistened before sowing and placed under polyethylene or mist will generally not require watering again until after the seedlings have emerged. If these methods are not used, the seed flats should be checked on a regular basis several times a day. The seed flats and young seedlings should never be allowed to dry out. When the seed flats do require irrigation, subirrigation is preferred to overhead watering. Overhead watering tends to pack the medium, reducing needed oxygen content, and may knock over or damage tender seedlings. If overhead watering must be done, a nozzle that produces a very fine mist should be used to reduce detrimental effects to the seedlings.

Increasingly, growers are using controlled environment chambers for germination and early growth. These chambers are totally enclosed and have the advantage of maintaining desired light, temperature, and humidity levels regardless of outside conditions. Germination tends to be quicker and more uniform. Although initial costs may be expensive, in many cases the investment is offset by the higher quality seedlings produced.

The increasing use of plugs to produce bedding plants has increased the use of supplemental lighting of seedlings. Seedlings in plug trays have more growing space and are less crowded than those in standard seed flats. As a result, growth can be accelerated with supplemental lighting, usually with high pressure sodium or metal halide lamps. These lamps are efficient and do not require as many fixtures as fluorescent lamps. To justify the expense of adding light fixtures, as many crops as possible must gain advantage from the light. Most seedlings gain the most benefits when given supplemental light for the first 4 to 6 weeks after germination. Beyond 6 weeks, supplemental lighting is not as beneficial. At this time, seedlings can be moved to the greenhouse to make room for a second or third planting to be placed under the lights.

To be useful, the lights must supply sufficient intensity. Usually, 300 ft-candles (fc) across the bench must be provided to be economically feasible. Lower intensities will not accelerate plant growth enough to justify the cost of installation and operation of the lamps. Optimum growth can be achieved if plants receive light, including sunlight, for 18 to 24 hours per day.

H. Fertilization

Fertilization requirements of seedlings are determined by plant species and the time the seedlings will remain in the flat. Fertilization may become necessary because germination media are often low in nutrients.

A well-balanced fertilizer such as 20–20–20 or the equivalent, should be applied immediately after seedlings emerge and weekly thereafter. For a small number of flats, a fertilizer solution is best applied with a sprinkling can that has been fitted with a fan-shaped flaring nozzle. A fertilizer proportioner in the water line is a good system for a large number of flats.

VI. PREPARATION FOR TRANSPLANTING

A. Containers

The type of container in which a bedding plant crop is grown will depend on the prospective market. Many types of containers are available for use, and each has its advantages and disadvantages.

Great advances have been made in the development of bedding plant containers. Originally, started plants were dug from the field and wrapped in newspaper or placed in wooden boxes. Some plants were grown in wooden flats. These methods were poor because the plants were often intertwined and suffered trans-plant shock when separated. Plastic flats were developed that contained open packs of about six plants. However, plant roots again became entangled and were hard to separate. This led to the development of the cell pack. Each plant was self-contained in an individual cell allowing for easy transplanting by the consumer. Depending on the market, plants are now marketed from cell packs, open packs, peat pellets, plantable pots, or 3½- or 4-inch pots or larger.

The bulk of the bedding plant crop is grown in flats using cell packs. A standard bedding plant flat may contain from 18 to 72 plants (Fig. 4). The trend has been to increase cell size, thus decreasing the number of plants grown per flat. When given this extra room, the plants tend to be of higher quality.

Some growers still use open packs to produce their crop. In most cases, a standard flat will contain 12 open packs. These packs are not subdivided into separate cells so the roots can become entangled. These packs contain more soil, which allows for less frequent watering.

Bedding plants are increasingly being produced in plastic pots. These pots are 3½ to 4 inches in diameter, and the standard flat will contain 15 to 18 pots. The large soil volume prevents the plants from drying out easily, and the extra room encourages greater plant development.

Other containers may include compressed peat pots in which the consumer transplants the entire pot with the plant, peat pellets that expand when moistened, and larger planters, pots, and hanging baskets that are intended not to be trans-

Fig. 4. Plug tray of seedlings ready for transplanting.

planted, but placed in the garden as is. The producer must select the proper container for each particular market.

B. Transplanting

The process of transplanting seedlings from the seed flat to the growing and sales container is one of the most time-consuming jobs in the bedding plant business. Attempts are constantly being made to reduce the time needed for transplanting. Time reduction may be achieved by a more efficient physical arrangement, by improving transplanting techniques, or by adapting mechanical systems, such as direct seeding or automatic transplanting.

Although much research has been devoted to mechanical transplanting equipment, hand transplanting is still the most commonly used method. There are two basic systems of hand transplanting. The first is to transplant in the greenhouse where the finished plants are to be produced. The second is to transplant in a centralized location, after which the flats are moved to the growing area.

Hand transplanting can be individualized or the combined efforts of several transplanters. In an individualized system, a single transplanter is responsible for planting a flat. The transplanter obtains the seedlings and a filled flat, dibbles the holes, plants the seedlings and removes the flat to the growing area or to a cart for later transport to the growing area. This system has been used for years and can be very successful when experienced transplanters are employed. Workers in this system are often paid a fixed amount per flat planted.

Combined transplanting is very efficient when an assembly line system is used. The flat moves along on a conveyor belt with several transplanters working on the same flat, each planting a specified portion. Other workers supply seedlings, place flats on the conveyor, and remove them. These workers usually receive an hourly wage instead of being paid per flat. Mechanical improvements are often added to an assembly line, such as an automatic flat filler, dibbler, or irrigator.

The biggest bottleneck in the transplanting process has always been removing the seedlings from the seed tray. This process was tedious because the roots were often intertwined and could be easily damaged. The use of plug-grown seedlings has greatly increased the efficiency of this process.

Because the seedlings are grown in individualized compartments, the roots cannot become tangled, resulting in little or no transplant shock. Also, the seedlings are easily removed from the plug tray by popping them up from the bottom. This can be done by hand or with any of several devices now on the market designed for this purpose. The plug tray is set over a series of pegs that line up with the drain holes on the bottom of the plug tray. Pressure is applied with a lever or foot pedal, and the seedlings pop out of the plug tray ready for transplanting.

Automatic transplanters are now being developed that reduce the number of transplanters necessary to complete the task. These machines plant seedlings by using a set of metal rods to push seedlings out the bottom of the plug tray into a flat below. Because of the high initial cost of the machines, this method has not yet gained widespread use.

All methods of transplanting use the same concept. A hole is made in the medium, a seedling is inserted, and the medium is settled around the roots to make a thorough contact. Though a simple idea, the process should be performed carefully to ensure successful growth.

Ideally, a medium is chosen that can be used from the time of germination through the production cycle. The flats or other containers should be loosely filled with the growing medium. Compacting the medium will restrict oxygen levels in the medium and reduce plant growth.

At the time of transplanting, the roots of the seedlings are very fragile and susceptible to desiccation. The medium should always be moistened before transplanting to prevent damage to the roots. On the other hand, if the medium is too wet, transplanting will be difficult and slow. The grower should be able to reach a compromise as to moisture levels so that the seedlings will not dry out but transplanting will still be efficient.

A general rule for transplanting has been to transplant the seedlings as soon as the first set of true leaves has appeared. Some plants, such as marigolds, tomatoes, and peppers, have been successfully transplanted with only the cotyledon leaves expanded. Usually, the sooner the seedling is transplanted, the less chance of checking growth in the early stages. This is especially true when plants are grown in seed flats instead of plug trays. If plants from seed flats are transplanted early, the roots will not tangle and break.

Seedlings grown in plug trays can be held much longer, because the plants are in individual compartments. Many growers may keep seedlings in plug trays for up to 6 weeks longer than they would if using seed flats, because less growing space is used than if the seedlings were transplanted out. Caution should be used when holding plug trays; however, the decreased soil volume can dry out quickly.

To ensure uniform growth in the flat after transplanting, many growers grade seedlings by size during the transplanting process. Even growth will result if all the small seedlings are planted together in a flat and all the large seedlings are planted in a separate flat.

The process of transplanting should be done carefully because the seedlings are delicate. When seedlings are removed from a seed flat, they should be scooped from the bottom and carefully separated. They should never be removed by pulling from the top. The holes in the medium should be large enough to avoid jamming the seedling into the hole, which risks breaking the roots or the stem. At this stage, the stem is easily crushed. If this happens the plant will usually die or be severely stunted. A damaged leaf, on the other hand, is more easily replaced. After transplanting, the medium should be pressed against the roots instead of simply closing the top of the hole leaving the rootball suspended in a cavity below the soil line. The root hairs must be in contact with the medium. If careful attention is paid to the transplanting process, growth should be quick and uniform.

Transplanting can be avoided all together by direct sowing the seed into the finishing flats either by hand or with a mechanical seeder. Some species of bedding plants are more suitable for this method than others. To ensure a full flat, usually more than one seed is planted per cell. Alyssum and portulaca can easily be grown by this method because the seed is cheap, and the plants grow well in clumps, which fill the flat quickly. This method has the advantages of eliminating the labor for transplanting and transplanting shock, but it has the disadvantage of increased seed use and the need for up to ten times the room in the germination area.

C. Posttransplanting Considerations

Advanced planning and a little extra attention after transplanting will ensure that the seedlings get off to a good start. After transplanting, the seedlings should immediately be watered in to prevent desiccation and to help settle the medium around the roots. The plants often benefit from the addition of a high phosphorus plant starter fertilizer to this first irrigation.

The newly planted seedlings should be kept off the greenhouse floor to prevent the plants from rooting into the ground as they grow (Fig. 5). Several methods have been used, such as raising the flats with plastic or wood risers, covering the floor with a material such as saran cloth, wood chips, or cement, or by growing the plants of raised benches.

Species of similar cultural needs should be grown together so that the necessary environmental conditions are easily supplied to the crop. Species known to

Fig. 5. Transplanted flats placed on plastic risers to keep the plants off the greenhouse floor.

develop insect or disease problems should be grown together to facilitate pesticide applications should they be necessary.

Seedlings can be especially vulnerable the first few days after transplanting. They should never be allowed to become water stressed, which will delay or stunt growth. They may also need to be protected from the sun if it is very bright by covering them with shade cloth for a few days. Once established, the seedlings should grow with few or no problems.

VII. MEDIA

The situation of the bedding plant grower is different from growers of other types of potted crops. The unique situation of the bedding plant grower is that of producing plants in very small media volumes, with most containers being only 2 to 3 inches deep. Therefore, it is essential that any mix used have good aeration. Many of the production problems associated with bedding plants can be traced to poor or improperly mixed media.

Historically, most bedding plants were grown in soil-based mixes. Mixes varied with geographic area of the country, depending on native soils and water supplies. The biggest problem was the variability in native soils and the extreme difficulty in obtaining a similar mix time after time. Also, many soils have been treated with persistent herbicides. These herbicides do little harm to field crops but can be devastating to bedding plant crops. These problems, as well as the increased cost of labor, encouraged many growers to switch to soilless mixes. One of the first was the U.C. mix developed by researchers at the University of California. This mix worked well in the western United States; however, one of the ingredients is a sharp, coarse, white sand not found in other parts of the United States. Unavailability of this sand was the primary reason for a lack of wide acceptance of this mix.

The Cornell peat-lite mix received a much wider acceptance. Its materials, sphagnum peat and horticultural vermiculite, are readily available and very competitively priced. Sphagnum-based mixes have advantages for seed germination as well as growing of the crop. They have been used with a very high success rate by many bedding plant growers.

More recently, mixes containing bark have been introduced. These mixes have found the greatest use in the southern and western United States. Research is now being conducted to explore possible uses of several other types of media and amendments.

Because soilless mixes are relatively insect, disease, and weed free, they are very convenient for the grower to use for germination and growing. Most growers do not pasteurize this mix; they merely open the bagged mix and fill their flats.

A typical soilless mix used in bedding plant production is outlined in Table VI. A number of commercial mixes similar to this type of Cornell peat-lite mix are available. Many greenhouse firms buy trailer-load lots of either bagged or bulk mix, thus eliminating the problems associated with mixing media on the premises.

Table VI

Typical Soilless Mix Used in Bedding Plant Production[a]

Amount	Component
½ cu yd	Sphagnum peat moss
½ cu yd	Horticultural vermiculite
5–10 lbs	Ground limestone
1–2 lbs	Superphosphate
1 lb	Potassium nitrate
2 oz	Fritted trace elements
3 oz	Wetting agent

[a]From Boodley and Sheldrake (1972).

VIII. THE GROWING ENVIRONMENT

A. Temperature

Temperature is the most controllable factor in the greenhouse. Thus, the grower is able to use temperature to control growth and to produce the type of plant desired.

Growing temperature recommendations are usually based on air temperature, although plant and air temperature may not be the same. On sunny days, the surface of the leaf may be several degrees warmer than the surrounding air. Medium temperature, too, usually differs from air temperature.

Raising the temperature increases growth rate, but most bedding plant species experience detrimental effects when the temperature exceeds 86° to 90°F. High temperatures encourage soft growth resulting in tall, thin-stemmed plants. The plants tend to primarily accumulate proteins and amino acids. Low temperatures slow plant growth and encourage the accumulation of carbohydrates. The resulting plants are shorter and stockier with thicker stems. These plants also tend to be of higher quality.

Night temperatures have traditionally been used to classify greenhouse crops because night temperatures were easier to control than day temperatures. With new developments in modern greenhouse technology, however, day temperature can now be easily controlled in most areas of the country. Night temperature classifications are no longer as important as they once were for bedding plants and other greenhouse crops.

There is no one optimum temperature for a given bedding plant crop. The optimum temperature depends on the species to be grown. For example, petunias grow better at a lower temperature than vinca or impatiens. The amount of light available also dictates the best temperature. Increasing the light level without changing temperature will result in earlier flowering in petunia by up to 2 weeks. The lower the amount of light energy available, the lower the temperature should be maintained to produce a high quality plant.

The stage of growth is also an important factor to consider. Germination will usually require the highest temperature. Seedlings and young plants have a higher optimum temperature for best growth than do older plants. When all plants in a greenhouse are the same age, adjusting temperature in relation to plant age might improve quality and yield of the crop.

In some bedding plants, flower bud initiation may depend on temperature, with plants initiating buds more rapidly when given the optimum temperature for the plant. Many taxa, such as several species of *Pelargonium*, must be maintained below a certain temperature for initiation to occur. Others must be grown above a minimum temperature, or only vegetative growth will occur. Also, optimum temperature for initiation may be different from the optimum temperature for floral development.

Research has shown that plant height for many greenhouse crops, including petunias, is influenced more by the difference between day and night temperatures and less by the actual temperatures themselves. A high day temperature, when combined with a low night temperature, encourages internode elongation resulting in taller plants. In the reverse situation, where day temperatures are cooler than night temperatures, internode elongation is limited, resulting in shorter plants.

Nutrition of the crop and fertilizer practices are highly dependant on the temperature regime used. The higher the temperature, the more frequent the need for watering and fertilization. Plants grown under cooler temperatures require less water and fertilizer. Temperatures below 50°F should be considered as holding temperatures and little water or fertilizer will be needed.

Quality bedding plants can be produced within a temperature range of 50°F to about 68°F. Plants grown at a constant 50°F will be short, compact, and well branched, but crop time will require an extra 2 to 4 weeks. If the night temperature is raised, compact growth will still result with a reduction in crop time. However, if both night and day temperature are high, the plants will flower quickly, but will be tall, have long internodes and be poor in quality.

Most species benefit from a temperature of 48° to 50°F during the last week of finishing. At these temperatures, flowering may be delayed, but the plants become acclimatized to outdoor conditions. If necessary, these temperatures can be used to hold back most species of bedding plants if the marketing date is delayed because of poor weather or other conditions that delay sales.

Temperature is a very powerful tool in controlling plant growth. Not all species, nor different stages of growth of the same species, respond the same to a given temperature. A grower must rely on observation and experience in order to provide the proper temperature at the proper stages to produce the desired crop.

B. Watering

Watering is one of the most critical cultural practices for producing bedding plants. If watering is not done properly, plants may be inferior in quality or completely destroyed. If a grower applies too much water, the plants become excessively lush and tall. These plants will have a poor shelf life and will be difficult to ship or transplant in the garden. On the other hand, if water is restricted excessively, plants will be short, have visible damage on the foliage, or may even die.

The water source should have a pH between 5.5 and 7.0 and a soluble salt level of potassium equal to 120 or less. There are many areas of the country where this quality of water does not exist. In these areas, the water must be treated to conform to these standards. Some growers have installed systems to remove dissolved solids from their water supply. Others find that the water pH is very high and resort to additives such as phosphoric acid or sulfuric acid to reduce the pH.

Growers must be cautious about their water supplies. Many problems can be encountered by using water directly from ponds or streams that are not checked or treated. The best source is usually from wells or municipal water supplies, though high levels of chlorine in many municipal systems may also cause problems.

1. Watering Seedlings

Seed may be germinated under a mist system. Alternatively, seed flats are completely saturated and allowed to drain, seed is sown, and then the entire flat is covered with polyethylene or glass to prevent evaporation and to keep humidity very high. If watering of the flat is delayed until after the seed is sown, finer seeds may be washed away and germination will be reduced. A very fine mist nozzle must be used to ensure that the force of the water will not uproot young seedlings or wash away any seed. Moisture is very important for germination. If the flat is allowed to dry out during this critical period, germination will be greatly reduced or completely inhibited. Once the seeds have completely germinated, the moisture level can be reduced to allow the seedlings to become sturdier for easier transplanting. The seedlings should never be allowed to wilt, however.

Plug production presents a special set of problems in the area of water management. The soil volume in plug trays is very small compared to regular seed flats. As a result, this small volume tends to dry out and be overwatered easily. Most plug losses are caused by improper watering. Also, as germination and seedling growth progresses, water regimes must be changed to meet the changing needs of the plants (Fig. 6A,B).

Boom systems outfitted with fine mist nozzles can be employed to maintain proper moisture levels in the plug flat. Passes of the boom can be increased or decreased accordingly as the moisture requirements of the seeds and seedlings change. Booms are an excellent tool for keeping even moisture levels across the flats.

Capillary mats and subirrigation can provide good results in situations where a boom system is impractical or impossible to install. Capillary mats can be used to subirrigate the seedlings without washing them away, while still maintaining even moisture across the flats.

The least desirable method of watering plug-grown seedlings is overhead with a hand-held nozzle. Coarse spray can wash out seeds and seedlings. Some growers have a problem applying water evenly to the flats. The result is wet and dry

Fig. 6. (*Top*) Large- and (*bottom*) small-scale bedding plant operations.

areas from flat to flat or even within the same flat. Seedling growth will then most likely suffer.

2. After Transplanting

Transplanted seedlings should not be allowed to dry out enough to wilt. They should be kept well watered for several days until new top growth indicates that roots are developing adequately. Thereafter the medium surface can be allowed to dry out each day to decrease chance for diseases and to allow adequate aeration in the medium. Watering should be done early in the day to allow the foliage to dry by nightfall.

For watering new transplants by hand, a nozzle should be used that will not knock them down, wash them out, or cover them with medium. A flaring, fan-shaped rose nozzle produces an even, gentle spray. Mechanized watering systems have been developed and are used primarily in the southern and western United States or in the northern areas when light is abundant and quick drying occurs. They are usually not used in the northern areas during the winter months under low light conditions.

3. Watering Established Bedding Plants

For the fastest growth of bedding plants, water stress must be minimal. Plants should receive water frequently enough to prevent wilting. Watering must be thorough, both in quantity and distribution, however, thorough does not necessarily mean frequent.

Water can be applied to a flat much faster than the flat can absorb it. Therefore, much of the water from a heavy watering may run over the top of the flat and soak into the ground. The medium in the flat may be left only partially watered. If water is not absorbed as fast as it is applied, several passes over the flats may be required for a thorough watering.

4. Finishing the Crop

Some growers claim that water is the best growth retardant. Just as adequate amounts are needed for unchecked growth, lack of water can hold back a crop that threatens to reach maturity before marketing time. Because excessive water loss may cause injury or death to plants, such practices should be handled with care.

Depending on general growing conditions in producing the crop, it may be desirable to water-stress plants prior to sending them to market. If this is not done, plant tissues may be too succulent. A test can be made by setting a well-watered flat of plants outside for an hour on a clear day. If the plants wilt, they should be better prepared for market by withholding some water for a few days. Care should be taken to avoid letting them wilt beyond recovery.

C. Fertilization

Historically, early bedding plant growers obtained rich field soils with good fertility because of previous farming practices or composting. They used this soil along with organic fertilizers to produce a crop. Many times the fertility level was not high enough, thus producing plants that were excessively stunted and hardened. At the other extreme, fertility levels could be so high that plant damage or death resulted.

Today, most growers use a high analyses, liquid-type fertilizer that is applied through the irrigation system after transplanting. The medium is tested, and the pH is adjusted to the range 5.5 to 7.0. Optimum pH would be 6.2 throughout the entire growing period for most bedding plants. Medium pH and soluble salts should be monitored on a regular basis. Nitrogen and potassium are applied at a rate of 100 to 250 ppm nitrogen and potassium as indicated by the soluble salts readings.

D. Daylength

Most bedding plant growers do nothing to manipulate daylength, but depend on natural daylength to produce the various bedding plants they grow. With certain species, daylength plays an important part in the time needed to flower.

Different bedding plant species cover all possibilities of photoperiodic combinations for flowering. Many are day neutral and will flower in relatively the same period of time under any reasonable photoperiod. Some species have been determined to be long-day plants while still others are short-day plants. Salvia may be found in all three categories, varying by cultivar. Petunias respond to long days in both time to flower and plant morphology between the temperatures of 55° to 70°F. Outside of this range, the plant responds to temperature instead of to photoperiod.

Photoperiodic responses of bedding plants can be expressed only in general terms. Cultivars of the same species may show varying effects of photoperiod. Seed suppliers are helpful in determining various photoperiodic effects on plants. Growers should attempt to select specific cultivars for specific times during the growing season (i.e., a short-day salvia cultivar should be grown early in the season when days are short, and a long-day cultivar should be selected for later in the season). Table VII details the general effects of photoperiod on selected bedding plant species.

E. Growth Retardants

Growers are interested in producing short, compact plants and, therefore, use cultural manipulations, as well as growth-regarding chemicals, to accomplish their goal. Cultural methods have long been used to keep plants compact. For example, lack of water may be the best growth retardant. Many growers carefully use this technique to keep their plants short. Another technique is to reduce or eliminate phosphorus from the media mix and to add it only as a liquid and at low levels. Until the introduction of growth-retarding chemicals, these cultural methods were the only ways possible to produce short plants.

Today, however, there are several chemicals available to control height. The most popular are daminozide (B-Nine), ancymidol (A-Rest), chlormequat (Cycocel), and paclobutrazol (Bonzi). Daminozide has the widest applications. It prov-

Table VII

**Classification of Selected Bedding Plant Species by Their
Response to Photoperiod**

Short day	Long day	Day neutral
Zinnia	Verbena	Alyssum
Basil	Snapdragon	Balsam
Coleus	Centaurea	Begonia
Celosia	Gaillardia	Gomphrena
Cleome	Gypsophila	Impatiens
Cosmos	Nicotiana	French marigold
Dahlia	Scabiosa	Pansy
Morning glory	Salpiglossis	Vinca
African marigold[a]	Ageratum	Carnation
Salvia[a]	Phlox	Pepper
	Geranium	Tomato
	African marigold[a]	Lobelia[a]
	Salvia[a]	Cabbage
	Lobelia[a]	Salvia[a]
	Petunia[b]	

[a]Response varies by cultivar.
[b]Long day between 55° and 70°F. Affected by temperature out of this range.

ides good height control at a reasonable cost for many species of bedding plants. Paclobutrazol is one of the newest growth retardants. It should be used with caution because it is highly effective at low concentrations, and plants do not readily grow out of its effects. Chlormequat is very popular for seed geranium production because of the height control it provides in addition to decreasing time to flower.

There are several advantages and disadvantages in using growth retardants. The advantages include (1) reduction in plant height (shorter internodes), (2) improved plant shape, (3) darker green foliage, (4) thicker leaves, and (5) more uniform flowering. Shelf life is usually greater because of the plants' ability to better withstand stress. The plant is more resistant to water stress and to heat, cold, or smog injury. Use of growth retardants allows the grower more flexibility in growing, because retardants keep plants short over a wide range of temperatures, fertilizer, and light levels. Usually they are safe and easy to apply.

The main disadvantage of using growth retardants is the varying response of different species to the chemicals. Some species are very responsive to growth retardants, while others are not. This varying response is even evident in different cultivars of the same species. Other disadvantages include delayed flowering if the chemicals are applied too late and injury from phytotoxic effects, in addition to the expense of purchasing and applying the chemicals. In most cases, however, the advantages usually outweigh the disadvantages.

Table VIII lists the effectiveness of available growth retardants on specific bedding plants. Before using any chemical, the grower should always check the label for registered uses and only use the chemical according to label directions.

Application methods are very important if success with growth retardants is to be achieved. When growth retardants are to be used, they should be applied early enough in the life cycle of the plant to retard future growth. No chemical is available that will make a plant shorter than it already is. Applications should be made to stress-free plants. The plants should be well watered and fertilized and at temperatures cool enough to avoid phytotoxic effects. The foliage should be dry in the case of foliar sprays, and the plants should not be watered for 24 hours so that the spray is not washed off the plants.

Growth retardants should be considered as one of the many tools available to the grower. They are no substitute for poor cultural practices. If used wisely, growth retardants are very helpful in producing a quality crop. They will not, however, make a poor grower into a good one, nor will they improve a poor crop.

Table VIII

Effectiveness of Currently Available Growth Retardants on Specific Bedding Plants[a]

Bedding plant	B-Nine	Arrest	Cycocel	Bonzi
Ageratum	+	+	−	+
Antirrhinum	+	+	−	+
Begonia (fibrous)	−	+	+	+
Begonia (tuberous)	z	−	+	−
Browallia	+	x	−	−
Callistephus	x	x	x	−
Catharanthus	+	x	x	+
Celosia	+	+	+	−
Cleome	+	+	+	−
Coleus	−	+	−	+
Dahlia	+	x	+	−
Dianthus	−	x	+	+
Garden mum	+	+	+	−
Impatiens	+	+	−	+
New Guinea impatiens	−	−	−	−
Pelargonium	−	+	+	+
Petunia	+	−	−	+
Salvia	+	+	+	+
Senecio	+	−	+	−
Tagetes	+	+	+	+
Verbena	+	+	+	−
Viola	−	−	+	+
Zinnia	+	+	−	−

[a] +, Effective; −, not effective or response not known; x, response varies by cultivar; and z, no retarding effect, but accelerates growth.

IX. PROBLEMS

A. Diseases

Disease problems, unless detected and treated, will usually devastate a bedding plant crop. At the very least, disease will reduce plant quality and lessen plant vigor. This low vigor is passed on to the consumer, who obtains unsatisfactory results. The dissatisfied consumer could be discouraged from buying plants in future years, thereby reducing sales.

Parasitic diseases of bedding plants are usually caused by bacteria or fungi. The following outline gives basic information on the parasitic diseases most commonly encountered by the bedding plant grower.

   I. Damping-off
      A. Usually caused by *Rhizoctonia* fungi or one of two water mold fungi, *Pythium* and *Phytophthora*.
      B. Symptoms or signs to consider
         1. Usually first detected in germination flats
         2. Weeds in flat may indicate unsterilized medium
         3. Poor stands of seedlings
         4. Plants fall over at the soil line
         5. Circles of dead and dying plants enlarge daily
         6. Threads of fungus can be seen on dead plants
  II. Sclerotinia crown rot
      A. Caused by *Sclerotinia sclerotiorium*
      B. Symptoms or signs to consider
         1. Cottony mold on medium or crown of the plant
         2. May be found on older flowering plants
         3. Causes rapid rot of plant tissue
         4. Its resistant, spreading structure (sclerotia) resembles large rodent droppings. They may be in rotted stems or on the medium near plants
         5. Spreads rapidly in warm, damp weather
         6. May also attack seedlings, causing damping-off symptoms
 III. Botrytis crown rot
      A. Caused by *Botrytis cinerea*
      B. Symptoms or signs to consider
         1. Soft decay of seedlings
         2. Fuzzy, gray growth
         3. Spores puff like dust when disturbed
         4. Works from top of seedling downward
         5. Starts on injured or dead parts of plants
         6. Restricted to cool, moist conditions
 IV. Rust diseases
      A. Caused by various species of *Puccinia*
      B. Symptoms or signs to consider
         1. Red leaf spots caused by fungus spores
         2. Red dust on hands
         3. May begin on lower leaves or on undersides of leaves
         4. Zones or rings of spores form as older leaf spots enlarge
         5. Most commonly seen on geranium, hollyhock, fuchsia, and snapdragons

V. Powdery mildew
    A. Caused by species of *Oidium*
    B. Symptoms or signs to consider
        1. White, fluffy, growth on leaf surfaces, may be in spots
        2. White spores may fly into air when leaf is flicked
        3. Usually seen on older zinnias and snapdragons
        4. Rarely seen on seedling plants

The greenhouse environment and cultural practices can aid in the spread of parasitic diseases. Threadlike parts of fungi can be found clinging to soil particles and decayed plant parts. Some of these can be transferred unknowingly to other flats, tools, and clothing. They can be carried through the air on dust particles or by splashing water. Some spores are capable of swimming through water in pore spaces between soil particles and on soil and plant surfaces. Some can get their start on a dead leaf or other part that falls onto a healthy plant, such as a fallen petal or leaf from a hanging basket to a plant below. Injured plants are especially vulnerable to infection.

B. Insects and Mites

A properly managed bedding plant greenhouse should have very few major insect problems. In the North, insects are usually not a problem at the time seed is sown or seedlings are transplanted. Only in late spring, when ventilators are open and insects are active outside, can they become a real problem, unless insects are already on other crops in the greenhouse.

An observant grower is the best key to insect control. Keeping a close watch on the crop will prevent a large insect population from developing.

There are only eight to ten pests that are common problems in bedding plant production.

Aphids (Aphididae). Commonly called plant lice or greenfly, they usually attack tender annuals in the spring of the year. They are found on the young, immature terminal areas and feed with piercing and sucking mouthparts. They vary in color from green to pink to black. A by-product of their feeding is a sticky honeydew that coats the foliage of infested plants. This honeydew will turn black under high humidity conditions. Aphids overwinter outside the greenhouse as eggs. In the spring, they hatch and develop into winged females that fly into the greenhouse, where they produce live, wingless female progeny. Occasionally a winged female is produced and flies to another plant, thus spreading the infestation. The higher the temperature, the greater the number of aphids produced. One female produces 1400 young at 70°F in 2 weeks.

Ants (Formicidae). They may live on the greenhouse floor, especially if a sandy fill is used. They have been known to carry aphids from one plant to another and can carry spores of *Botrytis* from one plant to another. Ants can be easily controlled by preventive spraying before flats are set on the greenhouse floor.

Whiteflies [*Trialeurodes vaporarionim* (Westwood)]. This insect infests many common bedding plants, as well as other crops grown by some bedding plant growers. The life cycle starts with small, oval eggs deposited on the underside of leaves. They hatch in 4 to 12 days into pale yellow, six-legged crawlers. These larvae or nymphs move about for a short time, avoiding direct light. They insert their heads into the plant and start sucking sap. In the next phase, they lose their legs and antennae and look like very small, flat, oval scales, often with a marginal fringe of white, waxy filaments, sometimes covered with plates or rolls of wax. They secrete large amounts of honeydew. After a second molt, the insect becomes a pupa and then the four-winged adult leaves the pupal skin, lays eggs, and the cycle starts over. The grower can usually see all stages of development on the underside of leaves of host plants. Chemical control on a regular application basis can eliminate this insect problem.

Moths (*Lepidoptera*). Larvae of several moths can cause considerable damage to bedding plants. They are usually noticed in the "worm" stage. Most go unnoticed when the eggs have been laid or just hatched. They are usually voracious eaters. Biological control with *Bacillus thringiensis* preparations (Dipel, Thircide, Biotrol) or chemical sprays are used. Foliage must be thoroughly covered. These worms are "chewers," so when they eat the treated foliage, they ingest the material and die.

Two-spotted mite (*Tetranychus telarius*). The two-spotted mite is the common spider mite or red spider mite found in greenhouses. They often go unnoticed until large populations develop because of their small size and because they feed on the undersides of leaves. They are greenish or reddish in color with two black spots on either side of their body. The female lays several hundred tiny, sphere-shaped eggs on the underside of leaves in 3 to 4 weeks. Eggs from unmated females only develop into males. The mites produce webs on the underside of leaves and from one leaf to another, sometimes entirely covering shoots and flower buds.

The number of spider mite generations increases with temperature. At 75°F, the adult stage is reached in 5 days; at 55°F, it takes 40 days. At 80°F, one female spider mite could theoretically give rise to over 13 million progeny in 4 to 6 weeks. There are many common pesticides used to control outbreaks of this pest. Environmental control can be obtained by reducing temperatures and washing mites from the leaves.

There are other mites that can cause problems, such as the broad mite (*Heimtarsonemus latus*), the false spider (*Brevipalpus*), and the "French fly" mites (*Tyrophagus*). These can be problems in popular annals such as fuchsias, begonias, and petunias.

Thrips (*Thysanoptera*). Thrips cause damage to foliage as well as to flowers by rasping surface cells and sucking up the content. Silver or white streaks, some turning brown, indicate thrips activity. Thrips are most commonly found in the shoot tips or flowers. The western flower thrip is of special concern. This insect was localized in California but has spread across the entire country over the past few years. The western flower thrip spreads tomato wilt spot virus, which can severely damage many species of bedding plants. This virus is easily spread by the thrips because of the many host plants which it infects. Because there is no cure for the virus, infected plants should be destroyed immediately to prevent spread to healthy plants.

C. Control

Two methods are available to control insect and disease problems. These are cultural and chemical. Cultural control is by far the safest and most economical method. Chemical controls should be used only in combination with good cultural practices to obtain the desired results.

The easiest way to prevent the spread of disease or insect problems is to reduce the chances of their occurrence. This can be done in many ways. As explained earlier, the grower should be very conscious of sanitation. All materials used should be free of any pathogens. Plant cultivars should be selected for their vigorous growth and natural resistance to insects or disease, if any. Areas in and around the greenhouses should be kept clear of weeds. Weeds are unsightly, and more importantly, provide host areas for insects and disease.

The grower must always be observant. The crop should be examined on a daily basis so that any problems can be detected early. Early detection makes control much easier and reduces losses of the crop. Workers should be trained to recognize plant problems so the number of people observing the crop is increased. Also, yellow sticky traps or boards coated with an adhesive, can be used to monitor insect infestations.

Good growing practices should always be used. Plants that are unstressed and vigorously growing are more resistant to disease and insects than are stressed plants. The plants should be properly watered and fertilized. The greenhouses should be well ventilated to reduce humidity.

Though the grower may use extreme caution in growing the plants, some problems will most likely develop. When one does, chemicals can usually be applied to control the problem. The list of available chemicals is constantly changing; the grower is advised to consult with the county cooperative extension agent for recommendations.

Extreme care should be used when applying chemicals. Most are toxic not only to the plants but to the applicator as well. Applicators should be trained in methods of chemical applications and how to protect themselves in the process. When using a chemical for the first time, only a small portion of the crop should be tested for phytotoxic reactions. Some species are sensitive to certain chemicals, and only specific cultivars of one species might be sensitive. New chemicals should be tested on all cultivars grown of each species before widespread applications are made.

Each state now requires that pesticide applicators be tested and licensed. Requirements vary from state to state. The federal government is restricting pesticide use more and more each year. Applicators should be aware of any new rules set forth by the government.

X. CARE AND HANDLING OF THE FINISHED PRODUCT

The methods used by the grower during the production of bedding plants greatly influence the garden performance of those plants. If the plants are healthy and vigorous when sent to market, the consumer has a better chance of success after the plants are in the garden. If poor practices are used, the consumer will likely fail, causing the reduction of future sales.

Before being sold, the plants must be properly acclimatized so they can easily adjust to garden conditions. Water and fertilizer levels should be reduced during the last 7 to 10 days of production. Temperatures should also be reduced to acclimate the plants to cool nights outdoors. Properly hardened plants are easier to ship and survive better at the retail and consumer levels.

The retail market has greatly expanded for the bedding plant industry. Garden centers were once the primary outlet for selling bedding plants, but over the past few years the retailing segment has expanded to include supermarkets, discount and hardware stores, and many other outlets. While increased retailing has helped the bedding plant industry grow, several new problems have developed. These new retailing outlets lack the experienced personnel common to garden centers. As a result, the plants suffer from lack of proper care. Plants damaged through water stress or temperatures that are either too hot or too cold are not uncommon. The producer should take some responsibility in educating retailers as to the proper care and handling of the finished product.

Several steps taken by both the producer and retailer will help ensure that the consumer receives a high quality product. The producer should adequately water the plants before shipment to ensure that they do not dry out in transit. The plants should be protected from mechanical damage while in shipment by using boxes or a rack system. The plants should not be allowed to sit for extended periods in the loading area prior to shipment. Increased ethylene levels and a prolonged dark period will reduce plant quality.

One of the responsibilities of the producer is to adequately label the product before it is shipped. Many seed companies and greenhouse suppliers make available picture labels that not only identify the plant, but also provide care instructions. The plants should be properly labeled with the correct species, and where applicable, cultivar, so the consumer is aware of the type of plant being purchased. The use of generic labels, such as those that identify a plant simply as a sunny annual or shady perennial, should be avoided because they do not help the consumer discern between different plant species.

The producer should provide retailers with instructions as to the proper care and handling of the plants. The retailer should unpack plants immediately on arrival and check to see if they need water. Preferably, the plants should be marketed from an enclosed structure to protect the plants from cold temperatures and drying winds. If such a structure is not available, the plants should be placed in a shaded area. Water is the important key in maintaining plant quality. Bedding plants should be checked several times each day to determine if they need water. If so, the plants should be watered thoroughly so that the soil ball is completely moistened. The retailer should also monitor temperatures and take necessary steps to prevent damage caused by cool or warm temperatures.

Some innovative producers are forming agreements with retailers in which the grower will provide a greenhouse structure to be erected in the retailer's parking lot or other accessible area and a person who will maintain the plants. The retailer would be responsible for selling the plants. Both parties benefit from such an arrangement. The producer has an outlet for the plants and is able to ensure plant quality through the retail process, while the retailer is able to offer the consumer a desired product without the problems of plant maintenance.

Many bedding plant growers think that finding a market is the last step in a bedding plant operation, however, the opposite is more logical. Finding a market should be the first step in a successful operation. A study by Ernest Dichter (1968) indicated people buy bedding plants for three reasons: creativity, excitement, and therapy. With the rise in food prices, perhaps necessity could be added as a fourth factor for producing vegetable transplants. When the product is promoted or sold, a need for the product must be established. This need can be established by appealing to one of these four reasons why people buy bedding plants.

Not every grower can grow all things for all people. Not every grower can meet the competition in all respects, but one can specialize. Instead of trying to grow all things for all people, the grower can become a specialist in one area and earn a good return on the investment and an adequate gross profit on the merchandise. Perhaps one can grow hanging baskets better than the competition, or grow different cultivars in different containers than what the competition will produce. Some producers have made a profitable venture out of growing potted annuals in 4- or 6-inch pots for late spring or early summer sales. If one is a small retail grower, an attempt to compete with large mass producers for the same market segment would most likely be unsuccessful. There are many profitable areas to be filled, if one is aware of what is needed in the marketplace.

REFERENCES

Agnew, H. H., and Koranski, D. S. (1986). "Standard versus refined impatiens seed." *Greenhouse Grower.* **4**(11): 29–30.

Anonymous. (1985). "Soil may determine whether bedding plant crop is successful." *Greenhouse Grower.* **4**(1): 7–10.

Anonymous. (1988). *Floriculture Crop 1987 Summary.* National Agric. Stat. Serv., Agric. Stats. Board. USDA, Washington, D.C.

Armitage, A. M. (1983). "Determining optimum sowing times of bedding plants for extended marketing periods." *Acta Hort.* **147**: 143–152.

Armitage, A. M. (1986). "Influence of production practices on post-production life of bedding plants." *Acta Hort.* **181**: 269–273.

Armitage, A. M. (1987). "Postproduction care of bedding plants." *Greenhouse Grower.* **5**(4): 44–45.

Armitage, A. M. (1988). "Supplemental lighting of plugs: Basic questions and answers." *Greenhouse Grower.* **6**(2): 48–49.

Ball, V. (1976). "Early American horticulture." *GrowerTalks*. **40**(3): 50.

Ball, V. (1985). *Ball Red Book*, 14th ed. Reston, VA: Reston.

Barret. J. E., and Nell, T. A. (1987). "Bonzi for bedding plants? At last a way to control runaway impatiens." *GrowerTalks*. **50**(9): 52, 54, 56, 58.

Bass, L. N. (1980). "Flower seed storage." *Seed Sci. Tech*. **8**(4): 591–599.

Boodley, J. W., and Sheldrake, R. S. (1972). "Cornell peat-lite mixes for commercial plant growing." *Cornell Univ. Plant Science Inform. Bull*. **43**.

Brown, W. (1985). "Keep plant growth under your control." *Greenhouse Manager* **3**(9): 114–116, 118, 120.

Bugbee, G. J. (1987). "Can chlorinated water harm your plants?" *Greenhouse Grower*. **5**(12): 62–63.

Carlson, W. H., and Johnson, F. (1985). "The bedding plant industry—Past and present." In *Bedding Plants III; A Penn State Manual* edited by J. W. Mastalerz and E. J. Holcomb. University Park: Pennsylvania Flower Growers. pp. 1–7.

Dichter, E. (1968). *To Buy or Not to Buy*. West Chicago, IL: Geo. J. Ball, Inc. p. 48

Erwin, J. E., Heins, R. D., and Karlsson, M. G. (1989). "Thermomorphogenesis in *Lilium longiflorum*." *Am. J. Bot*. **76**(1): 47–52.

Graper, D., and Healy, W. (1987). "Supplemental lighting benefits bedding plants." *Greenhouse Grower*. **5**(2): 34–36.

Kaczperski, M. P., and Carlson, W. H. (1989). "Producing salvia for profit, a commercial growers guide." *MSU Extension Bull*. **E-1661**.

Kaczperski, M. P., Carlson, W. H., Heins, R. D., and Biernbaum, J. (1988). "Petunias: Designed by cool days/warm nights." *GrowerTalks*. **52**(1): 37–39.

Kaczperski, M. P., Carlson, W. H., and Karlsson, M. G. (1990). "The influence of temperature and irradiance on the growth and development of *Petunia* × *hybrida* Vilm." *J. Am. Soc. Hort. Sci*. (In press).

Klingaman, G. L., and Jaster, S. W. (1982). "A comparison of seeding mixes for bedding plant production." *HortScience*. **17**(5): 735–736.

Koranski, D. S., and Laffe, S. R. (1985). "Plug production." In *Bedding Plants III: A Penn State Manual*, edited by J. W. Mastalerz and E. J. Holcomb. University Park, PA: Pennsylvania Flower Growers. pp. 126–140.

Laffe, S., and Koranski, D. S. (1988). "How to grow 32 plug crops: A pullout wall chart for growers." *Grower Talks*. **52**(8): 59–63.

Murray, G. E., Sanderson, K. C., and Williams, J. C. (1986). "Application methods and rate of ancymidol on plant height and seed germination of bedding plants." *HortScience*. **21**(1): 120–122.

Seeley, J. G. (1985). "Finishing bedding plants—Effects of environmental factors, temperature, light, carbon dioxide, growth regulators." In *Bedding Plants III: A Penn State Manual*, edited by J. W. Mastalerz and E. J. Holcomb. University Park, PA: Pennsylvania Flower Growers, pp. 212–244.

Seeley, J. G. (1989). "Temperature." In *Tips on Growing Bedding Plants*, edited by H. K. Tayama and T. J. Roll. Columbus, OH: The Ohio State University, pp. 30–33.

Stanwood, P. C. (1987). "Storage and viability of ornamental plant seeds." *Acta Hort*. **202**: 49–56.

Voigt, A. O. (1989). "Prices strengthen in '88 bedding plant season—'89 production prospects bright but maturing." *Prof. Plant Growers News*. **20**(1): 1–16.

White, J. W. (1984). "Keep 'em cool!" *Greenhouse Grower*. **12**(5): 12, 14, 16.

21

Hanging Baskets, Standards, Combination Pots, and Vertical Gardens

Virginia R. Walter

Introduction to Floriculture, Second Edition
Copyright © 1992 by Academic Press, Inc.
All rights of reproduction in any form reserved.

I. INTRODUCTION

Hanging baskets are grown by a variety of commercial producers. Some simply have baskets for spring sales, while others have a variety of plant material in hanging baskets for year-round sales. The plant material used is as diverse as the interests and imaginations of the growers. Equally diverse are the types of containers, growing mixes, and watering systems used by growers. Hanging baskets for spring or summer sales or those grown in mild climates can spend a significant amount of their growing life outdoors or under lath or plastic screen, making them an economical sales item. Even if the controlled environment of a greenhouse is necessary, baskets can use the volume of a greenhouse, which may be thought of as "free" space in terms of production costs for some growers, thus making baskets profitable items for greenhouse growers as well.

II. CONTAINERS

Many earlier references to hanging baskets describe the use of wire baskets lined with sphagnum moss, or wooden or clay containers designed to be suspended (Fig. 1 and 2). Ironically, many growers are finding that the appeal of these older containers is making a strong comeback and that many consumers seem willing to pay for the added costs. The vast majority of hanging baskets used today, however, are made of plastic. Various plastic container manufacturers have developed their own lines of baskets, which give growers more options than ever before. One can have either wire or plastic hangers and baskets with or without detachable saucers. The increasing demand for dripless baskets has created a basket style with an insert designed to create a false bottom. The water that drains out is captured in this reservoir. This reservoir can create watering difficulties for the grower, however much the consumer may like it. There are also more options available for baskets that come in several pieces. These rings allow easy insertion of plants into the side of the basket, thus creating instant hanging plants. This style of basket gives the growers more variety in the choice of plants that can be used and enable the growers to achieve a fuller basket in a shorter period of time. Plastic baskets are generally available in sizes from 5½ inches to 12 inches, in a wide range of colors, with dark green and white the most commonly used.

Use of unusual containers must be tempered by cost and salability. Although simple wire baskets may be as cheap as comparably sized plastic ones, the additional cost of manufactured liners or sphagnum moss to line the wire baskets plus the hangers can double the cost. Hangers for regular pots are available, but their use may also double the cost of the container. Ceramic containers and cedar or redwood containers are quite popular, but the cost of the container is generally four times or more that of a comparably sized plastic basket. All unusual containers should be checked for drainage holes or saucer attachment and ease of hanging. There should be room for sufficient soil volume, and the container must last through the growing and display period. The container should not be so heavy that when filled with soil and watered it is too cumbersome for safe, easy handling.

Fig. 1. Combination hanging basket with wire frame and sphagnum moss liner.

III. GROWING MEDIA

It is important that the growing medium be loose and coarse, because one of the biggest problems with many hanging containers is drainage. Soilless mixtures might be preferred to avoid the heavier weight of a soil mix. A standard potting mixture of 1:1:1 sphagnum peat moss, bark, and perlite (volume basis) will be light

Fig. 2. Combination hanging basket with wicker basket.

weight and also will hold sufficient water so that the baskets do not dry out too rapidly. Many of the commercially available bagged mixes are excellent for hanging baskets. The medium must be free of pathogens and other pests. Wetting agents, water absorbing additives, or both can be used when preparing the medium, but watering practices may have to be altered slightly.

IV. WATERING

Watering baskets by hand can be a time-consuming and dampening experience. Because most baskets are hung above eye level, it is also very difficult to use some of the more common visual methods for determining if the baskets need watering. Lifting a sample container will help determine if it is dry enough to water. A long, curved breaker makes handwatering easier. Tube irrigation is very common because once the baskets are in place overhead it can be very difficult to water by hand. Each container has its own drip tube. If the baskets are in a sales house it would be advisable to use tubes that can be shut off to protect customers when a pot has been removed. In-line restrictive flow irrigation systems are available that will deliver water to each container at a very slow rate (gallons per hour rather than gallons per minute). Each basket is placed at a predetermined place on overhead supports. Overhead moving hanging basket systems such as E.C.H.O. deliver water at only one location, and the baskets move to that spot at an appropriate rate.

Baskets should never be allowed to dry out, but overwatering will encourage root rot problems. The baskets should be watered thoroughly. High salt levels because of inadequate watering practices can be just as costly in plant losses as can soil-borne diseases encouraged by overwatering. Some growers may intentionally let their plants wilt occasionally to encourage a hanging appearance. Selecting plants with naturally hanging characteristics would be preferred because wilting can do harm to the plant as well. Baskets with fibrous liners will dry out much more frequently than plastic containers and must be watched more carefully. They can be plunged into a container of water to rewet if they dry out severely.

V. FERTILIZATION

Many growers use constant fertilizing/watering systems, and there is no reason that hanging baskets cannot be included on these systems as well. Nutritional needs for plants in hanging containers will vary with the plants being used, but standard nutritional practices appropriate for that plant material should be followed. Using slow release fertilizers in the growing medium may make fertilizing easier, and may also make the products last longer for the consumers.

VI. PLANTING

Although it is possible to plant any hanging container with small seedlings or newly rooted cuttings, it is generally much more feasible to use well-established plants. These plants may be established in cell packs or separate pots (from 2¼ to 4 inches) and then transplanted into the baskets. This will enable the grower to have a much higher percentage of success and a more uniform product. Planting with established plants is a necessity with mixed plant baskets, otherwise some plants that grow more rapidly will crowd out the others. Using containers that come in pieces that fit together will enable the grower to use plants that are practically of salable size, and just need sufficient time (from 1 to 3 weeks generally) to reorient themselves to their new location. The numbers of plants that will be used per basket is dependent on the size and style of the basket and the growth habit of the plant as well as the look one desires to create. Crowding of plants in the basket should be avoided. A much better basket will result from letting the plants fill out the basket as they grow. Enough space also should be allowed for proper watering. Many baskets are planted with one plant cultivar per basket, or different colors of the same plant species. Petunia or impatiens annual baskets or Boston fern foliage baskets are very common examples. Combination baskets are often sought by the consumer because they are different. These baskets often contain both cascading types of plants and upright plants mixed together to create unusual textural appearances. Most moss and wire baskets contain a mixture of different annuals such as alyssum, marigolds, or lobelia combined with plants that have different foliage textures such as ivy geraniums or dusty millers or even trailing vincas (Tables I, II). The plants are inserted into the side of the wire baskets (tucked through the moss at the time of planting) as well as on the top. Depending on its growth habit, the plant may then trail down or turn up and provide an unusual cover for the sides of the basket. Where the basket will ultimately be displayed determines whether the container has to be fully covered or not. Baskets seen only from below need not have a full center. Many trailing foliage baskets turn out this way. Baskets seen from all sides and at eye level (on patios, for example) should be as full as possible and their tops, as well as their sides, should be attractive.

Table I

Suggested Flowering Plants for Baskets

Botanical name	Common name	Botanical name	Common name
Abutilon	Flowering maple	Ipomoea	Morning glory
Acalypha pendula	Strawberry firetails	Lantana camara	Trailing lantana
Achimenes	Magic flower	Lathyrus odoratus	Sweet pea
Aeschynanthus	Lipstick plant	Lobelia erinus	Lobelia
Ageratum houstonianum	Floss flower	Lysimachia nummularia	Creeping Charlie
Antirrhinum majus	Snapdragon	Mandevilla	Mandevilla
Begonia	Begonia	Nemesia strumosa	Nemesia
Bougainvillea	Bougainvillea		Cup flower
Browallia speciosa	Browallia	Oxalis	Shamrock
Campanula isophylla	Italian bellflower	Pelargonium peltatum	Ivy geranium
Chrysanthemum morifolium	Chrysanthemum	Petunia hybrida	Petunia
Clerodendron thomsoniae	Glory bower	Portulaca grandiflora	Rose-moss
Coleus blumei	Coleus	Rhipsalidopsis gaertneri	Easter cactus
Columnea	Norse fire plant	Sanvitalia	Creeping zinnia
Convolvulus tricolor	Morning glory	Schlumbergera	Holiday cactus
Cuphaea platycentra	Firecracker plant	Tagetes	Marigold
Dianthus	Carnation	Thunbergia alata	Black-eyed Susan
Euphorbia pulcherrima	Poinsettia	Torenia fournieri	Wishbone flower
Fuchsia hybrida	Fuchsia	Tropaeolum majus	Nasturtium
Gazania	Gazania	Verbena elegans	Verbena
Heliotropium	Heliotrope	Viola tricolor	Pansy
Impatiens sultani	Sultana		

Table II

Suggested Foliage Plants for Baskets

Botanical name	Common name	Botanical name	Common name
Asparagus	Asparagus fern	Nephrolepis exaltata	Boston fern
Ceropegia woodii	Rosary vine	Peperomia rotundifolia	Yerba Linda
Chlorophytum	Spider plant	Pilea microphylla	Artillery plant
Cissus rhombifolia	Grape ivy	Plectrantus	Swedish ivy
Davallia fejeensis	Rabbit's foot fern	Saxifraga stolonifera	Strawberry begonia
Epipremnum aureum	Marble queen ivy	Sedum morganianum	Burro's tail
Ficus pumila	Creeping fig	Senecio	Various
Fittonia	Nerve plant	Soleirolia soleirolii	Baby's tears
Hedera helix	English ivy	Streptocarpus	Cape primrose
Helixine soleirolii	Baby's tears	Tolmiea menziesii	Piggyback plant
Hemigraphis alternata	Waffle plant	Tradescantia	Wandering Jew
Hypoestes	Polka-dot plant	Vinca major	Trailing vinca
Hoya carnosa	Wax plant		

VII. GROWING PROCEDURE

Although it certainly is not necessary to hang containers, most growers do (Fig. 3). Some will wait until the plants are fully established before hanging them so that there is less chance of losses created by visually ignoring them once they are hung.

Because most hanging baskets are grown hanging in the airspace in the growing structure rather than on benches, it is quite common for a grower to provide an internal frame on which to hang the containers rather than to use the growing structure (Fig. 4). Some greenhouse structures have collapsed under the weight of containers hanging from the structural members of the frame. Greenhouse structures are designed to support the glazing material and miscellaneous heating and cooling equipment, not pots heavy with soil and water. Internal frames also allow for unobstructed use of thermal blankets and sun screening curtains above the hanging containers. Properly designed irrigation systems can be attached to these frames and serve as automatic spacing indicators for the containers. Sufficient space must be provided for each basket to fill out properly. An 8-inch basket should have at least a foot of linear space to grow in. Most growers will allow up to 6 inches between baskets. Certainly another important consideration is the vertical space between the baskets and the plants on the bench below. Plants, such as foliage plants, that thrive in low light conditions can be layered in a greenhouse. Growers sometimes use up to seven layers and have even built greenhouses with exceptionally high gutters to accommodate these multiple layers of plants. Problems associated with layering are too heavy shading of the lowest plants, dripping and falling of plant debris onto the plants below, and disease contamination from one layer to another. These problems can be addressed by rotating the layers regularly. Plants from the top layers can be sold first and the

Fig. 3. Hanging baskets on bench top.

Fig. 4. Hanging baskets on specially constructed support in greenhouse.

younger plants moved up to fill those spaces. This requires a lot of hand labor unless the pots are on some sort of a conveyor. Also installing hooked wire stakes with spiraling hoops welded in place to hold the base of the container will allow the drainage water to drip harmlessly through to the floor (Fig. 5). Some growers intentionally hang the hook of the lower pot into the lip of the upper pot thus allowing the higher pot's leachate to water the pot below. Careful monitoring of both salts levels and diseases could enable this method to be successful.

In greenhouses where baskets share the environment with high light crops it is common to have the baskets hung over the walkways or along the side of the benches, or both. This will inconvenience normal worker movement but will enable the use of the volume of the structure with minimal harm to the primary crop in the house. An E.C.H.O. system that constantly moves and does not allow dripping on the crop below could be used over a high light intensity requiring crop during the

Fig. 5.   Specially made hoops to support hanging baskets in vertical fashion.

spring and summer months. It is common to start hanging baskets in a greenhouse and transfer them outside with minimal protection after the danger of frost has passed. Growers in warm climates may grow baskets in shade houses from the time of potting, especially if timing is not critical.

VIII. PINCHING

Unless the plants being used have a natural hanging form that is quite full (e.g., Boston ferns), it is likely that some sort of pinching, pruning, or shaping will be desired. Pinching will stimulate a plant to branch and will help to fill out the basket. Young plants can be pinched even before they are potted into the baskets. Although pinching is a labor-intensive process, the uniform, full shape that results proves worthwhile. Electric clippers or hedge shears will not damage the plants excessively if done carefully. Very full petunia, impatiens, browallia, fuchsia, and lobelia baskets, to name just a few, can be grown with proper pinching. Sufficient growth of sheared plants must be allowed before sales. Four weeks should be enough time for most herbaceous plants. Proper knowledge about plant response to both daylength and temperature will certainly help a grower make the right decisions about when to pinch during the growing period.

IX. SHIPPING AND HANDLING

A well-grown hanging basket is difficult to ship without some damage. Most growers will sleeve the plants using especially wide paper sleeves. Some will entwine the hanging shoots above the container so that they fit in the sleeve and do not hang out. Without some protection excessive bruising, especially of more succulent herbaceous flower plants, will occur. These plant types may not sleeve well, and certainly moss baskets with plants all around the sides will not, either. Shipping boxes designed to hang the plant inside can work quite well, but they are expensive and require proper handling during shipping to maintain them in an upright position. Growers making local deliveries can hang the baskets in their delivery trucks above other delivery items. Care must be taken to avoid excessive swaying of the baskets, as they could easily damage one another. Hanging the hooks between taut sets of wires will help prevent this swaying, or spacers between the baskets may be necessary. Placing the basket bottoms in holes cut into especially made flats or boxes will also stabilize them on the delivery van floor. Shipping and handling problems are reasons why only retail growers specialize in unusual baskets. Large wholesale growers simply cannot ship these products easily and generally grow the more common foliage hanging pots that sleeve well and are more convenient to ship.

Helping customers get the baskets home in good condition is important, too, and inexpensive slats or boxes that will keep the baskets stable often are used.

X. MARKETING

Because it is difficult for the large wholesale grower to ship unusual hanging basket shapes and sizes, this is a marketing niche for local growers. Supply and demand should be studied before large investments are made in production. The sales area should be well planned so the items can be readily seen. It is very important that faded flowers and other debris be removed regularly, and that the plants are adequately watered, so high quality can be maintained.

XI. CARE AND HANDLING BY THE CUSTOMER

The addition of a water-holding polymer to the soil in a hanging basket may be of value to the customer. This may be a valuable selling point, as watering is often a difficult task for the customer. Similarly, a slow release fertilizer might be

Fig. 6. *Helipterum anthemoides*, Everlasting, basket. Photo courtesy of Weidner's Begonia Gardens, Inc., Leucadia, CA.

beneficial. Occasional pruning, shaping, or pinching should help keep the basket looking better. Instructions or care sheets should accompany each basket. Customers should be told how quickly moss baskets and other densely planted baskets can dry out, and the value of proper watering procedures. The light and temperature requirements should also be explained.

XII. SELECTION OF PLANT MATERIAL

Many plant species are particularly suitable for hanging baskets. Tables I and II list some suggestions for flowering and foliage baskets and are by no means exhaustive in the selection of plants for hanging baskets. Seed and plant catalogs are excellent sources of ideas for baskets. Plants that are described as creeping, hanging, trailing, or pendulous will probably succeed in baskets if given proper care (Fig. 6). Various herbs, vegetables, and fruits can be tried as well. Chives,

Fig. 7.   Mixed seasonal pot.

mint, thyme, rosemary, lettuce, tiny fruited tomatoes, cucumbers, squash, and strawberries can be used. The Society of American Florists *Buyers' Guide* published biannually is a source of plant material, as well as containers.

XIII. INSECT AND DISEASE PROBLEMS

Like other plants grown in greenhouses, hanging baskets are susceptible to a variety of insect and disease problems. Proper monitoring of pests and safe preventive practices can help prevent serious problems from arising. Location of the baskets in the greenhouse may make early detection of pests difficult.

All plants are susceptible to a number of diseases, but good sanitation practices can help avoid many of them. Clean containers, plant material, soil, and tools are necessary, as well as maintaining good sanitation practices during production. Fungicide drenches also can be used. Often it is the crop beneath the hanging

Fig. 8.   Poinsettia-standard form. Photo courtesy of Paul Ecke Poinsettia Ranch, Encinitas, CA.

baskets that is most affected. For example, fallen petals can result in *Botrytis* infection on the foliage of the plant below.

Damage from insect populations can be excessive before the pests are actually detected. Spraying for pests is difficult when plants are overhead, and personnel must be properly protected and informed.

XIV. STANDARDS, WINDOW BOXES, COMBINATION POTS, AND VERTICAL GARDENS

Along with hanging baskets, plants can be grown in many other unusual forms (Figs. 7,8). These include standards or tree forms, window boxes, combination

Fig. 9. Sewer pipe cut to contain plants for vertical container.

pots or boxes, and vertical gardens. Because many of the same types of plants can be used for these forms and the marketing may be similar, a grower of baskets may find it convenient and profitable to produce these items as well.

Standards or tree forms of plants generally are upright, have a single stalk, and have been pruned to remove all shoots on the stalk below the head. Production time is quite long and may involve several years. Plants that grow rapidly and develop a strong, upright, main shoot are the easiest to train. The shoot should be staked to prevent damage and to keep it straight. The lowest side shoots are removed. Once the plant reaches the desired height, the main shoot is pinched. The side shoots that branch out are also soft pinched as necessary until the proper sized head is formed. Plants that are often used include azaleas, roses, fuchsias, geraniums, coleus, English ivies, herbs, chrysanthemums, and poinsettias.

Fig. 10. Vertical garden divider.

Window boxes and combination pots or boxes generally contain a variety of plants to give visual, textural, and flowering differences. Upright grasses can be used for height, trailing vincas for container coverage, and assorted flowering annuals for color accents. The containers vary according to use and may be replanted year after year. Containers with drainage holes are ideal in preventing waterlogged soil. Combination pots are especially common in some parts of the United States as grave decorations on Memorial Day.

Vertical gardens are a way to display plants attractively and at the same time provide screens, dividers, or backdrops in the garden. Frames can be constructed to any shape. The metal or wooden frame supports a chicken wire or welded wire fabric that is lined with plastic film. There should be horizontal cross-braces to prevent bulges and a sturdy base. A suitably light medium is placed inside the plastic film, and the film is cut where the plants are to be placed. Placing the plants relatively close together helps to prevent gaps. Watering can be difficult, as the lower part of the garden will be wetter than the top if water is only applied from the top of the frame. Watering from the outside occasionally will help keep plants moist without overwatering. Plant and water the frame on site, if possible, as moving the vertical garden later can be difficult. Alyssum, begonia, browallia, calendula, coleus, dianthus, exacum, ivy geranium, lobelia, marigolds, impatiens, herbs, pansy, petunia, portulaca, dwarf snapdragons, verbena, vinca, strawberries, and various vegetables are some of the plants that should do well in vertical gardens (Figs. 9,10).

REFERENCES

Anderson, T. (1977). "Vines for Hanging Baskets." *Southern Florist and Nerseryman.* March 18.

Anonymous. (1975a). *Container and Hanging Gardens.* Ortho Book Series.

Anonymous. (1975b). "Hanging baskets; What a way to grow." *TeleFlora Spirit.* March.

Anonymous. (1977). *Gardening in Containers.* Sunset Books, Lane Publishing Co., Menlo Park, CA.

Anonymous. (1984). "8 Simple steps to Super Baskets from Roger's Garden." *Southern Florist and Nurserymen.* March.

Anonymous. (1989). *Tips on Growing Bedding Plants.* Columbus, OH: Ohio State University.

Baumgardt, J. P. (1972). *Hanging Plants for Home, Terrace and Garden.* New York: Simon and Schuster.

Baxendale, M. (1986). *Gardening by Design-Window Boxes.* Manchester, NH: Salem House Publishers.

Brown, K., and Romain, E. (1987). *Creative Container Gardening.* Joseph, London, England.

Johns, L. (1974). *Plants in Tubs, Pots, Boxes, and Baskets.* Van Nostrand Reinhold Co.

Kramer, J. (1971). *Hanging Gardens.* Charles Schribner's Sons.

Maddux, R. (1978). Flowering Hanging Baskets for Mother's Day and Foliage Hanging Baskets-8 inch, Ohio Florists' Assoc. Bulletin #583.

Rathmell, J. K., Jr. (1975). "Hanging baskets, how to make and sell them. *Proc. Int. Bedding Plant Conf.,* 8th, Newport Beach, CA.

Whitman, A. (1990). "Extend your season and your cash flow with color bowls and containers." *GrowerTalks,* January.

22

Foliage Plants

Charles A. Conover

Introduction to Floriculture, Second Edition
Copyright © 1992 by Academic Press, Inc.
All rights of reproduction in any form reserved.

I. HISTORY AND BACKGROUND

Foliage plants, grown as ornamentals for centuries, are usually listed as pot plants, which include both foliage and flowering plants. Although it is sometimes difficult to separate the groups, foliage plants are generally grown for their attractive foliage that can be retained for long periods in an interior environment, rather than for their flowers.

A. Origin and Development of the Industry

The origin of the foliage plant industry is obscure because originally most plants used indoors were collected, and no organized commercial industry existed. During the late Victorian era foliage plants, including ferns, palms, *Aspidistra*, and *Sansevieria* were grown in greenhouses in Europe and the United States and sold for interior use in local areas. In the United States, Swanson (1975) traced the present foliage industry to Florida as early as 1906 when a northern Boston fern grower recognized the potential for growing foliage plants in Florida. However, greenhouse producers in northern states had already been growing limited numbers of ferns and other foliage plants for over 25 years.

Since the early 1900s, the foliage plant industry has developed dramatically in the United States as well as in Europe; the advent of central heating systems in homes undoubtedly contributed tremendously to this growth. In southern parts of Florida, Texas, and California the industry developed initially outdoors and under slat sheds, whereas more recently it has been moved into fiberglass and glass greenhouses. In northern areas, production in glass greenhouses has been centered in Ohio, Pennsylvania, New York, and nearby states. More recently (mainly since 1968), a number of companies have developed stock production units in the Caribbean region for shipment of propagative units to the United States and Europe. In 1977, it was estimated by Scarborough (1978) that at least 16 million dollars worth of propagative material was shipped into Florida for finishing. In northern Europe major increases in foliage plant production has occurred in glasshouses over the last 10 years, mainly in Denmark and The Netherlands. Most of the propagative units used in production are from tissue culture or from farms in southern Europe or Central America.

B. Production Areas

Major United States production areas are shown in Table I. The concentration of the foliage industry in Florida, California, Texas, and Hawaii is primarily due to reduced production costs associated with moderate winter temperatures and high light intensity throughout the year. Northern European production is concentrated in areas of moderate winter temperatures and specializes in smaller plants, while southern Europe specializes in larger plants. Major stock producing areas are located in Central America, Africa, Israel, and Asia.

Table I

Major United States Foliage Plant Production Areas and Estimated Acreage and Sales in 1988[a]

Area	Acres	Wholesale value (millions of dollars)
California	499	79
Florida	4132	254
Hawaii	128	12
Texas	159	15
United States—28 states total	5089	402

[a]Anonymous (1989a).

C. Economic Importance

Foliage crops were not of major economic significance in relation to other floriculture crops until the late 1960s. As late as 1970, foliage plants accounted for only 15 million dollars wholesale value in Florida. The most recent data from United States Department of Agriculture (Anonymous, 1989a), indicate that the total United States foliage plant wholesale value was nearly $402 million in 1988 (Table I). Foliage crops have become of major economic importance in a relatively short span of time.

The number of foliage plants used in Europe has also risen substantially during the past 20 years, with movement on the Aalsmeer auction increasing 1000-fold for some genera (Anonymous, 1987).

II. BOTANICAL INFORMATION

The best general sources of botanical information on foliage plants include those by Bailey and Bailey (1976) and Graf (1970).

Many definitions exist for foliage plants, but the plants are so diverse in habit and use that it is difficult to develop one definition that is inclusive.

A. Taxonomy

Anyone examining a representative group of foliage plants soon realizes that diversity among plants is enormous. In addition to both monocotyledonous and dicotyledonous members of the Angiospermae, foliage plants also contain representatives of the Gymnospermae—*Araucaria* (cone-bearing trees)—and Pteridophyta (ferns). Another problem with scientific classification of foliage plants is that the origin of many plants is uncertain, because many have been collected in the tropics and introduced into the industry without being identified taxonomically. The fact that many foliage plant cultivars are sports of the same species, although very different in appearance, confuses the matter further. Those interested in taxonomic details should consult references specializing in the subject.

B. Native Habitats

Most foliage plants are native to tropical and subtropical areas of the world although a few, such as *Aucuba* and *Pittosporum*, are also found in temperate zones. In tropical and subtropical zones, however, vast differences in climate may occur within relatively short distances because of the amount and distribution of rainfall, changes in temperature because of elevation, and variation in light intensity because of cloud cover. For these reasons, foliage plants are able to tolerate a wider range of soil moistures, light intensities, and temperatures than most people realize. However, the largest volume of foliage plants sold are adapted to areas where temperatures range between 55° and 100°F and where rainfall is somewhat evenly distributed throughout the year. Where dry periods are frequent, foliage plants often grow in the understory of trees or other vegetation, which reduces environmental stress and, therefore, water requirements of the foliage plants.

C. Flower Types

Most foliage plants are valued for their attractive foliage rather than their flowers although there are some exceptions, including *Aphelandra*, *Spathiphyllum*, bromeliads, Christmas cactus, and several others. Flowers of most foliage plants, however, are small, not very colorful, and hidden within the foliage. Some genera, such as *Epipremnum* or *Syngonium*, normally remain in the juvenile form when used indoors and thus never flower. Others, such as *Aglaonema* and *Dieffenbachia* within the Araceae, produce flowers indoors when grown under good conditions. These flowers, however, serve only as novelties and do not enhance the beauty of the plant.

D. Prominent Cultivars

Well over 1000 different foliage plant types are commonly sold in the foliage plant industry; thus, any discussion of a limited group is subject to inaccuracies. One of the best sources of information on numbers of any specific plant type produced in Florida is the list from the Florida Foliage Association (Anonymous, 1989b).

Information is unavailable on the total foliage plant product mix in the United States, but it has been compiled by McConnell *et al.* (1989) for Florida, which accounts for 63% of the total United States production.

The data in Table II are not listed by genera throughout, because many diverse genera may be included in plants sold as combinations. On a national basis, these statistics are probably low for hanging baskets and terrarium plants, which are often grown for local markets in northern areas of the United States. Also, larger

Table II

Estimated Relative Economic Importance (%) of Specific Genera, Species, Cultivars, and/or Categories of Florida Foliage Plants Sold in Specific Years[a]

Product	1975	1988
Aglaonema	2.0	4.0
Brassaia actinophylla	5.0	2.2
Dieffenbachia	5.0	5.9
Dracaena	11.0	9.4
Epipremnum	3.0	9.1
Ferns	3.0	1.2
Ficus	6.0	6.5
Hedera	—[b]	3.1
Palms	7.0	5.8
Philodendron (other)	6.0	1.7
Philodendron scandens oxycardium	14.0	3.0
Schefflera arboricola	—	1.3
Spathiphyllum	3.0	3.3
Syngonium	2.0	2.2
Combinations	2.0	1.0
Other	31.0	40.3

[a] From McConnell et al. (1989).
[b] —, Data not available.

foliage types are listed at percentages higher than the actual national percentage because Florida is responsible for well over 80% of the plants grown in 10-inch or larger containers.

European production figures for specific foliage plants can be obtained in part by reviewing sales on the auctions. An example is the annual listing of pot plants sold each month through the Aalsmeer auction in The Netherlands (Anonymous, 1987).

III. PROPAGATION

A. Stock Plant Culture

Foliage stock plants are grown under various cultural regimes. In tropical areas they are often grown under natural shade from trees, or in full sun, but more commonly under structures covered with polypropylene shadecloth. In areas where frost occurs, stock is most often grown in greenhouses, but it also may be grown in shadehouses covered with plastic in winter. On frosty nights, these plastic-covered structures can be heated or overhead irrigation water can be used for temperature control.

B. Field Cultural Systems

Growth of foliage stock plants in the field (outdoors) is restricted to areas where winter temperatures do not drop below freezing. Only a few stock plant species can be grown in full sun. Conover and Poole (1972, 1974b, 1974c) have shown that selection of the correct shade level is very important. Too little shade reduces the salability of propagative units because of yellowing or burning of edges of foliage and too much shade can also affect quality, as well as greatly reducing the number of propagative units available. For the most part, stock plants are grown under polypropylene shadecloth that provides the desired light levels (Table III).

Land selected for stock production should have good internal drainage as well as sufficient slope to allow surface water to drain off rapidly when excessive rainfall occurs. Temperature ranges should preferably be between 65°F minimum at night and 95°F maximum day for best quality and yield. Infrequent low temperatures of 50°F and high temperatures not above 105°F will not damage plants but will reduce yields. Location of a farm should also take wind speed, direction, and frequency into consideration because they may affect not only the types of structures that can be built, but also crop growth. Wind-induced problems include foliar mechanical damage from abrasion, tipburn due to wind, and drought caused by low humidity. Consideration should also be given to competition within the area for labor, willingness of individuals for "stoop-labor" employment, and ability of individuals to learn cultural aspects of production.

Table III

Light and Fertilizer Requirements of Foliage Plants Adapted to Stock Production in Full Sun or under Shadecloth

Botanical name	Suggested light level (fc)	Average fertilizer requirement (lbs/1000 sq ft/yr)		
		N	P_2O_5	K_2O
Aglaonema	2000–2500	34.5	11.5	23.0
Codiaeum variegatum cvs.	7000–8000	41.4	13.8	27.6
Cordyline terminalis cvs.	3500–4500	41.4	13.8	27.6
Dracaena deremensis cvs.	3000–3500	34.5	11.5	23.0
Dracaena fragrans cvs.	3500–10,000	41.4	13.8	27.6
Dracaena marginata	3500–10,000	48.3	16.1	32.2
Dracaena (not above)	3000–4500	41.4	13.8	27.6
Epipremnum aureum cvs.	3500–4500	41.4	13.8	27.6
Ficus elastica cvs.	3500–10,000	48.3	16.1	32.2
Ficus lyrata	3500–10,000	48.3	16.1	32.2
Monstera	3500–4500	41.4	13.8	27.6
Philodendron	3000–4500	41.4	13.8	27.6
Sansevieria	3500–10,000	27.6	9.2	18.4

Many foliage types will grow in ground beds that should have a north–south orientation with a width of about 4 feet and an elevation of about 6 inches above the aisles to facilitate drainage. Some plants well adapted to this type of culture are listed in Table III.

Structures used to support shadecloth are generally constructed of treated lumber with 4- × 4-inch posts 10 to 12 feet on center connected by 1- × 6-, 2- × 4-, or 2- × 6-inch stringers. Where wind is a problem, internal or external braces are often added between posts and stringers. More recently, some shade structures have been constructed with concrete or steel posts with cable stringers running at 90° and 45° angles to the posts. Height and size of the structure are important because they influence temperature at plant height. Because heat rises and air movement is slow through shadecloth, it is wise to provide a minimum of 8 feet and preferably 10 feet of clearance. Erection of several smaller 1- to 5-acre units with empty spaces between the units or elevated strips with open spaces, as shown in Figure 1, will help prevent excessive temperature buildup.

Foliage stock plants grown in full sun or under shadecloth are usually watered with an impulse or spinning sprinkler system, which is also commonly used for fertilization; therefore, it is very important to have a properly engineered system so that uniform coverage will be obtained. Use of low-angle, trajectory sprinklers will be necessary to prevent contact of water with shadecloth. Normally, foliage plants grown under shadecloth require 1 to 2 inches of water a week.

Unamended soils in tropical and subtropical areas are rarely satisfactory for foliage stock production. Sandy soils usually require organic components such as peat moss to improve water- and nutrient-holding capacities whereas heavy tropical soils require peat moss, bark, coarse sawdust, or rice hulls to improve internal aeration.

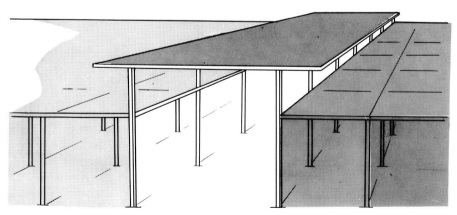

Fig. 1. Elevated sections in shade structures aid in cooling and provide the additional height required for vehicle movement.

Harvesting propagative units from stock plants requires an understanding of the sizes and forms required by the industry, as well as how harvesting influences future production. It is imperative that sufficient viable foliage always remain on the stock plant after cutting harvest to allow for extensive bud break. Constant removal of all cuttings of marketable size will reduce yields to the point that stock becomes unprofitable.

C. Greenhouse Cultural Systems

Although most foliage stock plants grown for cuttings or divisions may be grown in greenhouses, the economics of production limit the final plant selection. Size of plants in relation to yield of cuttings per square foot and need for specific types to be grown under cover for pest protection govern selection. Some of the most common greenhouse-grown foliage plant genera with their light and fertilizer requirements are shown in Table IV.

Foliage stock plants can be grown in any type of greenhouse that provides sufficient light, required temperatures, and proper moisture. In greenhouses where condensate drips on plants during the winter, there is the possibility of cold water damage (chilling damage) to foliage of some genera, such as *Aglaonema*, *Dieffenbachia*, and *Syngonium*. Additionally, lack of adequate air movement to help keep foliage dry can cause disease problems, such as *Botrytis*.

Greenhouse-grown stock plants in raised benches or in pots should have sufficient medium volume for good root growth, drainage, and adequate aeration.

Table IV

Light and Fertilizer Requirements of Selected Foliage Plants Adapted to Stock Production in the Greenhouse

Botanical name	Suggested light level (fc)	Average fertilizer requirement (lbs/1000 sq ft/yr)		
		N	P_2O_5	K_2O
Aglaonema	1500–2000	34.5	11.5	23.0
Aphelandra squarrosa cvs.	650–1000	41.4	13.8	27.6
Calathea	1500–2000	34.5	11.5	23.0
Dieffenbachia	2500–4000	41.4	13.8	27.6
Epipremnum aureum cvs.	3000–4000	41.4	13.8	27.6
Hoya carnosa cvs.	2000–3000	34.5	11.5	23.0
Maranta	1500–2000	27.6	9.2	18.4
Nephrolepis	2000–3000	34.5	11.5	23.0
Peperomia	2500–3500	27.6	9.2	18.4
Philodendron	2500–3500	41.4	13.8	27.6
Pilea	2500–3000	27.6	9.2	18.4
Schlumbergera truncata cvs.	3000–4000	27.6	9.2	18.4
Syngonium podophyllum cvs.	3000–4000	34.5	11.5	23.0

Benches with 6-inch sides serve this purpose best, but some smaller plants will grow well when 4-inch sides are used. Potted plants intended only for stock production should generally be in 8-inch or larger, containers.

Aluminum, cement, fiberglass, and wood benches can be used as long as an open, well-aerated medium is used for growing. Major advantages to bench growing versus using ground beds include elimination of stoop labor, increased growth because of higher soil and plant temperatures, and elimination of or reduced problems with soil-borne insects, nematodes, and disease organisms.

Selection of specific growing media will depend on local availability, cost, and personal preference. Some excellent media for foliage stock plant benches include (1) 50% peat moss and 50% pine bark, (2) 75% peat moss and 25% pine bark, (3) 75% peat moss and 25% sharp mason sand, (4) 75% peat moss and 25% perlite, and (5) rockwool. Because stock plants remain in benches for many years, it is important to select components that will not decompose rapidly.

Knauss (1971) has shown that selection of watering systems for stock plants grown in greenhouses is very important because foliar disease is less of a problem when the foliage is kept dry. Another advantage to keeping foliage dry and reducing pesticide usage is the reduction of foliar residues. The best method of irrigation that keeps foliage dry is irrigation tubing that allows water to seep through the entire surface wall or to be emitted at specific intervals. Spacing at 1-foot intervals across the bench will provide excellent coverage and will help prevent soluble salts buildup in the media between irrigation lines. Overhead sprinklers or spray stakes can also be used for stock plant production when foliar diseases are not a problem or can be easily controlled.

Temperatures needed for maximum yield from stock plants are 65°F minimum and 95°F maximum. Maintenance of the minimum temperature is expensive during winter months and if heat conservation is a problem, it is better to keep soil temperatures at 65°F minimum and allow air temperatures to drop slightly lower.

Light intensities listed in Table IV will maximize cutting yield, but they may be exceeded by as much as 25% if air temperatures are 80°F or less. Where excessive temperatures occur (100° to 110°F), the lowest suggested intensity should be used to improve plant appearance of *Aglaonema*, *Calathea*, *Maranta*, *Peperomia*, *Pilea*, and other plants requiring low-light intensity.

Fertilizer levels suggested are for listed light intensities; any change in light level above or below those shown in Table IV will require a change in the fertilizer level. Higher light levels require use of more fertilizer to maintain similar quality, whereas lower levels require less because of reduced total demand.

D. Economic Aspects of Stock Plant Culture

Growth of foliage stock plants in greenhouses is limited only by economics; valuable and costly space devoted to stock plant production may be better used for production of potted plants. It is generally conceded that bench space in northern greenhouses costs $10 to $15 a year per square foot, whereas in warmer areas the cost may range between $3 and $10. Only a few groups of foliage plants yield sufficient cuttings to make them profitable where heating costs are high. Before establishment of a stock production area, it is wise to compare the costs expected for an in-house stock production area to the costs of purchasing cuttings, tissue culture units, cane, or other propagative units.

E. Propagation Methods

Most foliage plants are relatively easy to propagate; consequently, producers are most interested in practices that enhance quality, percentage, and speed of rooting.

Propagation methods used for foliage plants include cuttings, seed, air layers, spores, division, and tissue culture. Conover and Poole (1970) have previously discussed propagation of specific foliage genera in some detail.

Propagation by cuttings is one of the most popular methods and can be tip, single- and double-eye leaf bud, leaf, or cane cuttings. Selection of a specific method depends on plant form (upright, vining) and availability of propagative material. Interest has developed in the effects of storage on cuttings because many producers are acquiring cuttings from Africa and Central America. Limited research by Poole and Conover (1988b, 1988c) has shown that storage of cuttings of several genera for up to 2 weeks does not severely impact rooting.

Seed propagation is increasing in popularity because costs are lower than for vegetative propagation and there is no need for stock production areas; however, seeds of many foliage plants are not available or the plant type is not stable from seed. Some of the more popular foliage plants grown from seed include *Araucaria*, *Brassaia*, *Coffea*, *Dizygotheca*, *Podocarpus*, and nearly all of the palms. Mikorski and White (1977) have conducted research on propagation requirements of some tropical seeds, but detailed information is lacking on many genera. Seed of tropical genera should be planted soon after harvest because the germination percentage decreases rapidly with increased time between harvest and planting.

Air layering is decreasing in importance as a propagation method because of high costs and the need for large stock plant areas. Plants that are most commonly air layered include *Citrus* (Calamondin), *Codiaeum*, *Ficus*, and *Monstera*. One of the drawbacks with air layers is their large size, which makes them difficult to ship without the occurrence of mechanical damage.

Division is the only method of propagation of some cultivars of *Sansevieria*. This is a labor-intensive method and presents problems of carrying disease, insect, or nematode pests to new plantings.

Spores are commonly used to propagate a number of fern genera although many ferns are grown from divisions or offsets and from tissue culture. Fern production from spores can take 1 to 2 years before marketable plants are available.

Tissue culture is becoming an important method of propagation for foliage plant producers (Fig. 2). Rapid multiplication of new cultivars is an important advantage

Fig. 2.   New cultivars of foliage plants can be rapidly increased through tissue culture.

of tissue culture, but some older cultivars such as Boston fern are also commonly propagated by this system. Some genera including *Dieffenbachia*, *Nephrolepis*, *Spathiphyllum*, and *Syngonium* are most commonly propagated through tissue culture than by any other method at the present time.

F. Propagation Systems

The usual propagation system is a mist area where cuttings are stuck, misted, and, when rooted, pulled and potted. During the last 10 to 15 years many producers have shifted to direct-stick propagation in which cuttings are placed directly in the growing pot, rooted, and finished without being moved. This system is especially adapted to plants that root quickly and grow rapidly, such as *Philodendron*, where three to five single-eye cuttings may be placed in each 4-inch pot. This type of propagation requires a high percentage of rooting viability and a large area available from stick-to-finish; this is not always practical in cooler regions where plant growth may be slower unless size of finished plant is kept small.

The frequency of misting depends on light intensity, temperature, humidity, and plant type and should be timed to keep some moisture on foliage at most times. In warmer climates where temperature and humidity conditions are extremely stressful, fine-spray misting or fog is set for very frequent intervals. This fine mist or fog keeps the foliage moist and temperatures down but does not saturate the media, thus it can be used frequently without serious disease problems. In cooler climates, growers often use tents over the propagation bench to provide 100% humidity and to eliminate misting. Seed propagation at temperatures of 80° to 90°F within germination chambers has also proven to be beneficial for many seed propagated genera, especially palms. Light requirements for propagation are the same as those required for plant production of the species.

G. Propagation Media, Fertilization, and Temperature

Major media requirements for most efficient rooting are good aeration, good drainage, proper pH, and cation exchange capacity so nutrients can be readily available to the cuttings as the roots develop.

Numerous media have been used for foliage plant propagation. Sphagnum peat moss is most commonly used either singly or amended with perlite, styrofoam, pine bark, or other organic components. Research has shown that dolomitic limestone and slow-release fertilizer are useful during propagation because, as soon as roots are initiated, nutrients are available to the plant. Nutrient mist has been used on foliage plant cuttings, but the high propagation temperatures are conducive to algal growth, especially on the slower-rooting cuttings, which can create a serious problem on leaves and reduce overall quality.

Mist application is very useful in lowering temperatures in summer months but can prolong rooting during periods when the propagation medium temperature drops below 65°F (Poole and Waters, 1971). For this reason, propagation beds or benches should be maintained at 70° to 75°F at all times within the medium. This can be accomplished by under-bench heating or by heating cables.

H. Breeding Programs

Only limited breeding work has been conducted on foliage plants. Some of the foliage plants receiving the most breeding attention to date include *Aglaonema*, *Anthurium*, *Dieffenbachia*, *Hoya*, *Philodendron*, and *Spathiphyllum* (Henny, 1988).

Breeding has concentrated primarily on cosmetics with the major objective being to produce a different-looking plant that will be bought in quantity by consumers. With *Philodendron*, another aim has been to produce a plant that is better adapted to interior environments by increasing tolerance to low light intensity and low humidity; with *Anthurium* the goal is to develop plants with both flower and plant size suitable for 4- to 8-inch pots (Fig. 3).

I. Plant Selection Programs

The majority of new foliage plants are selected from naturally occurring hybrids or "sports"; these are then introduced to the marketplace by observant, enterpris-

Fig. 3. New forms of anthurium are being developed through hybridization of *Anthurium* spp.

ing individuals or companies. Evaluation and selection of these plants (Henley and Poole, 1989; Ottosen, 1988) is especially valuable to the industry.

IV. CULTURE

A. Vegetative Stage

Systems for growing foliage plants vary between producers, which indicates that high quality foliage plants can be grown using different cultural systems. The key to any system depends on knowing how the various cultural factors interrelate so that logical decisions can be made if it becomes necessary to change one portion of the system.

1. Planting

Cultural variability among foliage plant genera and species prevents development of a detailed crop production guide, but factors influencing most foliage crops are discussed so that logical decisions may be made.

The potting medium used to grow foliage plants can range from 100% organic matter to approximately 50% organic and 50% inorganic matter. Key factors to consider in the selection of potting media include aeration (measured as capillary and noncapillary pore space), moisture retention (water-holding capacity), and nutrient retention (cation-exchange capacity). Table V provides an indication of the properties provided by some popular media components. Several other factors that must be considered when selecting potting media include uniformity, availability, weight, and cost.

Examples of the potting media used by commercial foliage growers are shown in the lower portion of Table V. The objective of foliage producers is to use a potting medium with excellent aeration that has good water and nutrient holding capacities but is not excessively heavy.

Normally, pH is adjusted at the time potting mixtures are developed. The best range for most foliage plants is between 5.5 and 6.5, but several genera, including *Maranta* and most ferns, grow most vigorously at a pH between 5.0 and 6.0. Dolomitic limestone is suggested for pH correction, but other calcium-containing materials can also be used. Superphosphate should not be incorporated into potting media unless foliage plants not sensitive to fluorides are being grown. Micronutrients are normally included in the fertilizer program, although they may be included in the potting medium.

Potting methods and systems fall into two main categories—hand potting and automatic planters. With hand potting there is more flexibility and systems vary. Hand-filled or machine-filled pots may be taken to the growing area and cuttings,

Table V

Physical and Chemical Characteristics of Potting Medium Components Commonly Used, and Selected Commercial Combinations[a]

Medium	Aeration	Water-holding capacity	Cation-exchange capacity	Weight
Composted pine bark	H	M	M	M
Perlite	H	L	L	L
Polystyrene foam	H	L	L	L
Sand	M	L	L	H
Sphagnum peat moss	M(V)	H	H	L
Vermiculite	M	M	M	L
Peat:bark (1:1)	H	H	H	L
Peat:bark:polystyrene foam (2:1:1)	H	M	H	L
Peat:bark:vermiculite (2:1:1)	H	H	H	L
Peat:perlite (2:1)	H	M	M	L
Peat:sand (3:1)	M	H	H	H

[a]H, High; M, average; L, low; and V, variable.

rooted or unrooted, planted in that location, or the cuttings may be brought to the pot-filling area, planted, and then planted pots taken to the growing area.

Automatic pot-filling machines are available that are also automatic planters. Rooted cuttings are brought to the central site, sent through the automatic planter, and the potted plants are then taken to the growing area.

The system of choice depends on the size and economics of the operation, as well as the method of propagation selected because unrooted cuttings or microcuttings from tissue culture cannot be sent through an automatic planter.

In the growing area plants that will be finished within 3 months are usually placed at their final spacing when placed on the bench with spacing distance varying from 0 to 3 times the pot diameter. Cavity trays are frequently used by growers to provide automatic spacing and pot support.

Plants grown in container sizes 6 inches or larger may take 6 months to 2 years to reach maturity. Such plants are often placed pot-to-pot until they become crowded and are then moved to their final spacing. Depending on growth form, spacing varies from 1 to 6 times the container diameter.

Pot spacing is important to producers (Christensen, 1976) because spacing of foliage plants directly controls final plant quality. Crowding of plants reduces the light reaching lower foliage and may cause it to abscise, or crowding may cause plants to grow tall without proportionate spread, which reduces value. Crowding also increases disease problems because plants are more difficult to spray, remain wet longer when watered, and have higher humidity around the foliage. Because crowded plants are more difficult to spray, the ability to control insect pests is also a problem.

Fertilizer directly influences growth rate and, therefore, profitability; fertilizer levels also inversely influence longevity indoors so it is important that it not be used in excess. Maximum growth rate of acclimatized foliage plants can be obtained with moderate levels of soluble, organic, or slow-release fertilizer applied constantly or periodically. Research by Conover and Poole (1982) has shown that the fertilizer requirement of a plant is related to light intensity. Suggested levels of fertilizer in Table VI are for plants growing under light intensities that will produce acclimatized, high-quality plants.

Poole and Conover (1978) found that fertilizer ratios for foliage plants need to be approximately 3:1:2 when potting mixtures listed in Table V are used. A ratio of 1:1:1 is also acceptable. Rates listed in Table VI can be easily calculated on a periodic basis and applied weekly or bi-weekly. Periodic fertilization less than every 2 weeks often results in decreased growth and quality. If a constant fertilization program is desired, the suggested level of nutrients at each application is 150 ppm nitrogen, 25 ppm phosphorus, and 100 ppm potassium. To aid producers in refining fertilizer programs, Poole et al. (1988) reported levels of macro- and micronutrients found in leaf tissue of 26 species of good-quality foliage plants.

Potting media used for foliage plant production are normally very low in micronutrients, thus most fertilizers used should contain at least the minimal micronutrient levels suggested in Table VII. Levels only slightly higher than desired,

Table VI

Suggested Light and Fertilizer Levels for Production of Potted Foliage Plants for Indoor Use

Botanical name	Light intensity (fc)	Average fertilizer requirement (lbs/1000 ft^2)		
		N	P$_2$O$_5$	K$_2$O
Aphelandra squarrosa cvs.	1000–1500	41.4	13.8	27.6
Brassaia	5000–6000	48.3	16.1	32.2
Chamaedorea (palms)	2500–3500	34.5	11.5	23.0
Cordyline terminalis cvs.	3000–4000	34.5	11.5	23.0
Dieffenbachia	2500–4000	34.5	11.5	23.0
Dracaena	3000–3500	34.5	11.5	23.0
Dracaena marginata cvs.	5000–6000	48.3	16.1	32.2
Ficus	5000–6000	48.3	16.1	32.2
Maranta	1000–2000	27.6	9.2	18.4
Peperomia	2500–3500	27.6	9.2	18.4
Philodendron	2500–3500	41.4	13.8	27.6
Pilea	2500–3000	27.6	9.2	18.4
Spathiphyllum	1500–2500	41.4	13.8	27.6

however, have been reported to be phytotoxic on *Aphelandra* and *Brassaia* (Conover *et al.*, 1975).

2. Environmental Control

Proper control of environmental factors is necessary for production of high-quality foliage plants. Too often producers fail to regulate the environment properly, with resultant loss of quality, plants, or both.

Temperature is very important because most foliage plants are tropical and require high night temperatures with 65°F the minimum. Soil temperatures are often

Table VII

Average Levels of Several Micronutrients Required for Foliage Crops

Element[a]	Spray application (oz/100 gal)	Soil drench (oz/1000 ft^2)	Soil incorporated (oz/yd^3)
Boron	0.13	0.03	0.010
Copper	1.60	0.30	0.100
Iron	16.00	3.00	1.000
Manganese	8.00	1.50	0.500
Molybdenum	0.13	0.01	0.001
Zinc	4.80	1.00	0.300

[a]One application is often sufficient for short-term crops, whereas reapplications are usually necessary for crops grown 6 months or more.

even more important, and, if they can be maintained at 65° to 70°F, the air temperature may drop as low as 60°F at night without significant crop damage (Conover and Poole, 1987; Poole and Conover, 1981).

The best temperature range for a wide number of foliage genera is 65°F minimum night and 75°F minimum day. Night temperatures as high as 80°F and day temperatures as high as 95°F will not be damaging; but they are unnecessary and often uneconomical to maintain in temperate climates.

Some foliage crops such as *Aphelandra, Aglaonema, Dieffenbachia, Epipremnum, Fittonia, Maranta,* and *Pilea* may be damaged by chilling temperatures of 50°F or below (McWilliams and Smith, 1978), whereas others such as *Ardisia, Brassaia, Hedera,* and *Pittosporum* tolerate temperatures as low as freezing without injury.

Information on light intensity levels for many foliage crops has been included in Tables III, IV, and VI. Light intensity is one of the most important factors to consider in culture because it directly controls quality factors, such as internode length, foliage color, carbohydrate level, growth rate, and acclimatization. Light green foliage or faded colors, such as in the case of *Codiaeum,* are indicative of excessive light and reduced chlorophyll levels. This problem can be corrected in most cases by increasing fertilization (Rodriguez and Cibes, 1977) or reducing light intensity. Because increases in fertilizer often cause excessive soluble salts levels, the proper corrective method is to reduce light intensity. In addition to color, many foliage plants change their leaf size, shape, and orientation in response to light intensity. For example, leaves of *Aglaonema* and *Dieffenbachia* will assume a nearly vertical position (20°F) under excessive light, whereas leaf orientation will be nearer 60°F under proper light; and *Ficus benjamina* leaves will be folded along the midrib under high light conditions.

Acclimatization of foliage plants for interior use is necessary to ensure that they perform well indoors. Acclimatization is the adaptation of a plant to a new environment—in this instance, a building with low light and humidity conditions. The most important aspect of the acclimatization process is development of "shade" foliage that is characterized by large, thin leaves with high chlorophyll levels that allows proper plant functioning in low light situations. A second important factor in the process is the nutritional level, which should be as low as possible while still producing a quality plant. Extensive research by Collard *et al.* (1977), Conover and Poole (1984a), and Fonteno and McWilliams (1978) has shown the importance of acclimatization.

Foliage plants, such as *Brassaia actinophylla, Chrysalidocarpus lutescens,* and *Ficus benjamina,* are often grown in full sun and then acclimatized by placing them under suggested shade conditions and lowering fertilizer levels for 3 to 9 months. Although chloroplasts and grana are capable of reorientation within sun-grown foliage, the leaf anatomy—small size and thick cross section—prevents them from being as efficient after acclimatization as shade leaves; therefore, such plants are less tolerant of low or medium light levels indoors than plants acclimatized during production.

Irrigation levels should be established that ensure that foliage plants receive sufficient water to remain turgid at all times. The water requirement during winter, when temperatures are low and growth is slowed, may be less than once a week; in spring or summer, daily irrigation may be necessary.

Use of irrigation systems that water the potting medium without wetting the foliage is desirable because this reduces foliar disease and residue problems.

The most common watering systems in use that meet this criteria are ebb-and-flow (mostly used in Europe), individual leader tubes to each plant, and capillary mats. Overhead application of water to foliage plants is most common, but it is strongly affected by plant canopy, which may cause much of the applied water to be deflected; thus, it is an inexact way to apply water or fertilizers.

Application of carbon dioxide to foliage plants is uncommon in the United States, but common in Europe, although growth increases of up to 25% have occurred with several foliage plants. Because temperature is also a key to increased growth during periods when greenhouses are closed, unless a range of 65° to 75°F minimum is maintained, the injection of carbon dioxide may not be beneficial.

Humidity requirements of foliage plants during production are not verified by research, but maintenance of 50% or higher relative humidity appears to be desirable. In areas where humidity falls below 25%, many growers install mist lines or raise humidity in some other way. Plant damage from low humidity has been observed on *Calathea*, *Ctenanthe*, and *Maranta*.

Wind can be a severe environmental problem if it causes physical damage to foliage, knocks over containerized plants, or increases the severity of cold damage to plants grown in outdoor shade structures. The best solution to excessive wind is to enclose structures, but use of windbreaks can also be beneficial. Many foliage plants will tolerate temperatures as low as 35° to 45°F, if protected from winds of 5 miles per hour or more.

3. Chemical Control

Regulation of vine length or plant height of foliage plants has been achieved by Henley and Poole (1974), McConnell and Poole (1974), and Poole and Conover (1988a) with ancymidol, daminozide, chlormequat, and paclobutrazol, but they generally increase production time without a significant increase in quality or sales price. Basal branching of *Peperomia* (Henny, 1985), *Spathiphyllum* (Henny and Fooshee, 1985), and *Dieffenbachia* (Henny, 1986), has been increased by spray applications of benzylamiopurine (BA) (N^6-benzyladenine), but propagation by tissue culture of specific cultivars with good breaking habits can decrease the reliance on labor-intensive and expensive chemical regulators; therefore, growth regulators have not received wide usage in the foliage industry.

Breaking bud dormancy of foliage plants through the use of chemicals has been reported by Joiner *et al.* (1978), but it is not of significance at this time. It could be very advantageous to producers if a way to induce bud break of leafless *Dracaena* or *Yucca* canes could be found.

B. Flowering Stage

For the most part, foliage plants are not grown for their flowers; consequently, the flowering of most genera is not desirable. There are, however, important factors that should be considered because flowers can improve or detract from foliage plant quality.

1. Environmental Control

A temperature near 45°F has been shown to be involved in flowering of *Aphelandra* (Christensen, 1969). Cool temperatures have also been used to aid in the induction of flowering in *Schlumbergera*, but temperatures higher than 80°F after flower initiation often cause bud drop. Influence of temperature on flowering of most foliage plants is unknown, but observations indicate it is often a factor.

Photoperiod, as shown by Poole (1971), controls flowering of *Schlumbergera*, but it has not been demonstrated to influence flowering of other foliage plants. This might be expected because most foliage plants are of tropical origin where photoperiod is relatively constant year-round. With *Schlumbergera*, 4 to 6 weeks of short days will initiate buds, which will then develop with either short or long days.

Both light intensity and duration have been shown to influence flowering of *Aphelandra*. This plant appears to be photoaccumulative, and flowers when light intensity is high or when lower intensity is received over longer periods (Kerbo and Payne, 1976). Light intensities of less than 1000 foot-candles (fc) for 8 to 10 hours a day will keep plants vegetative, whereas 1500 to 2000 fc for the same or shorter duration will result in flowering. A more complete listing of environmental effects on flowering has been reported by Henny (1988).

2. Chemical Control

Flower initiation in many bromeliads (including pineapple) is by chemical control in commercial operations. Research has shown that ethephon sprayed or poured into the vase, acetylene gas bubbled through water in the vase, or use of ethylene in an enclosed chamber will initiate flowering of many bromeliads (Adriansen, 1976). The plants must be of sufficient size to support a flower or fruit and usually take 2 or more years to grow to this size when started from seed. Henny (1988) reported effects of gibberellic acid (GA_3) on flower induction in numerous aroids for purposes of breeding new foliage plants.

3. Cultural Control

Influence of fertilization on flowering of foliage plants is mainly unknown at present. Observations on *Dracaena*, *Cordyline*, and many other foliage plants indicate that excessive nitrogen reduces flowering, but because commercial growers are mainly interested in vegetative growth, this area of research receives little attention.

Influence of soil moisture on flowering of *Schlumbergera* has been reported but seems to be of little importance provided photoperiod is controlled. When the correct photoperiod is partly received, but not precisely, drying plants down for 1 to 2 weeks is sometimes beneficial. Regulation of soil moisture for flowering of foliage plants is not a common commercial practice.

V. CONTROL OF INSECTS, DISEASES, NEMATODES, AND PHYSIOLOGICAL DISORDERS

Although most producers recognize the importance of producing high-quality plants, they often fail to recognize the importance of pest control. Good sanitation practices, both inside and outside production structures, will reduce many potential pest problems. Control of weeds that can harbor pests and disposal of infested or nonproductive plant specimens are two good management practices that can alleviate pest problems.

A. Insects and Mites

It takes a few uncontrolled insects or mites only a short time to increase their population to the point that they can severely damage plants.

Factors that affect pest populations include temperature, humidity, irrigation method, potting medium, and access to the structure. Temperatures above 80°F can cause rapid increases in mite populations, especially if humidity is low. Under such conditions, heavy dependence on chemical control is often necessary to produce mite-free plants. Cold or cool temperatures reduce pest problems in unheated production areas, whereas in temperature-controlled greenhouses they present year-round problems. Fungus gnats and shore flies are much more of a problem in greenhouses than outdoors, especially when organic potting media are kept too wet (Osborne et al., 1985). When a particular pest such as scale becomes a problem, cultural procedures must be checked carefully to see if stock plants are infested and if crawlers or adults (Osborne, 1986) have been carried through propagation to potted plant production areas. It is imperative that stock be kept as free of pests as possible to reduce the need for spraying potted materials, which are otherwise ready for sale. Frequent and continued spraying of a crop is undesirable because it increases potential for phytotoxicity and leaves persistent residues on foliage (Osborne and Oetting, 1989).

In areas open to pest movement, spraying or drenching with pesticides are primary methods of control at present. Mites present the most problems in spring, summer, and fall when temperatures are higher, whereas caterpillars and thrips are heaviest in spring and fall. Use of high-pressure sprayers and thorough coverage of both sides of foliage provide the best assurance of control. Air-blast sprayers, while providing high pressure and wide dispersal, often do not properly coat both sides of the foliage with pesticides. Major pests of foliage plants and some of the hosts they are most commonly found feeding on, as described by Hamlen et al. (1981), are shown in Table VIII. Chemicals registered for control of specific pests are currently being reviewed by EPA; thus, their future availability is in question. For recommendations on chemical pest control procedures on foliage plants, co-operative extension agricultural agents should be contacted.

Some of the pesticides used outdoors or under shade structures are also labeled for greenhouse use and are, therefore, available to the producers growing foliage in greenhouses. Smoke bombs and thermal fogs are also available for use in greenhouses, but they only work when greenhouses can be entirely closed for several hours. Because many greenhouses in warm climates are open on the pad side of the greenhouse for 6 months or more each year, this system is only practical in cooler climates.

B. Diseases Caused by Fungi, Bacteria, and Viruses

Diseases can be severe pests of foliage plants if proper cultural procedures are not followed. In some cases when diseases are systemic within tissue (*Erwinia, Xanthomonas*), they cannot be fully controlled even when indexed (disease-free) stock is available.

Fungal and bacterial diseases are more troublesome when wet foliage is combined with high temperatures and humidity. Therefore, these diseases are most prevalent in tropical and subtropical areas with high rainfall. Even in these areas, however, growing plants under cover and irrigating without wetting foliage will nearly prevent their occurrence. Soil-borne fungal diseases become more severe when poorer quality growing media without good aeration and drainage are used, or when plants are constantly overwatered. Fortunately, viral diseases are not commonly found on foliage plants.

Because outdoor production is usually in subtropical and tropical areas where high rainfall occurs, disease control is difficult except with preventative spray programs. During periods of frequent rainfall, and depending on disease pressure,

Table VIII

Major Foliage Plant Insect and Mite Pests and Crops Commonly Serving as Hosts

Pest	Hosts
Aphids	*Aphelandra, Brassaia, Gynura, Hoya, Dieffenbachia*
Broad mites	*Hedera, Aphelandra, Pilea*
Caterpillars	*Philodendron, Dracaena, Brassaia, Maranta, Aglaonema*
Fungus gnats	*Schlumbergera,* palms, *Peperomia*
Mealybugs	*Aphelandra, Ardisia, Dieffenbachia, Gynura, Asparagus, Maranta, Dracaena, Dizygotheca*
Scales	*Aphelandra,* bromeliads, *Ficus,* palms
Spider mites	*Brassaia, Codiaeum,* palms, *Cordyline, Calathea, Dieffenbachia, Maranta*
Thrips	*Brassaia, Ficus, Philodendron, Ctenanthe, Syngonium*

it may be necessary to spray weekly or more often; during dry seasons little, if any, pesticide application may be necessary. Irrigation applied directly to the soil or overhead irrigation only during the middle of the day when rapid drying can occur combined with proper plant spacing, which will lower the humidity, will aid in reducing disease problems. Directed high-pressure spray, rather than air-blast sprayers, provides the best control of plant pathogens and is, therefore, the most cost-effective method to use.

Foliar fungal and bacterial diseases generally can be controlled in greenhouses by keeping foliage dry. When this is not possible, fairly good control can be obtained with chemical sprays to the foliage. Soil drenches for control of most of the soil-borne diseases are fairly successful in raised benches and in containers off the ground. No control procedures are presently recommended for plants with virus except to rogue infected plants.

A listing of some of the major disease pests and primary hosts has been compiled by Chase (1987) and is shown in Table IX. As with insecticides and miticides, control information can be obtained from local cooperative extension agents.

C. Nematodes

These soil-borne pests are often found on foliage plants grown in the soil and propagated by division or when foliage plants are planted in contaminated potting mixtures.

Table IX

Major Foliage Plant Disease Organisms and Crops Commonly Serving as Hosts[a]

Organism	Plant parts affected	Common hosts
Fungal diseases		
Alternaria	Leaves, stems	Brassaia
Fusarium	Leaves, stems, roots	Aglaonema, Dieffenbachia, Dracaena
Myrothecium	Leaves, stems	Aglaonema, Anthurium, Dieffenbachia, Syngonium
Phytophthora	Leaves, stems, roots	Aglaonema, Aphelandra, Brassaia, Dieffenbachia, Peperomia, Philodendron
Pythium	Stems, roots	Aglaonema, Dieffenbachia, Epipremnum, ferns, Philodendron
Rhizoctonia	Leaves, stems	Epipremnum, ferns, Hedera, etc.
Sclerotium	Stems, roots	Brassaia, Dieffenbachia, Epipremnum, Philodendron, Syngonium, etc.
Bacterial diseases		
Erwinia	Leaves, stems, roots	Aglaonema, Dieffenbachia, Dracaena, Epipremnum, Philodendron, Syngonium
Pseudomonas	Leaves, stems	Brassaia, ferns, Ficus, Schefflera
Xanthomonas	Leaves	Aglaonema, Anthurium, Dieffenbachia, Hedera, Philodendron, Pilea

[a]From Chase (1987).

Use of clean planting stock, raised benches, and pasteurized growing media is the best way to prevent nematode infestations. Once nematodes are present in the media, they usually cannot be eradicated, and thus, partial control measures must be initiated on a continuing basis.

Once foliage plants are infested with nematodes, chemical control is the only way of preventing or reducing economic losses. Both granular and liquid nematicides are available that can be applied to the medium surface or as drenches. Hamlen (1976) has found that application to ground beds is beneficial but will require more frequent reapplication than those made to raised benches to get the same level of control.

D. Physiological Disorders

A wide number of problems of foliage plants occur because of excessive or insufficient amounts of chemicals applied to plants or the potting media. These management practices can cause serious economic losses if not recognized and remedied.

1. Nutrient Deficiencies

Within the macronutrient group, limited nitrogen availability results in light green foliage, similar to that seen with excessive light intensity, accompanied by a loss of lower foliage. Potassium deficiency occurs on lower foliage as a general marginal chlorosis, which may develop into necrosis, whereas magnesium deficiency is exhibited as chlorotic bands on each edge of older foliage. Deficiencies of phosphorus, calcium, and sulfur have not been described for any of the popular foliage plants probably because of the amendments used in the media and the use of complete fertilizer programs.

Within the micronutrient group, iron, manganese, and copper deficiencies on foliage plants have been described, whereas zinc, boron, and molybdenum deficiencies have not. Iron deficiencies are exhibited as general chlorosis to terminal foliage for most crops, as does manganese. Copper deficiency appears as cupped and dwarfed terminal foliage with some associated chlorosis (Fig. 4).

Quite a number of disorders that appear on foliage plants are commonly corrected with application of a complete fertilizer or a micronutrient application.

2. Nutrient Toxicities

Excessive application of macronutrients commonly causes soluble salts toxicities. Symptoms of excess soluble salts include dwarfed plants, marginal necrosis of foliage, and poor-quality roots. These symptoms may not appear simultaneously, and if all plants are treated the same, it may not be possible to determine that growth reductions are occurring. In production areas, salts levels ranging between 1000 and 3000 ppm total salts [as measured with a 2:1 (v/v) water:dry potting mix] should provide excellent growth. Because high soluble salts levels can be even more serious for the consumer, however, and because plants received from pro-

Fig. 4. Copper deficiency of *Aglaonema commutatum* 'Fransher.'

ducers with levels exceeding 1000 to 1500 ppm total salts will decline rapidly indoors (Conover and Poole, 1977), producers should make fertilizer program adjustments to lower soluble salts before marketing the crop.

Excessive applications of micronutrients to foliage plants have been reported to cause damage such as dwarfing, chlorosis, and necrosis. In time, specific micronutrient levels that cause problems will be published, but in the meantime, as long as levels listed in Table VII are not exceeded, toxicities should not occur.

3. Spray Material Toxicities

Foliage plants are rarely injured by pesticides despite careless or indiscriminate use. Because of the vast number of foliage plant genera, species, and cultivars, most pesticides have been tested only on a relatively few. Even though a specific pesticide lists foliage plants on the label, this labeling does not mean that it is entirely safe for all foliage plants (Chase, 1987). Symptoms of pesticide toxicity include dwarfed plants, chlorosis, necrosis, or both, of foliage, ring-spots on foliage, leaf drop, and dull-appearing foliage. Once damaged, the entire crop may be unsalable because attractive foliage is the selling point of a foliage plant. Some of the most important factors that influence the potential for spray material toxicities include (1) use of an excessive rate, (2) application at high temperatures, (3) too frequent application, (4) application to wilted plants, (5) application when pesticide may remain on foliage for many hours before drying, and (6) tank mixing.

4. Other Problems

It has been found that fluoride toxicity may result in severe damage to a number of foliage plant genera (Conover and Poole, 1974a; Poole and Conover, 1973). Fluorine may exist as a contaminant in water, air, potting media, fertilizer, or all of these. Some potting media identified with high fluoride content include German peat moss, perlite, and vermiculite. The most common source of fluoride in fertilizers is superphosphate, which may contain from 1 to 2% fluorine.

Foliage genera developing foliar chlorosis, tip or marginal, or necrosis from fluorine include *Calathea*, *Chlorophytum*, *Cordyline*, *Ctenanthe*, *Dracaena*, *Maranta*, *Spathiphyllum*, and *Yucca*. When growing sensitive genera, the producer will benefit from using potting media and fertilizers that do not contain fluorides.

Methods that will limit the uptake of fluoride present in water or the potting media include shading or cooling to reduce transpiration and, therefore, water usage and raising the pH so it is between 6.0 and 6.5 to tie up fluoride in the potting medium.

VI. HANDLING OF FINISHED PLANTS

A. Grading

At present there is no industry-wide grading system for foliage plants, although one has been proposed by Conover (1986). All producers, however, grade their crops; thus, only the most desirable plants should reach the market.

The most commonly used initial grading is based on plant size in relation to pot size. Essentially, this selection requires that in general appearance the overall symmetry is correct (i.e., the plant is not too small or too large in relation to pot size). If the overall size and form are acceptable, then other factors are considered (1) is the plant free from defects caused by insect feeding, disease attack, nutrient deficiency, and any physical or chemical damage and (2) is it well rooted but not pot-bound?

Because foliage plants are sold primarily for the beauty of their leaves and stems, any physical or chemical damage to foliage seriously affects marketability. In Europe, even one damaged leaf on a plant is cause for rejection by a consumer, but in the United States the damage must be fairly obvious before rejection occurs.

Residues on foliage, such as calcium, magnesium, or iron deposits from use of impure water sources or residues from pesticide sprays or fertilizer, can reduce marketability. It is unknown how seriously this factor affects the market because consumers do not readily explain reasons for not buying foliage plants. With the constant increase in labor costs, however, retailers cannot afford to remove residues by hand, which is the best method for cleaning the foliage. This problem is best controlled by producers, through prevention, by the wise selection of water sources, fertilizer materials, and sprays used, as well as by the method of application of each of these.

B. Packaging

Packaging systems in the foliage industry vary greatly and need further development. There are two main types of packaging systems: boxing and shipping loose in specially constructed racks and trucks. When one uses the boxing system, the boxes must meet interstate shipping regulations for weight of corrugated cardboard and must be either waxed or moisture-resistant. This latter requirement is necessary to prevent deterioration of boxes in transit because of moisture in the containers, because nearly all foliage plants are shipped in the containers in which they are grown (usually plastic).

Plants in containers up to 6 inches in size are usually placed in a waxed tray, and the tray is slid into a box of the proper height. These boxes usually do not contain dividers or other restraints. Some progressive producers have designed boxes with dividers and other restraints that hold pots in position and the potting medium in the containers. There is, however, usually an additional boxing charge when such boxes are used. Plants in 6-inch or larger pots are usually sleeved, in either plastic, fiberglass, or paper sleeves, and placed directly in boxes of the proper height.

Several large producers with their own truck fleets have built adjustable racks to accommodate variously sized plants, which they load directly into the trucks. Usually, no dividers or restraints are necessary with this system and physical damage is minimal.

Although several companies have developed special see-through packages of various materials, none are presently of any significance in the market. Several problems have occurred with these packages, including increased disease problems, ethylene buildup, and the inability of plants to adapt back to a low-humidity interior environment (Harbaugh et al., 1976).

C. Storage

Storage of foliage plants is not a normal practice, but with increased sales in mass market outlets that use central distribution points, there has been a need for information on how long plants can be held without a significant decrease in quality.

Extensive research on storage and shipping of potted plants has demonstrated that durations of up to 4 weeks are possible with many genera (Conover and Poole, 1984b; Poole and Conover, 1983; Poole et al., 1984).

Because several species of foliage plants experience chilling injury at 50° to 55°F, storage temperatures between 60° and 65°F are safest for mixed varieties. Storage durations of up to 4 weeks in the dark with no damage have been obtained with numerous genera, while other genera will not tolerate more than 7 days storage before leaf drop and chlorophyll destruction increase to the point that even if plants are salable, they will be of poor quality. With storage beyond 18 days, most foliage plants are not salable; if they do meet minimum salable requirements, the chance of providing consumer satisfaction is limited.

Boxes used for plant storage should have air vents for air exchange. It is possible that plants in nonvented boxes may experience 100% relative humidity, which may severely increase disease problems. Relative humidity should be maintained between 50 and 90%, if possible, for maintenance of best quality.

Foliage plants are not extremely sensitive to ethylene and other aerial contaminants, but they can be injured if placed in storage areas with fruits and vegetables. Generally, 1 or more ppm ethylene for several days will injure many foliage plants. Research to see if the use of low oxygen storage would improve quality of foliage plants showed that it was of no benefit (Poole et al., 1985).

D. Shipping

Foliage plants have become a world commodity and are shipped internationally by air, ship, and truck. Air shipments are generally limited to low weight cuttings, tissue culture plantlets, and small starter plants, while trucks and ships are used for plants of all sizes. Potted plants are shipped by truck in Europe and North America, while shipment from the Caribbean, Africa, and the United States to Europe move mostly in reefers on container ships.

Refrigerated trucks are used to ship approximately 90% of foliage plants within the United States. Truck transportation is reliable, relatively low in cost, and reaches almost all parts of the United States within 5 days. Foliage plants shipped in this manner are handled by specialty truckers rather than by those who haul general cargo. Trucks used for this purpose should have refrigeration and heating equipment that maintain a temperature between 60° and 65°F in all weather. In addition, trucks with air suspension are of value in limiting physical damage to foliage plants.

Because finished plants can enter the mainland United States from Hawaii and Puerto Rico and the cost of air containerization is so high, ship containerization is being used. Although the containers are the same as those used for trucks, it is important that they are sealed properly so desired temperature (60° to 65°F) and relative humidity (80 to 90%) can be maintained. Suggested long-term shipping temperatures have been listed (Table X) by Conover and Poole (1986).

E. Consumer Care of Products

A large number of books, bulletins, and other articles on foliage plant care indoors have been made available to consumers during the last 15 years. All too often there is conflicting information on care of specific plants, but all provide general information that can be valuable. Some of the main factors are mentioned here to aid in obtaining greatest satisfaction from foliage plants.

Light, or lack of it, is probably the main reason many foliage plants decline when used indoors. The three important factors to consider are light intensity, quality, and duration.

Light intensity required for growth indoors varies among species. Foliage plants are generally listed as requiring low (50 to 75 fc of light), medium (75 to 150 fc), or high (150 to 300 or more fc) light intensity. Although these are somewhat inexact because they fail to take into consideration light quality and duration, they are still useful (Table XI).

Light quality relates to the wavelengths of light that plants receive. Because natural light provides both red and blue wavelengths, and plants are responsive in both these bands, it was concluded that if artificial light were provided to plants it must provide both red and blue light. This has been proven untrue by Cathey and Campbell (1978) because foliage plants grown under blue light alone do as well as plants grown under a combination of wavelengths. For this reason, cool or warm white fluorescent lamps have proven to be one of the best indoor light sources.

Light duration is important because suggested intensities are based on light being received for at least 8 hours a day. Even better responses will be obtained when light is received for 12 or 16 hours daily if the intensity is above the light compensation point, but lighting beyond 18 hours per day can be detrimental (Conover et al., 1982).

Table X

Suggested Shipping or Storage Temperatures for Some Popular Acclimatized Foliage Plants[a]

Plant name	Temperature (°F) for duration[b]	
	1–14 days	15–28 days
Aglaonema 'Silver Queen'	60–65	60–65
Beaucarnea recurvata	55–60	55–60
Chamaedorea seifrizii	55–60	55–60
Dieffenbachia 'Tropic Snow'	55–65	55–65[c]
Dracaena fragrans 'Massangeana'	60–65	60–65
Epipremnum aureum	55–60	55–60[d]
Ficus benjamina	55–60	55–60
Ficus lyrata	55–60	55–60
Howea forsterana	55–65	55–65
Philodendron selloum	55–60	55–60
Spathiphyllum 'Mauna Loa'	50–55	55–60
Yucca elephantipes	50–55	50–55

[a]From Conover and Poole (1986).

[b]Plants shipped or stored for 1 to 7 days should be held at the highest temperature listed for that plant.

[c]Plants observed to have a loss in quality of about 25% per week beyond 2 weeks.

[d]Plants observed to have severe loss in quality beyond 2 weeks.

Table XI

Light Requirements of Some Popular Foliage Plants

Low light (50–75 fc)	Medium light (75–150 fc)	High light (150–300 fc)
Aglaonema commutatum	Brassaia actinophylla	Aphelandra squarrosa
Chamaedorea elegans	Calathea makoyana	Codiaeum variegatum
Dracaena deremensis 'Warneckii'	Chrysalidocarpus lutescens	Ficus benjamina
Maranta erythroneura	Dieffenbachia maculata	Ficus lyrata
Peperomia obtusifolia	Dracaena marginata	Hoya carnosa
Philodendron scandens oxycardium	Syngonium podophyllum	Saintpaulia ionantha
Sansevieria trifasciata	Epipremnum aureum	Nephrolepis exaltata

There is no secret to indoor watering of foliage plants. For most foliage plants, the potting medium should be kept moist but not wet, and it should not be allowed to dry out to the point that wilting occurs. At each irrigation, plants should be watered thoroughly so some leaching occurs, and then the medium allowed to become fairly dry before watering again. Amount of water and frequency of irrigation will depend on pot size, plant size, ability of the potting medium to hold water, temperature, light intensity, humidity, and rate of plant growth. A small plant in a large container under low light and high humidity might only need to be watered once a week, whereas a large plant in a small container under high light and low humidity might have to be watered every day.

Relative humidity indoors during the winter heating season often falls below 25% but will be about 50% when air conditioning is on. Most foliage plants do best at a humidity of 50% or higher. Foliage plants requiring high humidity can be maintained on a bed of wet pebbles, placed in a bathroom or kitchen where humidity is higher, or a humidifier can be installed into the heating system.

Normal interior temperatures between 65° and 80°F are satisfactory for most foliage plants. Basically, temperatures comfortable for most people will be satisfactory for foliage plants.

Foliage plants need very little fertilizer when grown indoors under low or medium light intensities. Under such conditions, fertilizer applications 3 or 4 times a year will be adequate. When plants are grown under high light intensity, fertilizer may need to be applied 6 or more times a year for best growth.

REFERENCES

Adriansen, E. (1976). "Induction of flowering in bromeliads with ethephon." *Tidsskr. Planteavl.* **80**: 857–868.

Anonymous. (1987). *Statistische Overzichten Snijbloemen Potplanten 1987.* VBA, Aalsmeer.

Anonymous. (1989a). *Floriculture Crops, 1988 Summary*. National Agricultural Statistics Service SpCr G-1 (89). USDA, Washington, D.C.

Anonymous. (1989b). *Florida Foliage Locator: 1989-1990*. Apopka, FL: Florida Foliage Association.

Bailey, L. H., and Bailey, E. Z. (1976). *Hortus Third*. New York: Macmillan.

Cathey, H. M., and Campbell, L. E. (1978). "Zero-base budgeting for lighting plants." *Foliage Dig.* **1**: 10–13.

Chase, A. R. (1987). *Compendium of Ornamental Foliage Plant Diseases*. St. Paul, MN: Am. Phytopath. Soc.

Christensen, O. V. (1969). "The influence of low temperature on flowering of *Aphelandra squarrosa*." *Tidsskr. Planteavl.* **73**: 351–366.

Christensen, O. V. (1976). "Planning of production—Timing and spacing for year-round production of pot plants." *Acta Hort.* **64**: 217–221.

Collard, R. C., Joiner, J. A., Conover, C. A., and McConnell, D. B. (1977). "Influence of shade and fertilizer on light compensation point of *Ficus benjamina* L." *J. Am. Soc. Hort. Sci.* **102**: 447–449.

Conover, C. A. (1986). "Quality." *Acta Hort.* **181**: 201–205.

Conover, C. A., and Poole, R. T. (1970). "Foliage plant propagation." *Grounds Maint.* **5**: 27–29.

Conover, C. A., and Poole, R. T. (1972). "Influence of shade and nutritional levels on growth and yield of *Scindapsus aureus*, *Cordyline terminalis* 'Baby Doll' and *Dieffenbachia exotica*." *Proc. Am. Soc. Hort. Sci. Trop. Reg.* **16**: 277–281.

Conover, C. A., and Poole, R. T. (1974a). "Fluoride toxicity of tropical foliage plants." *Florists' Rev.* **154**: 23, 59.

Conover, C. A., and Poole, R. T. (1974b). "Influence of shade and fertilizer source and level on growth, quality and foliar content of *Philodendron oxycardium* Schott." *J. Am. Soc. Hort. Sci.* **99**: 150–152.

Conover, C. A., and Poole, R. T. (1974c). "Influence of shade nutrition and season on growth of *Aglaonema*, *Maranta*, and *Peperomia* stock plants." *Proc. Am. Soc. Hort. Sci. Trop. Reg.* **18**: 283–287.

Conover, C. A., and Poole, R. T. (1977). "Influence of fertilization and watering on acclimatization of *Aphelandra squarrosa* Nees cv. 'Dania.'" *HortScience.* **12**: 569–570.

Conover, C. A., and Poole, R. T. (1982). "Slow-release fertilizers and light levels influence growth of *Araucaria heterophylla* and *Spathiphyllum* × Mauna Loa." *Proc. Am. Soc. Hort. Sci. Trop. Reg.* **25**: 73–76.

Conover, C. A., and Poole, R. T. (1984a). "Acclimatization of indoor foliage plants." *Hort. Rev.* **6**: 119–154.

Conover, C. A., and Poole, R. T. (1984b). "Influence of temperature and duration on simulated shipping of small potted foliage plants." *Proc. Fla. State Hort. Soc.* **97**: 280–282.

Conover, C. A., and Poole, R. T. (1986). "Shipping suggestions." *Int. Landscape Ind.* **3**: 32–37.

Conover, C. A., and Poole, R. T. (1987). "Growth of *Dieffenbachia maculata* 'Perfection' as affected by air and soil temperatures and fertilization." *HortScience.* **22**: 893–895.

Conover, C. A., Poole, R. T., and Nell, T. A. (1982). "Influence of intensity and duration of cool white fluorescent lighting and fertilizer on growth and quality of foliage plants." *J. Am. Soc. Hort. Sci.* **107**: 817–822.

Conover, C. A., Simpson, D. W., and Joiner, J. N. (1975). "Influence of micronutrient sources and levels on response and tissue content of *Aphelandra*, *Brassaia* and *Philodendron*." *Proc. Fla. State Hort. Soc.* **88**: 599–602.

Fonteno, W. C., and McWilliams, E. L. (1978). "Light compensation points and acclimatization of four tropical foliage plants." *J. Am. Soc. Hort. Sci.* **103**: 52–56.

Graf, A. B. (1970). *Exotica 3*. Rutherford, NJ: Roehrs Co.

Hamlen, R. A. (1976). "Efficacy of nematicides for control of *Meloidogyne javanica* on Maranta in ground bed and container production." *J. Nematol.* **8**: 287.

Hamlen, R. A., Dickson, D. W., Short, D. E., and Stokes, D. E. (1981). "Insects, mites, nematodes and other pests." In *Foliage Plant Production*, edited by J. N. Joiner. Englewood Cliffs, NJ: Prentice Hall.

Harbaugh, B. K., Wilfret, G. J., Engelhard, A. W., Waters, W. E., and Marousky, F. J. (1976). "Evaluation of 40 ornamental plants for a mass marketing system utilizing sealed polyethylene packages." *Proc. Fla. State Hort. Soc.* **89**: 320–323.

Henley, R. W., and Poole, R. T. (1974). "Influence of growth regulators on tropical foliage plants." *Proc. Fla. State Hort. Soc.* **87**: 435–438.

Henley, R. W., and Poole, R. T. (1989). "Evaluation of selected ornamental figs for interior use." *Proc. Fla. State Hort. Soc.* **102**: (in press).

Henny, R. J. (1985). "BA induces lateral branching of *Peperomia obtusifolia*." *HortScience.* **20**: 115–116.

Henny, R. J. (1986). "Increasing basal shoot production in a nonbranching *Dieffenbachia* hybrid with BA." *HortScience.* **21**: 1386–1388.

Henny, R. J. (1988). "Ornamental aroids: Culture and breeding." *Hort. Rev.* **10**: 1–33.

Henny, R. J., and Fooshee, W. C. (1985). "Induction of basal shoots in *Spathiphyllum* 'Tasson' following treatment with BA." *HortScience.* **20**: 715–717.

Joiner, J. N., Poole, R. T., Johnson, C. R., and Ramcharam, C. (1978). "Effects of ancymidol and N, P, K on growth and appearance of *Dieffenbachia maculata* 'Baraquiniana.'" *HortScience.* **13**: 182–184.

Kerbo, R., and Payne, R. N. (1976). "Reducing flowering time in *Aphelandra squarrosa* Nees with high pressure sodium lighting." *HortScience.* **11**: 368–370.

Knauss, J. F. (1971). "Nature, cause and control of diseases of tropical foliage plants." *Ann. Res. Rep. Fla. Agric. Exp. Stn.* 206.

McConnell, D. B., Henley, R. W., and Kelly, C. B. (1989). "Commercial foliage plants: Twenty years of change." *Proc. Fla. State Hort. Soc.* **102**: (in press).

McConnell, D. B., and Poole, R. T. (1974). "Influence of ancymidol on *Scindapsus aureus*." *SNA Nursery Res. J.* **1**: 13–18.

McWilliams, E. L., and Smith, C. W. (1978). "Chilling injury in *Scindapsus pictus, Aphelandra squarrosa* and *Maranta leuconeura*." *HortScience.* **13**: 179–180.

Mikorski, D. J., and White, J. (1977). "Foliage plants—Seed propagation and transplant research." *Florists' Rev.* **160**: 55, 99–102.

Osborne, L. S. (1986). "Dip treatment of tropical ornamental foliage cuttings in fluvalinate to prevent spread of insect and mite infestations." *J. Econ. Ent.* **79**: 465–470.

Osborne, L. S., and Oetting, R. D. (1989). "Biological control of pests attacking greenhouse grown ornamentals." *Florida Entomologist.* **72**: 408–413.

Osborne, L. S., Boucias, D. G., and Lindquist, R. K. (1985). "Activity of *Bacillus thuringiensis* var. *israelensis* on *Bradysia coprophila*." *J. Econ. Ent.* **78**: 922–925.

Ottosen, C. O. (1988). "Selection of fast-growing clones among visually unseparable genotypes of *Ficus benjamina* L." *Gartenbauwissenschaft.* **53**: 180–182.

Poole, R. T. (1971). "Flowering of Christmas cactus as influenced by nyctoperiod regimes." *Proc. Fla. State Hort. Soc.* **84**: 410–413.

Poole, R. T., and Conover, C. A. (1973). "Fluoride induced necrosis of *Cordyline terminalis* Kunth 'Baby Doll' as influenced by medium and pH." *J. Am. Soc. Hort. Sci.* **98**: 447–448.

Poole, R. T., and Conover, C. A. (1978). "Nitrogen and potassium fertilization of *Aglaonema commutatum* Schott. cvs 'Fransher' and *Pseudobracteatum*." *HortScience.* **12**: 570–571.

Poole, R. T., and Conover, C. A. (1981). "Growth response of foliage plants to night and water temperatures." *HortScience.* **16**: 81–82.

Poole, R. T., and Conover, C. A. (1983). "Influence of simulated shipping environments on foliage plant quality." *HortScience.* **18**: 191–193.

Poole, R. T., and Conover, C. A. (1988a). "Influence of paclobutrazol on foliage plants." *Proc. Fla. State Hort. Soc.* **101**: 319–320.

Poole, R. T., and Conover, C. A. (1988b). "Propagation of single eye cuttings of *Philodendron scandens oxycardium* after storage." *Acta Hort.* **II**: 587–590.

Poole, R. T., and Conover, C. A. (1988c). "Storage of philodendron and pothos cuttings." *Proc. Fla. State*

Hort. Soc. **101**: 313–315.

Poole, R. T., and Waters, W. E. (1971). "The influence of elevated medium temperature upon development of cuttings and seedlings of tropical foliage plants." *HortScience.* **6**: 463–464.

Poole, R. T., Chase, A. R., and Conover, C. A. (1988). "Chemical composition of good quality tropical plants." *CFREC-Apopka Research Report.* RH-88-6.

Poole, R. T., Conover, C. A., and ben-Jaacov, J. (1984). "Long-term storage of foliage plants." *Scientia Horticulturae.* **24**: 331–335.

Poole, R. T., Conover, C. A., and Staby, G. L. (1985). "Low oxygen storage allows foliage plants two weeks of rest." *Greenhouse Mgr.* **3**: 33–35.

Rodriguez, S. J., and Cibes, H. (1977). "Effect of five levels of nitrogen at six shade intensities on growth and leaf-nutrient composition of *Dracaena deremensis* 'Warneckii' Engler." *J. Agric. Univ. P. R.* **61**: 305–313.

Scarborough, E. F. (1978). *Fresh Foliage Plants Summary 1977.* Orlando, FL: Fed. State Mark. News. Serv.

Swanson, H. F. (1975). *Countdown for Agriculture in Orange County Florida.* Orlando, FL: Designers Press of Orlando.

Glossary

Abscission   Dropping of leaves, flowers or fruit, usually following formation of a layer of separate cells.

After-ripening   Physiological changes that must take place in a primary dormant seed before it can germinate.

Aldicarb (Temik)   Systemic insecticide.

Ancymidol-α-cyclopropyl-α-(ρ-methoxyphenyl)-5-pyrimidine methanol   The active ingredient in the commercial growth regulator A-Rest or Reducimol.

Annual   A plant that completes its life cycle in 1 year.

Antagonism   Situation in which the presence of one kind of nutrient ion interferes with the uptake of another.

Anthesis   Full flower; the period of pollination.

Apical meristem   The growing point of the shoot or root.

Artificial long days   Interruption of dark period or extension of natural daylength to prevent flower bud initiation of short-day plants, such as chrysanthemums and poinsettias.

Asexual propagation   Reproduction by vegetative means, such as cuttings, division.

Atrinal (Dikegulac)   Chemical pinching agent used to promote lateral shoot development on azaleas. Now sold as Atrimmec.

B-Nine (SADH) [(N-dimethylamino)succinamic acid]   Growth retardant used to control height of several floricultural crops and to promote flower initiation in others, such as azaleas.

Basal plate   Perennial, shortened, modified stem that has a growing point and to which bulb scales and roots are interjoined.

Benlate (Benomyl)   Systemic fungicide.

Biennial   A plant that lives through two growing seasons. It bears fruit and dies the second year.

Blackout system or blackclothing   A means of covering plants to shorten the photoperiod, to promote flowering in a short-day plant such as *Chrysanthemum morifolium*. Black sateen cloth or black polyethylene film generally used.

Blasting (flower)   The failure of a bulb to produce a marketable flower after the floral initiation has taken place.

Blindness (flower)   The failure of a bulb to produce any floral parts.

Blind shoots   Shoots that remain vegetative under conditions that normally stimulate formation of inflorescences.

Bract   Modified leaf, frequently associated with flowers.

Break   New lateral shoot, often developed following removal of apical dominance by pinching.

Breaker   A device on the end of a hose to permit application of water at low velocity.

Bud opening   Development of mature flowers in the wholesale or retail outlets; flowers cut prior to opening.

Bud scales   Leaflike structures that surround some flower buds, such as on azaleas.

Bulb   A specialized underground plant organ consisting of a greatly reduced stem (basal plate) surrounded by fleshy, modified leaves called scales.

Bulb dealer (jobber)   An individual or firm who stores, ships, and distributes bulbs.

Bulb forcer   A greenhouse operator who produces pot plants and cut flowers from bulbous species using artificial growing conditions.

Bulb grower (producer)   An individual or firm who propagates bulbs for wholesale purposes.

Bulb maturity   Measure of capacity of a healthy daughter stem axis to sprout without delay and to respond to flower-inducing treatments.

Bulb production phase   All aspects of bulb production that lead to the sale of forcing sized bulbs. The production phase may take 1 to 3 years depending on the species.

Bulb programming phase   All temperature treatments given to the bulbs from the time they are harvested until they are placed under greenhouse conditions.

Bulk density of soil   Weight per unit volume, such as grams per cubic centimeter.

Bullhead   Spherically shaped flower bud, usually resulting in a malformed flower.

Bullnose   A physiological disorder of *Narcissus* characterized by failure of the flower to open properly after it reaches the gooseneck stage of development.

Bypassing shoot   Vegetative shoot that develops immediately below the flower bed, often occurring on azaleas.

Case-cooled bulbs   Bulbs that are given low temperature treatment while still in the shipping container.

Caudicle   The stalk of the orchid pollinia.

Chimera   Plant part consisting of tissues of diverse genetic constitution, often observed in flowers.

Chlormequat [(2-chloroethyl)trimethylammonium chloride]   The active ingredient in the commercial product Cycocel.

Chlorosis   The yellowing of foliage due to loss or breakdown of chlorophyll.

Cold treatment   A cold-moist treatment prior to bulb planting that induces rapid shoot elongation and flowering.

Conidia   Asexual fungus spores.

Controlled temperature forcing (CTF)   Procedure for treating Easter lilies in which nonprecooled bulbs are potted, placed in a controlled temperature area at approximately 61°F for 2 to 4 weeks and, subsequently, cooled at 36°F to 45°F for 6 to 7 weeks prior to being placed in the greenhouse.

Corm   A specialized underground organ consisting of an enlarged stem axis with distinct nodes and internodes and enclosed by dry, scalelike leaves.

Cormels   Small corms arising on stolons that develop between the mother and daughter corms.

Corolla   Inner perianth (petals) of a flower.

Critical daylength   The daylength above or below which a plant will flower, depending on whether the plant is a short- or long-day plant.

Crown bud   Chrysanthemum inflorescence formed under adverse conditions, such as improper daylength. The bud may abort.

Cultivar   Horticultural or "cultivated" variety.

Cutting   Vegetative portion of a plant stem, which is rooted to produce a new plant.

Cyathium   Inflorescence of plants such as poinsettia. Relatively inconspicuous but bears pistils, stamens, and nectary glands.

Cyclic lighting   Intermittent illumination during the dark period, to stimulate a long day and prevent flowering of short-day plants.

Cyme   A relatively flat-topped determinate flower cluster (spray) with the central flowers or inflorescences first to open.

Daminozide (succinic acid 2,2-dimethylhydrazide)   The active ingredient in the commercial products B-Nine and Alar (SADH).

Damping off   Disease of seedlings caused by several organisms, such as *Pythium, Rhizoctonia*, and *Botrytis*.

Dark storage   A term used to describe the time foliage plants remain in darkness within shipping containers during transit or storage.

Daughter bulb   Scales and leaves initiated by and developing below and around the new daughter apex. This apex arises from a bud in the axil of a scale subtending the old or mother axis.

Determinant inflorescence   One in which the apical flower blooms first.

Devernalization   Negation of a vernalizing stimulus by temperatures above a critical level.

Dieback   Death of shoots, originating at the shoot tips.

Disbudding   The removal of lateral flower buds on stems of plant, such as the carnation and chrysanthemum.

Disc florets   Florets in center of chrysanthemum inflorescence, conspicuous in daisy- or anemone-flowered types.

Disorders (physiological)   Undesirable effects caused by a nonpathogenic agent or factor.

Division   Separation of root system of parent plant into several units; a method of asexual propagation.

Dormancy   The period of inactivity in buds, bulbs, and seeds when growth stops. Some change in environment usually required before growth will be resumed.

Dry-pack storage   The storage of cut flowers in vapor-proof containers with the stems not in water, usually at 32°F.

EC_e   Electrical conductivity of the saturated soil paste extract measured in millimhos per centimeter.

Endosperm   The seed tissue that contains stored food.

EPA   Environmental Protection Agency, authorized to regulate, among other duties, labeling of pesticides.

Epiphyte   An organism growing on a plant for support, without establishment of a parasite–host relationship.

Erwinia   Genus of bacterium that causes disease of succulent tissue, often called soft rot.

Ethylene   Colorless, odorless gas that hastens senescence of flowers. Often emitted by fruit, foliage, and incomplete combustion of oil and gas in heaters.

Ethylene scrubbers   Chemicals used to remove ethylene from the atmosphere.

Ethephon (2-chloroethylphosphonic acid)   The active ingredient in the commercial products Ethrel and Florel, which is metabolized to ethylene in the plant.

Eye   A lateral bud, as on a rose stem.

Flaccid   Wilted.

Floral bud   Immature flower that consists of petals, stamens, and pistil.

Floral preservative   Chemical added to water to extend vase life of cut flowers.

Floral primordium   Very early stage of flower bud.

Flower blasting   Phase of flower bud abortion occurring after flower differentiation is completed. When blasting has occurred visible signs of the floral organs are evident.

Flower bud abortion   Cessation of floral bud development at any stage of development.

Flower bud development   Progressive change in flower bud, from transition to flowering.

Flower bud initiation   Formation of floral primordium.

Flower differentiation   Complete morphological development of the floral organs following initiation.

Flower induction   An unobservable, preparatory step that occurs prior to visible flower bud initiation.

Flower initiation   Visible organization of flower primodia (buds) at the stem apex.

Foliage plant   Any plant grown primarily for its foliage and used for interior decoration or interior landscape purposes. While it may have flowers, these will be secondary compared to foliage features.

Foot candles (ft-c)   Illumination of a standard light source of one candle, one foot away; 1 ft-c = 10.76 lux.

Forcing   Acceleration of flowering by manipulation of environmental conditions.

Fumigun   A large "hypodermic needle" device for hand-injection of fumigant chemicals into soil.

Fungicide   Chemical used to control diseases caused by fungi.

Fungus gnats   Small black flies, damaging to plants when the maggots feed on plant roots and stems.

Fusarium   Genus of fungi that causes several plant diseases; can affect several portions of the plant.

Geotropic bending   Upward curvature of tips of spike flowers, such as snapdragons and gladiolus, when held horizontally, due to negative response to gravity.

Gibberellic acid (GA)   Chemical compound used to break flower bud dormancy or stimulate shoot elongation.

Grading   Classification of plants and flowers for market, based on size and quality.

Grassy growth   Excessive and noticeable production of axillary branches (e.g., on snapdragon stems).

Greenhouse phase   That portion of forcing that encompasses the time from placing the plants in the greenhouse until flowering.

Green pruning   Pruning of actively growing rose plants without benefit of a dormancy period.

Gooseneck   Term describing the proper stage of flower development to cut daffodils.

Gynandrium   The structure in the orchid flower that results from the fusion of the male and female portions of the flower.

Gypsum   Calcium sulfate, used to alter pH of the medium.

Hamper   Container used for shipping gladiolus.

Hanging basket   Container, usually plastic, used for bedding plants, foliage plants, poinsettias, and other plant material. Usually suspended from supports in greenhouse, and used to suspend plants in the home or garden.

Heat delay   A delay in the initiation of a flower bud because of an abnormally high temperature.

Herbicide   Weed killer, used to control weeds chemically.

Hypobaric storage   Storage at less than atmospheric pressure for long-term holding of plant material.

IBA [α(indole-3)-*n*-butyric acid]   An auxin commonly used to promote rooting of cuttings.

Indexed plants   Plants that have been tested by pathological methods and found to be free of known pathogens. Plants may also be indexed for one specific pathogen.

Inflorescence   A flower cluster.

Interveinal   Area between the veins of a leaf.

Interveinal chlorosis   Yellowing of leaf tissue between the veins.

Latex   Milky white fluid found in stems, foliage, and bracts of poinsettias.

Leaching   Applications of water to media to reduce soluble salts level.

Leaf counting   Procedure used to time Easter lily flowering.

Leaf scorch   Crescent-shaped necrotic areas that develop along the margin and tips of leaves as a result of physiological imbalances (also referred to as tip burn).

Light compensation point   That light intensity at which respiration and photosynthesis of the entire plant are balanced.

Light flux   The light intensity times the duration of light.

Long-day plant   Plant that flowers when the daylength is longer than the critical.

Marketing phase   The movement of the plants and/or flowers from the forcing facilities to the wholesaler and/or retailer at the proper stage of development so that the consumer receives the maximum possible enjoyment.

Mat watering (capillary watering)   Irrigation of potted plants by capillarity. Mats are composed of fabric, cellulose, or other water-absorbing materials.

Mealybugs   Small, sucking insects that feed on plants. Females, often wingless, are usually covered with a white mealy layer of wax.

Media   Substrates in which plants are grown. Can include soil, sand, peat moss, vermiculite, pine bark humus.

Meristem   Growing point where cell divisions occur. The undifferentiated plant tissue from which new cells arise by division.

Mesophyll   Large parenchyma cells located within the epidermal layers of a leaf.

Methyl bromide Chemical sterilant used to eliminate pests in growing media. Often contains chloropicrin (tear gas) for detection.
Mhos Measurement of conductivity; denotes amount of soluble salts in a medium. Measured with a Solubridge instrument.
Micronutrients Elements required in small amounts for plant growth and flowering.
Milliequivalent per liter (mEq/liter) One thousandth of an equivalent of an ion in a liter of solution.
Millimhos per cm (mmhos/cm) Unit used to express electrical conductivity of water or the soil solution. One millimhos equals 0.1 mhos.
Miticide Pesticide that is used to control mites.
Monopodial Having a strong terminal bud and forming a single upright stem (e.g., coconut palms). A plant with a primary upright stem that continues growth year after year (i.e., Vanda orchids).
Mother block (nucleus block) A group of selected, tested, and disease-free plants, used as basic stock for propagation of several species.
Mother bulb That portion of the bulb that is currently flowering and producing a daughter bulb in the axil of a scale subtending the mother axis. The old mother scales encompass the new daughter axis.
Multibranched plant One plant with several shoots and flowers. Achieved by pinching.
Multiflowering More than one flower or inflorescence produced by the branching of a single plant.
Mutation Change in genetic potential in a cell and subsequent change in growth of all other cells derived from the mutated cell.
Napthaleneacetic acid (NAA) Component of rooting hormone used to accelerate rooting.
Natural cooling Technique in which nonprecooled commercial bulbs are planted immediately on arrival and grown under cool natural conditions, but with frost protection, prior to being placed in the greenhouse.
Necrosis Symptom of plant injury, caused by spray damage, insect or disease injury, or other causes, which is characterized by dead, discolored cells and tissue.
Nematicide Chemical used to control nematodes.
Nematodes Wormlike organisms that can affect roots, stem, and foliage.
Node The point on the stem where the leaf is attached.
Noncooled bulb Bulb that is delivered direct to the forcer and has not received a cold treatment.
Nymph An immature stage of an insect.
Off-ShootO Methyl ester of a fatty acid that is used to kill shoot apices and promote lateral branching. Primarily used on azaleas.
Osmocote Slow-release, encapsulated fertilizer.
Osmunda fiber Potting medium for orchids; obtained from Osmunda fern.
Panning Transplanting or potting of rooted cuttings or bulbs.
Parts per million (ppm) Equivalent to milligrams per liter.
Pasteurization Process used to eliminate harmful pathogens. Temperature usually does not exceed 180°F, to distinguish pasteurization from sterilization at 212°F. Terms are often incorrectly used as synonyms.
Pathogen Infectious agent that causes disease.
Peat-lite Medium composed of peat moss and other ingredients, such as vermiculite and perlite.
Peat moss (moss peat) Partially decayed plant material often used as an ingredient in a growing medium. Usually acidic.
Pedicel Flower stalk.
Peduncle Stalk on which an inflorescence is borne.
Perched water table Concentration of soil moisture at the bottom of a plant growing container.
Perennial Plant that lives more than 2 years.
Perlite Volcanic rock heated to 1796°F. Expanded, porous aggregates are used in growing media to facilitate drainage and improve aeration.
Pesticide Chemical used to control undesirable organism. Growth regulators also are classified as pesticides, subject to EPA regulations and labeling.
Photoperiod The length of the day used in reference to its effect on growth and flowering.

Photoperiodic response Behavior of an organism to the length of day.

Phyllotaxy Arrangement of leaves on a stem.

Physiological disorders Undesirable effects caused by a nonpathogenic agent or factor.

Phytophthora Genus of fungi that causes plant diseases, such as root and crown rots. A water mold.

Pinch The removal of the shoot apex to overcome apical dominance and promote lateral shoot development.

Pine bark Ingredient in potting mix, usually used as pine bark humus. Obtained primarily from southern pine trees, and very popular in the Southeast.

Plant tissue culture Term applied to methods that allow growth and development *in vitro* of plant cells, tissues, or organs, in or on nutrient media, usually under aseptic conditions.

Plant vase A nontechnical term that describes the rosette of leaves that ascend forming a cup at the base of the plant. It is common to many genera within the Bromeliaceae family.

Plastochron The time interval between two successive and similar occurrences, for example, the rhythmic initiation of leaves by the apical meristem.

Ploidy (diploid, tetraploid, etc.) Refers to number of complete sets of chromosomes.

Precooling The dry storage of bulbs at temperatures between 35° and 48°F after floral initiation and development is completed but prior to planting.

Prepared (PR) hyacinths Hyacinths that have been harvested earlier than normal and given special temperature treatments by the bulb grower so that plants may be forced very early in the season.

Product mix A term common to the foliage plant industry that describes the ratio of different plants produced by individual producers or by the industry annually.

Propagation Any method used to increase plant populations, either sexually or asexually.

Pruning Removal of lateral vegetative shoots or the shaping of plants, such as azaleas, by trimming with shears.

Pythium Genus of fungi that can cause root rot of plants and damping-off of seedlings. A soil-borne water mold.

Quantitative long-day plant Plant that is not completely inhibited from flowering by short-day treatment but hastened by long-day treatment.

Ray florets The long, conspicuous florets that radiate from the center of an inflorescence; conspicuous on the standard or double chrysanthemum flowers.

Receptacle Portion of the flower stalk or axis that bears the floral organs.

Reefer Temperature-controlled, well-ventilated shipping containers used in overseas shipments.

Regular hyacinths Hyacinths that have been harvested at normal times and given temperature treatments for medium and late forcing periods.

Response group Classification of cultiars, based on response to environment; for chrysanthemums, indicates number of weeks required to flower after the start of short days. For snapdragons, indicates time of year cultivar should be in flower.

Retarded iris Dutch iris that has been stored at 86°F to prevent flower development.

Rhizoctonia Genus of fungi that can cause root rots and damping-off.

Rhizome A horizontal stem either on the ground or just below the surface of the soil.

Rogueing The elimination of undesirable plants, which might be diseased, inferior, or nontypical.

Rooting hormone Compound such as IBA or NAA that stimulates rooting.

Rooting room A controlled temperature facility used to root and satisfy the cold requirement of bulbs.

Rostellum A gland, literally a small beak, on the orchid stigma.

Saran cloth Material installed in greenhouse or field to reduce light intensity and radiant heat.

Scaling Technique used to propagate foundation stock, wherein the scales of a bulb are removed and planted. This produces numerous scale bulblets from a single mother scale.

Scape Peduncle originating at the base of the plant and bearing one or more flowers at the apex.

Self-branching Axillary buds may initiate growth without pinching.

Senescence Aging of plant parts, such as the flower. Usually the stage from full maturity to death.

Sepal A unit of the calyx, the first formed series of floral parts that are often green but may be other colors.

Shattering Abscission of snapdragon and seedling geranium florets, caused by ethylene, pollination, or senescence.

Shoot   Upright stem, often arising from axillary position following a pinch.

Short-day plant   Plant that flowers when the daylength is shorter than the critical.

Single-stemmed plant   Plant confined to one stem, in contrast to a multistemmed, pinched plant.

Slab-side   A carnation flower where the petals elongate and expand first on one side of the calyx; a lopsided flower.

Slipping of bark   Period of extreme cambial activity in spring when the bark–phloem section of a stem separates readily from the xylem at the cambial layer.

Slipping stage   The time when the inflorescence emerges from the leaf sheathes.

Slow-release fertilizer   Fertilizer not immediately soluble or readily available to plant roots because of coating of the granule, relative insolubility, or need for bacterial breakdown.

Sodium hypochlorite (Clorox)   Common household bleach, used for surface sterilization of greenhouse benches and tools.

Sodium methyl dithiocarbamate   The active ingredient in the commercial product Vapam, used in soil fumigation.

Solubridge   An instrument used to measure the electrical conductivity of the soil solution. When used with a saturated paste extract (EC_e), the units are in millimhos/e_{cm}; with a 1:2 or 1:5, soil:water extract, the units are mhos $\times$ 10^{-5}.

Spadix   A succulent stalk bearing an inflorescence, as on plants in the Aroid family.

Special cooling of tulips   A special technique used to program tulips for cut flower forcing. When used, the entire cold requirement is given as a dry cold treatment and rooting takes place during the greenhouse phase of forcing. Also known as "41°F forcing" and "direct forcing."

Spider mites   Various species of mites, often parasitic, on plants.

Spike   An elongated inflorescence, exemplified by gladiolus, stock, and snapdragon.

Spitting   In hyacinths, the abscission or release of the entire floral stalk and inflorescence from the basal plate during forcing.

Spores   Reproductive units of fungi, equivalent to seeds of higher plants.

Sport   A mutant that is inherited and transmitted to progeny.

Spray treatment (chrysanthemums)   Removal of apical flower bud to stimulate development of lateral flowers.

Spring-flowering bulbs   The broad classification given to those bulbous plants that are planted outside in the fall, overwintered under low temperatures, and flower in the spring months.

Stage "G"   The term used to indicate that the gynoecium (pistil) has formed in the flower of the tulip. The same stage occurs in other Liliaceae species.

Standard forcing   A technique used to program spring-flowering bulbs in which all or most of the cold requirement is given after planting of the bulbs. Rooting takes place during the programming phases of forcing. Also, known as container forcing, normal forcing, or conventional forcing.

Stem topple   In tulips, the physiological disorder that is characterized by the collapse of a small portion of the internode of the floral stalk located just underneath the flower. It can occur either just before or during flowering.

Stock plants   Plants from which cuttings are taken for propagation.

Sun scald   Injury caused by excessively high light intensity and radiant heat; often associated with standard chrysanthemum flowers in summer.

Supplementary illumination (lighting)   A means of supplying additional light (usually in the middle of the dark period) to lengthen the photoperiod; used to keep chrysanthemums, poinsettias, and other short-day plants vegetative.

Sympodial   Plants with a main stem or axis that ceases growth each year, and new growth arises from the base (i.e., Cattleya orchid).

Syringing   Application of water, sprayed on the foliage. Used to reduce transpiration, control some insect pests, reduce leaf temperature, or as a means of watering.

Systemic   Usually used to indicate that a pesticide can be translocated throughout the plant. Generally longer lasting than a spray application.

Taxonomy   The science of classification of plants and animals.

Tensiometer   Instrument used to measure tension with which water is held in the growing medium.

Tuber   A thickened, often short, subterranean stem, enabling the plant to be asexually propagated.

Tunic   The dry, papery scales that surround the fleshy organs of a bulb or corm.

Understock   The lower portion of a budded or grafted plant that develops into the root system.

Vase life   Longevity of cut flowers.

Vermiculite   Mica platelets, formed by heating to about 1360°F and used as an ingredient in growing media.

Vernalization   Cold-moist treatment applied to a seed, plant, or bulb to induce or hasten the development of the capacity for flowering.

Verticillium   Genus of fungi that causes vascular disorders of plants.

Viaflo (twin-wall tubes)   Commercial designations for irrigation systems composed of tubes through which water is applied to media at very slow rates.

Wardian case   A glass-topped enclosure usually supplied with bottom heat and used for propagation of plants.

Xanthomonas   Genus of bacterium that can cause leaf spots on plants.

Year-round flowering   Control of daylength and temperature to produce flowering plants throughout the year. Often used to describe azalea and chrysanthemum programs.

Zygomorphic   An irregular flower capable of being divided in half, but in only one plane. Used to describe some orchid flowers.

Index